Communications in Computer and Information Science 2970

Series Editors

Gang Li, *School of Information Technology, Deakin University, Burwood, VIC, Australia*
Joaquim Filipe, *Polytechnic Institute of Setúbal, Setúbal, Portugal*
Zhiwei Xu, *Chinese Academy of Sciences, Beijing, China*

Rationale

The CCIS series is devoted to the publication of proceedings of computer science conferences. Its aim is to efficiently disseminate original research results in informatics in printed and electronic form. While the focus is on publication of peer-reviewed full papers presenting mature work, inclusion of reviewed short papers reporting on work in progress is welcome, too. Besides globally relevant meetings with internationally representative program committees guaranteeing a strict peer-reviewing and paper selection process, conferences run by societies or of high regional or national relevance are also considered for publication.

Topics

The topical scope of CCIS spans the entire spectrum of informatics ranging from foundational topics in the theory of computing to information and communications science and technology and a broad variety of interdisciplinary application fields.

Information for Volume Editors and Authors

Publication in CCIS is free of charge. No royalties are paid, however, we offer registered conference participants temporary free access to the online version of the conference proceedings on SpringerLink (http://link.springer.com) by means of an http referrer from the conference website and/or a number of complimentary printed copies, as specified in the official acceptance email of the event.

CCIS proceedings can be published in time for distribution at conferences or as post-proceedings, and delivered in the form of printed books and/or electronically as USBs and/or e-content licenses for accessing proceedings at SpringerLink. Furthermore, CCIS proceedings are included in the CCIS electronic book series hosted in the SpringerLink digital library at http://link.springer.com/bookseries/7899. Conferences publishing in CCIS are allowed to use our online conference service (Meteor) for managing the whole proceedings lifecycle (from submission and reviewing to preparing for publication) free of charge.

Publication process

The language of publication is exclusively English. Authors publishing in CCIS have to sign the Springer CCIS copyright transfer form, however, they are free to use their material published in CCIS for substantially changed, more elaborate subsequent publications elsewhere. For the preparation of the camera-ready papers/files, authors have to strictly adhere to the Springer CCIS Authors' Instructions and are strongly encouraged to use the CCIS LaTeX style files or templates.

Abstracting/Indexing

CCIS is abstracted/indexed in DBLP, Google Scholar, EI-Compendex, Mathematical Reviews, SCImago, Scopus. CCIS volumes are also submitted for the inclusion in ISI Proceedings.

How to start

To start the evaluation of your proposal for inclusion in the CCIS series, please send an e-mail to ccis@springer.com

Hamid R. Arabnia · Leonidas Deligiannidis ·
Farzan Shenavarmasouleh ·
Farid Ghareh Mohammadi · Soheyla Amirian
Editors

Emerging Trends in Computational Science, Computational Biology, Parallel & Distributed Computing, and Internet of Things

12th International Conference on Computational Science and Computational Intelligence, CSCI 2025
Las Vegas, NV, USA, December 3–5, 2025
Proceedings Part IV

Springer

Editors
Hamid R. Arabnia
University of Georgia
Athens, GA, USA

Leonidas Deligiannidis
Wentworth Institute of Technology
Boston, MA, USA

Farzan Shenavarmasouleh
Medialab Inc.
Lawrenceville, GA, USA

Farid Ghareh Mohammadi
Verify Radiology Images Consultants
Houston, TX, USA

Soheyla Amirian
Pace University
New York, NY, USA

ISSN 1865-0929 ISSN 1865-0937 (electronic)
Communications in Computer and Information Science
ISBN 978-3-032-28465-5 ISBN 978-3-032-28466-2 (eBook)
https://doi.org/10.1007/978-3-032-28466-2

This Springer imprint is published by the registered company Springer Nature Switzerland AG
The registered company address is: Gewerbestrasse 11, 6330 Cham, Switzerland

Preface

It is our great pleasure to present this collection of research papers from the 12th International Conference on Computational Science and Computational Intelligence (CSCI 2025). This volume, titled "Emerging Trends in Computational Science, Computational Biology, Parallel & Distributed Computing, and Internet of Things", comprises selected papers accepted in the Research Tracks on Parallel & Distributed Computing (CSCI-RTPD); Computational Science (CSCI-RTCS); Internet of Things & Internet of Everything (CSCI-RTOT); and Computational Biology (CSCI-RTCB). The conference was held in Las Vegas, Nevada, USA, from December 3 to December 5, 2025.

The CSCI 2025 International Conference brought together papers from a diverse array of communities, including researchers from universities, corporations, and government agencies. Accepted papers are published by Springer Nature, and the proceedings showcase solutions to key challenges in various critical areas of Computational Science and Computational Intelligence.

Computational Science (CS) is the study of addressing problems that are impossible to solve (or difficult to solve) without computers. CS can be considered to be the bridge between computer science and other sciences. The field is interdisciplinary by nature and includes the use of advanced computing capabilities to understand and solve complex problems. In short, CS is the science of using computers to do science. Computational Intelligence (CI) is the study of computational methods that exhibit intelligence. These methods adapt to changing environments and changing goals. There is a significant overlap between the fields of CI and Artificial Intelligence (AI). However, there is also a difference: AI techniques often involve top-to-bottom methods (i.e., methods to reach solutions are imposed from the top) whereas CI techniques often involve bottom-up methods (i.e., solutions emerge from unstructured beginnings). An important part of CI includes a set of Nature-inspired computational approaches to address complex problems for which traditional methods are infeasible. Computational Science and Computational Intelligence both share the same objective: finding solutions to difficult problems. However, the methods to reach the solutions are different. The main objective of the CSCI Conference is to facilitate increased opportunities for cross-fertilization across CS and CI.

Considering the above broad outline, the CSCI 2025 International Conference was composed of the following focused Research Tracks:

Artificial Intelligence (CSCI-RTAI); Big Data and Data Science (CSCI-RTBD); Computational Science (CSCI-RTCS); Computational Intelligence (CSCI-RTCI); Computational Biology (CSCI-RTCB); Cyber Warfare, Cyber Defense, & Cyber Security (CSCI-RTCW); Signal & Image Processing, Computer Vision & Pattern Recognition (CSCI-RTPC); Smart Cities and Smart Mobility (CSCI-RTSC); Education - CS & CE (CSCI-RTED); Health Informatics and Medical Systems (CSCI-RTHI); Mobile Computing, Wireless Networks, & Security (CSCI-RTMC); Software Engineering (CSCI-RTSE); Internet of Things & Internet of Everything (CSCI-RTOT); Social Network

Analysis, Social Media, & Mining (CSCI-RTNA); Cloud Computing and Data Centers (CSCI-RTCC); Parallel & Distributed Computing (CSCI-RTPD); and Large Language Models (LLM). The scope of each track can be found at: https://www.american-cse.org/csci2025/topics and https://www.american-cse.org/LLM2025/topics.

An important objective of the CSCI 2025 International Conference and its associated Research Tracks was to foster opportunities for cross-fertilization between the fields of Computational Science and Computational Intelligence. The CSCI Conference is deeply committed to promoting diversity and eliminating discrimination, both in its role as a conference organizer and as a service provider. Our goal is to create an inclusive culture that respects and values differences, promotes dignity, equality, and diversity, and encourages individuals to reach their full potential. We are also dedicated, wherever possible, to organizing a conference that represents the global community. We sincerely hope that we have succeeded in achieving these important objectives.

The Steering Committee and the Program Committee would like to extend their gratitude to all the authors who submitted papers for consideration. CSCI 2025 received submissions from 48 countries, with approximately 48% of them coming from outside the United States. Each submitted paper underwent a rigorous peer-review process, with at least two experts (an average of 2.5 referees per paper) evaluating the submissions based on originality, significance, clarity, impact, and soundness. In cases where reviewers' recommendations were contradictory, a program committee member was tasked with making the final decision, often consulting additional referees for further guidance. The CSCI Conference followed the guidelines of COPE (Committee on Publication Ethics):

- Typical submissions underwent a single-blind peer review process, in which the authors remain unaware of the identities of the reviewers, while the reviewers are informed of the authors' identities.
- Papers authored by one or more members of the program committee, including co-chairs, were subjected to a double-blind peer review process, ensuring that neither the authors nor the reviewers were aware of each other's identities or affiliations.

The Research Track on Parallel & Distributed Computing (CSCI-RTPD) received 142 submissions, of which 28 papers were accepted, resulting in a paper acceptance rate of 20%; we selected only 4 papers for this volume. The Research Track on Computational Science (CSCI-RTCS) received 154 submissions, of which 34 papers were accepted, resulting in a paper acceptance rate of 22%; we selected only 20 papers for this volume. The Research Track on Internet of Things & Internet of Everything (CSCI-RTOT) received 58 submissions, of which 14 papers were accepted, resulting in a paper acceptance rate of 24%; we selected only 9 papers for this volume. The Research Track on Computational Biology (CSCI-RTCB) received 120 submissions, of which 18 papers were accepted, resulting in a paper acceptance rate of 15%; we selected only 10 papers for this volume. A total of 43 papers are included in this volume.

We are deeply grateful to the many colleagues who contributed their time and effort to organizing the CSCI 2025 Conference. In particular, we extend our thanks to the members of the Program Committee, the Steering Committee, and the referees. We would also like to express our appreciation to the primary sponsor of the conference, the American Council on Science & Education. The list of members of the Program

Committee for each track can be found at: https://www.american-cse.org/csci2025/committees.

We extend our heartfelt gratitude to all the speakers and authors for their valuable contributions. We would also like to thank the following individuals and organizations for their support: the staff at the Luxor Hotel (an MGM property, conference/meeting department) and the staff of Springer, for their assistance in various aspects of the event.

We are pleased to present the proceedings of CSCI 2025 (selected papers of CSCI-RTPD, CSCI-RTCS, CSCI-RTOT, and CSCI-RTCB Research Tracks). These proceedings represent a collection of outstanding research contributions that reflect the diversity and depth of work in Computational Science, Computational Biology, Parallel & Distributed Computing, and Internet of Things.

Co-editors: Research Tracks on Parallel & Distributed Computing (CSCI-RTPD); Computational Science (CSCI-RTCS); Internet of Things (CSCI-RTOT); and Computational Biology (CSCI-RTCB) of CSCI 2025 International Conference

December 2025

Hamid R. Arabnia
Leonidas Deligiannidis
Farzan Shenavarmasouleh
Farid Ghareh Mohammadi
Soheyla Amirian

Organization

Steering Committee – Co-chairs (CSCI 2025)

Hamid R. Arabnia	University of Georgia, USA
Leonidas Deligiannidis	Wentworth Institute of Technology, USA
Fernando G. Tinetti	Universidad Nacional de La Plata, Argentina
Quoc-Nam Tran	Southeastern Louisiana University, USA

Publication Co-chairs and Co-editors

"Emerging Trends in Computational Science, Computational Biology, Parallel & Distributed Computing, and Internet of Things"

Research Track: Parallel & Distributed Computing (CSCI-RTPD)
Research Track: Computational Science (CSCI-RTCS)
Research Track: Internet of Things & Internet of Everything (CSCI-RTOT)
Research Track: Computational Biology (CSCI-RTCB)

Part of Proceedings of The 2025 International Conference on Computational Science and Computational Intelligence (CSCI)

Hamid R. Arabnia	University of Georgia, USA
Leonidas Deligiannidis	Wentworth Institute of Technology, USA
Farzan Shenavarmasouleh	Medialab Inc., USA
Farid Ghareh Mohammadi	Verify Radiology Images Consultants, LLC, USA
Soheyla Amirian	Pace University, USA

Members of Steering Committee (CSCI 2025)

Baak Akhgar	Sheffield Hallam University, UK
Abbas M. Al-Bakry	University of Information Technology & Communications, Iraq
Nizar Al-Holou	University of Detroit Mercy, USA
Soheyla Amirian	Pace University, USA
Hamid R. Arabnia	University of Georgia, USA
Rajab Challoo	Texas A&M University-Kingsville, USA
Chien-Fu Cheng	Tamkang University, Taiwan
Hyunseung Choo	Sungkyunkwan University, South Korea

Kevin Daimi	University of Detroit Mercy, USA
Leonidas Deligiannidis	Wentworth Institute of Technology, USA
Eman M. El-Sheikh	University of West Florida, USA
Mary Mehrnoosh Eshaghian-Wilner	University of California, Los Angeles, USA
David L. Foster	Kettering University, USA
Ching-Hsien (Robert) Hsu	ChungAsia University, Taiwan
Farid Ghareh Mohammadi	Verify Radiology Images Consultants, LLC, USA
James J. (Jong Hyuk) Park	SeoulTech, South Korea
Mohammad S. Obaidat	University of Jordan, Jordan
Gerald Schaefer	Loughborough University, UK
Farzan Shenavarmasouleh	Medialab Inc., USA
Fernando G. Tinetti	Universidad Nacional de La Plata, Argentina
Quoc-Nam Tran	Southeastern Louisiana University, USA
Shiuh-Jeng Wang	Central Police University, Taiwan
Layne T. Watson	Virginia Polytechnic Institute & State University, USA
Chao-Tung Yang	Tunghai University, Taiwan
Mary Yang	University of Arkansas for Medical Sciences, USA

Program Committee – Computational Science and Computational Intelligence (CSCI 2025)

Afrand Agah	West Chester University of Pennsylvania, USA
Bharat Bhushan Agarwal	IFTM University, India
Omaima Nazar Ahmad AL	Al-Zaytoonah University of Jordan, Jordan
Wasim A. Al – Hamdani	Kentucky State University, USA
Ismail Khalil Al Ani	Ittihad University, United Arab Emirates
Mehran Asadi	Lincoln University, USA
Travis Atkison	University of Alabama, USA
Azita Bahrami	IT Consult, USA
Mehdi Bahrami	Fujitsu Laboratories of America, Inc., USA
P. Balasubramanian	Nanyang Technological University, Singapore
Petra Saskia Bayerl	Erasmus University Rotterdam, The Netherlands
Jane M. Binner	University of Birmingham, UK
Juan-Vicente Capella-Hernandez	Universitat Politècnica de València, Spain
Juan Jose Martinez Castillo	Universidad Nacional Abierta, Venezuela
Rui Chang	Mount Sinai School of Medicine, USA
Dongsheng Che	East Stroudsburg University of Pennsylvania, USA

Jianhung Chen	Chung Hua University, Taiwan
Mu-Song Chen	Da-Yeh University, Taiwan
Xin (Thomas) Chen	University of Hawaii, USA
Steve C. Chiu	Idaho State University, USA
Mark Yul Chu	University of Texas Rio Grande Valley, USA
Jose Alfredo F. Costa	Universidade Federal do Rio Grande do Norte, Brazil
Arianna D'Ulizia	National Research Council of Italy, Italy
Zhangisina G. Davletzhanovna	Central Asian University, Kazakhstan
Wesley Deneke	Western Washington University, USA
Noel De Palma	Grenoble Alpes University, France
Mohsen Doroodchi	University of North Carolina at Charlotte, USA
Levent Ertaull	California State University, East Bay, USA
Mahmood Fazlali	Shahid Beheshti University, Iran
Boyuan Feng	University of California, Santa Barbara, USA
George A. Gravvanis	Democritus University of Thrace, Greece
Gheorghe Grigoras	"Gheorghe Asachi" Technical University of Iaşi, Romania
Ray Hashemi	Georgia Southern University, USA
Houcine Hassan	Universitat Politècnica de València, Spain
Abdolreza Hatamlou	Islamic Azad University, Iran
Bing He	Cisco System Inc., USA
Henry Hexmoor	Southern Illinois University at Carbondale, USA
Gahangir Hossain	Texas A&M University - Kingsville, USA
Guofeng Hou	AQR Capital Management, USA
Ren-Junn Hwang	Tamkang University, Taiwan
Naseem Ibrahim	Penn State Erie, The Behrend College, USA
Rabia Jafri	King Saud University, Saudi Arabia
Shahram Javadi	Azad University, Iran & University of Lisbon, Portugal
Young-Sik Jeong	Dongguk University, South Korea
Aleksandr Katkow	University of Computer Sciences & Skills, Poland
Byung-Gyu Kim	Sun Moon University, South Korea
Taihoon Kim	University of Tasmania, Australia
Sang-Wook Kim	Hanyang University, South Korea
Dattatraya Vishnu Kodavade	D.K.T.E Society's Textile & Engineering Institute, India
Elena B. Kozerenko	Russian Academy of Sciences, Moscow, Russia
Guoming Lai	Sun Yat-sen University, China
Hyo Jong Lee	Chonbuk National University, South Korea
Bo Liu	NEC Labs China, China
Eleanor Lockley	Sheffield Hallam University, UK

Weiqiang Wang	Opera Solutions LLC, USA
Yin Wang	Lawrence Technological University, USA
Alicia Nicki Washington	Duke University, USA
Wei Wei	Xi'an University of Technology, China
Yong Wei	University of North Georgia, USA
Santoso Wibowo	Central Queensland University, Australia
Alexander Woehrer	Vienna Science and Technology Fund, Austria
Bernd E. Wolfinger	University of Hamburg, Germany
Jay Xiong	NSEsoftware, LLC, USA
Hyun Yoe	Sunchon National University, South Korea
Jane You	Hong Kong Polytechnic University, China
Ismail Yusuf	Lamintang Education and Training Center, Indonesia
Wei Zhong	University of South Carolina Upstate, USA

Contents

Internet of Things and Applications

Computational Biology

Poster Papers

Parallel and Distributed Computing and Applications

Understanding CPU Limitation Mechanisms in Cloud-Native Parallel Applications

Samuel Rac and Mats Brorsson(✉)

Interdisciplinary Center for Security. Reliability and Trust, University of Luxembourg, Esch-sur-Alzette, Luxembourg
mats.brorsson@uni.lu

Abstract. Containerized applications can be deployed on cloud platforms, such as a Kubernetes cluster, with limits on how many resources they are allowed to use. These limitations ensure that an application cannot use all available resources on a server. Setting limits correctly facilitates resource allocation by enabling explicit resource requests. However, the resulting performance significantly depends on the CPU limitation mechanism, and the default mechanism used by most users may not yield optimal results.

Containers have two main CPU limitation mechanisms: Time division and Core division. In this article, we present the differences in performance that we observe between these two mechanisms when running parallel applications. Choosing the right mechanism for a given application can significantly impact the execution time, even for the same number of CPUs allocated. Based on these insights, we developed a method that automatically selects the optimal CPU limitation mechanism at runtime. We implemented this method to evaluate it on a Kubernetes cluster, and it demonstrates faster execution time and lower energy consumption compared to the default approach using the Time division mechanism. This method supports heterogeneous clusters comprising different server types.

Keywords: Container · Kubernetes · Degree of parallelism · Auto-scaling

1 Introduction

Containerization has greatly simplified software deployment. Containers package all dependencies into a single artifact, ensuring consistent execution across container runtimes, provided the underlying architecture matches. This also applies to parallel applications, which can use multiple CPU cores to solve a particular problem faster or a larger problem at the same time. Most container runtimes allow resource usage limitations to facilitate multi-tenancy and assist orchestration systems in selecting appropriately sized servers.

H. R. Arabnia et al. (Eds.): CSCI 2025, CCIS 2970, pp. 3–25, 2027.
https://doi.org/10.1007/978-3-032-28466-2_1

An example of resource limitations is *CPU limits*, which play a critical role in CPU resource allocation. They ensure that the CPU resources allocated to an application can be used only by this application. Explicit resource limitations enable efficient scheduling policies and maximize resource utilization. However, it is a challenging task to define CPU limits: over-allocating resources leads to waste, while under-allocation degrades application performance and Quality of Service.

Time division and *Core division* are the two main mechanisms to limit container CPU usage. Time division gives access to all cores in a CPU to the application based on a fraction of a time slot, which provides a very flexible and fine-grained resource allocation. It is also the default CPU resource limit mechanism in Docker and Kubernetes, therefore used by the majority of users. In contrast, Core division divides a CPU vertically and gives an application the entire core(s) allocated to it. As discussed later, Core division can improve performance for certain parallel applications by dedicating cores and ensuring predictable execution.

Some parallel applications (described in the next section) have a faster execution time when using Core division instead of Time division, even if the same number of CPUs is allocated to the application. Since Time division is the default CPU limitation mechanism, it is essential to understand how it works and what impacts its performance to avoid wasting resources. Furthermore, we wanted to investigate the possibility of automatically choosing the right CPU limit mechanism as it is not always easy for developers to understand what would be the best depending on the architecture of the available nodes in a Kubernetes cluster, for instance.

The key contributions of this article are:

1. Explain the differences in performances according to the CPU limitation mechanism
2. Provide guidelines to set CPU limitations: using the most efficient limitation mechanism
3. Present a method to automatically set CPU limitations and its evaluation on a Kubernetes cluster.

The rest of this article is organized as follows: Sect. 2 presents the differences in performances we can observe on parallel applications when using Time division or Core division. It provides some guidelines for setting CPU limitations. Section 3 describes our CPU Limitation Setter (CLS). CLS is a methodology for automatically choosing the best CPU limitation mechanism. It also updates the number of allocated CPUs to avoid waste of resources, it allocates more CPUs only if they provide a significant speedup of the execution time. For example, allocating one more CPU core to an application if it provides a speedup of execution time of only 1% could be considered as a waste of resources. This section also describes the CLS implementation and evaluation on a Kubernetes cluster. Then, Sect. 4 describes related work. Finally, we present our conclusions and future work directions in Sect. 5.

2 Setting Limits

This section describes the main container CPU limitation mechanisms and the performance difference we observe between the two when running parallel applications.

2.1 Container CPU Limitation

Containers have two main mechanisms to limit CPU usage: Time division and Core division. *Control group* (*cgroup*) [12] is the Linux kernel mechanism that limits container resource usage. Cgroups can be used to restrict CPU usage, in addition to memory, IO, and other resources.

The default mechanism to limit CPU usage is Time division. This mechanism gives an application access to the whole CPU for a fraction of the time. There are two main scheduling strategies to allocate time between the different tasks: Completely Fair Scheduler (the default scheduler) and Real-Time scheduler. *Period* and *quota* are the two parameters to use for defining a Time division rule for the CFS scheduler. A containerized application limited by Time division can use the whole CPU for *quota* microseconds each *period* microseconds. Once the quota is spent, the application needs to wait until the period to access the CPU again. The default *period* is 100000 microseconds. The maximum *quota* is equal to 100000 microseconds times the number of vCPU on the machine. Allocating 1.2 vCPU to an application sets the *period* to 100000 microseconds and the *quota* to 120000 microseconds.

The *cpu-set* is the Core division implementation of cgroups. This mechanism limits the container to use only one or many CPU cores. For example, it can limit an application to access only the cores 0 and 1 of a CPU. In addition, it is possible to restrict many applications simultaneously to use cores 0 and 1. However, Kubernetes offers exclusivity guarantees. When using Kubernetes *static* CPU management policy, pods are getting exclusive usage of the CPUs restricted with cpu-set [8]. If a pod is given core 1, the Kubernetes scheduler will not allocate this core to any other pod on the node. Nevertheless, this exclusivity is only for other pods on the node; other processes, like the container runtime, can continue to run on these reserved cores.

Time division and the CFS is the default approach for most containerized workloads, it is the default configuration of Docker and most Kubernetes container runtimes. It allows fine-grain CPU allocation; allocating a fraction of a CPU core is possible. The notation mCPU refers to one-thousandth of CPU time. For example, allocating 500 mCPU to a pod means allocating half a core to the pod. On the contrary, Core division is coarse-grain. It is only possible to allocate full cores, not a fraction of a core.

Time and Core divisions perform similarly most of the time. However, we note that performances are different for parallel applications.

2.2 Time Division and Parallel Applications

We observe that containerized parallel applications can be significantly slower when using Time division. A parallel application's default number of threads varies from one framework or programming language to another. The default value can be set to the number of cores of the machine or a pre-defined setting. This is problematic when running parallel applications with the Time division mechanism. When the number of CPUs allocated with Time division is below the number of threads created by the parallel application, we observe performance degradation. That degradation is higher as the difference between the number of allocated CPUs and the number of threads created increases. A solution is to set that number of threads lower, but there is no general API or a global variable to set it for every parallel framework. Therefore, deploying a parallel application on a Kubernetes cluster with limits and default configuration impacts performance.

However, running the same application and configuration but using Core division instead of Time division changes the application performance. When using Core division, the application does not create more threads than the CPU limits; this improves execution time.

The rest of this section evaluates the impacts of choosing time or Core division on execution time, power consumption, and energy consumption. It also explains the difference in performance between time and Core division.

Experimental Methodology. Table 1 describes our experimental setup. We are using Docker v27.1 and Linux v6.8.

Table 1. Setting limits experimental setup

	Intel Xeon
Microarchitecture	Skylake 8175M
Architecture	x86_64
#Sockets	2
#Phys. Cores	48 (24 + 24)
#Threads	96
Clock speed	2.5 GHz
L1 Cache I/D	1.5 MiB/1.5 MiB
L2 Cache	48 MiB
L3 Cache	66 MiB
Main memory	384
Node name	intel

We measure the execution time, power consumption, and energy consumption for several benchmark applications. We use the RAPL (Running Average Power

Limit) sensor available in intel processors for power and energy consumption. For the initial benchmarks, we use the NAS Parallel Benchmark (NPB), C++ implementation [14] parallelized using OpenMP.

In the first experiment, we compare execution time, power, and energy consumption when we have static CPU limits but varying the number of threads, and with different CPU limitation mechanisms: No limit, Time division, and Core division. Figure 1 and Fig. 2 present the results for the Embarrassingly Parallel (EP) and Conjugate Gradient (CG) benchmarks from NPB. OpenMP programs will, by default, use as many threads as there are cores on the host CPU, and it will not consider multithreading (hyperthreading), so it is very easy to overcommit the processor with threads. In this case, we use the environment variable `OMP_NUM_THREADS` to control the number of threads.

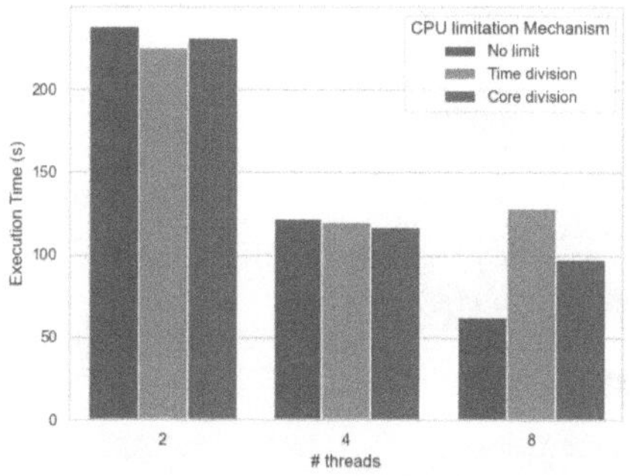

(a) Execution time

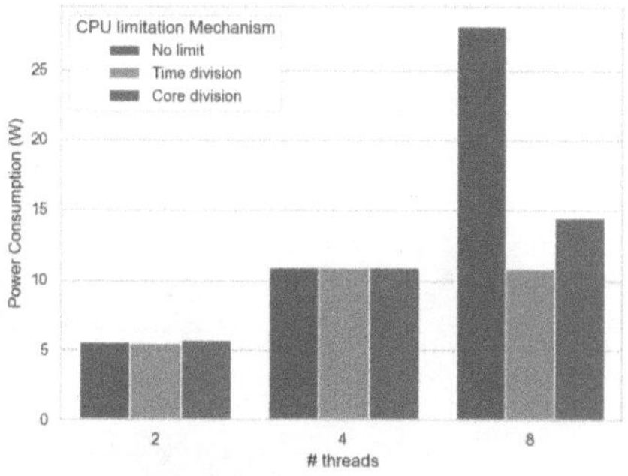

(b) Power Consumption

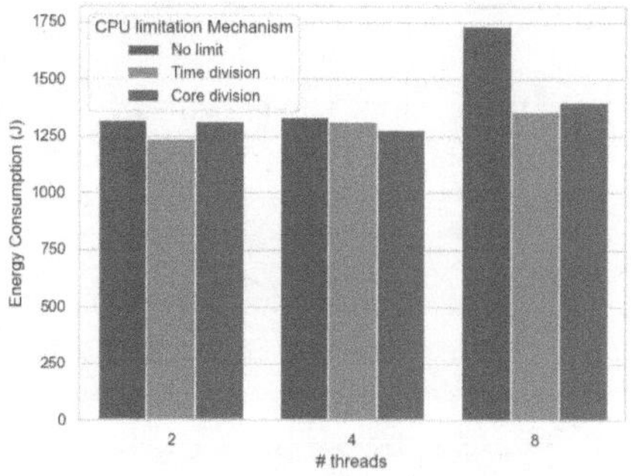

(c) Energy Consumption

Fig. 1. Results for a limit of 4 CPU - Embarrassingly Parallel

For these two applications, we note that the average execution time is similar when the number of threads is lower or equal to the CPU limits. However, it is different when the number of threads is strictly higher than the CPU limit. Execution is then considerably slower when using Time division compared to

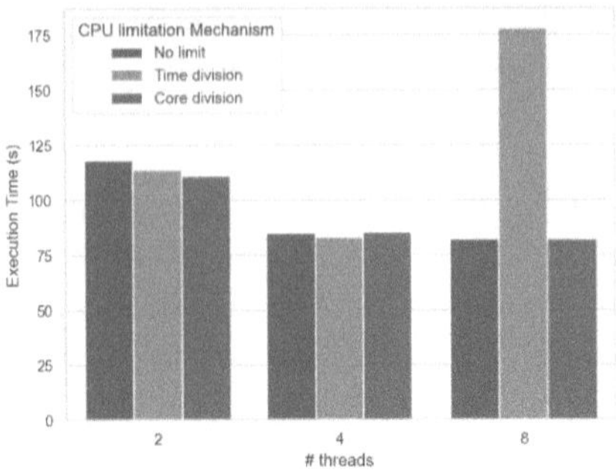

(a) Execution time

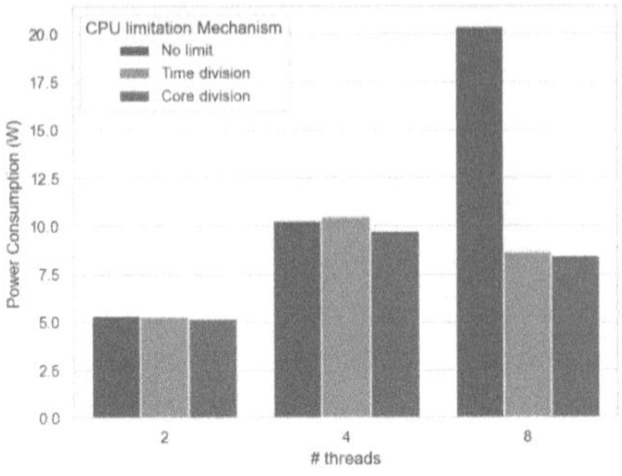

(b) Power Consumption

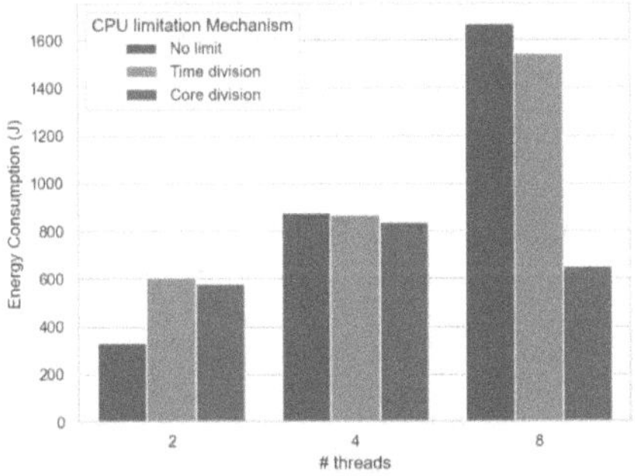

(c) Energy Consumption

Fig. 2. Results for a limit of 4 CPU - Conjugate Gradient

Core division. For the EP application, the execution is faster when there are no limitations. That is the expected behavior; without limitations, more resources are available. For the CG application, the execution is the same when using the Core division or when there are no limitations, indicating that we have reached the practical limit of parallelism. For both of these programs, we observe a slower execution time when using Time division when the number of threads is higher than the CPU limit.

When the number of threads is below the CPU limit, the power consumption is similar for all mechanisms and also with no limit set. However, as with execution time, the power consumption differs when the number of threads is larger than the CPU limit. Without limitation, the power consumption is almost twice as high as it is for the Time division. For Core division, power consumption is higher than Time division for EP, but it is similar for CG. We also observe that the power consumption increases when more threads are added.

The energy values are similar when using fewer threads than the CPU limit. Using no limitations leads to higher energy consumed in every case. The energy consumed is much lower in Core division than in Time division for CG. Longer execution times lead to higher energy consumption.

Image Processing Workload. We also experiment with an image processing workload to confirm our observations on the NAS parallel benchmarks. We build this image processing workload using Yolov8 [24], a computer vision model. This model implementation uses PyTorch [20], one of the main machine learning frameworks. The application processes images from a video stream of a driving scene. The application identifies elements and their locations in the images. These can be helpful for autonomous driving; an embedded camera could, therefore, extract elements around the car and their locations. We want to observe if this parallel application using a framework different from the NAS benchmark presents similar performance variations when changing the CPU limitation mechanism.

Table 2 presents the execution times of the Image processing workload for various CPU Limitation mechanisms. We run these experiments on the Intel node.

Table 2. Image Processing average executions time for Time division (TD), Core division (CD), and No limitation.

	Average Inference Time (ms)		
CPU Limit	TD	CD	No Limitations
1	798	93.2	46.2
2	251	60.9	
3	52.4	46.6	
4	42.5	41.6	
5	43.7	43.0	
6	43.8	43.7	
7	42.6	44.1	
8	44.1	44.2	

We observe a significant difference in performance between Time division and Core division when the CPU limit is equal to 3 vCPUs or less. From 4 vCPUs to 8, there is no significant difference. By default, we observe that the image-processing application creates 4 threads. When using the Time division approach, there are more threads than CPU allocated. The application adjusts the number of threads created when using the Core division approach. Bellow 4 vCPUs allocated, it creates a number of threads that matches the number of allocated vCPUs. For 4 vCPUs and above, it allocates only 4 threads. Without limitations, the application creates only 4 threads. It cannot use all the processing power of the Intel node.

Linux Kernel Preempt Events. This section investigates why execution is slower when using the Time division mechanism. The two previous experiments have shown that parallel applications are slower when using more threads than the number of allocated CPUs and selecting Time division instead of Core division. To explain the performance difference, we study the Linux kernel preempt events. We trace the execution of the LU application from the NAS parallel benchmark suite.

Figure 3 shows the duration and the number of the preempt events and the execution time of the LU application when using different CPU limitation mechanisms and a limit of 2 vCPU. We compare the results when using the default number of threads to the case where we manually set the number of threads to 2. We study Time division (TD), Core division (CD), and No-limit set (No). Figure 3a presents the duration of the preempt events, and Fig. 3b their number of occurrence. Figure 3c shows the total execution time for the different CPU limitation mechanisms.

The duration is similar when the number of threads is set to the same value for each approach. This confirms our previous observations. The pre-empted duration is the same for all CPU limitation mechanisms, and so are the execution times. However, the number of preempt events differs, even when the same number of threads is set.

When the number of threads is set to default, we observe different behaviors for each limitation mechanism.

Execution time is much faster for the No-limit approach as it can use the 96 available CPUs. Then, using Core division over Time division gives an execution time that is 13 times faster.

Preempted duration corresponds to the sum of each preempt event duration for each thread. The No-limit approach has a longer preemption duration than the Core division approach because it uses 96 threads, while the Core division uses only 2. Then, when all threads are preempted, the total preempt duration is longer, but the execution time is still lower. However, that preempt duration is very long for Time division. First, this duration is longer because of the 96 threads, like the No-limit approach. Then, all of the 96 threads are sharing the same time budget of 2 vCPU. That means that each thread has only a very limited time to run. When the time budget is spent, the Linux kernel scheduler runs other processes. Therefore, the threads spend most of their time switching contexts and processing them with limited time. Finally, we think that small measurement inaccuracy might inflate the total duration of preemption. The application and the tracing software are running in the same pod with a limit of 2 CPUs. That might affect the precision of the measures. However, we run all these experiments in the same conditions, with the tracing software always running with the application in a CPU-limited container.

Conclusion. We observe similar effects between these experiments. When the default number of threads is higher than the CPU limit, there is a negative impact on performance when using Time division. Using Core division adjusts the number of allocated threads. Then, we do not observe such a significant impact on performances. When the default number of threads is lower or equal

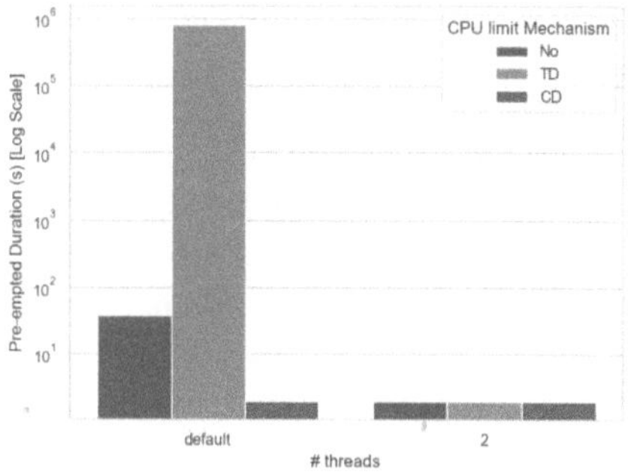

(a) Duration of preempt events

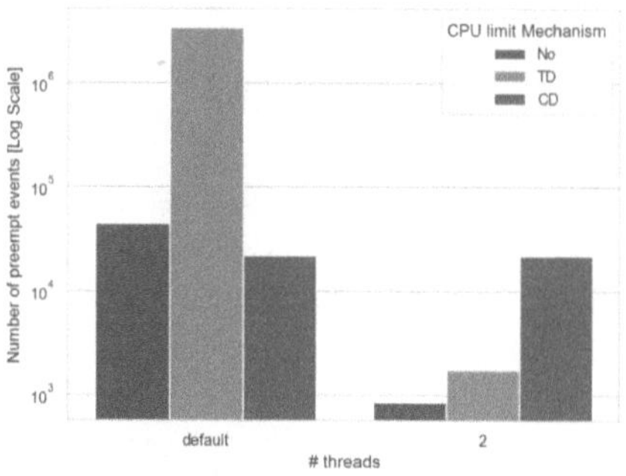

(b) Number of preempt events

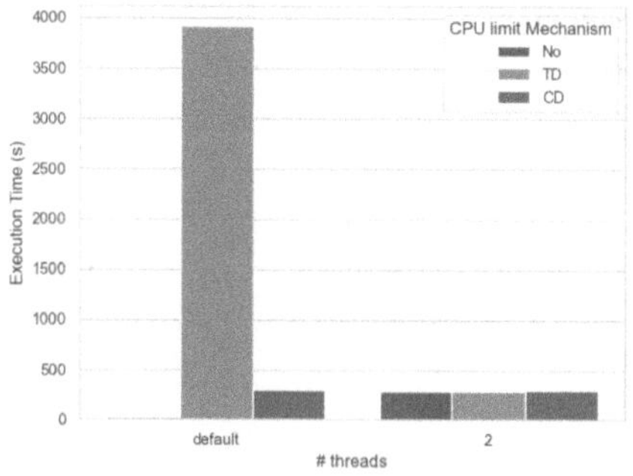

(c) Execution time

Fig. 3. Duration and number of preempt events and execution time for the LU application for a limit of 2 vCPU

to the CPU limit, Time division and Core division perform similarly. In this case, we recommend using Time division. Time division enables more flexibility in resource allocation. Also, Time division supports the in-place resize feature of the CPU limits. That helps update CPU limits without restarting the application. Core division also supports this feature in cgroup and Docker containers but not in Kubernetes. However, using in-place resize for Core division does not update the number of created threads. Therefore, the application needs to be restarted nonetheless to update the number of created threads.

In summary, using Core division helps improve performance when deploying parallel applications on Kubernetes. It is primarily essential when deploying applications on processors with a large number of cores and allocating only a few of them to the application.

3 CPU Limitation Setter (CLS)

This section describes our CPU Limit Setter (CLS). It is a tool that automatically adjusts the CPU limits (i.e., the number of allocated CPUs and the CPU limit mechanism). It selects limits based on the results of the previous section. Setting resource limitations is helpful for scheduling, even if it is not mandatory in Kubernetes. It helps maximize resource utilization while ensuring enough resources for applications to run.

We designed the CLS to be integrated into a Kubernetes cluster. Kubernetes is the de facto standard for container orchestration. Setting limits can be a tedious task. Setting limits too low leads to performance issues, and setting limits too high results in the waste of resources. We design the CLS to adjust resources automatically to find a tradeoff between performance and resource utilization. According to the previous section, having well-set resources also helps reduce energy consumption. Therefore, a tool that automatically sets limits for Kubernetes deployment can be beneficial for application developers.

CLS can determine how many CPUs to allocate and which limitation mechanism to use, realizing a quick application profiling. It creates a copy of the deployment to evaluate its performances with various parameters (e.g., number of CPU allocated, limit mechanism). Then, profiling the application does not disturb the execution of the original application. In addition to ad hoc profiling, the CLS records the actual application's performances to increase the profiling's accuracy without additional costs. CLS makes sure not to profile the same application twice. It checks the hash of the container image to make sure there are no existing profiles about it. Profiles can be reused for different types of nodes at the cost of a lower accuracy. It helps reduce the number of resources required to profile an application.

The CLS proceeds as follows: it first detects which limitation mechanism to use and then profiles the application using the selected mechanism. Finally, it updates the number of CPUs allocated to the application and its CPU limitation mechanism. It only increases the number of allocated CPUs if there is an actual speedup.

Applications are profiled once and run multiple times. Then, it is possible to use application profiles multiple times. Spending a few resources to benchmark the application once and get the benefits of well-adjusted limits for all instances is an acceptable tradeoff. We could even extend this approach to share profiles between multiple clusters, but it is outside of the scope of this study.

3.1 CLS Methodology

This section describes our CPU Limitation Setter (CLS) methodology, its main algorithm, and its integration into the Kubernetes ecosystem. The main idea of this approach is to profile applications while they run and execute the additional tests in the background.

First, the CPU Limitation Setter (CLS) watches each workload running in the cluster. It maintains a list of profiled applications and already optimized pods. CLS uses the hash of the container image and the command of containers to recognize which applications are already profiled. Therefore, if an application has already been profiled, it is unnecessary to do it again. Different instances are automatically detected even if they are using different deployments. When the application is updated, the container image hash will be different, and the CLS will detect it as a new application to profile.

Profiles are stored in a database, using the hash of the container and its command as the key for the entry. A profile is a set of performance measurements (i.e., execution times, power consumption) with different labels (i.e., type of node, CPU limitation mechanism used). Once they are recorded, this measurement can be used for any application deployment. The more instances an application has deployed, the lower the cost of profiling is. We aim to profile applications that are deployed multiple times to achieve economies of scale.

We use a copy of the production deployment for profiling applications without disturbing the production pods. The idea is to create one or more pods similar to the production pod in the background. We can then change the CPU limitation mechanism and CPU limit of the application without disturbing the production environment. Updating the CPU limitation mechanism requires restarting the pod. Restarting the application multiple times would negatively impact the production environment.

We profile the application to detect the right CPU limitation mechanism. It runs two additional deployments, one using Core division and the other using a Time division. Then, it measures performance. These new deployments run in the background and do not affect the production deployment. CLS uses, by default, Time division unless it is more than ϵ percents (e.g., 5%) slower than Core division. It is better to use Time division when it does not negatively affect the performance because it can allocate fractions of a CPU. It helps maximize resource usage and set finer grain limits. However, we use Core division if there is a performance gap between the limitation mechanisms. In this case, the Core division can perform better with the same resources reserved.

The CLS starts benchmarking the application once the CPU limitation mechanism is set. It deploys new pods and measures performances similarly to the CPU limit selection process. Then, once the profiling is done, the CLS stores all the metrics in a database. We do not want to benchmark the application multiple times. The CLS can reuse this profile the next time it detects another instance of the same application. These profiles could even be available for different clusters.

When the application profile is available, the autoscaler can adjust the CPU limit and limitation mechanism of the deployment. CLS ensures that allocating more CPUs to the application really improves the execution time. The rule is to add one more CPU to the limit only if the execution time speedup is greater than η percent (e.g., 5%). This rule avoids wasting CPU resources.

Finally, the CLS continues iterating about the workloads in the cluster, adjusting limits to the objectives.

Algorithm 1 presents the algorithm of the CLS. The CLS watches the workload deployments in the cluster. First, it checks if the deployment pods are already profiled. If not, it starts selecting the proper limitation mechanism. Then, it profiles the application for various CPU values. To select the proper limit mechanism and benchmark the application, CLS creates a copy of the deployment to avoid disturbing its execution. After the measurements, it deletes all the deployments it used for profiling the application. All the profiles are stored in a dedicated database. The pod must be restarted when changing the CPU limitation mechanism (e.g., from Time division to Core division). However, updating the number of CPUs allocated using the Time division *in-place resize* feature is possible without restarting the application. There is no similar method available for the Core division mechanism.

Algorithm 1. CPU limit setter

```
Require: ε > 0, η > 0
  for each deployment in cluster do
    if deployment not profiled then                          ▷ check container image hash
      profile ← CreateDeploymentCopy(deployment, td, 1)      ▷ Use Time division and set CPU limit to 1
      profile.save()
      profile ← CreateDeploymentCopy(deployment, cd, 1)      ▷ Use Core division and set CPU limit to 1
      profile.save()
      if |profile.tdPerformances−profile.cdPerformances| / max(profile.tdPerformances,profile.cdPerformances) < ε then
        profile.limitationMechanism ← TD
      else
        profile.limitationMechanism ← CD
      end if
      cpuLimits ← 2
      while cpuLimit ≤ MaxCPULimit AND profile.lastSpeedup > η do
        profile ← CreateDeploymentCopy(deployment, profile.limitationMechanism, cpuLimit)
        profile.save()
        cpuLimits ← cpuLimits + 1
      end while
    end if
    profile ← getProfile(deployment)
    updateLimits(deployment, profile.LimitMechanism, profile.NumberCPU)
  end for
```

Kubernetes Implementation. Kubernetes is the *de facto* standard for container orchestration. That is why we choose to integrate the CLS into the Kubernetes ecosystem.

The CLS is a Go application deployed in a Kubernetes cluster in a dedicated pod. It can access the Kubernetes APIs to create, update, and delete deployments. It can also create copies of deployments, modifying only some parameters like CPU limits. CLS can pull monitoring data from open source software like Prometheus [19]. Then, it stores the software profiles in a dedicated database. Each profile is identified with the hash and the command of the container of the profiled application. A profile contains information about the execution time, CPU characteristics, CPU limitation mechanism, and allocated resources. The

CLS uses the profiles to set appropriate limits when deploying a new application. Profiles are created when the first instance of the application is deployed. Then, profiles are updated continuously while application instances are running.

Figure 4 presents an overview of the integration of the CLS into the Kubernetes ecosystem.

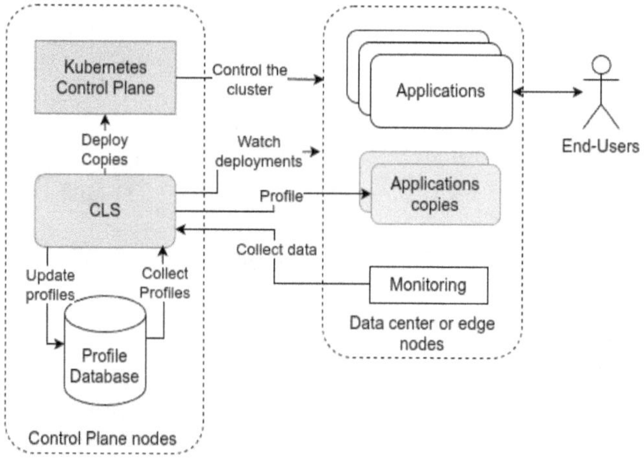

Fig. 4. Overview of the CPU Limitation Setter (CLS)

3.2 CLS Evaluation

Experimental Setup. We use the NAS benchmark C++ implementation [14] for these performance experiments. It uses OpenMP APIs.

Table 3 describes our experimental setup. All nodes use the same software configuration: Docker v27.1, Kubernetes v1.28, and Linux v6.8.

We use the RAPL (Running Average Power Limit) sensor to measure power and energy consumption. This sensor is only available on x86 processors from Intel and AMD. Therefore, there is no energy data available for the ARM processor.

During these benchmark experiments, we compare the execution time, power, and energy consumption.

Results. Table 4 presents the CLS results. It compares the execution speedup, the energy consumed, and the resource allocated for six applications from the NAS benchmark suite [14]: CG - Conjugate Gradient, EP - Embarrassingly Parallel, FT - discrete 3D fast Fourier Transform, IS - Integer Sort, LU - Lower-Upper symmetric Gauss-Seidel, MG - Multi-Grid on a sequence of meshes.

We compare the performance of each application using different CPU limitation strategies, using the Time division mechanism (the default Kubernetes

Table 3. CLS evaluation experimental setup

	Intel Xeon	AMD EPYC	AWS Graviton
Microarchitecture	Skylake 8175M	EPYC 7R13	Graviton 2
Architecture	x86_64	x86_64	aarch64
#Sockets	2	2	1
#Phys. Cores	48 (24 + 24)	96 (48 + 48)	64
#Threads	96	192	64
Clock speed	2.5 GHz	2.6 GHz	2.5 GHz
L1 Cache I/D	1.5 MiB/1.5 MiB	3 MiB/3 MiB	4 MiB/4 MiB
L2 Cache	48 MiB	48 MiB	64 MiB
L3 Cache	66 MiB	384 MiB	32 MiB
Main memory	384	768	256
Node name	Intel	AMD	ARM

approach) and comparing it to our CPU Limit Setter (CLS). The CLS can choose either Core division or the Time division mechanism. We present execution time speedup normalized relatively to the execution time of the application with only one CPU allocated.

We reproduce this experiment on three different CPUs (Intel, AMD, and ARM). We use the RAPL sensor to record power and energy consumption. This sensor is only available on Intel and AMD x86 chips. Therefore, we have energy values to discuss for the ARM node. We report the total energy consumed during the execution of each application.

For these applications, the CLS chooses the Core division mechanism every time.

The CLS has the highest execution time speedup for every application and each CPU. The higher the execution time speedup, the better. CLS gets a higher speedup than the default approach of Kubernetes, even when they are using the same number of CPUs. That can be explained by the fact that CLS chooses to use Core division instead of the Time division mechanism. The Core division mechanism offers better performances for these parallel applications.

The CLS uses the maximum CPU for most of the cases. The lower the resources allocated, the better. It uses the most CPU possible, but there is still a significant speedup when adding more. For the Conjugate Gradient application, the AMD node gets the faster speedup with only five CPUs. For the Multi-Grid application, the AMD node gets the faster speedup with only three CPUs. The CLS allocates fewer CPUs because the relative speedup between five and six, and between three and four CPUs was less than 5%.

The CLS consumes less energy than any other approach. The lower the energy consumption, the better. The power consumption is similar when using Time division or Core division, as described in the previous section. However, the execution time when using Core division is much lower; therefore, the energy consumption is lower.

Transferable Profiles. In this section, we study how to reuse application profiles when deploying new application instances on a different CPU type (e.g., with a different clock speed, generation, constructor, or even architecture). It is important to support heterogeneous clusters to be able to use the CLS in the Compute Continuum. The idea is to avoid profiling the application on every type of node in the cluster, using an approximation based on the initial profiles. Over time, profiles become increasingly complete as they also include information about the new instances running (and sometimes running on different types of nodes). Reducing the number of nodes to profile enables economies of time, costs, and energy.

To understand how similar profiles can be between two different types of CPU, we study i) the decision of the CLS for the CPU limitation mechanism and ii) the relative execution time speedups when increasing the number of allocated vCPUs. We compare these decisions and speedups for the Intel, AMD, and ARM nodes.

Table 5 presents the chosen mechanisms between Time division (TD) and Core division (CD) for each application and CPU type. The decision of the CLS is the same for all CPU types and applications except for the EP application on the ARM node. There is no significant difference between TD and CD execution time. That is why CLS chose TD over CD. However, with similar performances between TD and CD, choosing CD instead of TD is not problematic. For these six applications, the choice of the CPU Limitation Mechanism can be the same for every node. There are no negative impacts if the decision of the CLS from another node regarding the CPU Limitation Mechanism is reused.

Figure 5 and Fig. 6 present the relative speedups when allocating one more CPU. Results are different for Core division and Time division.

For Core division, speedups are mainly similar from one CPU to another. The results of the CLS would be the same except for two applications (CG and MG) on the AMD node.

For Time division, speedups are different from one node to another. It would be difficult to predict the results for a different node. However, these are parallel applications; CLS chose to use Core division because of the performance difference between the two mechanisms. If CLS had chosen the Time division, we could expect performances to be similar to or better than the Core division (otherwise, the Core division would have been chosen). If performance is similar between the two, we could expect to have results similar to Core division where reusing CLS decisions for another node is not problematic.

Using a profile from a different type of node seems to be a promising tradeoff. In most cases, the CLS decision does not change from one node type to another. The error due to the profile transfers could have a more negligible impact on resource utilization than systematically profiling applications on every node. In addition, profiles are updated with production data. That means that transfer

Table 4. CLS Results

	Execution Time Speedup			Resource utilization (vCPU)			Energy Consumed (J)		
CG	Intel	AMD	ARM	Intel	AMD	ARM	Intel	AMD	ARM
TD 1	1,00	1,00	1,00	1	1	1	1,01 10^3	1,47 10^3	-
TD 2	2,08	1,08	1,87	2	2	2	9,73 10^2	2,72 10^3	-
TD 4	3,76	1,82	3,91	4	4	4	1,08 10^3	3,24 10^3	-
TD 8	23,3	4,17	8,03	8	8	8	3,48 10^2	2,83 10^3	-
CLS	**141**	**235**	**13,7**	8	5	8	**5,77** 10^1	**3,14** 10^1	-
EP	Intel	AMD	ARM	Intel	AMD	ARM	Intel	AMD	ARM
TD 1	1,00	1,00	1,00	1	1	1	1,59 10^2	1,41 10^2	-
TD 2	2,19	2,93	2,01	2	2	2	1,45 10^2	9,62 10^1	-
TD 4	4,63	6,27	4,01	4	4	4	1,37 10^2	8,99 10^1	-
TD 8	9,50	14,0	**8,09**	8	8	8	1,34 10^2	8,07 10^1	-
CLS	**11,2**	**17,9**	**8,09**	8	8	8	**1,14** 10^2	**6,29** 10^1	-
FT	Intel	AMD	ARM	Intel	AMD	ARM	Intel	AMD	ARM
TD 1	1,00	1,00	1,00	1	1	1	2,83 10^2	2,49 10^2	-
TD 2	2,30	2,33	1,95	2	2	2	2,46 10^2	2,14 10^2	-
TD 4	5,12	5,61	3,57	4	4	4	2,21 10^2	1,77 10^2	-
TD 8	14,0	11,3	7,15	8	8	8	1,62 10^2	1,76 10^2	-
CLS	**45,5**	**62,9**	**8,99**	8	8	8	**4,98** 10^1	**3,17** 10^1	-
IS	Intel	AMD	ARM	Intel	AMD	ARM	Intel	AMD	ARM
TD 1	1,00	1,00	1,00	1	1	1	1,44 10^1	2,04 10^1	-
TD 2	1,00	1,61	1,89	2	2	2	2,88 10^1	2,53 10^1	-
TD 4	1,94	2,93	3,65	4	4	4	2,98 10^1	2,79 10^1	-
TD 8	12,8	5,53	6,20	8	8	8	9,00 10^0	2,96 10^1	-
CLS	**40,5**	**127**	**13,1**	8	8	8	**2,85** 10^0	**1,29** 10^0	-
LU	Intel	AMD	ARM	Intel	AMD	ARM	Intel	AMD	ARM
TD 1	1,00	1,00	1,00	1	1	1	2,66 10^3	7,69 10^3	-
TD 2	0,71	4,33	1,15	2	2	2	7,51 10^3	3,55 10^3	-
TD 4	0,39	5,82	5,40	4	4	4	2,73 10^4	5,29 10^3	-
TD 8	13,8	32,8	19,2	8	8	8	1,54 10^3	1,87 10^3	-
CLS	**115**	**697**	**33,0**	8	8	8	**1,85** 10^2	**8,83** 10^1	-
MG	Intel	AMD	ARM	Intel	AMD	ARM	Intel	AMD	ARM
TD 1	1,00	1,00	1,00	1	1	1	3,37 10^2	9,04 10^2	-
TD 2	2,11	1,83	1,87	2	2	2	3,19 10^2	9,88 10^2	-
TD 4	3,22	4,89	3,91	4	4	4	4,18 10^2	7,40 10^2	-
TD 8	7,20	6,14	8,03	8	8	8	3,74 10^2	1,18 10^3	-
CLS	**41,17**	**99,26**	**13,73**	8	3	8	**6,54** 10^1	**2,73** 10^1	-

Table 5. CPU Limitation Mechanism chosen by the CLS.

	Chosen CPU Limitation Mechanism		
Application	Intel	AMD	ARM
CG	CD	CD	CD
EP	CD	CD	TD
FT	CD	CD	CD
IS	CD	CD	CD
LU	CD	CD	CD
MG	CD	CD	CD

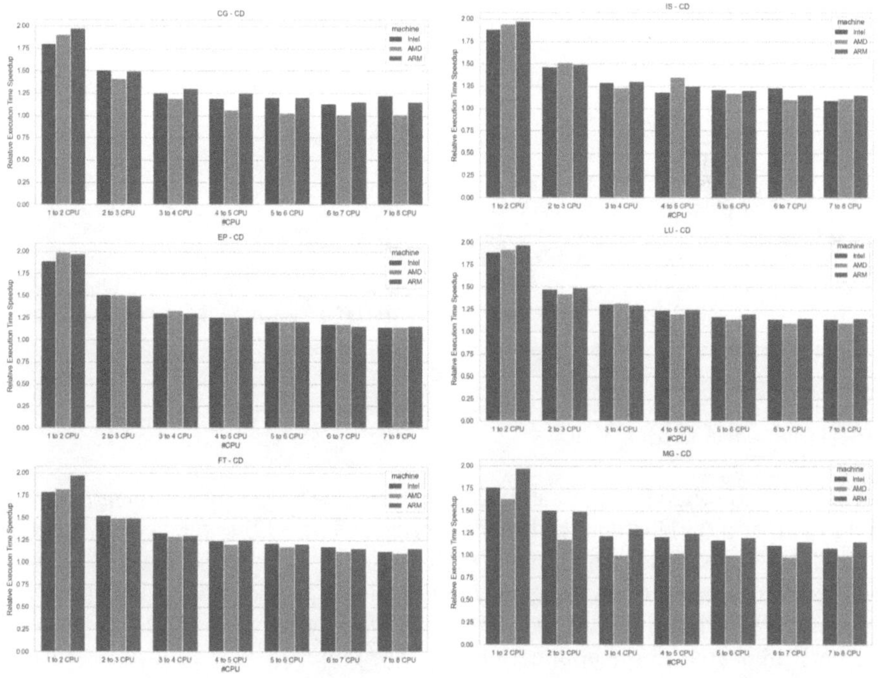

(a) Relative speedups for CG, EP, and FT - Core division

(b) Relative speedups for IS, LU, and MG - Core division

Fig. 5. Relative speedups - Core division

profile approximations are only necessary until one instance runs or has run on every node type.

CLS Overhead. Fig. 7 shows the CPU time total usage of the Kubernetes Control Plane, Monitoring, and CLS pods. These values correspond to a 20-minute experiment. We observe that the total CPU time for the CLS was 1 s

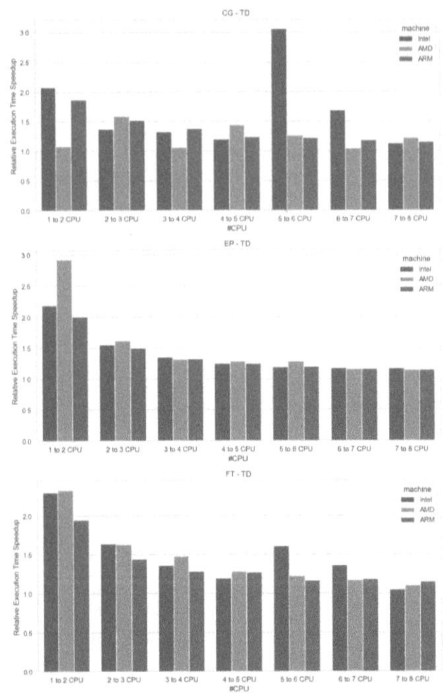

(a) Relative speedups for CG, EP, and FT - Time division

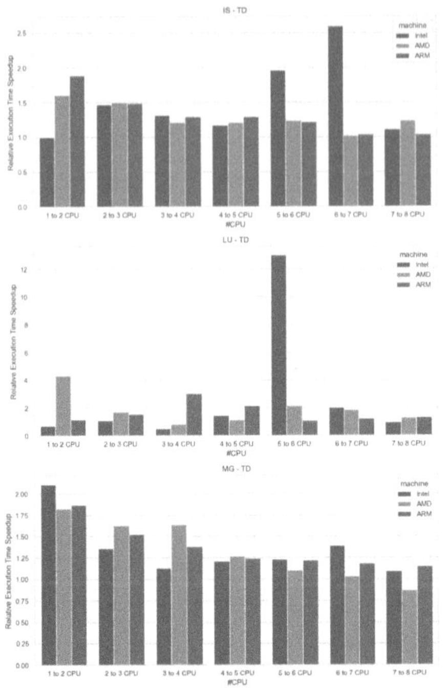

(b) Relative speedups for IS, LU, and MG - Time division

Fig. 6. Relative speedups - Time division

during the experiment. It is significantly lower than other components we can find in Kubernetes clusters like the Prometheus. The CLS application has a very limited impact on resource usage.

However, the copies of the deployments have a larger impact. It depends on the number of CPU limits tested, but in the worst case, it doubles the resources currently allocated to the application for a few minutes. This additional usage of resources is not negligible, but applications are only profiled once and deployed multiple times often.

In summary, using the CLS has a cost associated with using additional resources to profile the applications. This cost should be observed in relation to the number of instances of the application that will be running. Profiling applications can save resources overall if they are deployed multiple times.

4 Related Work

Auto-scaling is a method for automatically adjusting allocated resources. There are two main approaches for auto-scaling: horizontal scaling (i.e., adding more machines or application instances), the most common approach, and vertical

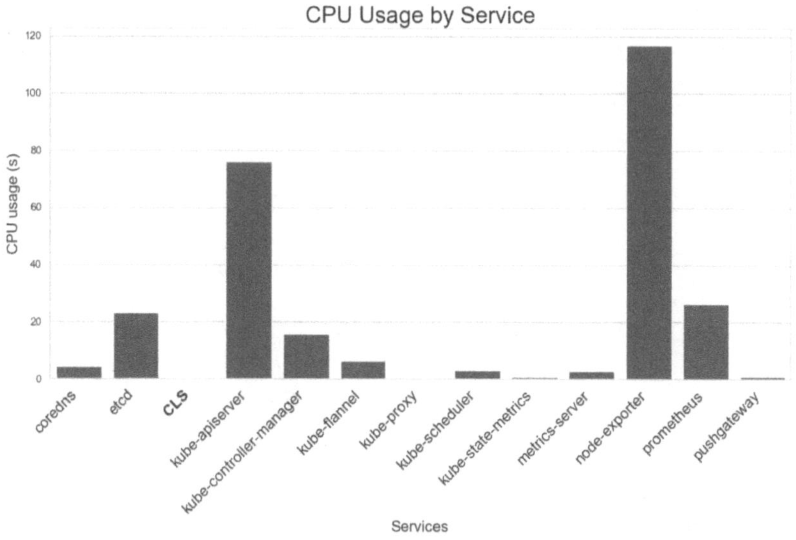

Fig. 7. Total CPU time of the Kubernetes Control Plane and Monitoring

scaling (i.e., increasing the resources allocated to one machine) is another option. Kubernetes has a built-in auto-scaler [9]. It is a reactive auto-scaler that offers a horizontal and a vertical approach. The Horizontal Pod Autoscaler (HPA) deploys more or less pod replicas. The Vertical Pod Autoscaler (VPA) adjusts the resources allocated to a pod. It is only available in *beta*; it is not fully operational like the HPA. The *Cool Down Time* parameter sets the duration between two scaling operations up or down. It prevents the auto-scaler from making new decisions before observing the effects of the previous actions. The cluster scaling adjusts the number of nodes in the Kubernetes cluster. It offers API compatible with most cloud providers. The default mode of the VPA and the HPA is to monitor resource usage (CPU and memory) to make auto-scaling decisions. However, it is also possible to use custom metrics. Holoscale presents an approach that mixes horizontal and vertical scaling to reduce delays when creating new Virtual Machines [17]. Quattrocchi et al. compare their two autoscaling methods to existing methods of major cloud providers [21]. One of their autoscaler is based on control theory, and the second is based on queuing theory. [18] uses machine learning to predict how many pod replicas should be deployed to achieve the expected response time. [3] predicts the pod resource limits using decision tree regression.

Recent research has extended auto-scaling approaches to include energy considerations. [15] presents an energy-aware scheduling algorithm for microservices in Kubernetes clusters that reduces energy consumption while maintaining service level agreements. This approach complements traditional CPU and memory-based auto-scaling by incorporating power consumption metrics into scheduling decisions. [7] provides container-level energy observability in Kubernetes clusters, enabling fine-grained energy monitoring that can inform auto-scaling decisions. The integration of energy awareness with auto-scaling represents a growing trend toward sustainable computing in containerized environments.

There are many approaches to finding a **suitable number of threads**. De Lima et al. use Artificial Neural Networks to predict thread count combinations that provide optimal energy-delay product [11]. Da Silva et al. present an autoscaler that first tunes the number of threads during the runtime to find a tradeoff between energy consumption and performance and then reallocates freed resources (if any) to other containers [22]. One limitation is that this approach only supports OpenMP workloads.

Balla et al. have created an adaptative autoscaler to adjust CPU limits to have just enough processing power to be below execution time requirements [2]. [4] uses machine learning to automatically tune the degree parallelism for query optimization.

Machine learning approaches are increasingly being applied to optimize parallel program performance in containerized environments. [1] uses machine learning to predict software defects in hybrid MPI and OpenMP parallel programs, which can help identify optimal thread configurations before deployment. This predictive approach complements runtime thread tuning by preventing performance issues related to incorrect parallelization patterns. Such ML-driven optimization techniques are particularly relevant for automated CPU limit setting as they can predict application behavior under different resource constraints.

Performances of containers and the Linux CFS has been studied in different studies [5,6,13,25]. [5] proposes to extend the notion of namespace to include more detailed information about available resources to the containers. This could improve resource management for parallel runtimes and programming languages.

Zhang et al. propose a joint optimization of CPU scaling and core sharing to address a limitation of the CFS [25]. The CFS forces threads from different containers to share cores, which can waste resources.

Liu et al. observe performance degradations in multi-tenant environments [13]. They find that containers are not getting all the resources they requested due to the CFS forcing the threads of different containers to share CPU cores. [6] studies limitations of the CFS for burstable containers. The authors present two OS mechanisms to mitigate the performance issues. Recent comprehensive studies have evaluated container performance across different platforms and measurement approaches. [10] provides performance evaluation of containerized virtualization platforms including Docker, Podman, and LXC/LXD, showing performance overhead of approximately 8% in containerization environments.

In [16] the authors present a comprehensive analysis of open-source power measurement tools in containerized clouds, comparing various approaches for monitoring power consumption in container environments.

These studies complement our RAPL-based measurements by providing broader context for container performance evaluation. Real-world deployment experiences, such as [23], demonstrate the practical challenges and benefits of container adoption in HPC environments, highlighting the importance of proper CPU resource management for achieving expected performance gains.

5 Conclusion

This article analyzes the performances of different Container CPU limitation mechanisms. We observe that the Core division limitation mechanism performs better than Time division for some containerized parallel applications. Using Core division can provide lower execution time and energy consumption for the same CPUs allocated.

We propose a CPU Limitation Setter (CLS) method for automatically selecting the CPU limitation mechanism that offers the best performance. Our Kubernetes implementation of this method demonstrates faster execution time and lower energy consumption compared to the default approach using the Time division mechanism.

Profiling applications with the CLS has an overhead worth paying when deploying many instances of the same application. In addition, profiles are updated over time with already deployed applications to reduce this overhead. Also, using a profile for different machines is a reasonable approximation. Transferring profiles from one machine type to another helps support heterogeneous clusters with different kinds of servers.

Future work can investigate ways to update CPU limitation decisions during application runtime to react to workload variation. New CPU limitation decisions could rely on data acquired during runtime to improve the accuracy of the decision.

Acknowledgements. This research has been partly funded by the Luxembourg National Research Fund (FNR) under contract number 16327771 and has been supported by Proximus Luxembourg SA.

References

1. Alenezi, M., Almomani, M.A.: Predicting software defects in hybrid MPI and OPENMP parallel programs using machine learning. Electronics **13**(1), 182 (2024). https://doi.org/10.3390/electronics13010182
2. Balla, D., Simon, C., Maliosz, M.: Adaptive scaling of Kubernetes pods. In: NOMS 2020 - 2020 IEEE/IFIP Network Operations and Management Symposium, pp. 1–5 (2020). https://doi.org/10.1109/NOMS47738.2020.9110428, https://ieeexplore-ieee-org.proxy.bnl.lu/abstract/document/9110428
3. Bouflous, Z., Haraka, F., Ouzzif, M., Bouragba, K.: Enhanced vertical pod auto scaling with decision tree regressor-max in Kubernetes. In: 2024 11th International Conference on Wireless Networks and Mobile Communications (WINCOM), pp. 1–5 (2024). https://doi.org/10.1109/WINCOM62286.2024.10654970, https://ieeexplore.ieee.org/abstract/document/10654970

4. Fan, Z., Sen, R., Koutris, P., Albarghouthi, A.: Automated tuning of query degree of parallelism via machine learning. In: Proceedings of the Third International Workshop on Exploiting Artificial Intelligence Techniques for Data Management, pp. 1–4. aiDM '20, Association for Computing Machinery (2020). https://doi.org/10.1145/3401071.3401656
5. Huang, H., Rao, J., Wu, S., Jin, H., Suo, K., Wu, X.: Adaptive resource views for containers. In: Proceedings of the 28th International Symposium on High-Performance Parallel and Distributed Computing, pp. 243–254. HPDC 2019, Association for Computing Machinery (2019). https://doi.org/10.1145/3307681.3325403
6. Huang, H., et al.: Adapt burstable containers to variable CPU resources **72**(3), 614–626 (2023)
7. Huber, S., Becker, S.: Container-level energy observability in Kubernetes clusters (2024). https://arxiv.org/abs/2504.10702
8. Kubernetes: Control CPU management policies on the node (2024). https://kubernetes.io/docs/tasks/administer-cluster/cpu-management-policies/#static-policy. Accessed 9 May 2024
9. Kubernetes: Kubernetes autoscaler (2024). https://github.com/kubernetes/autoscaler?tab=readme-ov-file. Accessed 9 May 2024
10. Kumar, A., Singh, R.: Performance evaluation of containerized virtualization platforms. SN Comput. Sci. **6**, 1–20 (2025). https://doi.org/10.1007/s42979-025-04297-3
11. de Lima, E.C., Rossi, F.D., Luizelli, M.C., Calheiros, R.N., Lorenzon, A.F.: A neural network framework for optimizing parallel computing in cloud servers. J. Syst. Architect. **150**, 103131 (2024)
12. Linux Kernel: Control group (2024). https://www.kernel.org/doc/html/latest/admin-guide/cgroup-v2.html. Accessed 9 May 2024
13. Liu, L., Wang, H., Wang, A., Xiao, M., Cheng, Y., Chen, S.: Mind the gap: broken promises of CPU reservations in containerized multi-tenant clouds. In: Proceedings of the ACM Symposium on Cloud Computing, pp. 243–257. SoCC 2021, Association for Computing Machinery (2021). https://doi.org/10.1145/3472883.3486997
14. Löff, J., et al.: The NAS parallel benchmarks for evaluating C++ parallel programming frameworks on shared-memory architectures. Futur. Gener. Comput. Syst. **125**, 743–757 (2021)
15. Mahmud, A.R., Ullah, A., Hassan, M.M.: Energy-aware scheduling algorithm for microservices in Kubernetes clouds. J. Grid Comput. **22**(3), 1–25 (2024). https://doi.org/10.1007/s10723-024-09788-w
16. Martinez, C., Johnson, D., Smith, E.: Impact of power consumption in containerized clouds: a comprehensive analysis of open-source power measurement tools. Comput. Netw. **245**, 110232 (2024). https://doi.org/10.1016/j.comnet.2024.110232
17. Millnert, V., Eker, J.: HoloScale: horizontal and vertical scaling of cloud resources. In: 2020 IEEE/ACM 13th International Conference on Utility and Cloud Computing (UCC), pp. 196–205 (2020). https://doi.org/10.1109/UCC48980.2020.00038, https://ieeexplore-ieee-org.proxy.bnl.lu/abstract/document/9302812
18. Pramesti, A.A., Kistijantoro, A.I.: Autoscaling based on response time prediction for microservice application in Kubernetes. In: 2022 9th International Conference on Advanced Informatics: Concepts, Theory and Applications (ICAICTA), pp. 1–6 (2022). https://doi.org/10.1109/ICAICTA56449.2022.9932943, https://ieeexplore.ieee.org/abstract/document/9932943
19. Prometheus (2024). https://prometheus.io/. Accessed 9 May 2024

20. PyTorch foundation: Pytorch (2024). https://pytorch.org/. Accessed 9 May 2024
21. Quattrocchi, G., Incerto, E., Pinciroli, R., Trubiani, C., Baresi, L.: Autoscaling Solutions for Cloud Applications under Dynamic Workloads, pp. 1–17 (2024). https://doi.org/10.1109/TSC.2024.3354062, https://ieeexplore.ieee.org/abstract/document/10419899
22. da Silva, V.S., et al.: Synergistically rebalancing the EDP of container-based parallel applications. IEEE Trans. Parallel Distrib. Syst. (2024)
23. Thompson, M., Wilson, S.: Container adoption in campus high performance computing at texas a&m university. In: Practice and Experience in Advanced Research Computing 2024: Human Powered Computing, pp. 1–8. PEARC 2024, ACM (2024). https://doi.org/10.1145/3626203.3670550
24. Ultralytics: Ultralytics YOLOv8 (2024). https://github.com/ultralytics/ultralytics. Accessed 9 May 2024
25. Zhang, Y., Zhang, H., Wu, Y., Ma, H.: Joint optimization of CPU scaling and core sharing in container clouds. In: 2023 IEEE 29th International Conference on Parallel and Distributed Systems (ICPADS), pp. 1031–1038 (2023). https://doi.org/10.1109/ICPADS60453.2023.00152, https://ieeexplore.ieee.org/abstract/document/10476265#full-text-header

AI-Cache: A Caching Method Applying Application Inference Targeting File Sharing Servers

Keiichiro Higashihara[1(✉)], Daigo Michishita[2], Koji Kida[3], and Hitoshi Kamei[4]

[1] Graduate School of Science for Creative Emergence, Kagawa University, Takamatsu, Japan
s25g362@kagawa-u.ac.jp
[2] NEC Corporation, Tokyo, Japan
[3] Faculty of Engineering and Design, Kagawa University, Takamatsu, Japan
[4] Integrated Center for Informatics, Kagawa University, Takamatsu, Japan

Abstract. The performance of file sharing servers, such as NFS server, is highly dependent on caching. Meanwhile, the page cache which is a general-purpose caching mechanism implemented in the Linux kernel is not optimized for the I/O accesses of specific applications. Thus, the cache efficiency decreases for certain workloads. To address this problem, we propose a caching method that infers applications running on clients by a server based on received NFS I/O patterns. To verify the effectiveness of the proposed method, we build a Proof-of-Concept (PoC) system that integrates the application inference method as an external module with an NFS server and it performs read-ahead. The results of evaluation show that the proposed method eliminates the penalty of a cache miss, especially in large-scale file access. It improves I/O throughput; thus, it can reduce 42% of file transfer times in case of 3.2 GB file. This result demonstrates the effectiveness of caching with application inference.

Keywords: Caching · File Sharing Server · Application Inference · File Access

1 Introduction

Server performance relies heavily on effective caching mechanisms to bridge the speed gap between memory and storage. The page cache implemented in general-purpose operating systems like Linux provides stable performance for various workloads [1, 2]. However, this generic approach is not optimized for the specific I/O patterns of applications. As modern applications become increasingly diverse, this general-purpose cache mechanism often leads to suboptimal performance and may not fully leverage workload-specific characteristics that could significantly improve cache efficiency.

Conventionally, a common solution to this problem has been manual cache parameter tuning by system administrators. However, this approach is static and cannot be fully utilized in environments with changing workloads. Therefore, an automated approach is required. We hypothesize that if we can automatically identify the application accessing the server, we can dynamically apply caching method optimized for that specific workload.

H. R. Arabnia et al. (Eds.): CSCI 2025, CCIS 2970, pp. 26–38, 2027.
https://doi.org/10.1007/978-3-032-28466-2_2

Various studies have shown that methods exist that leverage file system I/O patterns [3–6]. In some of these studies, methods for inferring client applications based on I/O patterns are proposed [3, 6, 7]. These are proposed for security purposes such as ransomware detection. Thus, applying these methods becomes a challenge.

In this paper, we propose Application Inference Cache (AI-Cache) which is a caching method for NFS that applies these inference methods. This paper evaluates the effectiveness of AI-Cache. To evaluate this effectiveness, this paper describes the design and implementation of Proof-of-Concept (PoC) of AI-Cache. Through a series of experiments, this paper evaluates the feasibility of this approach and demonstrates its effectiveness in improving I/O throughput and reducing file transfer times, particularly for large-scale files such as 3.2 GB files.

2 Background

2.1 File Sharing Server

A file sharing server is an appliance for storing and sharing files among clients via a network. The server exports a local file system to multiple clients. The clients mount the file system exported by the server and manipulate files on the server in the same way as regular files stored in a client locally. Protocols used for file sharing are CIFS/SMB for Windows OS and NFS for UNIX-like OS such as Linux [8, 9].

According to RFC 1094, NFS has the following specifications [8]. Network Filesystem (NFS) protocol provides transparent remote access to shared files via networks. The NFS protocol is designed for portability across different machines, operating systems, network architectures, and transport protocols. This paper focuses on NFS.

When a client reads a file stored on an NFS server, the process of NFS is executed as follows:

1. The client verifies access permissions and attributes.
2. The client sends a "read call" procedure to a NFS server. In this request, the client specifies the file handle of a target file, access offset, and length to read.
3. NFS server returns the data in a "read reply" according to the client's instructions.
   This "read call" and "read reply" process is repeated as many times as necessary.

Therefore, NFS servers only accept requests from clients regarding file access. NFS server does not know what applications the client is running.

2.2 Cache Methods for NFS

The I/O performance of NFS relies on caching mechanisms, which exist on both the server and client side in an NFS environment. Each is described below:

1. Server Cache: The server-side cache is not NFS-specific, but rather a page cache provided by the OS. Whenever a file is read, the data is put into the page cache memory, to avoid disk access on the subsequent reads [1]. This significantly reduces I/O latency and improves the throughput of the NFS server.

2. Client Cache: NFS clients store data returned from the server as cache. The first reason is to access data faster. The second reason is to reduce network bandwidth and server load. File data is almost always considered to be cacheable. In addition, if caching data can reduce RPC requests to the server, the results of NFS operations such as lookups and directories are also stored in the cache [10].

2.3 Application Inference Method

File sharing servers only accept requests related to file access. Thus, it is difficult to directly detect client behavior. Michishita et al. proposed a method to infer multiple applications running on clients based on the hypothesis that file access patterns observed on file sharing servers exhibit temporal periodicity [7]. This method consists of three processes: acquisition, analysis, and inference. It is described below.

1. Acquisition process: This process uses tcpdump to capture packets of NFS requests on the NFS server and extract the NFS procedures and their issuance times.
2. Analysis process: In this process, the acquired data is transformed. After sampling the acquired data into file access data at regular time intervals, an STFT (Short-Time Fourier Transform) is performed. This allows us to obtain a spectrogram showing the frequency distribution at different times.
3. Inference process: This process uses fully supervised NMF (non-negative matrix factorization) to predict the STFT results of known applications. It uses the STFT results obtained in Step 2 as the input data. The training data consists of a frequency basis matrix extracted by NMF from the STFT of a single application run. Then, for each STFT of a predicted known application, the power of each frame is compared to the threshold to infer whether the target application was running during that frame.

2.4 Challenge

This paper focuses on server caches. The page cache in the Linux kernel is designed for various workloads. This design provides stable performance across various situations. However, it is not optimized for any specific workload. Furthermore, applications have become remarkably diverse in recent years, and their I/O patterns and characteristics are also diverse. Consequently, a generic caching method is often suboptimal. It applies to a uniform strategy and misses opportunities for significant performance improvements that a more flexible approach could provide. Meanwhile, NFS server only accepts requests from clients regarding file access. NFS server does not know what applications the client is running. Therefore, it is a challenge to identify the applications.

If a server identifies the application generating the I/O, it would be possible to apply a caching policy specifically optimized for that workload. By identifying applications and specializing in the caching strategy, we can expect to achieve more effective caching and unlock further performance gains.

3 System Design and Implementation

3.1 AI-Cache System Design

We will implement a system that externally implements application inference functionality for the NFS server and performs caching based on the inference results. When it is inferred that a specific application has accessed the server, caching is performed by reading ahead the accessed files.

Figure 1 shows Application Inference Cache (AI-Cache) system design. Proposed method integrates two components: a caching method and the application inference method described in Sect. 2.3. The sequence of AI-Cache is described below.

1. The NFS server receives an I/O request from a client application.
2. The server forwards information about this I/O access pattern to the Application inference (AppInf) module.
3. The AppInf module analyzes the pattern, infers the application, and calls the Read-Ahead (RA) module accordingly.
4. Based on the inferred application, the RA module reads data previously.
5. The NFS server returns data in response to client requests. When doing so, if read-ahead has been performed as described in step 4, or if the data already exists in the cache, allowing the server to respond immediately.

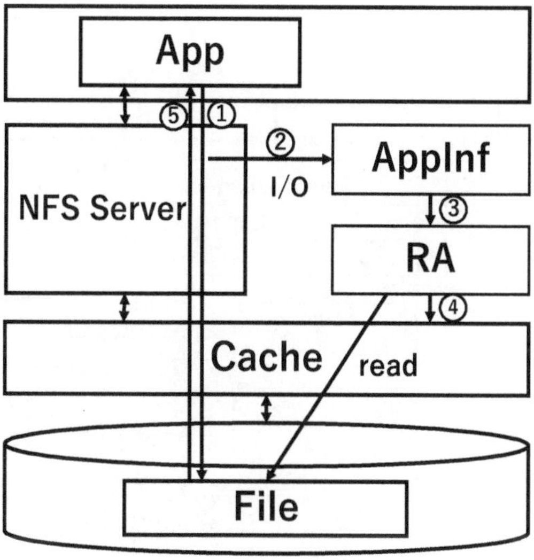

Fig. 1. System design

3.2 Implementation Details

This subsection describes the implementation based on the system design explained in Sect. 3.1. Figure 2 shows the system overview. It consists of two modules described below.

1. NFS Server: This system targets files exported via the Network File System. For this implementation, NFS-Ganesha is used. NFS-Ganesha is a user-mode file server for NFS v3, 4.0, 4.1, and 4.2 [11]. Additionally, the following processes were added.
   a. *FSAL MDCACHE*: NFS-Ganesha includes FSAL (File System Abstraction Layer). Within this, a hook was added to FSAL MDCACHE to launch the application inference module. FSAL MDCACHE is a stackable module for FSAL. It provides metadata cache. FSAL MDCACHE is placed on top of other modules. Hooking into this module ensures the application inference module launches when a read call is invoked.
   b. *RA thread*: This thread receives results from the application inference module. If RA is required based on the results received, it reads the target file to its end.
2. Application Inference Module: This module infers applications attempting to read files exported via the Network File System when it receives start-up instruction. It then returns the result to NFS Ganesha. It consists of three processes described as follows.
   a. *Acquisition process*: This process acquires file access data for the file sharing server when it receives a startup request from the file system. The system uses tcpdump to capture packets and extracts NFS protocol procedures and their issuance times.
   b. *Analytical process*: This process transforms the acquired data. First, the process counts NFS procedure calls within a fixed period. These calls are generated by applications executed against the file sharing server. It then performs an STFT (Short-Time Fourier Transform) on this data to convert it into frequency components.

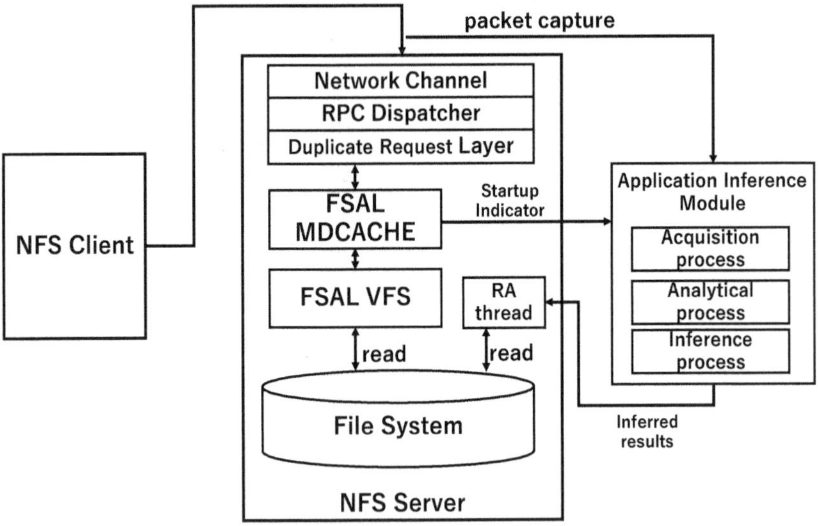

Fig. 2. System implementation

   c. *Inference process*: This process infers whether an application was executed. It predicts the STFT result of known applications from the data obtained in the "analytical process." It uses fully supervised NMF for prediction. The training data utilizes the STFT results obtained when running the known applications.

First, an application is executed on the NFS server from the NFS client. During this process, NFS read calls are continuously issued against files on the NFS server. When the first NFS read call is issued, the launch function hooked into the FSAL MDCACHE launches the application inference module.

Next, upon receiving the launch instruction, the application inference module begins with packet capture. After acquiring packets required for inference, it determines whether the application was executed. It then returns the inference result to the NFS server. The NFS server performs read-ahead based on the received inference results. This ensures the target file is loaded into the page cache.

4 Evaluation

This section describes the experimental evaluation. First, the evaluation environment is described in Sect. 4.1. Next, a preliminary evaluation to verify the feasibility of our application inference method is presented in Sect. 4.2. Section 4.3 verifies the effectiveness of AI-Cache. Finally, Sect. 4.4 concludes with a discussion of our findings.

4.1 Evaluation Environment

This subsection describes the evaluation environment. Table 1 describes the evaluation environment, and Fig. 3 depicts the experimental environment.

Table 1. Evaluation environment

	NFS Server	Client
CPU	Intel(R) Core(TM) i5–7500 CPU @ 3.40GHz	Intel(R) Core(TM) i5–8500 CPU @ 3.00GHz
Memory	8GB	8GB
Storage	HDD 1 TB	
File System	ext3	
I/O Scheduler	none	
Merge	/sys/block/sd*/queue/nomerges= > 2	
Application		vim, emacs
Network	1Gbps LAN	
NFS	ver3, rsize = 8192	
OS	Ubuntu 24.04 LTS	Ubuntu 24.04 LTS

The evaluation assumes an environment where the NFS server mounts the disk. To prevent I/O scheduling, the I/O scheduler was set to 'none,' which implements the FIFO

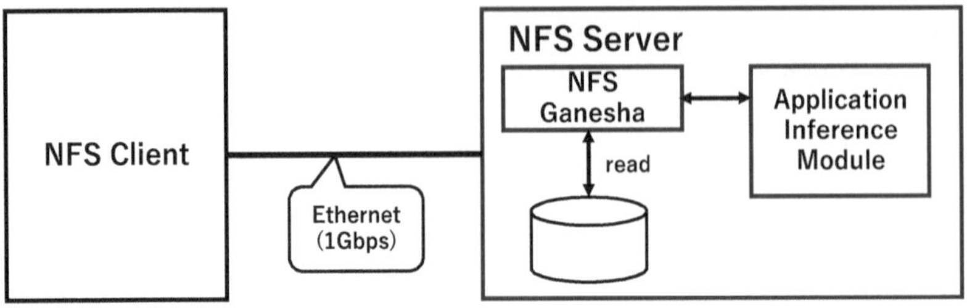

Fig. 3. Experimental environment

scheduling algorithm. Additionally, settings were modified to prevent I/O merging. In this evaluation, the applications executed by the client are the editor applications vim and emacs as in previous studies.

4.2 Preliminary Evaluation

This subsection describes a preliminary evaluation to verify the feasibility of our application inference method. AI-Cache must infer the applications that clients are running against NFS. Thus, in this evaluation environment as well, it is necessary to conduct evaluations similar to those in the previous study. This is to verify the feasibility of inference.

Section 4.2.1 evaluates the similarity of I/O temporal characteristics across applications, while Sect. 4.2.2 assesses the accuracy of the inference.

4.2.1 Application Similarity

This subsection describes the similarity of temporal periodicity of file access for each application. For this verification, we employ DTW (Dynamic Time Warping). DTW is a method used to measure the distance or similarity between time series data. The DTW distance indicates how similar waveforms are to each other; a smaller DTW distance signifies greater similarity between waveforms.

To use DTW, several preprocessing steps are performed. First, the applications are run, and data on "NFS read call" is collected during file reading. Second, a short-time Fourier transform is then performed on this collected data to extract periodicity features. Finally, these extracted features are used with DTW to compare the similarity between each application.

Figure 4 depicts the evaluation results. The horizontal axis represents the test data, while the vertical axis shows the ratio of the DTW shortest distance relative to the baseline application. (a) shows the results when using emacs as test data and comparing it with emacs and vim. For all test data, a DTW distance difference of at least 1.4 times was observed. (b) shows the results when using vim as test data and comparing it with emacs and vim. For all test data, a DTW distance difference of at least 1.3 times was observed. These results indicate that each application exhibits temporal periodicity and possesses distinct characteristics.

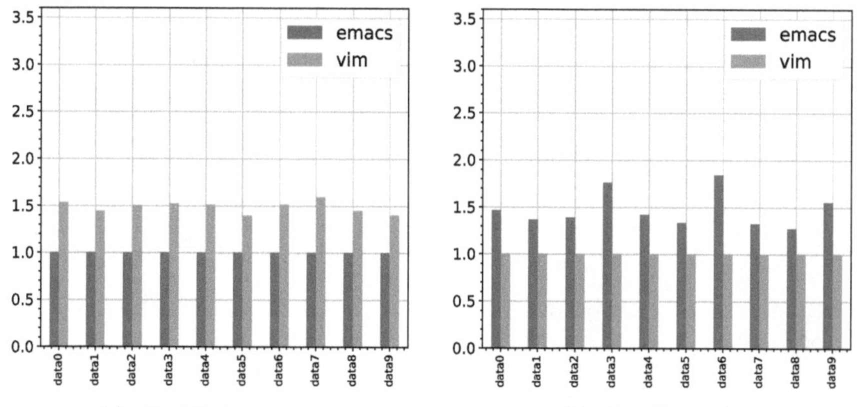

(a) Test Data: emacs (b) Test Data: vim

Fig. 4. DTW distance

4.2.2 Application Inference Accuracy

This subsection describes the results of inferring Applications by having the NFS client run application against files on the NFS server, respectively. The applications for evaluation were emacs and vim. The inferred results are shown in Table 2.

The execution patterns for the two applications, emacs and vim, fall into four categories: "both running," "emacs only," "vim only," and "not running." When loading a single file in emacs, 3.06% of cases were inferred to be "both running." For 91.84% of cases, the inference result was "emacs only." For 4.08% of cases, the result was "vim only," and for 1.02% of cases, it was "not running." When loading a single file in vim, 6.12% of cases were inferred to be "both running." For 3.06% of cases, the inference result was "emacs only." For 85.71% of cases, the result was "vim only," and for 5.10% of cases, it was "not running." This shows that inference is possible on the NFS server.

4.3 Experimental Results

This subsection evaluates the effectiveness of the AI-Cache. We compared the performance of our proposed method which utilizes predictive reading ahead against a baseline case without reading ahead. Furthermore, to confirm that AI-Cache was operating as intended, the read I/O from disk was measured on the NFS server and the overall throughput was evaluated. AI-Cache is expected to perform read-ahead. While this may cause a temporary increase in read I/O operations, the time required to read the entire file is expected to be shorter. Consequently, the overall throughput is expected to improve compared to the cache miss case without read-ahead.

Figures 5 and 6 plot the overall throughput (MB/s) against file size. The horizontal axis represents file size, while the vertical axis shows the throughput (MB/s). The file size was set to 100 MB as the baseline and increased by powers of two up to 3.2 GB. Following this, Figs. 7 and 8 present the transfer time as a ratio relative to the cache miss scenario. The horizontal axis represents file size, while the vertical axis shows the

Table 2. Application inference result

	inference result			
	both	emacs	vim	none
emacs	3.06%	91.84%	4.08%	1.02%
vim	6.12%	3.05%	85.71%	5.10%

transfer time as a ratio relative to the cache miss. The file size was set to 100 MB as the baseline and increased by powers of two up to 3.2 GB.

As shown in Figs. 5 and 6, the throughput for a cache hit is approximately 110 MB/s. For a cache miss, the throughput is lower, ranging from 60 to 90 MB/s, and decreases for files larger than 800 MB. When AI-Cache is used, the throughput for a 100 MB file is similar to that of a cache miss. However, as the file size increases, its throughput approaches the level of a cache hit.

Figures 7 and 8 show the transfer time as a ratio relative to a cache miss. For a 100 MB file, a cache hit case reduces the transfer time by approximately 30% compared to the cache miss case. The AI-Cache case reduces it by approximately 5% compared to the cache miss case. As the file size increases, the time for a cache hit case becomes shorter relative to the baseline, decreasing by 44% for the 3.2 GB file. The transfer time for the AI-Cache case remains close to that of the cache hit. For a 3.2 GB file, it reduced transfer time by 42% compared to a cache miss case.

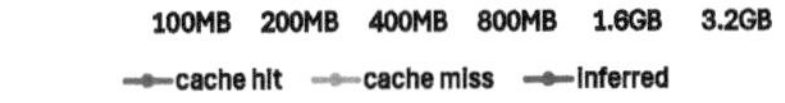

Fig. 5. vim throughput

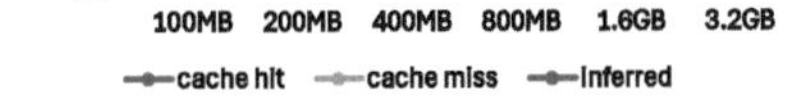

Fig. 6. emacs throughput

Figures 9, 10, 11, 12, 13 and 14 plot the disk I/O throughput (MB/s) over time as captured by blktrace. The horizontal axis represents elapsed time, while the vertical axis shows the throughput. In a cache miss case, the throughput remained stable at approximately 60 MB/s, with a maximum of around 90 MB/s. In contrast, with the AI-Cache, the throughput was initially like the cache miss case. Meanwhile after a brief period, it increased and stabilized at a rate slightly above 135 MB/s. Furthermore, a comparison of the total I/O duration shows that disk read operations are shorter with the AI-Cache than in the cache miss case. These results, specifically the increase in throughput mid-transfer and the shorter overall read duration, confirm that the AI-Cache was functioning effectively.

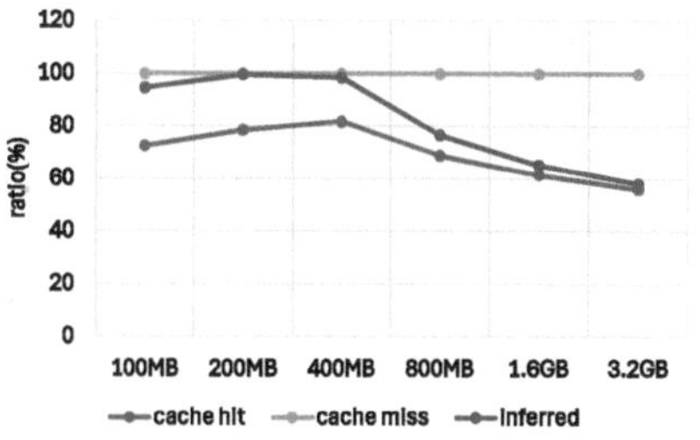

Fig. 7. vim transfer time ratio based on cache miss

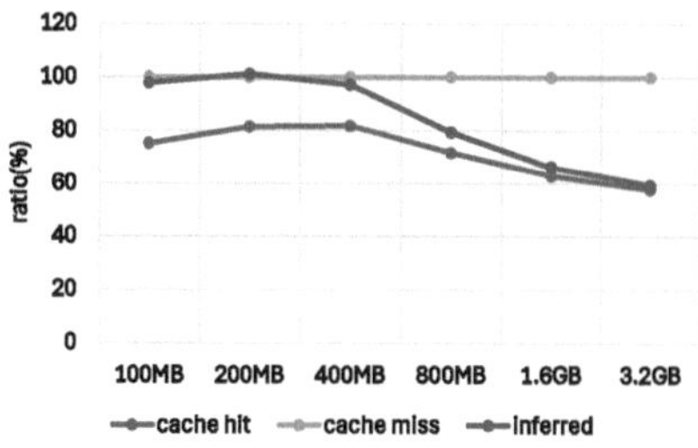

Fig. 8. emacs transfer time ratio based on cache miss

4.4 Discussion

Figure 5, 6, 7 and 8 show improvements in throughput and reduced transfer times. Additionally, the blktrace results in Figs. 9, 10, 11, 12, 13, and 14 show increased disk I/O throughput during file data transfers. These findings indicate that AI-Cache successfully inferred the application and read-ahead the file data into the page cache before the application requested it via NFS read calls. This saved time required for reading from disk, improving overall read efficiency.

AI-Cache has a limitation. When inferring applications, if the time required to read the file is less than the time required for inferring, read-ahead cannot be performed. Therefore, no performance improvement can be expected from AI-Cache in this case. This result is observed for small file sizes. Comparing the results for 100 MB to 400 MB files, there is little difference between the cache miss case and AI-Cache. This is because it took approximately 4 s for the inference module in our PoC system. Furthermore, if the prediction result is incorrect, unnecessary data may be cached. In the worst case, data that was needed could be evicted from the cache. It potentially leads to performance degradation.

However, the effectiveness of AI-Cache was particularly pronounced for large files. For a 100 MB file, the overhead of inference and read-ahead processing was significant, and no performance improvement was observed. However, as file size increased, the time saved by preemptively performing disk I/O likely outweighed the overhead of inference. This is supported by the results for the 3.2 GB file: our method completed the transfer in 58% of the time required for a cache miss. This performance level is nearly as effective as cache hit which took 56% of the cache miss time.

Based on these results, proof-of-concept for AI-Cache demonstrated improved read performance, indicating AI-Cache is effective.

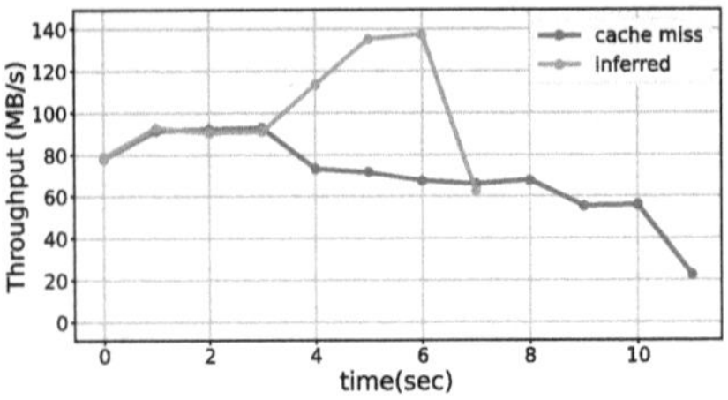

Fig. 9. Disk I/O throughput (vim, 800 MB)

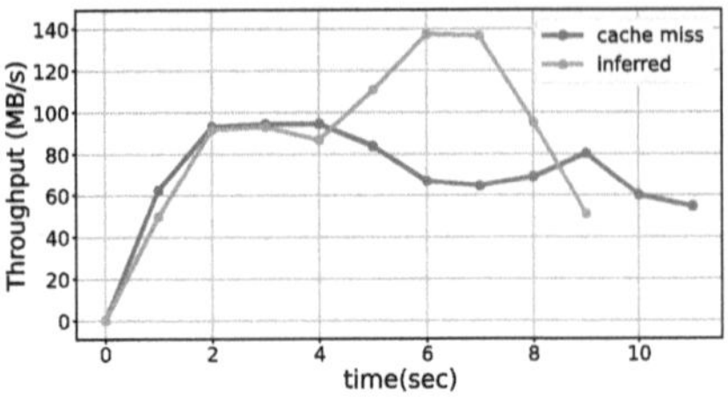

Fig. 10. Disk I/O throughput (emacs, 800 MB)

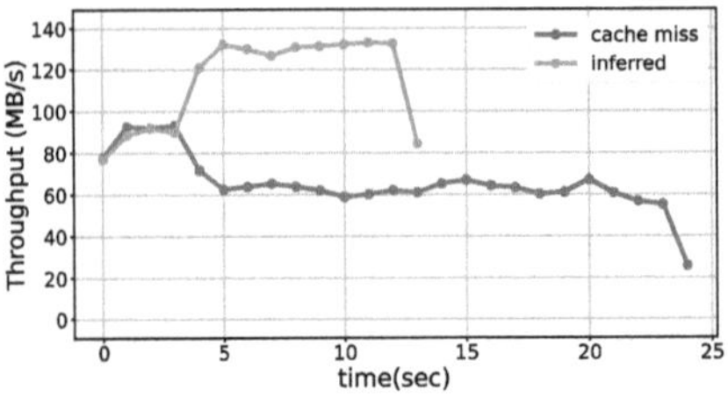

Fig. 11. Disk I/O throughput (vim, 1.6 GB)

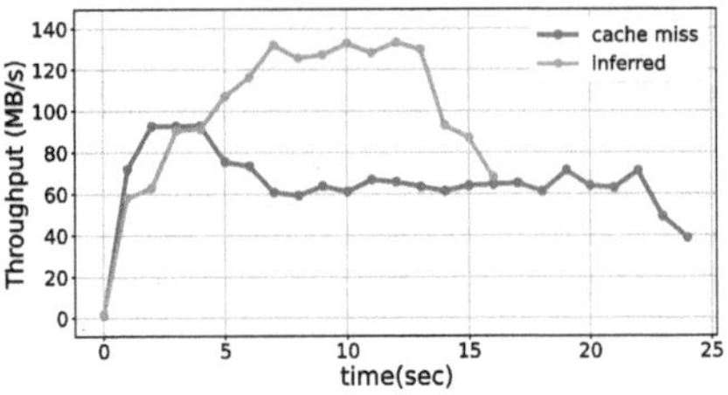

Fig. 12. Disk I/O throughput (emacs, 1.6 GB)

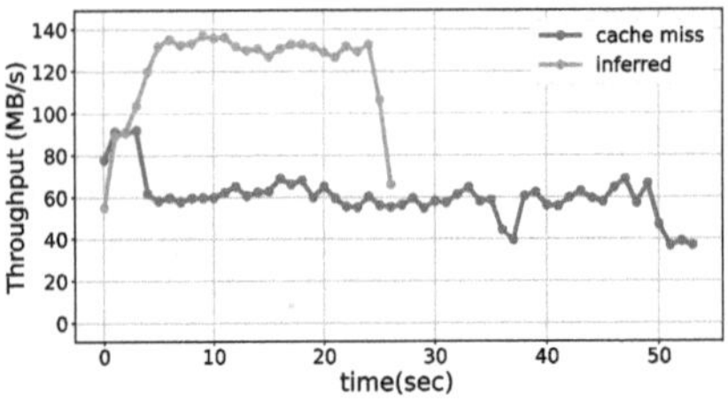

Fig. 13. Disk I/O throughput (vim, 3.2 GB)

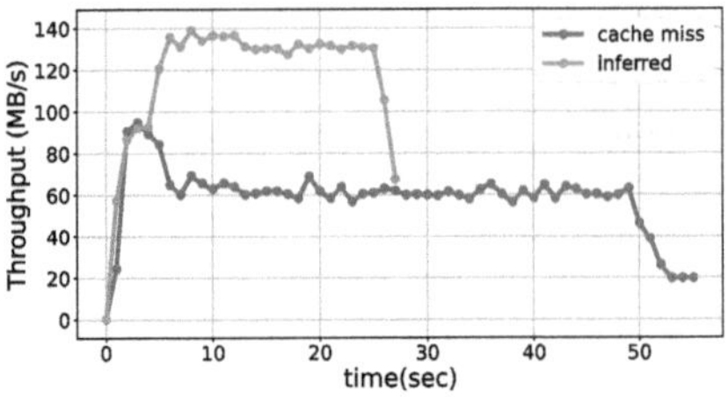

Fig. 14. Disk I/O throughput (emacs, 3.2 GB)

5 Related Work

Neeraja et al. proposed a method to infer applications running on a client from an NFS server [3]. It uses the results of analyzing NFS procedure issuance as features. Specifically, it pre-trains a transition model using the issuance of NFS procedures during application execution as states. It uses the number of lines in NFS procedures as input and infers the hidden application state using a hidden Markov model. The results showed that when trained on single-application executions and then used for inference, it achieved over 70% accuracy. This research has something in common with the application inference method applied by the proposed method in that it infers client behavior from file accesses on a file sharing server. However, this research infers the file access of each client separately from the client IDs when multiple clients operate a file.

Hododa et al. conducted research on classifying ransomware and benign programs using machine learning based on their storage device access patterns [6]. Against the backdrop of increasing ransomware variants, it has become increasingly difficult to create signatures for all of them. To address this issue, they investigated a new behavioral detection method that uses ransomware storage access patterns as features. Building upon classical machine learning algorithms, they trained deep learning models and evaluated their ransomware detection performance. The results showed an accuracy of approximately 98% for the two-class classification.

Herodotou proposes "AutoCache" which uses machine learning to automate caching (which files to put in memory and which to delete) in distributed file systems such as HDFS [12]. Specifically, a gradient boosting tree (XGBoost) model is pre-trained using the access history of individual files as features. This model is used to predict the probability that a file will be accessed in the near future or not for a while. Evaluation experiments show that the model produced has high prediction accuracy (AUC > 0.96), and the proposed method improves job completion time by up to 25% and cluster resource efficiency by up to 39% compared to existing policies such as LRU. Like our proposed method, this work automates caching strategies based on file access patterns on file-sharing servers. Our proposed method infers the type of client applications. In contrast, this study predicts the access probability of individual files.

6 Conclusion

This paper described that application inference methods toward Network File Systems can be applied not only to ransomware detection but also to caching. Our proposed method employs application inference method to perform caching based on the inferred results. We conducted Proof-of-concept (PoC) to validate its effectiveness. The results demonstrated that the effectiveness of caching increases for larger files, confirming its validity. Building on this validation, we will proceed with further design and implementation, while also exploring more effective caching methods utilizing application inference methods.

7 Acknowledgments

This research was supported by the Kagawa University Research Promotion Program in fiscal year 2025 (Grant No. 25K0D019).

References

1. The Linux Kernel documentation: https://www.kernel.org/doc/html/latest/admin-guide/mm/concepts.html#page-cache. Accessed 20 Oct 2025
2. Reed, B., Long, D.D.E.: Analysis of Caching Algorithms for Distributed File Systems. In: ACM SIGOPS Operating Systems Review (1996)
3. Yadwadkar, N.J., Bhattacharyya, C., Gopinath, K.: Discovery of application workloads from network file traces. In: 8th USENIX Conference on File and Storage Technologies (FAST '10), USENIX Association, USA (2010)
4. Tarasov, V., Hildebrand, D., Kuenning, G., Zadok, E.: Virtual MachineWorkloads: the case for new benchmarks for NAS. In: 11th USENIX Conference on File and Storage Technologies (FAST' 13), USENIX Association, pp. 307–320, USA (2013)

5. Tarasov, V., et al.: Extracting flexible, Replayable models from large block traces. In: 10th USENIX Conference on File and Storage Technologies (FAST' 12), USENIX Association, USA (2012)
6. Hodota, R., Hirano, M., Kobayashi, R.: A deep-learning-based ransomware detection system using disk access patterns. Comput. Secur. Symp., 26–29 (2021)
7. Michishita, D., Kida, K., Kamei, H.: A proposal of method for inferring client behavior by using file access features on file sharing servers. Information Processing Society of Japan, 5ZC-05 (2023)
8. RFC 1094 NFS Network File System Protocol Specification: https://www.rfc-editor.org/rfc/rfc1094. Accessed 20 Oct 2025
9. RFC 1813 NFS Version 3 Protocol Specification: https://www.rfc-editor.org/rfc/rfc1813. Accessed 20 Oct 2025
10. Callaghan, B.: NFS Bible, 1st edn. Kenichi Suzuki, Tokyo (2001)
11. NFS-Ganesha Homepage.: https://github.com/nfs-ganesha/nfs-ganesha/wiki. Accessed 13 Oct 2025
12. Herodotou, H.: AutoCache: employing machine learning to automate caching in distributed file systems. In: 2019 IEEE 35th International Conference on Data Engineering Workshops (ICDEW), pp. 133–139 (2019)

Fault Tolerant Reconfigurable ML Multiprocessor

Tangrui Li(✉), Justin Y. Shi, Matteo Spatola, and Hongzheng Wang

Temple University, Philadelphia 19102, PA, USA
{tuo90515,shi,matteo.spatola,hongzheng.wang}@temple.edu

Abstract. This paper reports three computational experiments for a von Neumann inspired reconfigurable fault tolerant multiprocessor for neural network (NN) training workflows. The experiments are intended to prove the feasibility of the proposed reconfigurable multiprocessor architecture for non-regular workflows on robustness of adaptability. A potential integration with MLIR compilers is also discussed for integrating diverse accelerator hardware for existing practical applications.

Keywords: Reconfigurable Multiprocessors · Fault Tolerance · Automatic Load Balance · Distributed Neural Net Training · Scalability

1 Introduction

The rapid growth of NN models has led to a surge in computational demands. While high-performance GPU clusters seem to meet these needs, such devices are becoming cost-prohibitive due to thermal management difficulties and checkpoint overheads. These costs grow as infrastructure expands. A renewed research interest has shifted toward utilizing multiple devices rather than centralized GPU clusters. However, scaling the NN models across multiple heterogeneous devices introduces three major challenges: 1) **Single Point of Failure (SPF)**: As the system grows, the probability of a single device failure bringing down the entire application increases. 2) **Inter-device Communication Overhead**: Frequent data transfers between devices introduce substantial latency and bandwidth pressure, especially as model size and number of devices increase. 3) **Load Balancing**: Achieving high memory utilization and compute efficiency requires partitioning of the computation graph to ensure prompt termination without processors idling at the same time.

Eliminating SPF in multiprocessor clusters has been an open challenge. Even the latest GPU clusters require frequent checkpoints. The checkpoint time and storage overheads grow as the processing infrastructure expands. Checkpoint is like a "shotgun approach" to defend against potential loss of computational results. Creating reconfigurable multiprocessors is another open challenge that requires mapping arbitrary workflows (NN models in this paper) to processors to ensure high processing and communication efficiencies. Recently, a von Neumann

H. R. Arabnia et al. (Eds.): CSCI 2025, CCIS 2970, pp. 39–52, 2027.
https://doi.org/10.1007/978-3-032-28466-2_3

principle inspired [10] fault-tolerant reconfigurable multiprocessor was reported to have solved both challenges at the same time [18]. Due to the theoretical impossibility of reliable failure detection [3], all application programs are required to handle timeout/retransmission discipline as an integral part of the distributed application. A timeout event signals a potential component failure, a slow network, or a busy or slow component. A simple retransmission logic with redundancy elimination can solve the application fault tolerance challenge without explicit checkpoints. Carefully developed the timeout/retransmission discipline can result in significant load-balancing savings for non-regular workflows by tuning task distribution parameters.

This paper reports a computational simulation result for such a fault-tolerant reconfigurable processor for NN training applications with a focus on robust load balancing over heterogeneous processors.

The fault tolerant reconfigurable multiprocessor architecture is enabled by the ACAN data representation and program-program synchronization semantics in order to map arbitrary workflows to any multiprocessor cluster or clusters. This framework naturally compliments traditional optimizing compiler efforts, such as MLIR (Multi-Level Intermediate Representation) [6] programs generated by the compilers assume unrealistic component reliabilities. The compilers provide optimized program mapping for different applications by enabling dialect-based abstraction and transformation on specific accelerator hardware. Optimized program mapping at the global processing graph level is difficult. The proposed method can thus solve the broader scalability challenge of removing significant barriers to the adoption of MLIR-based compilers in large-scale AI/ML projects.

While a complete system would involve systematic dialect specification and backend-specific lowering, this work focuses on a proof-of-concept implementation using a simplified NN class composed of linear layers only. At this stage, we do not provide a full dialect design or device lowering pipelines, as those require extensive engineering efforts. Instead, our approach introduces a hardware-agnostic abstraction layer via the TS model, which can conceptually align with any MLIR-compatible NN and be lowered flexibly in future accelerator implementations.

The remainder of this paper is organized as follows. We first review related work on parallel training of neural networks, followed by an introduction to the key components of our system: Tuple Space and Active Content Addressable Networking. Then, we detail how these components interact to support efficient NN parallelization, and present experiments evaluating feasibility, adaptability, and robustness. A discussion on MLIR integration and potential lowering strategies is also included.

2 Related Work

Data Parallelism (DP)

The most straightforward distributed training strategy, where each device holds a full copy of the model and processes distinct data batches. Gradients are computed locally and must be aggregated on a primary device [1,16]. This synchronization becomes a scalability bottleneck. Moreover, memory utilization is inefficient, as all devices redundantly store the same model copy.

Distributed Data Parallelism (DDP)

To mitigate the communication bottleneck of DP, DDP [7] was proposed to decentralize gradient aggregation. The most widely used strategy is Ring AllReduce [4,11], where each device exchanges partial gradient tensors with its neighbors in a ring network topology. Over several rounds of communication, all devices converge to the same gradients. High-performance libraries like Nvidia NCCL[1] and accelerator hardware like Tenstorrent[2] can satisfy such requirements.

Model Parallelism (MP)

Unlike DP focus on data, MP [2] distributes the model across devices. This is often necessary when the model is too large to fit into a single device. MP is generally implemented in two main categories.

Pipeline Parallelism (PP) By partitioning modules sequentially across devices, pipeline parallelism enables concurrent processing of multiple micro-batches in different stages [5,8]. However, pipeline bubbles and imbalances between stages can lead to low resource utilization. On the other hand, such pipelines are usually manually designed, although recent approaches, such as Alpa [21], enable automated pipeline partitioning; such methods still face limitations, such as: 1) They are primarily effective for a limited set of model architectures with sequential or layered structures. 2) The underlying combinatorial search for optimal partitions is computationally expensive. 3) When the external hardware configuration changes, the pipeline often needs to be recomputed from scratch.

Tensor Parallelism (TP) This finer-grained strategy partitions the computation within individual layers across multiple devices. Megatron-LM [9,19], a widely adopted implementation of tensor parallelism, requires manual modification of the model code to integrate its custom parallel primitives. This constraint poses challenges when adapting large-scale or pre-existing codebases.

ZeRO (Zero Redundancy Optimizer)

Instead of replicating all parameters, gradients, and optimizer states across devices, ZeRO [13] partitions these elements among devices and reconstructs only the necessary parts during computation, significantly improving memory efficiency and scalability. As implemented in DeepSpeed [14] through progressive stages (ZeRO-1, 2, 3), this approach trades off communication overhead for a reduced memory footprint.

However, while ZeRO excels at memory optimization, it does not alter the computation graph and can not resolve issues where individual layers exceed device memory limits. The benefit is that ZeRO does not require manual source code change, but if the original model contains a layer that exceeds the memory capacity of a single device, ZeRO will not be able to handle it.

---

[1] https://github.com/NVIDIA/nccl.

[2] https://github.com/tenstorrent/.

Guided Self-scheduling

It has long been recognized that the task size assigned to a processor will have a significant impact on the overall performance and efficiency of computation. Most practical application task durations are unpredictable and vary significantly. Guided Self-Scheduling (GSS) aims to automatically determine parallel tasks for all applications, minimizing idle times and maximizing processor efficiency by evaluating task response times [12]. For non-regular applications using heterogeneous processors, GSS uses a "small->large" heuristic tuning strategy. Faster processors would complete the small tasks quickly while the slower processors became busy most of the time. Eventually, the faster processors would get more work done than slower ones automatically.

3 Tuple Space and ACAN

Active Content Addressable Networking (ACAN) departs from Content Addressable Networking (CAN) [15] in two key ways: 1) networked data representation, and 2) networked data access protocols. Traditional CAN is typically implemented using a Distributed Hash Table (DHT) for quick data retrieval. The data contents are linked to "buckets" inside a DHT [20]. Therefore, the network data representation is <data, bucket ID>. The data accesses protocols are "read" and "write". The data-bucket ID binding violated von Neumann's principle of reliable system synthesis [10], since the "bucket ID" may not be available sometimes. It is the cause for single-point failures.

ACAN's networked data representation is <key, value>. It supports three data access methods: 1) put(key, value), 2) read(&pattern, &buffer) and 3) get(&pattern, &buffer), where the read(•) and get(•) are blocking calls for synchronization between programs. These "Tuple Space" semantics are supported by all participating nodes as an extension to the TCP/IP stack. Nodes within the "Tuple Space" communicate via automatic peer-to-peer discovery protocols. Together, they form an application-aware computing cloud.

ACAN completely decouples application programs and data from physical and virtual components. The program/data decoupling from devices and the tuple space communication with synchronization semantics enable runtime systems to perform automatic fault isolation and load balancing. At runtime, the application's workflow will automatically form SIMD (Single Instruction Multiple Data), MIMD (Multiple Instruction Multiple Data) pipelines using all available resources. The following section explores this idea in-depth.

For NN training applications, ACAN can improve the performance and fault tolerance of compute. First, ACAN significantly reduces single point of failure risk by enabling run-time fault isolation and task rerouting unless all devices fail at the same time. Second, as discussed in timeout/retransmission discipline, the TS semantics enable efficient communication and synchronization between heterogeneous devices without requiring manual directions. Third, the proposed method requires the application programs to handle timeout events in support of automatic graph partitioning, eliminating the need for manual scheduling or slicing. Finally, this method enables dynamic device participation, allowing new devices to join or leave at run-time while maintaining service continuity, making the system suitable for non-deterministic environments.

4 Using a Custom ACAN Cloud

An ACAN application typically consists of a single manager and multiple handlers. Both the manager and the handler can modify the tuple space. For mission-critical applications, all updates can be logged in an immutable blockchain, ensuring traceability and accountability.

The manager is responsible for publishing tasks to tuple space while handlers pull tasks from it. After the manager publishes a task, each handler will get(•) the tasks that it can handle according to the content of the task and its own capabilities. A get(•) request will populate the custom cloud until a match is found. When a task is taken after a match, other handlers will no longer see it. The handler who receives this task can choose whether to "process" or "store" the task. A stored task may be matched later by a different handler. When multiple handlers are competing for the same get() pattern, SIMD parallelism is formed automatically. When multiple handlers are asking for different patterns, MIMD will form. The natural dependencies between the input and output tuples will form task pipelines seamlessly.

In the ACAN architecture, the manager and handlers operate without direct communication, enabling the runtime to perform global resource multiplexing efficiently. Once the manager dispatches a task, a timer (timeout) is set to allow handlers to process it. Upon timeout, the system evaluates the completion status of all dispatched tasks. If a task remains uncompleted, the manager can re-send it. Similar to GSS, we initially use a small fixed task size, that is, we try to test if the processing difficulty of each task is close enough, so that the higher-performance handler can process the task and complete at approximately the same time without knowing the processor details. The ideal situation involves tasks being released at one time can be completed at the same time regardless of the performance of the handlers. Considering that the service of the handler is not necessarily reliable, the manager can adjust 1) the task size, 2) the size of the task released at one time (called pouch size), and 3) the timeout for waiting for the completion of the released task to finetune the deliverable application performance.

The custom ACAN cloud allows any handler to fail at any time. The loss of a Handler will only affect the task pouch size and timeout. The Manager has the option to store the partial result in a stable storage or leave them in TS. And the crash of Manager should not damage the partially computed results. If needed, the Manager restart can be programmed to read the tuple space state and continue. Since the Manager uses a single processor, the current industry record of single processor MTBF (Mean Time Between Failure) is more than

11 years[3]. Unless the application requires that long running time, the Manager checkpoints may not be necessary.

5 Method

This section focuses on deploying NNs using ACAN. To that end, we first introduce the deployment content and its TS representation, then we will introduce the dispatching and conflict resolution details.

5.1 Individual ACAN Task

Considering that the current framework focuses solely on NNs composed of linear layers, we first obtain the necessary parameters (input and output dimensions) either through manual input or by parsing the MLIR. Based on this information, three prototype task types are generated for each layer: **forward**, **backward**, and **update**. If the layer is not the last layer, an additional **activation** task is created (since we assume a regression scenario). For the last layer, a **loss** task is created instead.

Each task type has specific execution preconditions:

- Forward task: Requires the output of the previous layer (or the input data, if it is the first layer) to be available in the tuple space. Otherwise, the task fails upon timeout and is discarded.
- Activation task: Requires the hidden variable to be activated.
- Loss task: Requires both the final output and the ground-truth label.
- Backward task: Depends on the availability of the gradient from the next layer (or the gradient derived from the loss function, if it is the last layer). This task computes the first-order derivative based on the chain rule (without involving advanced optimization techniques, though such extensions are possible).
- Update task: Requires the corresponding gradient for the parameter to be updated.

TS is queried using variable names as keys to determine task readiness. Although this leads to frequent accesses, the design characteristics of TS mitigate performance concerns.

Notably, when tasks are created, they are represented as declarative descriptions (e.g., a string of the task type and parameters) rather than instantiated objects. This design enhances the decoupling between the Manager and Handler in ACAN. It is important to emphasize that these are merely prototype tasks. They require further processing before being submitted to the tuple space.

Together, these tasks define the core execution pipeline for NN training. The associated execution states, such as current data items or active layers, are also recorded in TS, serving as the foundation for task dispatching, execution, and validation by the Manager and Handler.

[3] https://chatgpt.com/share/6876cffe-0d6c-800b-908a-e4d47d94ca4b.

5.2 Prototype Task Partition

As discussed earlier, all tasks are of the same size for ease of tuning to achieve optimal utilization of system computational resources in a Handler-agnostic manner. This allows for consistent control of execution granularity through the adjustment of pouch size and timeout. However, since NN layers vary in size, the prototype tasks derived from the network structure must be partitioned to satisfy the above constraints.

Since tasks are represented as descriptions rather than instantiated objects, the designated size can be adjusted dynamically when the Manager inserts them into TS. The system could adapt the task size at runtime to accommodate significant changes in computational capacity. For example, when so many low-performance devices join (favoring smaller tasks) or when high-performance devices dominate (favoring larger tasks to reduce I/O overhead). However, in this paper, to make it simple we assume a fixed task size, which is determined in advance and remains unchanged during execution.

Task Size: The size of a task is influenced by multiple factors, such as the memory required to store parameters and the number of multiplications/additions involved. In particular, the loss task typically involve more complex computations and are better to be assigned a proportionally larger size.

Task partitioning is performed by reducing the input and output sizes of a task under the same type. For example, a forward task with input dimension m and output dimension n can be split into **four** smaller forward tasks, each corresponding to one formed by dividing the input and output in half. Specifically, these sub-tasks cover: (first $m/2$ inputs & first $n/2$ outputs), (first $m/2$ & last $n/2$), (last $m/2$ & first $n/2$), and (last $m/2$ & last $n/2$). Since tasks are represented as descriptions, the Handler can independently retrieve the necessary weights and biases from TS during execution.

Backward tasks follow the same logic, as they also depend on both input and output dimensionality to retrieve the weights/biases. For other task types, splitting is more straightforward:

- An activation task operating on an m-dimensional vector is split into two tasks, each handling $m/2$ elements.
- An m-dimensional update tasks are similarly split into two tasks, each updating $m/2$ parameters.
- An m-dimensional loss tasks are divided into two smaller tasks, each responsible for computing loss over $m/2$ outputs. The Handler automatically retrieves the corresponding ground truth labels during processing.

After generating the prototype tasks for each NN layer, the Manager process them according to the predefined task size and inserts the resulting task descriptions into TS, where they await propagating by the Manager.

5.3 Task Dispatching

Following the design of ACAN, the Manager retrieves from TS both the data required for the current stage of NN training and a pouch of partitioned tasks (up to the predefined pouch size). These tasks are then inserted into TS and remain available for processing for a duration defined by the timeout.

Each Handler continuously attempts to process tasks by taking one available task from TS under the key *task*. If the preconditions are met, the Handler performs the computation and writes the results back to the corresponding TS entries. Additionally, it marks the associated keys for task completion, enabling the Manager to track progress. The above process is repeated continuously by all Handlers.

After the timeout, the Manager checks the status of all previously issued tasks. Based on how many tasks were completed, it adjusts the timeout to adapt to the current system computational power. The Manager then collects all uncompleted tasks (and possibly new tasks from TS if it is not enough to give a pouch) and re-issues them. If a particular data item or network layer has been fully processed, the Manager updates the relevant TS entries as a checkpoint, so as to reset some flags.

5.4 Conflict Resolution

Once a task is published by the Manager, the Handler that takes this task temporarily gains the right to access the related variables. For example, in a forward task, the Handler reads the input data, weights, and biases, and writes the resulting output.

However, due to ACAN's handler-agnostic nature, a Handler may take a task but fail to complete it before the timeout. From the Manager's perspective, the task is incomplete and will be reissued. Meanwhile, the original Handler may still proceed, potentially leading to duplicate execution of the same task.

For tasks that do not write the read variables, such as forward (it does not need to overwrite the inputs, weights and biases), activation, backward, and loss tasks, duplicate execution poses no correctness issue. As long as each execution is valid (a property ensured by the blockchain-like consistency of TS), redundant computation is okay.

In contrast, update tasks involve overwriting the read variables. Redundant execution in this case may result in unintended cumulative updates. Checking task ID for completion and eliminate concurrent updates can be easily done following the same sliding-window discipline in TCP/IP protocols. Once all update tasks for a given layer are complete, the Manager can safely overwrite them.

6 Experiments

Before orchestrating a large ACAN cloud, we simulate our method using multithreading on a single machine. The target model is a neural network comprising two linear layers, with the first layer sized $N \times N$ and the second $N \times 1$, where $N = 4^4$. Each Handler is constrained to a maximum capacity of 4^4, which is also the upper bound for task size in ACAN. Consequently, the overall model is about 256 times larger than what a single device (Handler) can accommodate, highlighting the necessity of task-level parallelism.

Instead of adopting the Gossip protocol or implementing ACAN on an actual blockchain infrastructure, we simulate the tuple space with a daemon thread which maintains a Python dictionary. For simplicity, we also assume that all Handler threads behave correctly and do not return malicious results.

Our simulation involves **one Manager thread and four Handler threads**, all of which may crash during execution. The daemon thread continuously monitors the system and revives failed Manager thread using the latest checkpoint. Although ACAN is designed to be Handler-agnostic and does not require active fault recovery for Handlers, in our simulation we still recreate crashed Handler threads. Instead, to emulate fluctuating computational resources, we dynamically vary the processing speed of Handler threads during runtime. To highlight the robustness of our approach, we fix both the task size (4^4) and the pouch size (100) throughout the experiment and only adjust the timeout.

We conduct three experiments in total:

- Experiment 1 serves as a basic feasibility check. Both the Manager and Handlers remain stable, and the processing speed is fixed. We perform a regression task to demonstrate that the NN can be successfully trained under our method.
- Experiment 2 introduces dynamic changes in Handler processing speeds while still assuming no crashes. This allows us to evaluate whether ACAN's adaptive timeout mechanism can effectively track performance shifts across Handlers.
- Experiment 3 builds upon Experiment 2 by introducing probabilistic failures for both the Manager and Handlers. We analyze regression loss and timeout dynamics to assess the method's resilience under fault-prone conditions.

6.1 Feasibility Test

To construct a learnable target function, we randomly generate a set of parameters that define a mapping from a 4^4-dimensional input to a 1-dimensional output. Based on this function, we synthesize 100 data points and train the neural network using the Mean Squared Error (MSE) loss. The goal is to verify whether the proposed method enables the network to train properly.

Given that the primary focus of this experiment is to validate the operational feasibility rather than assess generalization performance, we do not generate a large dataset or include a separate test set. For the same reason, we forego mini-batch training and instead use one data point at a time (i.e., stochastic gradient descent with batch size = 1).

Figure.1 below shows the training loss curve over 2 epochs.

Observing an effective decrease in the loss function is enough to show that the method is working properly.

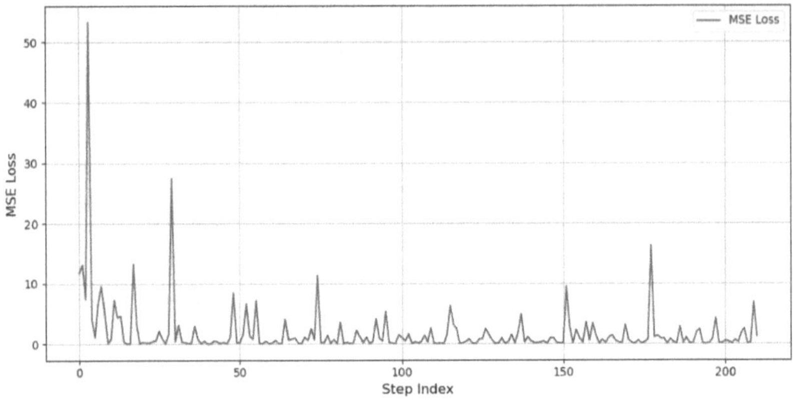

Fig. 1. The MSE loss curve in the feasibility test.

6.2 Adaptability Test

We set 3 different performance levels for Handlers, with a processing speed ratio of 1:5:10. During training, each Handler has a 100% chance of random changing every 5 s. After processing a task, each Handler will sleep for a period of time based on the processing speed and the size of the task being processed.

We processed a total of 20 data (about 3,500 pouches) and obtained Fig. 2 of Handler computing power and ACAN timeout, which showed an obvious inverse proportional relationship, indicating that the system can use lower timeout to speed up processing when computing power is high, and vice versa.

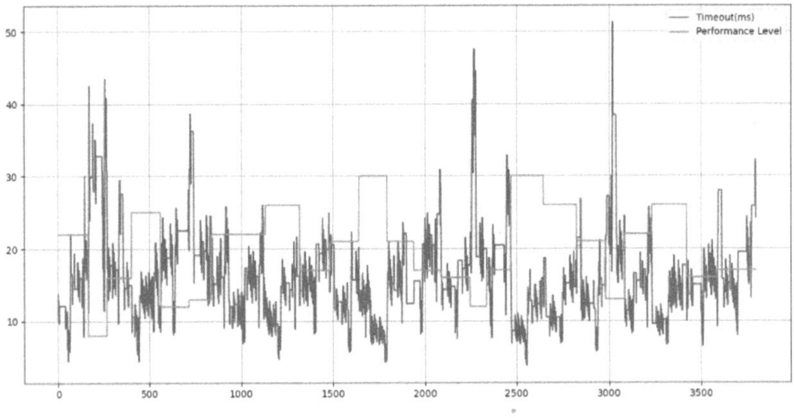

Fig. 2. The relation between the timeout and the computational power in the adaptability test.

6.3 Robustness Test

Based on the adaptability test, we make the Manager 100% crash every 5 s, all Handlers 100% crash every 5 s, and 100% change the processing speed every 5 s. We also tested on 20 data and get the following curve.

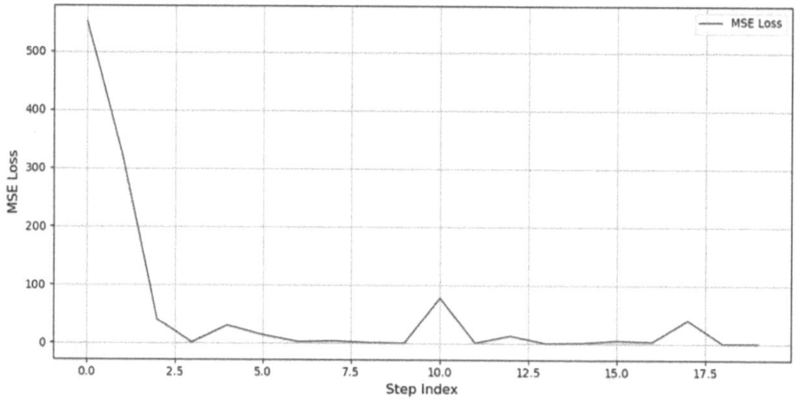

Fig. 3. The MSE loss curve in the robustness test.

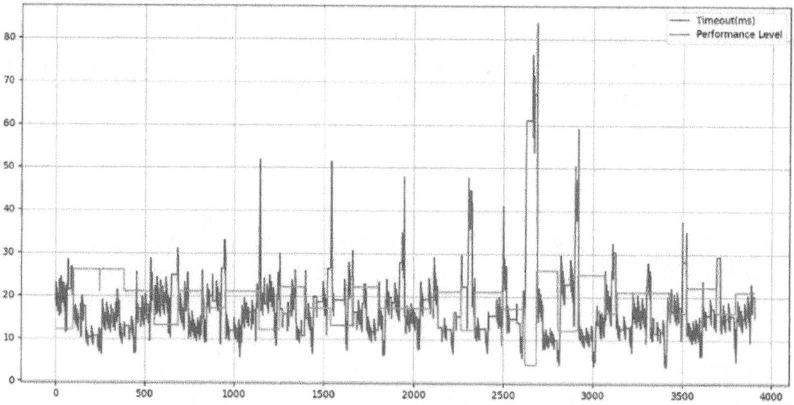

Fig. 4. The relation between the timeout and the computational power in the robustness test.

The loss curve in Fig. 3 still shows that the method is working properly, while the performance curve in Fig. 4 shows that the overall relationship between timeout and performance level is still inversely proportional, indicating that this method can still show stability under high crash conditions. The number of pouches to be processed increases due to the extremely high crash probability.

7 Significance

The scale of neural network training continues to grow rapidly due to intense AI/ML competition. The reconfigurable von Neumann multiprocessor architecture promises to deliver sustainable performance and reliability regardless of scale. The reported experiments used extreme cases to prove the feasibility of these claims.

8 Design Tradeoffs

The reported experiments used a dictionary to simulate the ACAN Tuple Space. It has identical data retrieval semantics, without the program-to-program synchronization semantics. Therefore, the performance may be better due to the absence of delays.

Using a tuple space for indirect program-to-program communication will incur higher communication overhead relative to direct program-to-program communication (nearly 2x). However, the savings in minimized synchronization overheads and checkpoint delays with the custom cloud's natural parallelization effects, the proposed method can still out-perform direct program-program communication counterparts even without checkpoints. The energy savings would allow training of even larger models.

The ACAN Tuple Space service is implemented in reality as an extension to operating systems communication stack for each participating node. Unless all nodes fail at the same time, the application can still continue. However, data corruption during transmission or storage requires special application programming to ensure the quality of data delivery. These small probability errors are typically negligible in scientific simulations and NN training applications.

9 Conclusion

The reported study proves the feasibility of a reconfigurable multiprocessor architecture for NN workflows. The next step is orchestrating a large custom ACAN for targeted applications. In particular, experiments with Tenstorrent processors can possibly shed more light on the proposed programming paradigm.

The proposed ACAN concept and protocols [17] depart from the traditional client-server programming paradigm in its complete program and data decoupling features. Rather than relying on compilers to map applications onto multiprocessors, the proposed framework enables runtime formation of computational graphs dynamically. It is also possible to compliment compiler generated programs to overcome the scaling challenges without checkpoints.

A salient feature of the proposed framework is "unlimited scalability" due to the resolution of the SPF challenge. Without SPF threats, all applications can scale indefinitely, limited only by available resources. Although security was not discussed in this report, the scalability feature can compensate for any security measure overhead by simply expanding the size of the infrastructure.

ACAN further restricts the threat vectors within the application boundary. Therefore, the proposed framework is suitable for all mission-critical applications.

Acknowledgment. The reported work is supported in part by NSF Chameleon resource grants #CH-817746 and #CHI-251439, Temple University Technology Transfer Office, Temple AGI Team and the SMC Labs. LLC.

References

1. Dean, J., et al.: Large scale distributed deep networks. In: Pereira, F., Burges, C., Bottou, L., Weinberger, K. (eds.) Advances in Neural Information Processing Systems, vol. 25, Curran Associates, Inc. (2012)
2. Dean, J., et al.: Large scale distributed deep networks. In: Proceedings of the 26th International Conference on Neural Information Processing Systems - Volume 1, pp. 1223–1231. NIPS 2012, Curran Associates Inc., Red Hook, NY, USA (2012)
3. Fekete, A., Lynch, N., Mansour, Y., Spinelli, J.: The impossibility of implementing reliable communication in the face of crashes. J. ACM **40**, 1087–1107 (1993). https://doi.org/10.1145/174147.169676
4. Gibiansky, A.: Bringing HPC techniques to deep learning (2017). https://andrew.gibiansky.com/blog/machine-learning/baidu-allreduce/. Accessed 14 July 2025
5. Huang, Y., et al.: GPipe: efficient training of giant neural networks using pipeline parallelism. Curran Associates Inc., Red Hook, NY, USA (2019)
6. Lattner, C., et al.: MLIR: scaling compiler infrastructure for domain specific computation. In: Proceedings of the 2021 IEEE/ACM International Symposium on Code Generation and Optimization, pp. 2–14. CGO 2021, IEEE Press (2021). https://doi.org/10.1109/CGO51591.2021.9370308
7. Li, S., et al.: Pytorch distributed: Experiences on accelerating data parallel training (2020). https://arxiv.org/abs/2006.15704
8. Narayanan, D., et al.: Pipedream: generalized pipeline parallelism for DNN training. In: Proceedings of the 27th ACM Symposium on Operating Systems Principles, pp. 1–15. SOSP 2019, Association for Computing Machinery, New York, NY, USA (2019). https://doi.org/10.1145/3341301.3359646
9. Narayanan, D., et al.: Efficient large-scale language model training on GPU clusters using MEGATRON-LM. In: Proceedings of the International Conference for High Performance Computing, Networking, Storage and Analysis. SC 2021, Association for Computing Machinery, New York, NY, USA (2021). https://doi.org/10.1145/3458817.3476209
10. von Neumann, J.: Probabilistic logics and the synthesis of reliable organisms from unreliable components. In: Automata Studies. Princeton University Press (1956)
11. Patarasuk, P., Yuan, X.: Bandwidth optimal all-reduce algorithms for clusters of workstations. J. Parallel Distrib. Comput. **69**(2), 117–124 (2009). https://doi.org/10.1016/j.jpdc.2008.09.002
12. Polychronopoulos, C., Kuck, D.: Guided self-scheduling: a practical scheduling scheme for parallel supercomputers. IEEE Trans. Comput., 1425–1439 (1987)
13. Rajbhandari, S., Rasley, J., Ruwase, O., He, Y.: Zero: memory optimizations toward training trillion parameter models. In: Proceedings of the International Conference for High Performance Computing, Networking, Storage and Analysis. SC 2020, IEEE Press (2020)

14. Rasley, J., Rajbhandari, S., Ruwase, O., He, Y.: DeepSpeed: system optimizations enable training deep learning models with over 100 billion parameters. In: Proceedings of the 26th ACM SIGKDD International Conference on Knowledge Discovery & Data Mining, pp. 3505–3506. KDD 2020, Association for Computing Machinery, New York, NY, USA (2020). https://doi.org/10.1145/3394486.3406703
15. Ratnasamy, S., Francis, P., Handley, M., Karp, R., Shenker, S.: A scalable content-addressable network. SIGCOMM Comput. Commun. Rev. **31**(4), 161–172 (2001). https://doi.org/10.1145/964723.383072
16. Sergeev, Balso, D.: HOROVOD: fast and easy distributed deep learning in tensorflow (2018). https://github.com/horovod/horovod, https://arxiv.org/abs/1802.05799
17. Shi, J.Y.: Statistic multiplexed computing system for network-scale reliable high performance services. U.S. PTO #11,588,936B2 (2023)
18. Shi, J.Y.: Network-scale fault tolerant computing - a von Neumann multiprocessor architecture. In: 25th International Conference on Foundations of Computer Science (2025)
19. Shoeybi, M., Patwary, M., Puri, R., LeGresley, P., Casper, J., Catanzaro, B.: Megatron-LM: training multi-billion parameter language models using model parallelism (2020). https://arxiv.org/abs/1909.08053
20. Stoica, I., Morris, R., Karger, D., Kaashoek, M.F., Balakrishnan, H.: Chord: a scalable peer-to-peer lookup service for internet applications. ACM SIGCOMM Comput. Commun. Rev. **31**, 149–160 (2001)
21. Zheng, L., et al.: ALPA: automating inter- and Intra-Operator parallelism for distributed deep learning. In: 16th USENIX Symposium on Operating Systems Design and Implementation (OSDI 22), pp. 559–578. USENIX Association, Carlsbad, CA (2022). https://www.usenix.org/conference/osdi22/presentation/zheng-lianmin

Performance and Consumption Analysis of LU Matrix Decomposition on Heterogeneous and Traditional Multicore CPUs

Francisco J. Rosa(✉), Alberto Valderas-González, José M. Aragón-Jurado, Juan Carlos de la Torre, and Bernabé Dorronsoro

Department of Computer Engineering, University of Cádiz, Cádiz, Spain
{franciscojavier.rosavega,alberto.valderasgonzalez}@mail.uca.es,
{josemiguel.aragon,juan.detorre,bernabe.dorronsoro}@uca.es

Abstract. Modern processors increasingly adopt heterogeneous architectures implementing both high-performance and energy-efficient cores, at the expense of traditional homogeneous designs. This shift breaks the simple link between core count and throughput, making classical performance models unreliable. This work compares three LU factorization variants—Crout, Doolittle, and LUP—implemented in Java across two representative CPU platforms (Intel Core i9-12900HK and AMD Ryzen 7 7800X3D) and two operating systems (Windows and GNU/Linux). Execution time and energy consumption are measured for sequential and parallel implementations, revealing that performance and efficiency depend strongly on the interaction of algorithmic memory-access patterns, processor microarchitecture, and OS/runtime policies rather than following simple, hardware-agnostic trends. These findings highlight the need to evaluate algorithms within their target hardware and software environments and to revisit optimization strategies for emerging heterogeneous processors, where scheduling and energy trade-offs are increasingly complex.

Keywords: LU decomposition · Heterogeneous multicore processors · Energy efficiency · Java high-performance computing

Acronym: CSCI-RTPD

1 Introduction

In recent years, the landscape of processor design has begun to change significantly. Whereas traditional CPUs offered many identical cores and predictable scaling characteristics, a new generation of heterogeneous architectures—most visibly the hybrid designs that combine high-performance (P-cores) with energy-efficient (E-cores)—has emerged as a prominent trend. These architectures break

H. R. Arabnia et al. (Eds.): CSCI 2025, CCIS 2970, pp. 53–67, 2027.
https://doi.org/10.1007/978-3-032-28466-2_4

the simple mapping between core count and throughput by introducing cores with different microarchitectural properties, instruction-level performance, and power envelopes, with the aim of improving their energy efficiency. As a result, algorithms originally developed and tuned for homogeneous multicore chips must be re-evaluated to understand how their performance and efficiency behave on this new class of hardware.

Historically, the performance of dense linear algebra routines and other high-intensity kernels has been characterized using well-established models that assume uniform processing elements, consistent memory hierarchies, and predictable contention behavior. Under these assumptions, parallel scalability, cache blocking, and load balancing strategies could be designed with relatively stable and reliable outcomes. However, the heterogeneity introduced by mixed core types alters these assumptions: execution latency, peak FLOPS, and memory subsystem interactions can differ between cores, and the cost of migrating work between core types or scheduling fine-grained tasks becomes nontrivial. Consequently, the classical performance models and optimizations need to be revisited for modern hybrid processors.

Matrix multiplication and related dense linear algebra operations remain as main examples of compute-bound workloads: for very large matrices, they dominate both execution time and memory traffic in scientific and engineering applications. Because of their cubic computational complexity and high arithmetic intensity, naive implementations quickly become bottlenecked by either compute performance or memory bandwidth. For solving linear systems and many other numerical tasks, LU decomposition is a widespread alternative to repeated matrix inversion or naive algorithms; it reduces problems to triangular solves and matrix operations, enabling high-performance implementations and benefiting from decades of optimization research.

Numerous studies have examined LU decomposition across platforms and applications: Özcan and Şen [20] showed that GPUs outperform CPUs in simple LU factorization, while other works highlight diverse uses such as watermarking in images [25], copyright protection in Internet of Things environments [18], and domain-specific uses in engineering and security [9,17,28]. Parallel CPU implementations, such as the OpenMP-based Crout algorithm proposed by Rana et al. [22], demonstrate effective scaling. However, despite this broad literature, little is known about the performance of parallel LU algorithms on emerging hybrid processors compared with conventional homogeneous multicore systems. This gap raises important questions about task granularity, scheduling, data placement, memory contention, and the trade-offs between throughput, scalability, and energy efficiency.

This study presents an overview of parallel LU decomposition algorithms from the literature, focusing on the Crout [10], Doolittle [8], and LUP [26] methods, and introduces a Java implementation that employs concurrency and parallel processing technologies. Computational performance is evaluated by measuring execution time and energy consumption on both modern heterogeneous processors—combining performance cores (P-cores) and efficiency cores

(E-cores)—and conventional homogeneous multicore systems, under two operating systems (Windows and GNU/Linux), with the aim of informing future software optimization.

This work makes three main contributions. First, it develops parallel Java implementations of three widely used LU decomposition algorithms: Crout, Doolittle, and LUP. Second, it empirically evaluates their execution time and energy consumption on both conventional homogeneous multicore processors and modern heterogeneous architectures, considering two operating systems (Windows and GNU/Linux). Third, it offers a comparative analysis of performance and energy-efficiency trade-offs, providing guidance for optimizing parallel LU decomposition on emerging processor designs.

The remainder of this paper is organized as follows. Section 2 reviews the most relevant work on LU matrix decomposition as well as studies on parallelism in homogeneous and heterogeneous architectures. Section 3 then describes the methodology followed in this study, while Sect. 4 introduces the configuration of the experiments conducted. After that, Sect. 5 presents and discusses the obtained results. Finally, Sect. 6 concludes the paper and outlines possible directions for future work.

2 Related Work

2.1 LU Decomposition Algorithms

Several classical algorithms exist for LU decomposition, such as Crout [10], Doolittle [8], or Cholesky [11] (the latter for symmetric positive-definite matrices). Each method decomposes a given matrix A into the product of a lower triangular matrix L and an upper triangular matrix U, differing in how L and U are constructed and updated. Early research focused on numerical stability, pivoting strategies, and reduction of floating-point errors [12,13]. Foundational theoretical work established the conditions for existence and uniqueness of the factorization, while later contributions addressed practical aspects such as blocking strategies, cache-friendly kernels, and robust implementations in high-performance libraries like LAPACK and ScaLAPACK [2,5].

Algorithmic variants (e.g., Crout vs Doolittle), blocking choices, and pivoting policies have a direct impact on both numerical robustness and computational performance. Crout's algorithm is widely used for dense matrices because its data-access pattern and dependency structure are well suited to blocked implementations and task-based decomposition, facilitating parallelization on shared-memory systems [15]. As a result, Crout is often selected when evaluating how factorization algorithms map onto emerging novel hardware architectures.

Beyond algorithmic design, platform-specific implementations also play a critical role in performance. In the Java ecosystem, several numerical libraries (e.g., EJML, Parallel Colt) provide blocked LU factorizations; performance studies on these implementations highlight the influence of array layout, primitive data types, and just-in-time (JIT) compilation effects when porting high-performance LU kernels to the JVM [1,27].

2.2 Parallelism in Homogeneous and Heterogeneous Architectures

Parallelism in modern computing encompasses a variety of programming models and runtime strategies, including thread- and process-based approaches (e.g., POSIX threads, MPI), task-based models (e.g., OpenMP tasks, task graphs), and dataflow/runtimes that manage dependencies and task placement. Task-based parallelism—where computations are expressed as fine-grained tasks with explicit data dependencies—has gained significant traction for shared-memory and heterogeneous systems because it exposes parallelism while delegating scheduling and data movement to the runtime [19].

Efficient scheduling is central to task-based execution. Work-stealing schedulers provide provable load-balancing guarantees for dynamically generated task graphs and have been widely adopted in high-performance runtimes; they perform well when task shapes are irregular or when parallelism is discovered at runtime [6]. For heterogeneous platforms, runtime systems add further complexity: they must consider device capabilities, data transfer costs, and topology-aware placement. Systems such as StarPU [3] and PaRSEC [7] offer unified programming models that allow the same task graph to be scheduled across CPUs and accelerators while providing scheduling policies and automated data movement to maximize throughput and locality. Beyond these native runtimes, the Java ecosystem also offers task-parallel frameworks—such as Fork/Join, parallel streams, and CompletableFuture—that have been used to express fine-grained task graphs for dense linear algebra on shared-memory systems [16].

2.3 Applications and System-Level Studies

At the system level, both hardware heterogeneity and operating system policies strongly affect measured performance. The emergence of hybrid core designs challenges the assumptions behind classical performance models: such CPUs integrate cores with differing microarchitectural properties and power envelopes, making simple linear scaling models based on identical cores no longer valid. As Intel notes in its guidance on hybrid (P-core/E-core) architectures:

> "Achieving the best performance on a hybrid architecture is harder because load balancing with heterogeneous cores is more complicated." [14]

Vendor guidance, along with several empirical studies, therefore recommends careful thread-to-core binding and topology-aware scheduling. Żytko and Badurowicz [29] show that P-cores can significantly outperform E-cores on floating-point workloads, while E-cores may offer energy savings on memory-bound tasks, thus highlighting trade-offs between raw throughput and energy efficiency.

Operating systems introduce another source of variability. Comparative studies examining the same hardware under different OSs report conflicting results: some identify Windows as faster for compute-bound kernels [23], while others favor Linux [4]. These differences are attributable to scheduler policies, default affinity behavior, timer resolution, power-management defaults, and runtime library interactions. Consequently, a comprehensive performance characterization for LU decomposition must jointly consider algorithmic variants, runtime scheduling policies, operating system behavior, and energy/performance trade-offs on both homogeneous and heterogeneous platforms.

In summary, prior work spans algorithmic foundations, parallel runtimes, application-driven constraints, and system-level performance measurements. Yet there remains a clear gap in integrated studies that simultaneously evaluate parallel LU variants, OS-level scheduling/affinity, and energy measurements across homogeneous and heterogeneous CPUs. Adressing this gap is the primary motivation for the experiments and analysis presented in this paper.

3 Methodology

Several well-known theorems describe how to perform *LU* decomposition, including those underlying the Crout [10], LUP [26], and Doolittle [8] algorithms. However, the focus of this work is on *LU* decomposition without permutation. This is feasible because the permutation matrix is unnecessary when the leading principal minors of the matrix satisfy the following condition:

$$\det(A_k) \neq 0 \quad \text{for } k = 1, 2, \ldots, n$$

Under this assumption, the factorization considered in this work is:

$$A = L \times U$$

for algorithms that do not require permutation, where A is the initial matrix, and L and U are the lower and upper triangular matrices, respectively.

Custom Java implementations were developed for the parallel versions of the Crout, Doolittle, and LUP algorithms[1]. The code uses `ThreadPool` executors and the `Runnable` interface, with synchronization managed through barriers. An exception-handling mechanism was added to prevent the experimental suite from terminating abruptly.

To ensure the correctness of these implementations, black-box testing was performed. Various matrices were provided as input, and outputs were verified without inspecting internal behavior. Correctness was confirmed by multiplying L and U; if their product equaled the original matrix A, the factorization was considered correct.

After verifying correctness, computational performance was assessed. The main metric is *speedup*, defined as the ratio between the execution time of the best sequential implementation and that of the parallel version executed with multiple threads [21]:

$$Speedup = \frac{t_{sequential}}{t_{parallel}}$$

[1] All source code and experimental scripts are publicly available at https://github.com/goal-group-uca/Performance-and-Consumption-Analysis-of-LU-Matrix-Decomposition-in-Java.

In addition to performance, power consumption is measured across various hardware platforms and operating systems (Windows and GNU/Linux) to characterize the energy efficiency of the Java implementations and to analyze algorithm–architecture trade-offs. Experiments compare performance and energy consumption across three factors—hardware architecture, operating system, and matrix size—for the Crout, Doolittle, and LUP algorithms.

Because execution times are inherently variable [24], each experiment was repeated 30 independent times, and the mean of these repetitions was taken as the representative value. The resulting average runtime was then used to compute the speedup of each algorithm under all tested configurations. We assume that any impact of the JVM on the measurements is constant across all experimental configurations, and therefore does not affect the comparative analysis.

4 Experimental Setup

The experimental setup was designed to evaluate the performance and energy consumption of sequential and parallel LU decompositions implemented in Java across two modern CPU architectures with different designs (namely, Intel Core i9-12900HK and AMD Ryzen 7 7800X3D), two operating systems (GNU/Linux and Windows), and three well-known algorithms: Crout, Doolittle and LUP.

Table 1 summarizes the components of the experimental setup and the configurations tested. The Intel Core i9 processor features a heterogeneous architecture with P-cores and E-cores, whereas the AMD Ryzen has a homogeneous design with only P-cores. Both systems were evaluated under Windows 11 and the Linux distribution Ubuntu 24.04, using square matrices ranging from $n = 3000$ (3000×3000) to $n = 10000$ (10000×10000 elements). Execution times for the algorithms were measured within the scripts using Java's built-in timing utilities. Power consumption, however, was recorded by an external device that reports real-time power usage for each hardware system at one-second intervals. This device, the industrial-grade *NETIO PowerCable REST 101F* external power outlet, was used to measure the total energy consumption E as:

$$E = \sum_{i=1}^{n} P_i dt$$

where n is the total number of samples and dt is the sampling interval (fixed at 1 s in this study), and $\overline{P}$ denotes the mean power consumption:

$$\overline{P} = \frac{E}{\Delta t}$$

where Δt is the duration of the algorithm, in seconds, for each configuration.

Table 1. Summary of experimental configurations

Hardware	Type	OS	Matrix sizes (n)
AMD Ryzen 7 7800X3D	Homogeneous	Linux, Windows	3000–10000
Intel i9-12900HK	Heterogeneous	Linux, Windows	3000–10000

5 Results and Discussion

This section presents the evaluation results, focusing on execution time and energy consumption. The discussion is organized into three subsections: the first two analyze each hardware architecture separately, and the last one highlights notable anomalies and their plausible causes.

5.1 Multi-core Heterogeneous Architecture

Figure 1 shows the speedup values, computed as the mean of 30 independent runs, for the studied algorithms on both operating systems using the Intel Core i9-12900HK processor. The number of threads was varied from 2 to 32, twice the number of logical cores available on this processor.

On the i9-12900HK under Linux (see Fig. 1a), Crout's implementation scales steadily up to 6 threads (matching the number of high-performance cores). Activating a seventh thread causes a marked drop in speedup: performance with 7+ threads decreases substantially compared to the 1–6-thread regime. Under Windows, the scaling pattern differs: speedup continues to increase up to about 12 threads, with only marginal gains beyond that point (see Fig. 1b). Examining the other results for the i9, the Doolittle (see Figs. 1e and 1f) and LUP (see Figs. 1c and 1d) implementations in Java follow the same qualitative trend—a platform-dependent knee followed by flattened or reduced returns—with modest, algorithm-specific differences in stability and magnitude.

Regarding energy performance, Fig. 2 illustrates the energy consumed by each algorithm on the Intel Core i9-12900HK running both operating systems. On Linux, executions on the i9 tend to achieve their best energy efficiency near 6 threads (matching the number of high-performance cores). Small matrices (e.g., size 3000) result in a higher mean power draw than medium or large matrices in the Linux measurements (see Figs. 2a, 2c, and 2e). On Windows (see Figs. 2b, 2d, and 2f), the measured total energy shows a consistent ordering across algorithms: *LUP* consumes the least total energy, *Crout* is intermediate, and *Doolittle* consumes the most, although the absolute differences remain relatively modest.

5.2 Multicore Homogeneous Architecture

Figure 3 depicts the average speedup obtained from 30 independent executions of the three LU decomposition algorithms on the AMD Ryzen 7 7800X3D under both Linux and Windows. Thread counts range from 2 up to 32, representing twice the number of logical cores provided by this processor.

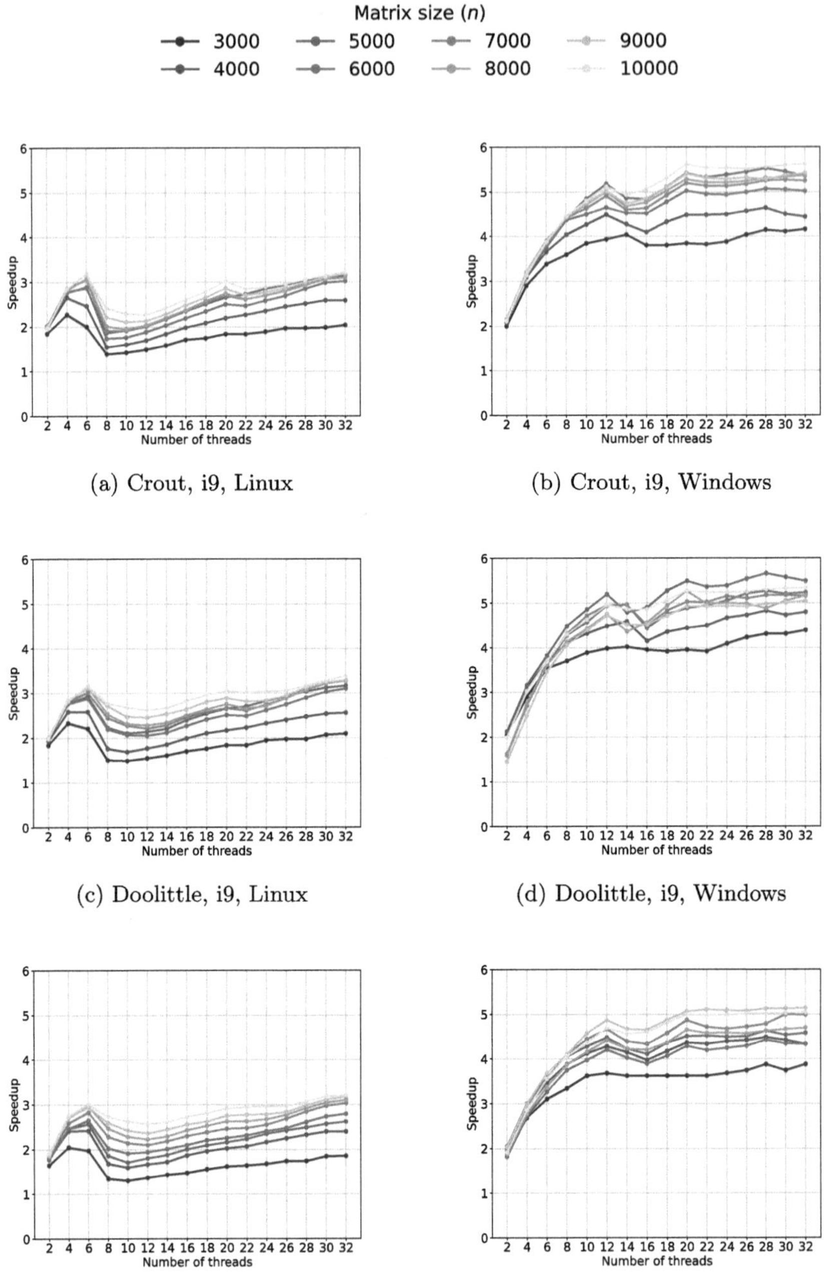

(a) Crout, i9, Linux

(b) Crout, i9, Windows

(c) Doolittle, i9, Linux

(d) Doolittle, i9, Windows

(e) LUP, i9, Linux

(f) LUP, i9, Windows

Fig. 1. Speedup curves for Crout, Doolittle, and LUP implementations on the Intel Core i9-12900HK under Linux and Windows.

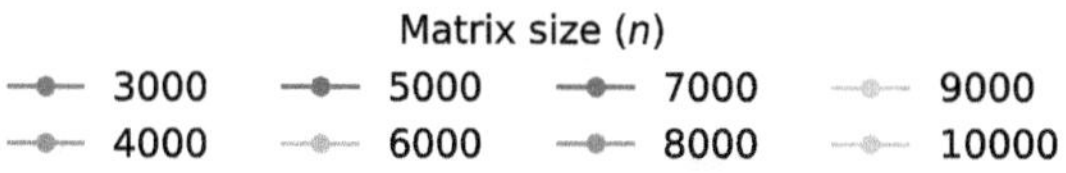

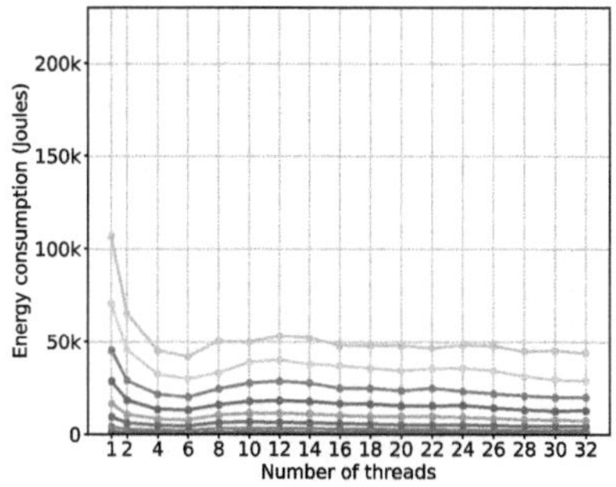

(a) Crout, i9, Linux

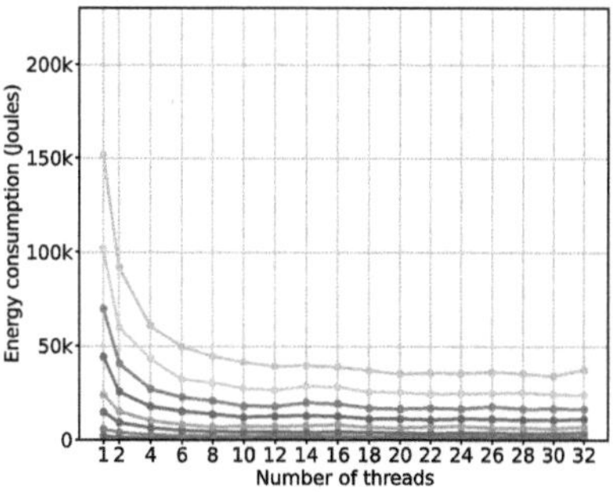

(b) Crout, i9, Windows

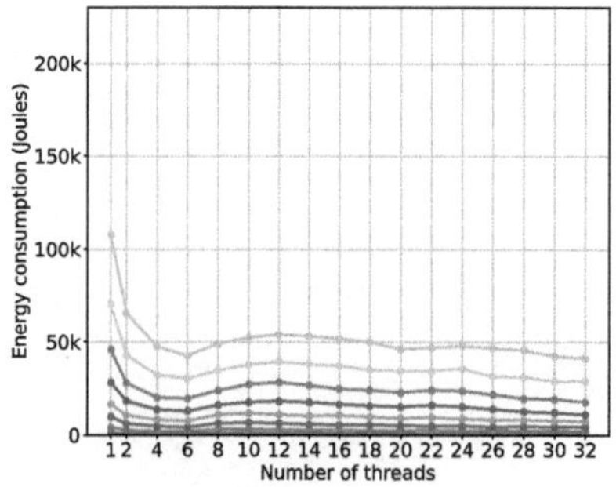

(c) Doolittle, i9, Linux

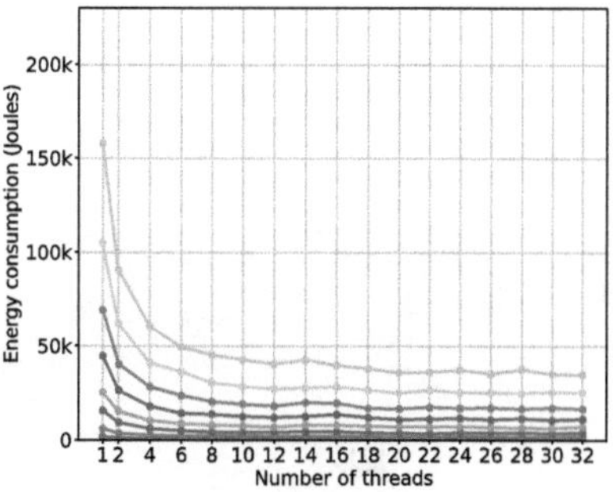

(d) Doolittle, i9, Windows

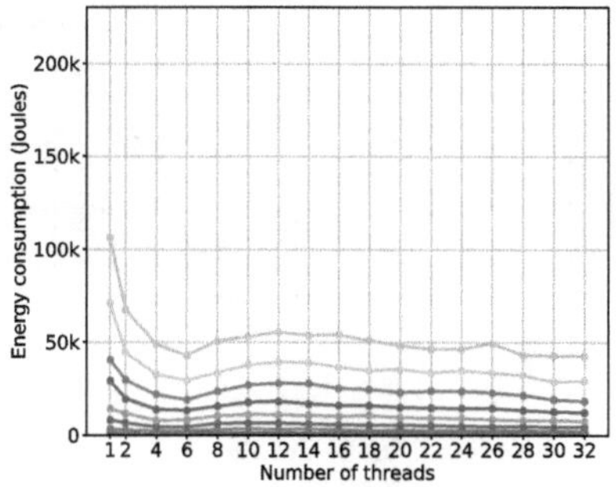

(e) LUP, i9, Linux

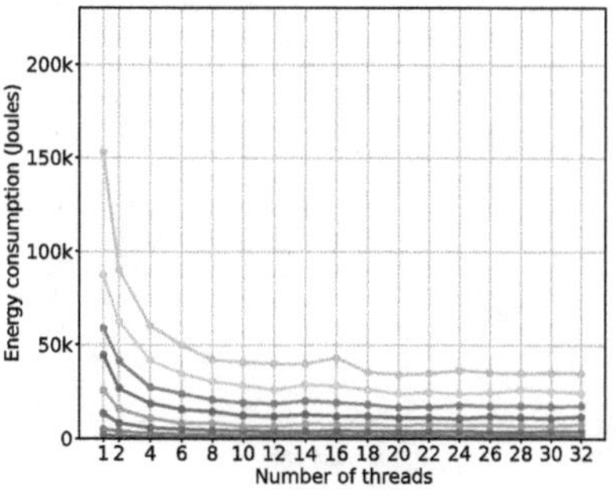

(f) LUP, i9, Windows

Fig. 2. Energy consumption curves for Crout, Doolittle, and LUP implementations on the Intel Core i9-12900HK under Linux and Windows.

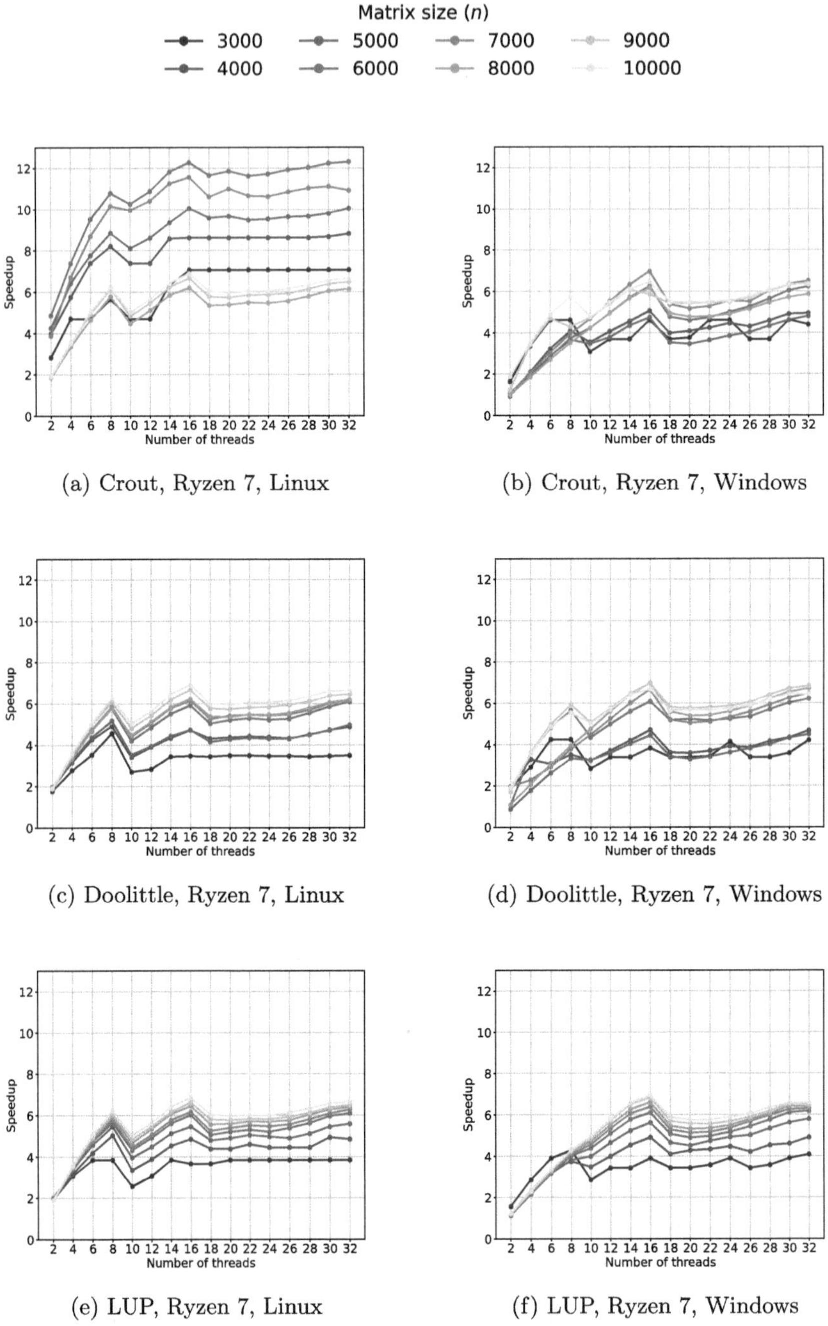

(a) Crout, Ryzen 7, Linux

(b) Crout, Ryzen 7, Windows

(c) Doolittle, Ryzen 7, Linux

(d) Doolittle, Ryzen 7, Windows

(e) LUP, Ryzen 7, Linux

(f) LUP, Ryzen 7, Windows

Fig. 3. Speedup curves for Crout, Doolittle, and LUP implementations on the AMD Ryzen 7 7800X3D under Linux and Windows.

The Ryzen 7 7800X3D exhibits a distinct performance profile, compared with the previously studied one. Under Linux, Crout's implementation shows superlinear speedups for matrix sizes roughly in the 4000–7000 range, where the measured speedup exceeds the nominal number of threads (see Fig. 3a). For sizes greater than 8000, incremental per-thread gains diminish. On Windows, Crout displays a similar initial increase in speedup up to about 16 threads, after which performance declines (see Fig. 3b). Doolittle (Figs. 3d and 3c) and LUP (Figs. 3f and 3e) follow a broadly similar trend but achieve better speedups for larger matrices and do not exhibit the same post-8000 decrease observed in Crout.

Results on the energy performance study, displayed in Fig. 4, show a strong dependency on both the operating system and the matrix size. On Ubuntu, total energy consumption varies considerably with matrix size (see Figs. 4a, 4c, and 4e). On Windows, the observed ranking of total energy consumption is: *Crout* is the most efficient, followed by *Doolittle*, with *LUP* consuming the most energy, although the gaps remain relatively modest (see Figs. 4b, 4d, and 4f).

5.3 Notable Anomalies and Plausible Causes

Several unexpected behaviors emerged during the experiments and warrant closer analysis. First, the sharp drop in speedup on the i9 when using more than seven threads under Linux is likely caused by a shift in CPU scheduling and core composition. Once additional threads are launched, hyper-threading and the activation of E-cores can lower the effective per-core frequency to the E-core maximum, reducing single-thread throughput. This effect may be reinforced by increased context switching and by Java runtime events such as JIT compilation or garbage collection, which are triggered under higher concurrency levels.

Second, the superlinear speedups observed on the Ryzen are explained by memory-hierarchy effects. The large on-chip caches in this processor family can substantially reduce memory traffic for medium-sized working sets (roughly 4000–7000). When the workload is partitioned across threads, cache hit rates improve and lead to more-than-linear performance gains.

Third, the differences in energy ordering (e.g., LUP consuming less energy than Crout on the i9) likely stem from subtle interactions between algorithmic memory-access patterns and the Java runtime's code generation. Pivoting affects data locality, while different JIT compilation paths can produce distinct instruction mixes. These effects are further influenced by measurement variability and by power-state transitions occurring between runs.

Overall, these anomalies show that performance and energy efficiency cannot be generalized across platforms: the interplay of hardware heterogeneity, cache behavior, and runtime code generation causes each LU algorithm to behave differently depending on the processor and operating system. This underscores the importance of evaluating algorithms individually on the specific hardware and software environments where they will run, rather than relying on results from other architectures and operating systems.

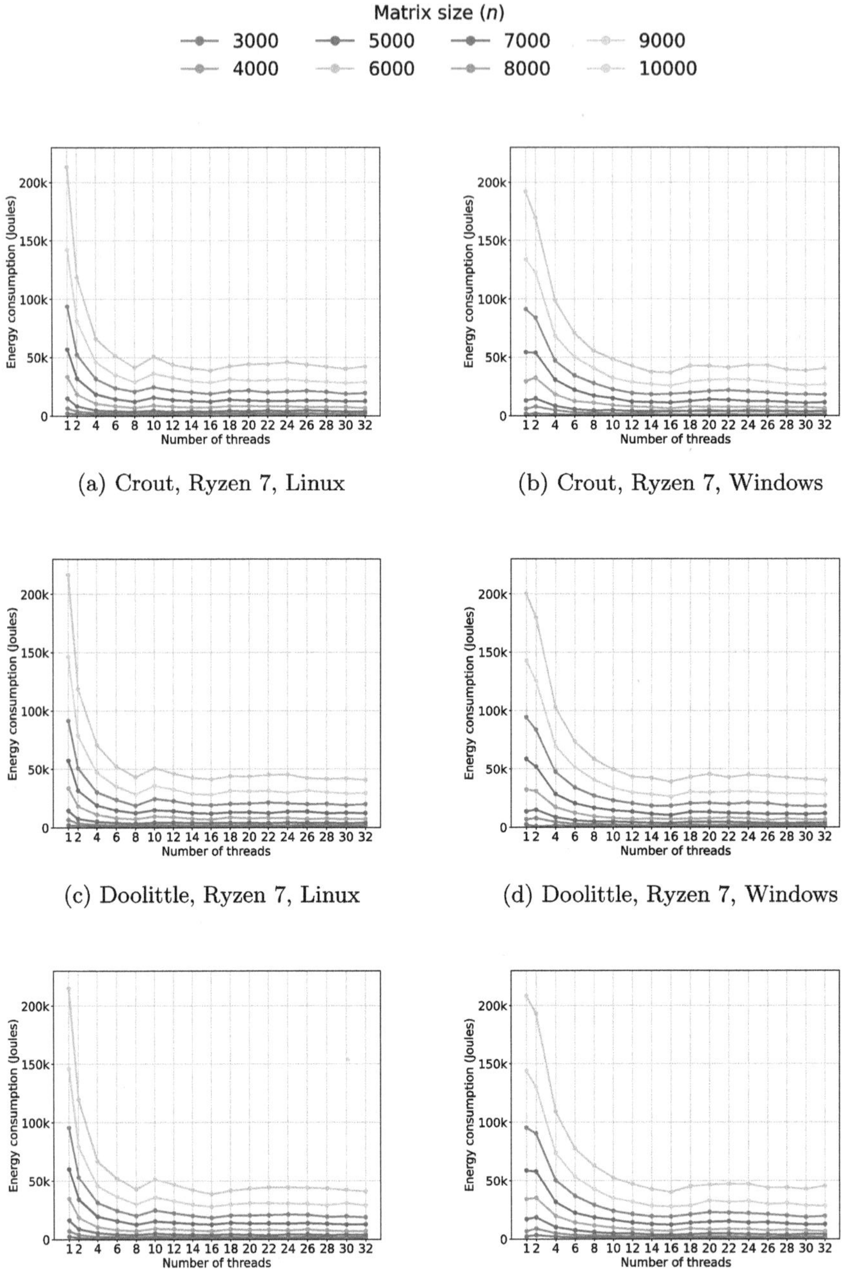

(a) Crout, Ryzen 7, Linux

(b) Crout, Ryzen 7, Windows

(c) Doolittle, Ryzen 7, Linux

(d) Doolittle, Ryzen 7, Windows

(e) LUP, Ryzen 7, Linux

(f) LUP, Ryzen 7, Windows

Fig. 4. Energy consumption curves for the Java implementations of Crout, Doolittle, and LUP on the AMD Ryzen 7 7800X3D under Linux and Windows.

Limitations

This empirical study is limited to Java implementations on two CPUs, two operating systems, and a single JVM configuration; therefore, its results cannot be broadly generalized beyond these specific hardware, OS, and runtime settings. Power measurement granularity (1-second sampling) and the lack of microarchitectural traces (e.g., performance counters, detailed JIT logs) constrain causal attribution. Although we used warm-up runs and 30 repetitions to reduce uncertainty, scheduling nondeterminism and background activity may still affect some results.

6 Conclusions and Future Work

This work has compared the execution time and energy consumption of three LU factorization variants (Crout, Doolittle and LUP) implemented in Java across two representative CPU platforms—an Intel Core i9-12900HK (heterogeneous P/E cores) and an AMD Ryzen 7 7800X3D (homogeneous cores)—and under two operating systems (GNU/Linux and Windows). The experiments show that performance and energy behavior result from a complex interaction among algorithmic memory-access patterns, processor microarchitecture, and OS/runtime policies, rather than from simple, hardware-agnostic trends.

The experiments demonstrate that LU factorization performance in Java depends strongly on hardware and operating system. On the i9, scaling is effective on the six P-cores but drops when E-cores or Hyper-Threading are used under Linux, while Windows sustains higher thread counts. On the Ryzen, medium matrices (4,000–7,000) sometimes achieve superlinear speedups due to cache effects. Energy efficiency also varies with OS and algorithm (e.g., LUP uses less energy than Crout on the i9). Overall, heterogeneous cores and OS scheduling can alter scalability and energy trade-offs, and no single algorithm–OS–hardware combination is optimal; the best choice depends on matrix size, performance versus energy goals, and the target platform.

Future research should further investigate the causes of the observed performance and energy behaviors. Controlled experiments with explicit CPU affinity and power policies could isolate scheduling and frequency effects, while finer timing and high-resolution power telemetry would reduce noise and separate core, uncore, and DRAM contributions. Microarchitectural counters (e.g., cache misses, LLC hit rates) could confirm whether superlinear speedups arise from cache and memory effects. In addition, JVM configurations with explicit warm-up, fixed heap sizes, and deterministic garbage collection would clarify runtime impacts. More repetitions with robust statistics and significance testing would improve reliability. Together, these steps would enable clearer causal insights and provide more robust guidance for optimizing LU factorization on both heterogeneous and homogeneous CPUs.

Acknowledgments. This publication is part of GRASS (FEDER-UCA-2024-A2-23) and gCODE (PIDI_2024_01307) projects, both funded by European Regional Development Funds (ERDF) and the Consejería de Universidad, Investigación e Innovación de la Junta de Andalucía. J.M. Aragón-Jurado would like to acknowledge the Spanish Ministerio de Ciencia, Innovación y Universidades for the support through FPU21/02026 grant. Juan Carlos de la Torre would like to acknowledge the University of Cadiz for its support through "Tecnicas avanzadas de Inteligencia Artificial aplicadas al Tratamiento de Imágenes en entornos reales" project (I+i 2024/03).

References

1. Abeles, P., Contributors: EJML – efficient java matrix library. https://ejml.org. Accessed 2 Oct 2025
2. Anderson, E., et al.: LAPACK Users' Guide, 3rd edn. SIAM, Philadelphia (1999)
3. Augonnet, B., Thibault, S., Namyst, R., Wacrenier, P.A.: STARPU: a unified platform for task scheduling on heterogeneous multicore architectures. In: Proceedings of the 18th International European Conference on Parallel Processing (Euro-Par) Workshops, pp. 863–874 (2011). https://doi.org/10.1007/978-3-642-29737-3_94
4. Baggs, R., Delgado, H., Bakhtiani, P.: Performance analysis of *i/o*-intensive & CPU-intensive benchmarks on windows 7, 8.1 & ubuntu 14.04 LTS. In: Proceedings of the IEEE International Conference on Electro/Information Technology (EIT), pp. 415–420 (2015). https://doi.org/10.1109/EIT.2015.7293429
5. Blackford, S., et al.: ScalaPack users' guide. Lawrence Berkeley National Laboratory/University of Tennessee, Tech. rep. (1997)
6. Blumofe, R.D., Leiserson, C.E.: Scheduling multithreaded computations by work stealing. J. ACM **46**(5), 720–748 (1999)
7. Bosilca, G., Bouteiller, A., Herault, T., Dongarra, J.: Parsec: a runtime system for dynamic task-based parallelism on heterogeneous architectures. Parallel Comput. **37**(9), 521–540 (2011). https://doi.org/10.1016/j.parco.2011.08.005
8. Burden, R.L., Faires, J.D.: Numerical Analysis. Cengage Learning (2010)
9. Coon, A., Stadtherr, M.: Parallel implementation of sparse LU decomposition for chemical engineering applications. Comput. Chem. Eng. **13**(8), 899–914 (1989). https://doi.org/10.1016/0098-1354(89)85063-X
10. Crout, P.D.: A short method for evaluating determinants and solving systems of linear equations with real or complex coefficients. Electr. Eng. **60**(12), 1235–1240 (1941)
11. Golub, G.H., Van Loan, C.F.: Matrix Computations, 4th edn. Johns Hopkins University Press, Baltimore, MD (2013)
12. Higham, N.J.: Accuracy and Stability of Numerical Algorithms, 2nd edn. SIAM, Philadelphia, PA (2002)
13. Higham, N.J., Higham, D.J.: Large growth factors in gaussian elimination with pivoting. SIAM J. Matrix Anal. Appl. **10**(2), 155–164 (1989)
14. Intel corporation: Managing performance with heterogeneous cores (2023). https://www.intel.com/content/www/us/en/docs/onemkl/developer-guide-windows/2023-0/managing-performance-with-heterogeneous-cores.html. Accessed 12 May 12 2025
15. Kaya, D., Wright, K.: Parallel algorithms for LU decomposition on a shared memory multiprocessor. Appl. Math. Comput. **163**(1), 179–191 (2005). https://doi.org/10.1016/j.amc.2004.01.027
16. Lea, D.: A Java fork/join framework. In: Proceedings of the ACM 2000 Conference on Java Grande, pp. 36–43. ACM, San Francisco, CA, USA (2000). https://doi.org/10.1145/337449.337465

17. Li, X., Niu, J., Khan, M.K., Wang, Z.: Applying LU decomposition of matrices to design anonymity bilateral remote user authentication scheme. Math. Probl. Eng. **2013**, 910409 (2013). https://doi.org/10.1155/2013/910409
18. Mohammed, A.O., Hussein, H.I., Mstafa, R.J., Abdulazeez, A.M.: A blind and robust color image watermarking scheme based on DCT and dwt domains. Multimedia Tools Appl. **82**(21), 32855–32881 (2023). https://doi.org/10.1007/s11042-023-14797-0
19. OpenMP architecture review board: OpenMP application program interface, version 5.0 (2018). https://www.openmp.org/specifications/
20. Ozcan, C., Sen, B.: Investigation of the performance of LU decomposition method using CUDA. Procedia Technol. **1**, 50–54 (2012). https://doi.org/10.1016/j.protcy.2012.02.011, first World Conference on Innovation and Computer Sciences (INSODE 2011)
21. Palma Méndez, J.T., Garrido Carrera, M.D.C., Sánchez Figueroa, F.: Quesada Arencibia, A.: Programación Concurrente. Ediciones Paraninfo (2003). SA
22. Rana, V.S., Lin, M., Chapman, B.: A scalable task parallelism approach for LU decomposition with multicore CPUS. In: 2016 Second International Workshop on Extreme Scale Programming Models and Middleware (ESPM2), pp. 17–23 (2016). https://doi.org/10.1109/ESPM2.2016.008
23. Ristov, S., Gusev, M.: Performance vs cost for windows and Linux platforms in windows azure cloud. In: Proceedings of the 2013 IEEE 2nd International Conference on Cloud Networking (CloudNet), pp. 194–200 (2013). https://doi.org/10.1109/CloudNet.2013.6710581
24. Sestoft, P.: Numeric performance in C, C# and java. Tech. rep., IT University of Copenhagen (2010). https://www.itu.dk/~sestoft/papers/numericperformance.pdf, 14 pages
25. Su, Q., Wang, G., Zhang, X., Lv, G., Chen, B.: A new algorithm of blind color image watermarking based on LU decomposition. Multidimension. Syst. Signal Process. **29**(3), 1055–1074 (2018). https://doi.org/10.1007/s11045-017-0487-7
26. Gauhati university institute of science and technology: Numerical methods: LUP decomposition (1968-1996). https://lampz.tugraz.at/~hadley/num/ch2/2.3a.php
27. Wendykier, P., contributors: Parallel colt – multithreaded version of colt (JAVA) (2010). https://sourceforge.net/projects/parallelcolt/. sourceForge project / code mirrors on GitHub; see also Wendykier et al., 2010
28. Yang, A., Liu, C., Chang, J., Guo, X.: Research on parallel LU decomposition method and its application in circle transportation. J. Softw. **5**(11), 1250–1255 (2010). https://doi.org/10.4304/jsw.5.11.1250-1255
29. Żytko, D., Badurowicz, M.: Analysis of performance and energy efficiency of processors with hybrid architecture. J. Comput. Sci. Inst. **34**, 54–59 (2025). https://doi.org/10.35784/jcsi.6690

Computational Science - Theories, Algorithms, and Applications

Differentially Private Synthetic Health Data

Joseph Pedersen[1(✉)] and Kristin P. Bennett[2]

[1] Rensselaer Polytechnic Institute Troy, USA 12180, NY, USA
pederj4@rpi.edu
[2] Rensselaer Polytechnic Institute Troy, USA 12180, NY, USA
bennek@rpi.edu

Abstract. Generative adversarial networks (GANs) have been successfully used to generate synthetic health data to preserve patient privacy. We develop a differential private (DP) Wasserstein GAN with gradient penalty (WGAN-GP) that empirically demonstrates high utility, resemblance, and privacy on three health care problems. We derive tight bounds on the gradient of the loss for a WGAN for certain architectures. Computational tests show the noise added to achieve DP results in convergence to better local minima. The proposed DP WGAN-GP generates synthetic data that performs better in classification tasks than the original WGAN with no noise. We hypothesize that adding differential privacy reduces the problem of mode collapses that frequently occurs in GANs, and perhaps the regularization that it induces may benefit other optimization problems.

1 Introduction

Governments, companies, and other organizations have a wealth of data, but the need to protect individual privacy can often hinder its use. For example, laws protecting health information, such as the Health Insurance Portability and Accountability Act (HIPAA) in the United States, impose severe fines and the possibility of imprisonment for violation. Thus, there is a need to have strong guarantees of privacy before releasing data. Unfortunately, this impoverishes data science education and research in all disciplines by forcing reliance upon "toy" data sets or a small number of de-identified data sets.

Synthetic data generation offers a potential solution to enable research while preserving privacy. Generative adversarial networks (GANs), represent the state-of-the-art for synthetic health data modeling and generation [1]. However, the model produced by a GAN is not necessarily private; the weights can memorize the training data [2], so stronger privacy guarantees are required.

A common choice is differential privacy (DP), which matches an intuitive notion of privacy. Any algorithm that uses stochastic gradient descent (SGD) can be made DP by: clipping each gradient to be bound in ℓ_2 norm, adding noise based on that bound, and taking a step in that noisy direction. [3] The added noise can impact output quality, so the tightest bound possible is desirable.

H. R. Arabnia et al. (Eds.): CSCI 2025, CCIS 2970, pp. 71–81, 2027.
https://doi.org/10.1007/978-3-032-28466-2_5

Therefore we derive tight bounds on the gradient of the loss for a WGAN for a wide class of neural architectures. We use this bound as a basis for a novel DP WGAN with gradient penalty (DP WGAN-GP) that empirically demonstrates high utility, resemblance, and privacy on three health care problems.

The rest of this paper is organized as follows. First, we briefly review relevant facts about Wasserstein GANs and discuss the need for privacy. Next, we derive our bounds on the gradients of WGAN and WGAN-GP. Then, we describe our algorithm which uses those bounds for a differentially private WGAN-GP. Finally, we analyze the results of using DP-WGAN-GP on some real datasets by examining the utility, resemblance, and privacy of the synthetic data for different levels of privacy. We conclude with discussion and directions for future research.

The specific contributions of this paper include:

- An improved tight bound on gradient of WGAN over published results,
- A novel bound on the gradient penalty term of a WGAN-GP,
- An algorithm which uses the bounds for a differentially private WGAN-GP,
- Code for above at https://github.rpi.edu/RensselaerIDEA/synthetic-data
- Empirical analysis of DP-WGAN-GP on real datasets, and demonstrating that differential privacy can help reduce convergence to poor local minimums and improve performance while maintaining privacy.

2 Generative Adversarial Networks

A technique developed to improve upon the training of GANs is the **Wasserstein GAN** (WGAN) of Arjovsky et al.(2017). Their architecture includes a generator G and *critic* D, which solve the following bi-level optimization problem, where $\mathcal{D}$ is the set of all 1-Lipschitz functions [4]:

$$\min_{G} \max_{D \in \mathcal{D}} \mathbb{E}_{x \sim p_r}[D(\mathbf{x})] - \mathbb{E}_{\tilde{x} \sim p_g}[D(\tilde{x})] \tag{1}$$

Arjovsky et al.(2017) show that this optimization problem is sound and the training is more stable than the original GAN formulation. To satisfy the Lipschitz constraint, they clip the weights of their critic network, although they note that weight clipping imposes various difficulties in the training. [4]

We adopt the improved WGAN-GP of Gulrajani et al. [5], that uses a *gradient penalty*, instead of weight clipping, to enforce the Lipschitz constraint. Their proposed gradient penalty would bound the norm of the gradient by 1, making the critic 1-Lipschitz. However, they use a soft version of the constraint by adding a penalty term to the loss function, computed on the gradient norm for random samples $\hat{x} \sim p_{\hat{x}}$.

$$L = \mathbb{E}_{\tilde{x} \sim p_g}[D(\tilde{x})] - \mathbb{E}_{x \sim p_r}[D(x)] + \lambda \mathbb{E}_{\hat{x} \sim p_{\hat{x}}}\left[(\|\nabla_{\hat{x}} D(\hat{x})\|_2 - 1)^2\right] \tag{2}$$

3 Private Synthetic Data

HealthGAN, an implementation of this WGAN-GP architecture, has been used to generate synthetic databases of electronic medical records (EMR) [6,7]. Their methodology is designed to meet regulatory requirements for privacy, while retaining instructional utility. [6] They developed novel metrics for measuring utility, resemblance, and privacy loss; and used those to compare their method to several others. Although HealthGAN demonstrated good privacy empirically by those metrics, no theoretical privacy guarentees have been established for HealthGAN.

Differential privacy (DP), developed by Dwork et al.(2006) offers a rigorous guarantee of a certain notion of privacy, and has been widely adopted and well studied. The intuition for the method is that, the probability of a given outcome should not change much by changing one record in the data.

Definition 1. *(Differential Privacy) A randomized algorithm $\mathcal{M}$ with domain $\mathbb{N}^{|\mathcal{X}|}$ is (ε, δ)-differentially private if $\forall \mathcal{S} \subseteq Range(\mathcal{M})$ and $\forall x, y \in \mathbb{N}^{|\mathcal{X}|}$ such that $\|x - y\|_1 \leq 1$, the following holds [8]:*

$$Pr[\mathcal{M}(x) \in \mathcal{S}] \leq \exp(\varepsilon) Pr[\mathcal{M}(y) \in \mathcal{S}] + \delta \tag{3}$$

Dwork et al.(2006) show that any algorithm can be made DP if the "sensitivity" of each query of the private data is bounded, and an amount of noise determined by that bound is added to the true query result. [9] For SGD, the query is the sum of the gradients computed on a batch, and the ℓ_2-sensitivity is the maximum possible ℓ_2 norm of any of the individual gradients. As noted, one common technique for bounding the norm is to simply clip the gradients to be within the desired bound.

4 Bounds on Gradient of Wasserstein GANs

We derive a stronger bound (tight in fact) for the gradient of a WGAN with all layers fully connected and only ReLU or LeakyReLU activation functions. Xie et al.(2018) created a DP WGAN, called DPGAN, and showed the norm of the gradients could be bound by only clipping the weights. [10]:

$$\begin{aligned}\left\|\nabla_W \left[D_W\left(\mathbf{x}^{(i)}\right) - D_W\left(G_\Theta(\mathbf{z}^{(i)})\right)\right]\right\| &\leq 2\left\|\nabla_W \left[D_W\left(\mathbf{x}^{(i)}\right)\right]\right\| \\ = 2\sum_l \sum_{ij} \left|\frac{\partial C}{\partial W_{ij}^{(l)}}\right| &\leq 2C_p B_\sigma B_{\sigma'}^2 \sum_{l=1}^{L-1} m_l m_{l+1} = C_D\end{aligned} \tag{4}$$

with bounds: C_p on the weights of the critic, B_σ on the output of the activation function, and $B_{\sigma'}$ on the derivative of the activation function. They note that, although activation functions like ReLU or Leaky ReLU are not bounded, if the input is from a compact set (true for HealthGAN) the bound will exist.

We bound each component of the gradient, rather than just the norm. We also derive a bound on the norm of the gradient of the gradient penalty term of the WGAN-GP. These bounds are computed each training iteration, obviating the need to clip the weights, which is not part of the WGAN-GP algorithm. The code to compute the bounds is available at our github repo.[1]

4.1 Derivation of Tight Bound on Gradient of WGAN

We note by the linearity of differentiation, the gradient can be decomposed into two terms, $\nabla_W \left[D_W\left(\mathbf{x}^{(i)}\right)\right] - \nabla_W \left[D_W\left(G_\Theta(\mathbf{z}^{(i)})\right)\right]$, and only the first term is a query on the data. This enables us to take a different approach than Xie et al.(2018) which added noise calibrated to a bound on the norm of the gradient of the entire loss function. Rather than derive our bound from a constant bound on the weights, C_p, we use the actual weights of the critic each iteration, t_c. Since the weights are initialized to small values early in training, this yields a much tighter bound. The bounds for component j of the gradient are of the form:

$$C_{0,j}^{\nabla_{W^{(t_c)}}} \leq \left[\nabla_{W^{(t_c)}} D_{W^{(t_c)}}(\mathbf{x})\right]_j \leq C_{1,j}^{\nabla_{W^{(t_c)}}} \tag{5}$$

The bounds $C_0^{\nabla_{W^{(t_c)}}}$ and $C_1^{\nabla_{W^{(t_c)}}}$ can not be given by a single closed form expression, but instead are computed recursively, similar to the gradients, using the same forward and back propagation equations. Our notation for the bounds is that, for some quantity z, C_0^z is the lower bound and C_1^z is the upper bound, with the superscript indicating what quantity is being bound. For critics that use only fully connected layers with ReLU or LeakyReLU activation functions (very common activation functions in the literature), the bounds we derive are tight, in the sense that when the component-wise upper and lower bounds of the input training data are the same, then the computed component-wise bounds on the gradients are the same (i.e. the value of the gradient for that input).

Our critic, $D(x)$, is a neural network with L hidden layers, all fully connected. The activation functions are all ReLU, LeakyReLU, or identity, which can be expressed as a family parameterized by $\alpha \in [0,1]$, defined by $\phi_\alpha(x) = \max\{x, \alpha x\}$. The entire network is described by the following well known neural network equations. Superscripts denote layers, and subscripts denote components. The input layer is x^0, and the output layer is $D(x) = D(x^0) = x^{L+1}$.

$$s_j^\ell = \left(\sum_i W_{ij}^\ell x_i^{\ell-1}\right) + b_j^\ell \quad \text{and} \quad x_j^\ell = \phi_{\alpha^\ell}\left(s_j^\ell\right) \tag{6}$$

For continuous activation functions, a bound $C_{0,i}^{x^0} \leq x_i^0 \leq C_{1,i}^{x^0}$ on the input (ours is bounded by $0 \leq x_i^0 \leq 1$) propagates forward. For non-decreasing activation

[1] https://github.rpi.edu/RensselaerIDEA/synthetic-data.

functions, for k in $\{0,1\}$, the bound is given by:

$$C^{s^\ell}_{k,j} = \sum_{i:W^\ell_{ij}\geq 0} W^\ell_{ij} C^{x^{\ell-1}}_{k,i} + \sum_{j:W^\ell_{ij}<0} W^\ell_{ij} C^{x^{\ell-1}}_{1-k,i} \tag{7}$$

$$C^{x^\ell}_{0,j} = \phi_{\alpha^\ell}\left(C^{s^\ell}_{0,j}\right) \leq x^\ell_j \leq \phi_{\alpha^\ell}\left(C^{s^\ell}_{1,j}\right) = C^{x^\ell}_{1,j} \tag{8}$$

These forward propagation bounds are clearly tight in that, if $C^{x^{\ell-1}}_0 = C^{x^{\ell-1}}_1$, then the left hand side equals the right hand side in (8).

Denote the loss function for the critic of the WGAN for a single training example by $e(x,z)$. The gradient penalty term of the WGAN-GP will be treated separately shortly. The gradients from back propagation are given by:

$$\nabla_{b^\ell}\, e = \phi'_{\alpha^\ell}(s^\ell) \odot \nabla_{x^\ell}\, e \tag{9}$$

$$\nabla_{x^{\ell-1}}\, e = W^\ell \nabla_{b^\ell}\, e \tag{10}$$

$$\nabla_{W^\ell}\, e = (\nabla_{x^{\ell-1}}\, e)\,(\nabla_{b^\ell}\, e)^T \tag{11}$$

Here, $\nabla_z\, e$ has the dimensions of z, with $[\nabla_z\, e]_i = \partial e/\partial z_i$, and $\odot$ represents the Hadamard product. Denote the function min by $g_0(\cdot)$ and max by $g_1(\cdot)$. We need the following facts about the products of vectors y and z: if $a_i \leq y_i \leq b_i$ and $c_j \leq z_j \leq d_j$, and denoting $m^{(k)}_{ij} = g_k\{a_i c_j, a_i d_j, b_i c_j, b_i d_j\}$, then we know: $m^{(0)}_{ii} \leq [y \odot z]_i \leq m^{(1)}_{ii}$ (if $y \odot z$ is defined), and $m^{(0)}_{ij} \leq [yz^T]_{ij} \leq m^{(1)}_{ij}$. These are tight: $a_i = b_i$ and $c_j = d_j$ implies $m^{(0)}_{ij} = m^{(1)}_{ij}$. From forward propagation:

$$C^{\phi'_{\alpha^\ell}}_{0,j} = \phi'_{\alpha^\ell}\left(C^{s^\ell}_{0,j}\right) \leq \phi'_{\alpha^\ell}(s^\ell) \leq \phi'_{\alpha^\ell}\left(C^{s^\ell}_{1,j}\right) = C^{\phi'_{\alpha^\ell}}_{1,j} \tag{12}$$

The gradient of the output is $\nabla_{x^{L+1}}\, e = -1$ for the WGAN (without gradient penalty term), but during actual computations we find that numerical errors require adding some tolerance, such as $\nabla_{x^{L+1}} \in (-0.999999, -1.000001)$. Using these bounds with back propagation gives, for k in $\{0,1\}$:

$$C^{\nabla_{b^\ell}}_{k,j} = g_k\{C^{\nabla_{x^\ell}}_{0,j} C^{\phi'_{\alpha^\ell}}_{0,j}, C^{\nabla_{x^\ell}}_{0,j} C^{\phi'_{\alpha^\ell}}_{1,j}, C^{\nabla_{x^\ell}}_{1,j} C^{\phi'_{\alpha^\ell}}_{0,j}, C^{\nabla_{x^\ell}}_{1,j} C^{\phi'_{\alpha^\ell}}_{1,j}\} \tag{13}$$

$$C^{\nabla_{x^{\ell-1}}}_{k,j} = \sum_{i:W^\ell_{ij}\geq 0} W^\ell_{ij} C^{\nabla_{b^\ell}}_{k,j} + \sum_{j:W^\ell_{ij}<0} W^\ell_{ij} C^{\nabla_{b^\ell}}_{1-k,j} \tag{14}$$

$$C^{\nabla_{W^\ell_{ij}}}_{k,j} = g_k\{C^{\nabla_{x^{\ell-1}}}_{0,j} C^{\nabla_{b^\ell}}_{0,j}, C^{\nabla_{x^{\ell-1}}}_{0,j} C^{\nabla_{b^\ell}}_{1,j}, C^{\nabla_{x^{\ell-1}}}_{1,j} C^{\nabla_{b^\ell}}_{0,j}, C^{\nabla_{x^{\ell-1}}}_{1,j} C^{\nabla_{b^\ell}}_{1,j}\} \tag{15}$$

4.2 Derivation of Bound on Norm of Gradient of WGAN-GP

Theorem 1. *The norm of the gradient of the gradient penalty is bounded by*

$$\|\nabla_{W^\ell}\left(\|\nabla_x D(x)\|_2 - 1\right)^2\|_F \leq \\ 2\max\{(\|\nabla_x D(x)\|_2 - 1)\,, 1\}\prod_{k\neq\ell}\|W^k\|_F \prod_{1\leq k\leq L}\|\phi'_{\alpha^k}\|_F \tag{16}$$

Proof. Using the chain rule, with a dummy variable z, and ignoring the non-differentiability at the origin, we have:

$$\nabla_z\left(\|\nabla_x D(x)\|_2 - 1\right)^2 = 2\left[(\|\nabla_x D(x)\|_2 - 1)\right]\nabla_z\|\nabla_x D(x)\|_2 \tag{17}$$

$$\nabla_z\|\nabla_x D(x)\|_2 = \nabla_z\left(\nabla_x D(x)\right)\nabla_x D(x)/\|\nabla_x D(x)\|_2 \tag{18}$$

$$\|\nabla_z\left(\|\nabla_x D(x)\|_2\right)\|_F \leq \|\nabla_z\left(\nabla_x D(x)\right)\|_F \tag{19}$$

The gradient term $\nabla_x D(x)$ is computed using standard back propagation:

$$\frac{\partial D(x)}{\partial x_i^{\ell-1}} = \sum_j \left(\frac{\partial x_j^\ell}{\partial x_i^{\ell-1}}\right)\left(\frac{\partial D(x)}{\partial x_j^\ell}\right) = \sum_j \phi'_{\alpha^\ell}\left(s_j^\ell\right)W_{ij}^\ell\left(\frac{\partial D(x)}{\partial x_j^\ell}\right) \tag{20}$$

$$\nabla_{x^{\ell-1}} D(x) = \left(W^\ell\right)\left(\phi'_{\alpha^\ell}\right)\left(\nabla_{x^\ell} D(x)\right) \tag{21}$$

$$\nabla_{x^0} D(x) = \left(W^1\right)\left(\phi'_{\alpha^1}\right)\ldots\left(W^{L+1}\right)\left(\phi'_{\alpha^{L+1}}\right) = \prod_{\ell=1}^{L+1}\left(W^\ell\right)\left(\phi'_{\alpha^\ell}\right) \tag{22}$$

Instead of Hadamard products with vectors $\phi'_{\alpha^\ell}(s^\ell)$, here we are letting ϕ'_{α^ℓ} denote the corresponding diagonal matrices, and $\phi'_{\alpha^{L+1}}$ is the identity and may be dropped when convenient. Although ϕ'_α is technically not differentiable at zero, we adopt the convention that $\phi''_\alpha(x) = 0$ (constant), making the gradient with respect to any of the bias vectors equal to zero. With respect to the weight matrices, we have $\nabla_{W^\ell}\left(\nabla_x D(x)\right) =$

$$\left(\prod_{1\leq k<\ell}\left(W^k\right)\left(\phi'_{\alpha^k}\right)\right)^T\nabla_x D(x)\left(\prod_{\ell<k\leq L+1}\left(\phi'_{\alpha^{k-1}}\right)\left(W^k\right)\right)^T \tag{23}$$

Empty products are identity transformations. We are bounding the ℓ_2-norm, which when considering weights as matrices is the Froebenius norm. Combining all equations yields the desired bound (without the factor of λ). □

Algorithm 1. Differentially Private WGAN-GP

Ensure: data in $[0,1]^d \subset \mathbb{R}^d$; generator (gen.) G_Θ and critic D_W fully connected with ReLU or LeakyReLU activation functions.
Require: ϵ, δ: DP parameters; n_{g}: gen. iterations; n_{c}: critic iterations (per gen. iter); m: batch size; λ: gradient penalty coefficient; α, β_1, β_2: Adam hyper-parameters

Initialize: $\Theta^{(1)}$ and $W^{(1)}$
$\sigma_{\mathrm{dp}} \leftarrow \left(\sqrt{2\log(1.25/\delta)}\right)/\epsilon$
for $t_g = 1, \ldots, n_{\mathrm{g}}$ **do**
  for $t_c = 1, \ldots, n_{\mathrm{c}}$ **do**
    Compute $C_0^{\nabla_{W^{(t_c)}}}, C_1^{\nabla_{W^{(t_c)}}} : C_0^{\nabla_{W^{(t_c)}}} \le \nabla_{W^{(t_c)}} D_{W^{(t_c)}}(\mathbf{x}) \le C_1^{\nabla_{W^{(t_c)}}}$
    Compute $C^\lambda = \max_{\mathbf{x}\in[0,1]^d} \left\| \nabla_{W^{(t_c)}} \left(\|\nabla_{\widehat{\mathbf{x}}} D_W(\widehat{\mathbf{x}})\|_2 - 1 \right)^2 \right\|_2$
    Compute total ℓ_2 sensitivity $\ell_2^{(t_c)} = \sqrt{\left\| C_1^{\nabla_{W^{(t_c)}}} - C_0^{\nabla_{W^{(t_c)}}} \right\|_2^2 + \lambda^2 \|C^\lambda\|_2^2}$
    for $i = 1, \ldots, m$ **do**
      Sample real data $\mathbf{x} \sim \mathbb{P}_r$, latent variable $\mathbf{z} \sim \mathbb{P}_z$, random number $u \sim U[0,1]$
      $\widetilde{\mathbf{x}} \leftarrow G_\Theta(\mathbf{z})$
      $\widehat{\mathbf{x}} \leftarrow u\mathbf{x} + (1-u)\widetilde{\mathbf{x}}$
      $g_W^{(i)} \leftarrow \nabla_{W^{(t_c)}} \left[D_W(\widetilde{\mathbf{x}}) - D_W(\mathbf{x}) + \lambda \left(\|\nabla_{\widehat{\mathbf{x}}} D_W(\widehat{\mathbf{x}})\|_2 - 1 \right)^2 \right]$
    Sample $n \sim N(0, 4\sigma_{\mathrm{dp}}^2 \mathbf{I}_W)$
    $\hat{g}_{W^{(t_c)}} \leftarrow \frac{1}{m}\left(\sum_{i=1}^m g_w^{(i)} \right) + \frac{1}{m} n$
    $W^{(t_c+1)} \leftarrow \mathrm{Adam}\left(\hat{g}_{W^{(t_c)}}, W^{(t_c)}, \alpha, \beta_1, \beta_2 \right)$
  Sample a batch of latent variables $\{\mathbf{z}^{(i)}\}_{i=1}^m \sim \mathbb{P}_z$.
  $\Theta^{(t_g+1)} \leftarrow \mathrm{Adam}\left(\frac{1}{m} \sum_{i=1}^m \nabla_{\Theta^{(t_g)}} \left[-D_{W^{(t_c+1)}}\left(G_{\Theta^{(t_g)}}(\mathbf{z}^{(i)}) \right) \right], \Theta^{(t_g)}, \alpha, \beta_1, \beta_2 \right)$
Return: W and Θ, and $\ell_2^{(t_c)}$ to compute total privacy cost $(\varepsilon_T, \delta_T)$.

5 DP WGAN-GP with Analysis on Health Data

We have used the derived bounds with Algorithm 1 to modify HealthGAN of Yale et al. [6] to be DP. We compared performance of the original HealthGAN (HGAN) to DP WGAN-GP on three datasets from the MIMIC-III (Medical Information Mart for Intensive Care) database on patients admitted to critical care units at a large tertiary care hospital. [11]. These datasets were designed to reproduce published medical studies. Each dataset has an associated task: task 1 predicts mortality of diabetics [12], task 2 predicts mortality of acute kidney injury patients [13], and task 3 predicts readmission to ICU within 30 days [14] (Table 1). In our experimental design, we used a 50/50 training/test data split, and trained the synthetic data model only on the training data. To assess the effect of DP, the same algorithm was used on all three datasets, with the level of noise varied: zero noise for HealthGAN, and four levels of noise σ_{dp} (10^{-7},10^{-6},10^{-5}, 10^{-4}).

The models were trained using Google's TensorFlow. Both the generator and critic of the WGAN-GP had all fully connected layers. The generator had 100

Table 1. Characteristics of three MIMIC-III Datasets. + Rate refers to percentage of people who died/were readmitted in the training data.

Dataset	Size	Features	Label	+ Rate
1	2085	6 continuous, 5 categorical	Mortality	6.3%
2	2119	31 continuous, 2 categorical	Mortality	23.4%
3	1719	6 diagnostic values	Readmission	3.03%

nodes in the input (uniform random variables $\sim U(0,1)$), with other layer sizes: $2d, \lfloor 1.5d \rfloor, d$, where d is the number of features in the dataset. The activation function was ReLU in the hidden layers and sigmoid for the output. The critic layer sizes were: $d, 64, 128, 256, 1$. The activation function was LeakyReLU with $\alpha = 0.2$ in the hidden layers and none in the output. The batch size used was $\lfloor N/10 \rfloor$. We trained the generator for 200,000 iterations, and the critic for 5 iterations between each generator iteration. Before training, the data is preprocessed to be in $[0, 1]$. For categorical variables, we use the formulation of the synthetic data vault [15]. After training, synthetic data is converted to unnormalized values using the inverse transformation.

The optimization problem solved in WGANs is highly non-convex and algorithms frequently produce poor local minima. Algorithm 1 produced models with positive class rate (+ rate) much lower in the synthetic data than the training data, commonly known as "mode collapse". To combat this, we trained a fixed number of models per noise level, and chose the one with the highest positivity rate (least mode collapse). For datasets 1 and 2, we trained two models per noise level, and created two synthetic datasets for each. For dataset 3, which had the rarest class and suffered the most from convergence to local minima, we trained 10 models per noise level, and created two synthetic datasets for each. Table 2 shows the + rate for the best synthetic dataset for each noise level, and in the training and test sets, for each of the datasets. We assessed utility, resemblance, and privacy, comparing HealthGAN (HGAN) to the WGAN-GP model with the noise level resulting with highest (+ rate), with results summarized in Table 3.

Utility. To assess the utility of the synthetic data, which is how well it can be used for a specific purpose, we used the "gold standard" of creating predictive models on the synthetic data and using them to classify the test data, quantified by the area under the curve (AUC) of the classifiers. Although the goal of creating a DP WGAN-GP was to have the rigorous privacy guarantee, with the hope that utility would not be substantially decreased, we found that differential privacy actually slightly improved the utility for all three datasets.

Resemblance. We also want the synthetic data to resemble the real data. We measure resemblance in two ways. Since the class labels are imbalanced in the training data, and GANs tend to suffer from mode collapse, we specifically check for mode collapse by checking that the proportion of the rare class is not under-represented in the synthetic data. For each noise level, we trained a fixed number

Table 2. Comparison of highest positivity rates (%) of the synthetic data from random restarts of Alg 1 for different noise levels, to those in the training and test sets

	Dataset 1	Dataset 2	Dataset 3
training set	6.28	23.4	3.03
test set	6.04	25.9	2.97
$\sigma_{\text{dp}} = 0$ (HGAN)	4.46	17.8	2.60
$\sigma_{\text{dp}} = 10^{-7}$	**5.37**	19.1	2.39
$\sigma_{\text{dp}} = 10^{-6}$	5.23	**20.1**	2.68
$\sigma_{\text{dp}} = 10^{-5}$	4.22	19.6	**3.20**
$\sigma_{\text{dp}} = 10^{-4}$	4.89	18.9	2.97

Table 3. Comparison of results on three datasets between HGAN and DP WGAN-GP. Best results are bolded, but differences may not be significant

	Dataset 1		Dataset 2		Dataset 3	
Measure	HGAN	w/ DP	HGAN	w/ DP	HGAN	w/ DP
Utility – AUC	0.723	**0.724**	0.614	**0.657**	0.713	**0.720**
+Rate	4.5%	**5.4%**	17.8%	**20.2%**	2.6%	**3.2%**
TrResemblLoss	0.545	**0.539**	**0.856**	0.860	0.628	**0.627**
TeResemblLoss	**0.534**	0.550	**0.868**	0.872	**0.648**	0.649
PrivacyLoss	**-0.011**	0.012	**0.012**	**0.012**	0.022	**0.020**

of models and selected the one with the highest positivity rate. We found that the higher noise levels tended to have the highest maxima for positivity rate, corresponding to the models with the least mode collapse.

We also measured overall resemblance using the nearest-neighbors based metrics defined by Yale et al. (2019) [6]. Intuitively, these measure whether points from the synthetic data and real data are closer to neighbors in the same set or the other set, with 0.5 corresponding to both being equally likely (same distribution). TrResemblLoss measures the resemblance loss of the synthetic data to the training data, and TeResemblLoss measures the resemblance loss to the test data. We found that the differentially private WGAN-GP had comparable resemblance to HealthGAN with these metrics.

Privacy. The above bounds prove that the algorithm is technically DP, however the sensitivity of the gradient penalty term did not permit a practical value of ε. We also investigated the empirical privacy using the PrivacyLoss metric introduced by Yale et al.(2019), which is the difference of the resemblance loss on the test set and the resemblance loss on the training set. More resemblance loss on the test set is an indication of overfitting the training set. We did not find any significant difference between the value of this metric on HealthGAN and DP WGAN-GP. Our interpretation is that, although it is possible for a GAN to memorize the training data, it does not seem like HealthGAN is memorizing these datasets.

In addition to privacy, we find DP optimization has the ability to improve performance. As noted previously, our goal is to achieve an optimal solution in which the generator is approximating the true distribution of the data. However, since we are learning from a finite sample, we have to try not to overfit the training data. Differential privacy, which makes it unlikely for any one training point to have too large an effect on the optimal value found, intuitively may reduce the likelihood of overfitting, leading to better generalization.

6 Conclusion

In this work, we developed a DP version of the WGAN-GP that empirically demonstrates high utility, resemblance, and privacy on three health care problems. We derived strong bounds on the gradient of the loss functions of WGAN and WGAN-GP, and show how the latter can be used for a differentially private WGAN-GP. We find that the noise added for differential privacy reduced mode collapse and improved the utility of the synthetic data generated.

In future work, we plan to incorporate our tight bound for WGAN with other methods to enforce the Lipschitz constraint, and explore recent variations of differential privacy. Also, instead of keeping the noise constant across training iterations, more noise toward the beginning of training and less toward the end could potentially improve convergence. Our bounds also have the potential to be used with non-spherical Gaussian noise, which may improve performance. Since GANs are extremely popular, and mode collapse is one of their main problems, an in-depth examination of the use of differential privacy to mitigate mode collapse could be very useful.

Acknowledgments. This work was partially supported by United Health Foundation. The authors thank the contributions of Karan Bhanot, Isabelle Guyon, and John Erickson.

References

1. Goodfellow, I.J., et al.: In: Proceedings of the 27th International Conference on Neural Information Processing Systems - vol. 2 (MIT Press, Cambridge, MA, USA), NIPS'14, pp. 2672–2680 (2014)
2. Zhang, C., Bengio, S., Hardt, M., Recht, B., Vinyals, O.: Understanding deep learning requires rethinking generalization (2017). arXiv:1611.03530 ArXiv preprint
3. Abadi, M., et al.: Proceedings of the 2016 ACM SIGSAC Conference on Computer and Communications Security (Association for Computing Machinery, New York, NY, USA), CCS '16, pp. 308–318 (2016). https://doi.org/10.1145/2976749.2978318
4. Arjovsky, M., Chintala, S., Bottou, L.: In: Proceedings of the 34th International Conference on Machine Learning (2017)
5. Gulrajani, I., Ahmed, F., Arjovsky, M., Dumoulin, V., Courville, A.: Improved training of Wasserstein GANs (2017). arXiv:1704.00028 ArXiv preprint

6. Yale, A., Dash, S., Dutta, R., Guyon, I., Pavao, A., Bennett, K.P.: In: ESANN 2019 - European Symposium on Artificial Neural Networks, Computational Intelligence and Machine Learning, Bruges, Belgium (2019). URL https://hal.inria.fr/hal-02160496
7. Yale, A., Dash, S., Dutta, R., Guyon, I., Pavao, A., Bennett, K.P.: Neurocomputing **416**, 244 (2020)
8. Dwork, C., Roth, A.: Foundations and trends in theoretical computer science **9**(3–4), 211 (2014). https://doi.org/10.1561/0400000042
9. Dwork, C., McSherry, F., Nissim, K., Smith, A.: Theory of Cryptography Conference (Springer), pp. 265–284 (2006)
10. Xie, L., Lin, K., Wang, S., Wang, F., Zhou, J.: Differentially private generative adversarial network (2018). ArXiv preprint arXiv:1802.06739
11. Johnson, A.E., et al.: Scientific data **3**(1), 1 (2016)
12. Anand, R.S., et al.: AMIA Summits on Translational Science Proceedings **2018**, 310 (2018)
13. Fernandes, M.P., Silva, C.F., Vieira, S.M., Sousa, J.M.: 2014 IEEE International Conference on Fuzzy Systems (FUZZ-IEEE) (2014)
14. Celi, L.A.G., Tang, R.J., Villarroel, M.C., Davidzon, G.A., Lester, W.T., Chueh, H.C.: Journal of Healthcare Engineering **2**(1), 97 (2011)
15. Patki, N., Wedge, R., Veeramachaneni, K.: 2016 IEEE International Conference on Data Science and Advanced Analytics (DSAA), pp. 399–410 (2016). https://doi.org/10.1109/DSAA.2016.49

"Rules of the Road" Navigation with Logic, Theorem Proving and a Large Ontology

James Timberlake and Adam Pease(✉)

Naval Postgraduate School, Monterey, CA, USA
{james.timberlake,adam.pease}@nps.edu

Abstract. The deployment of maritime autonomous surface ship (MASS) is accelerating across various domains, driven by an expanding range of use cases. In non-military applications, MASS are utilized for oceanographic research, marine resource exploration, environmental monitoring, search and rescue operations, and commercial shipping [1]. In the military domain, MASS are employed for missions such as intelligence gathering, surveillance, deception, protection, and offensive operations targeting maritime infrastructure. A critical concern shared across these domains is the ability of an autonomous vessel on the high seas to adhere to the Convention on the International Regulations for Preventing Collisions at Sea (COLREGs), or the "Rules of the Road" [2]. Our work employs typed first order logic with arithmetic and a large ontology along with automated theorem proving to create a decision support system for navigation that is compliant with the COLREGs.

Keywords: COLREGs · ontology · logic · decision support system · theorem proving

1 Introduction

We used the Suggested Upper Merged Ontology (SUMO) [3,4], and associated inference technology [5–7] to develop a decision-aid framework for autonomous vessels that ensure COLREGs-compliant navigation.

SUMO is a comprehensive upper-level ontology and collection of domain ontologies that provides a framework for integrating knowledge across domains. It captures foundational concepts such as time, space, events, and objects, and offers a platform for structuring domain-specific data. SUMO employs higher-order logic [8], enabling knowledge engineers to declare constraints on actions through formal rules [3]. These declarative rules allow for validation through theorem proving [9], creating systems that can reason and justify decisions autonomously. By embedding decision-making processes within the logic itself, SUMO eliminates the need for external procedural implementation, streamlining the development of decision-support systems. SUMO's large inventory of

H. R. Arabnia et al. (Eds.): CSCI 2025, CCIS 2970, pp. 82–95, 2027.
https://doi.org/10.1007/978-3-032-28466-2_6

concepts and definitions provides for reuse, lessening the burden of formalizing concepts. By creating new content within the context of a large ontology, it ensures that changes in scope or context of the project do not expose hidden assumptions that would break existing models [10]. SUMO will be explained further in Sect. 2.4.

The primary objective is to formalize the decision-making logic necessary for autonomous vessels to identify and navigate close-quarters scenarios in compliance with applicable COLREGs rules. By representing these rules in a formal logic, the research aims to bridge the gap between theoretical maritime regulations and practical autonomous navigation, enabling MASS to reason about and execute collision-avoidance maneuvers with transparency and traceability.

This research relies on several key assumptions:

- **Situational Awareness**: The autonomous vessel is assumed to have situational awareness (SA) of its own position, course, speed, and that of surrounding vessels in its vicinity.
- **Sensor Inputs**: It is presumed that the vessel's radar and sensor systems provide accurate real-time inputs, such as closest point of approach (CPA), course over ground (COG), speed over ground (SOG), range and relative bearings of nearby vessels.
- **Focus on Decision-Making**: The scope of this research is confined to formalizing the decision-making process using ontology and automated reasoning tools, specifically for identifying COLREGs scenarios and determining compliant actions. Path replanning and maneuver execution are outside the scope of this work.

If integrated with onboard radar systems, the ontology and reasoning engine could evaluate navigational scenarios in real-time, offering several benefits:

- **Rule Applicability Assessment**: The system identifies which COLREGs rule applies to a given situation, reducing ambiguity for operators.
- **Collision Avoidance Recommendations**: By analyzing vessel dynamics and relative positioning, the system suggests actionable maneuvers, such as altering course or speed, to avoid collisions.
- **Enhanced Situational Awareness**: The tool provides clear, rule-based insights to operators, enabling more informed and confident decision-making.
- **Training**: This tool, coupled and integrated with simulation software, could provide direct Rules of the Road training for maritime navigators.

Unlike traditional procedural software testing, which relies on partial coverage through test cases, formal logic allows for rigorous, static validation, ensuring that the reasoning processes are provably error-free. If a logical contradiction is found, theorem proving provides a proof that precisely describes the inconsistency. This guarantees a higher degree of reliability and robustness, essential for the high-stakes environment of maritime navigation.

2 Background

2.1 COLREGs

The crux of almost every close-quarters situation between two vessels at sea is that of COLREGs. The International Maritime Organization (IMO) formalized The Rules of the Road in 1972 and they came into effect in 1977, and since then there have been over 60 amendments to these rules. The United States adopted the International Rules in 1977 and later developed the Inland Rules in 1981; this research will focus on the International Rules [2]. COLREGs consist of 38 rules split between five sections (A-E): General, Steering and Sailing Rules, Lights and Shapes, Sound and Light Signals, and Exemptions. This research will focus on a decision aid for portions of Part B, Steering and Sailing Rules. COLREGs are built around two vessels interacting in a one-on-one encounter, leaving ambiguity for a situation involving more than two vessels. This responsibility normally falls on the Ship's Master to safely apply the rules to avoid collisions. Part of this research will be to both aid in the COLREGs-compliant decision making as well as investigate how an ontology can adjudicate complex and abstract situations.

2.2 Ontology Languages

"An ontology defines a set of representational primitives with which to model a domain of knowledge" [11]. The primitives in this case are the relationships between vessels on the high seas that can be modeled to represent the knowledge of COLREGs. The current research in this field primarily utilizes the Semantic Web standards such as Resource Description Framework (RDF) and Web Ontology Language (OWL) to assist autonomous vessels. These standards are effective for certain applications but employ a limited logic (description logic) that is less expressive than well-known formalisms like first order logic. An expressive logic can better represent the definitions of abstract concepts such as time, space, events, and objects whereas the Semantic Web operates at a more granular level of machine-readable metadata. This research aims to improve on prior work by presenting a novel approach to collision avoidance decision aids for autonomous vessels by use of SUMO.

2.3 Ontologies and MASS

Prior work in this area has typically provided a set of concepts with a type structure and relationships. That approach does not fully capture rules that define the concepts. If the logic used is not as expressive as first order logic, definitions of terms must be either expressed informally as English comments, omitted entirely and left to intuition about the term names, or encoded in a separate procedural language that is not subject to static proof of consistency.

The research described in [12] proposes an ontology-based approach to enhance MASS navigation safety and collision avoidance decision-making. The authors construct an ontology model for navigation scenarios, integrating structured representations of scene elements, such as autonomous ships, environmental factors, and obstacles. The model captures entity attributes and relationships to facilitate semantic understanding of navigation scenes. [13] explores the use of ontologies as a foundational framework to enhance autonomous ship functionality, focusing on semantic data integration, improved SA, and effective decision-making. [14] explores the integration of ontologies within the 4D/RCS (Real-Time Control System) architecture to enhance decision-making in autonomous vehicles and improve path planning and collision avoidance. The study introduces an ontology-based model to represent environmental objects and their attributes, such as rigidity, movability, and crushability. These attributes support the system in collision damage assessment, guiding path replanning, and real-time decision making. The work in [15] explores the application of ontology-based methods to enhance semi-automatic communication in collision avoidance scenarios at sea. The authors developed ontologies tailored to navigational and communication processes to facilitate automated generation and interpretation of messages, incorporating elements of IMO's Standard Marine Communication Phrases.

2.4 SUMO and Reasoning

The SUMO[1] is a comprehensive ontology that includes an domain-independent upper ontology and dozens of domain-specific ontologies that are all mutually compatible. Originally proposed by the IEEE-sanctioned Standard Upper Ontology (SUO) Working Group, SUMO was developed to address the need for a shared, formal representation of knowledge across diverse domains, including engineering, philosophy, artificial intelligence (AI), and information science [3].

One of the core original purposes of SUMO was to provide a formalized foundation for domain-specific extensions. These extensions enable the ontology to adapt to specialized fields while maintaining interoperability with its general structure. SUMO has since been extended for applications in biology, transportation, military logistics, medicine and dozens of other domains [3].

The development and application of SUMO is supported by the Sigma Knowledge Engineering Environment (SigmaKEE), a platform designed to facilitate ontology browsing, editing, and reasoning [16]. SigmaKEE integrates with external resources such as WordNet [17] to enhance its natural language processing capabilities, enabling semantic interpretation and linguistic reasoning.

SigmaKEE automatically converts SUMO into the Thousands of Problems for Theorem Provers (TPTP) family of logical languages, making it compatible with standard automated reasoning tools. These transformations, explained in [5,18,19], enable SUMO to support logical verification, decision support, and system validation, expanding its applicability to domains requiring rigorous formal reasoning.

[1] https://www.ontologyportal.org.

We employ the Vampire theorem prover [20] for reasoning with SUMO. Vampire is an automated theorem prover (ATP) designed to automate the proof-search process for logical theories. It supports first-order logic (FOL) as well as several different logics beyond first order. It is typically the winner in most of the divisions it competes in at Computer Automated Systems Competition (CASC) [20]. Vampire supports a range of reasoning tasks, including theorem proving, model generation, and satisfiability checking, making it particularly suitable for applications in formal verification, artificial intelligence, and knowledge representation [21].

3 Implementation

The formalization process for the COLREGs is structured into three key components. First, foundational definitions and attributes are established, providing the essential building blocks for the system. Next, specific rules from COLREGs, such as *head-on*, *crossing*, and *overtaking* situations, are translated into formal logic to enable systematic reasoning. Finally, helper functions and auxiliary mechanisms are introduced to facilitate the application of these rules in dynamic and complex maritime scenarios.

Building on this principle, it is assumed that the vessels in the generated scenarios are *mutually visible at time X* and *concurrently translocating at time X*. These assumptions are represented in the code snippets by the predicates:

```
(mutuallyVisibleAtTime ?VES1 ?VES2 ?TIMENOW)
(concurrentTranslocationAtTime
    ?VES1 ?VES2 ?TIMENOW)
```

These assertions are currently hard-coded to establish that, at time `?TIMENOW`, `?VES1` and `?VES2` are visible to one another and underway.

3.1 Radar Inputs

The reasoning software is presumed to receive a range of inputs from an onboard radar system, enabling real-time decision-making. These inputs pertain to both the user vessel (denoted with the attribute `OwnVessel`) and the other vessel involved in the interaction (denoted with the attribute `TargetVessel`). Since maritime conditions are dynamic, all input data is tied to specific vessels and evaluated as snapshots in time. The formal definitions of various terms summarized below can be found in [22].

The radar inputs include:

- **SOG**: The speed of the vessel in relation to the ground.
- **COG**: The course the vessel is following with respect to true north.
- **CPA**: The minimum distance between two vessels if they continue on their current courses and speeds.

- **CPAt**: The time until CPA (CPAt) is the time remaining until the vessels reach their closest point. It is common for radar systems to display a negative value for CPAt if it as already passed.
- **Relative Bearing**: The angle between the two vessels, relative to own vessel.
- **Range**: The distance between the two vessels.

The user input includes:

- **Captain's Thresholds for CPA and CPAt**: Defined as `cpaThresh` and `cpaTThresh`, these parameters determine what constitutes a close-quarters situation based on the Captain's judgment. For example, a `cpaThresh` and `cpaTThresh` of 2nm and 10 min respectively would indicate that risk of collision exists if there's another vessel with a CPA of less than 2nm in less than 10 min.

These inputs form the foundational dataset for the reasoning engine, enabling it to assess compliance with COLREGs and to recommend or execute appropriate maneuvers.

3.2 Vessel Types

The types of vessels that are defined in COLREGs that this work considers are power-driven vessels (PDVs), Sailing Vessel, Vessel Engaged in Fishing, Vessel Constrained by Draft, Vessel Restricted in Her Ability to Maneuver, and Vessel Not Under Command. The relationship between these vessels and who has the right-of-way is mostly defined by Rule 18. The exception to Rule 18 is Rule 13 (overtaking), and the handling of that will be discussed shortly.

In SUMO, there is a class `WaterVehicle`; all the vehicle types mentioned above are instances of this class and the different vehicle types are modeled as attributes. In SUMO, entities may not have different class membership during their lifetimes. The vehicle types above are not their own class because a vessel could at one point in time be a PDV, but as soon as it puts up a sail and turns off its engine it becomes a Sailing Vessel. At all times, these vessels are instances of the class of water vehicles, and, at a certain time they have only one vessel type `Attribute`.

An PDV is represented in SUO-KIF as:

```
(=>
  (and
    (instance ?VES1 WaterVehicle)
    (attribute ?VES1 PowerDrivenVessel))
  (not
    (or
      (attribute ?VES1 NotUnderCommand)
      (attribute ?VES1 RestdInAbilToManeuver)
      (attribute ?VES1 ConstrainedByDraft)
      (attribute ?VES1 VesselEngagedInFishing)
      (attribute ?VES1 SailingVessel))))
```

This can be read in plain English as, *if there is an instance of a Water Vehicle, and that vehicle has the attribute of a PDV, then it does not have the attribute of a vessel not under command, restricted in her ability to maneuver, constrained by draft, engaged in fishing, or sailing.*

3.3 Hierarchy

Rule 18 establishes the hierarchy between vessels and the only exception to this rule is overtaking. For example, a PDV is to keep out of the way of a Sailing Vessel, however, if the Sailing Vessel is overtaking the PDV, the roles reverse and the PDV is to maintain course and speed and the Sailing Vessel shall give-way and pass at a safe distance.

Below is the formalization for a Sailing Vessel in SUO-KIF. A similar definition exists for all vessel types. Some term names are shortened due to space constraints.

```
(=>
  (and
    (instance ?VES1 WaterVehicle)
    (attribute ?VES1 SailingVessel)
    (instance ?VES2 WaterVehicle)
    (mutuallyVisAtTime ?VES1 ?VES2 ?TIMENOW)
    (or
      (attribute ?VES2 NotUnderCommand)
      (attribute ?VES2 RestdInAbilToManeuver)
      (attribute ?VES2 ConstrainedByDraft)
      (attribute ?VES2 VesselEngagedFishing))
    (not
      (overtakingSitn ?VES2 ?VES1 ?TIMENOW)))
  (and
    (giveWayVessel ?VES1 ?VES2 ?TIMENOW)
    (standOnVessel ?VES2 ?VES1 ?TIMENOW)))

(=>
  (and
    (instance ?VES1 WaterVehicle)
    (attribute ?VES1 SailingVessel)
    (instance ?VES2 WaterVehicle)
    (attribute ?VES2 PowerDrivenVessel)
    (mutuallyVisAtTime ?VES1 ?VES2 ?TIMENOW)
    (not
      (overtakingSitn ?VES1 ?VES2 ?TIMENOW)))
  (and
    (giveWayVessel ?VES2 ?VES1 ?TIMENOW)
    (standOnVessel ?VES1 ?VES2 ?TIMENOW)))
```

In an effort to not explain code line-by-line, this section handles a hierarchy situation for a sailing vessel, where either the target vessel is not overtaking the sailing vessel, or the Sailing Vessel is not overtaking the target vessel. By explicitly indicating an overtaking situation does not exist, the reasoner will land on a refutation for who gives-way to who in accordance with Rule 18 (hierarchy).

3.4 Conclusive Terms

Several terms used throughout these axioms are considered conclusive, meaning they provide the operator (or route-planning software of a MASS) with the information needed to make a decision. These terms represent key questions, such as *am I the give-way vessel?* or *What is my next obligatory action?*

The conclusive terms include:

- **Obligatory Next Action**: Specifies the baseline action a vessel is expected to take in accordance with COLREGs. Examples include:
  - Alter course to starboard
  - Slow down
  - Back-down engines
  - Pass port-to-port
  - Pass at a safe distance
  - Maintain course and speed
- **Give-way Vessel**: Identifies the vessel required to take proactive measures to avoid a collision, as per Rule 16.
- **Stand-on Vessel**: Identifies the vessel expected to maintain its course and speed, as per Rule 17, unless circumstances necessitate otherwise.

Each rule is articulated through multiple definitions to account for all possible variations of the scenario. For instance, an overtaking situation can manifest as *I am overtaking another vessel* or *I am being overtaken by another vessel.* Additionally, it is essential to explicitly define the negation of each rule, such as *I am NOT overtaking them* or *I am NOT being overtaken by them.* Formalizing these negations is critical to ensure the reasoning engine can evaluate scenarios comprehensively and determine when a rule does not apply.

It should be mentioned that classical logic does not have the default method of negation by failure, such as is found in the logic programming language Prolog. In Prolog and similar languages, if a system cannot conclude X then it will conclude (not X). This is expedient, since proving that something is not true is not guaranteed to terminate in first order logic. By taking this shortcut, Prolog optimizes performance, but this is at the potential cost of reaching unjustified conclusions.

4 Evaluation

The evaluation of the developed ontology and reasoning framework follows a structured and iterative process to ensure logical consistency and validate scenario-based rule compliance. The steps undertaken are outlined below:

1) **Base Ontology Conversion**
   - Convert the foundational ontologies into typed first-order form (TFF) format.
   - This initial TFF file does not contain any scenario-specific data but includes the defined axioms and the minimal SUMO ontology necessary for reasoning.
2) **Scenario Generation**
   - Construct an itemized database of test scenarios covering a range of probable maritime situations.
   - Format the necessary scenario-specific parameters into TFF syntax.
   - Generate a corresponding visual representation of each scenario for verification and interpretability.
3) **Scenario Integration**
   - Append the scenario-specific TFF data to the base TFF file.
4) **Consistency Checking**
   - Execute Vampire on the combined TFF file, with no conjecture, to check for logical consistency.
5) **Conjecture Execution**
   - Append one conjecture at a time to the scenario.
   - Run Vampire for each query individually.
   - Catalog the output, including Vampire refutation outcomes and execution times.
6) **Iteration Across Scenarios**
   - Repeat steps 3–5 for all test scenarios to ensure comprehensive evaluation.
7) **Result Compilation**
   - Aggregate all recorded results into a structured `.csv` file.
   - The compiled results serve as the basis for performance evaluation, statistical analysis, and validation of the reasoning framework.

5 Model Consistency and Satisfiability

Ensuring the validity of the model requires evaluating both its consistency and satisfiability. A set of axioms is satisfiable if there exists at least one interpretation in which all statements hold true [23]. Conversely, a model is considered consistent if no contradiction can be derived from its axioms. While these concepts are closely related, they are distinct: a consistent model may still be unsatisfiable, and proving satisfiability is often significantly more challenging than demonstrating consistency.

5.1 What Can Be Said: Local Consistency

This research establishes that the model exhibits *local consistency*—meaning within the scope of tested conjectures, no contradictions arose. Specifically, every conjecture formulated as a theorem to be proven yielded the expected results, with Vampire providing refutations and correct logical consequences. This indicates that the subset of axioms involved in each proof was internally consistent.

Furthermore, at no point did the model produce a counter-satisfiable result. That is, no refutation was found that negated a previously established theorem. This reinforces the notion that the axioms, at least within the bounds of individual proofs, maintain local consistency.

The following conjectures were applied to each scenario:

1) **Crossing Situation:** Is there an `X` and `Y` in a crossing situation?
2) **Head-on Situation:** Is there an `X` and `Y` in a head-on situation?
3) **Overtaking Situation:** Is there an `X` overtaking a `Y`?
4) **Give-way Vessel:** Is there an `X` that is required to give-way to `Y`?
5) **Stand-on Vessel:** Is there an `X` that is required to stand-on to `Y`?
6) **Own Vessel's Obligation:** What is the obligatory next action of own vessel?
7) **Target Vessel's Obligation:** What is the obligatory next action of the target vessel?

It should be noted that not every conjecture will yield an answer for every scenario. Only one of the first three conjectures (1–3) may return a result, as a scenario can only represent one type of situation at a time. COLREGs does discuss special situations with more than one vessel, however, that is beyond the scope of this work. The remaining conjectures (4–7) generally return results for all applicable scenarios, provided the situation necessitates a give-way or stand-on action.

5.2 Example Crossing Situation: Give-Way

The scenario shown in Fig. 1 represents another type of crossing situation between Blue and Red. The parameters defining the scenario are as follows:

- **Blue:** COG = 293°, SOG = 16 knots
- **Red:** COG = 245°, SOG = 6 knots
- **Relative Bearing of Red from Blue:** 45°
- **CPA:** 2 nm
- **CPAt:** 18 min
- **Range:** 4 nm

The relative bearing of 45° and aspect angle indicate that Red is approaching Blue from its starboard side. This scenario aligns with a crossing situation where the vessel on the starboard side (Red) is the stand-on vessel, while the vessel on the port side (Blue) is the give-way vessel.

Although from a birds-eye view this can be considered the same type of situation as the previous crossing example. However, in first-person point of view, specifically from the perspective of the Blue vessel, this is different in that own vessel is now to give-way instead of stand-on.

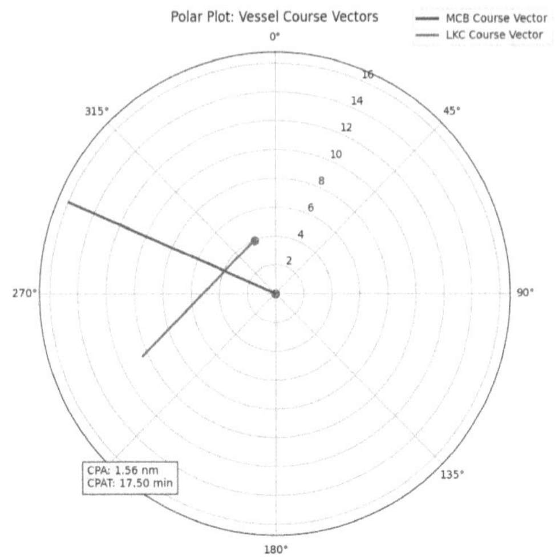

Fig. 1. Example of a Crossing Situation as the Give-way.

5.3 Summary of Validating Requirements

The final test set included 288 distinct scenarios, with the distribution of scenario types summarized in Table 1. The distribution was not perfectly balanced across different scenario types.

Table 1. Summary of Scenario Tests

Situation Type	Count	Situation Variation	Count
Crossing Situations	88	Blue Stand On	38
		Red Stand On	50
Overtaking Situations	68	Blue Stand On	3
		Red Stand On	65
Head On Situations	16		
No Situation Found	116		
Total Scenarios	288		
Average Time to Refutation	0.531 s		

The model demonstrated a high level of accuracy, with 171 of the 172 scenarios that resulted in a refutation yielding the expected results—representing a 99.4% true positive rate. Only one scenario produced an unexpected outcome, corresponding to a false positive rate of 0.58%. The cause of this anomaly is currently unknown.

It is important to note that the evaluation of results involves a degree of subjectivity. Unlike traditional classification tasks with an absolute ground truth, maritime scenarios often involve context-dependent judgment. The model's results are closely tied to the author's interpretation of COLREGs, which may differ from another mariner's perspective. This subjectivity is especially relevant in scenarios where no situation was identified. While many of these outcomes were objectively correct—such as scenarios with a negative CPAt, indicating that the vessels had already passed their closest point of approach—others could be open to interpretation depending on the evaluator's experience and judgment.

This work aims to reduce such ambiguity by formalizing COLREGs in a machine-interpretable framework. However, this does not imply universal consensus among maritime professionals regarding all outcomes.

6 Conclusion

The validation testing demonstrated that the model effectively adheres to the Rules of the Road. No evidence of logical contradictions was observed, confirming local consistency within the tested scenarios. While the model successfully identified crossing, overtaking, and head-on situations, it also correctly determined when no formal COLREGs situation existed.

This research demonstrates the feasibility and effectiveness of formalizing COLREGs within an ontology-based framework to support autonomous vessel decision-making. By leveraging the Vampire ATP and the SUMO ontology, the model successfully identifies and classifies maritime situations with a high degree of accuracy, while adhering to the logical principles of COLREGs. The findings of this study lay the foundation for future advancements in transparent autonomous maritime navigation, contributing to safer and more efficient operations at sea.

References

1. Yang, P., Xue, J., Hu, H.: A bibliometric analysis and overall review of the new technology and development of unmanned surface vessels. J. Marine Sci. Eng. **12**(1) (2024). Available: https://www.mdpi.com/2077-1312/12/1/146
2. United States Coast Guard: Rules of the Road. United States Department of Transportation, United States Coast Guard, Washington, D.C. (1992)
3. Niles, I., Pease, A.: Towards a standard upper ontology. In: Proceedings of the International Conference on Formal Ontology in Information Systems - vol. 2001, ser. FOIS '01. New York, NY, USA: Association for Computing Machinery, pp. 2–9 (2001). Available: https://doi.org/10.1145/505168.505170
4. Pease, A.: Ontology: A Practical Guide. Articulate Software Press (2011). Available: https://books.google.com/books?id=Mvs3twAACAAJ
5. Pease, A., Schulz, S.: Knowledge engineering for large ontologies with sigma KEE 3.0. In: Automated Reasoning. Demri, S., Kapur, D., Weidenbach, C., (eds.) Cham: Springer International Publishing, pp. 519–525 (2014)

6. Kovács, L., Voronkov, A.: First-order theorem proving and vampire. In: Proceedings of the 25th International Conference on Computer Aided Verification, ser. CAV 2013, vol. 8044. New York, NY, USA: Springer-Verlag New York, Inc., pp. 1–35 (2013)
7. Schulz, S.: E - a brainiac theorem prover. J. AI Commun. **15**(2/3), 111–126 (2002)
8. Benzmüller, C., Pease, A.: Progress in automating higher-order ontology reasoning. In: Workshop on Practical Aspects of Automated Reasoning (PAAR-2010), Konev, B., Schmidt, R., Schulz, S., (eds.) Edinburgh, UK: CEUR Workshop Proceedings (2010)
9. Pease, A., Schulz, S.: Contradiction detection and repair in a large theory. In: The International FLAIRS Conference Proceedings (2022)
10. Pease, A., Benzmüller, C.: Ontology archaeology: mining a decade of effort on the suggested upper merged ontology. In: The ECAI-10 Workshop on Automated Reasoning about Context and Ontology Evolution (2010)
11. Ozsu, M.T., Liu, L. (eds.): Encyclopedia of Database Systems. Springer, New York (2009)
12. Lan-xuan, A., Xin-yu, Z., Zhi-guo, D., Cheng-bo, W.: Research on ontology-based situation understanding and decision-making approach for mass. In: IOP Conference Series: Materials Science and Engineering, vol. 929, p. 012029 (2020)
13. Sumić, D., Maleš, L., Marušić, T., Rosić, M.: Ontology-based information infrastructure for autonomous ships. Trans. Maritime Sci. **13**(1) (2024)
14. Provine, R., Schlenoff, C., Balakirsky, S., Smith, S., Uschold, M.: Ontology-based methods for enhancing autonomous vehicle path planning. Robot. Autonomous Syst. **49**(1), 123–133 (2004). knowledge Engineering and Ontologies for Autonomous Systems 2004 AAAI Spring Symposium. [Online]. Available: https://www.sciencedirect.com/science/article/pii/S0921889004001423
15. Hatlas-Sowinska, P., Wielgosz, M.: Ontology-based approach in solving collision situations at sea. Ocean Eng. **260**, 111941 (2022)
16. Pease, A.: The Sigma Manual: A Guide for Users and Developers of the SUMO Toolset (2024). Accessed 14 Jan 2025. https://github.com/ontologyportal/sigmakee/blob/master/doc/manual/SigmaManual.pdf
17. Fellbaum, C.: Semantic network of English verbs. In: WordNet: An Electronic Lexical Database, ser. Language, Speech, and Communication, Fellbaum, C., (ed). MIT Press, pp. 69–104 (1998). Available: http://books.google.com.hk/books?id=Rehu8OOzMIMC
18. Pease, A.: Converting the suggested upper merged ontology to typed first-order form (2023). Available: https://arxiv.org/abs/2303.04148
19. Benzmüller, C., Pease, A.: Higher-order aspects and context in SUMO. In: Special issue on Reasoning with Context in the Semantic Web, I. J. V. Jos Lehmann and A. Bundy, Eds. Science, Services and Agents on the World Wide Web, vol. 12-13 (2012)
20. Sutcliffe, G., Desharnais, M.: The CADE-29 automated theorem proving system competition (CASC-29). AI Commun. **37**(4), 485–503 (2024). Available: https://doi.org/10.3233/AIC-230325
21. Team, V. D.:"Vampire: Theorem prover." https://vprover.github.io/ (2025). Accessed 03 Jan 2025

22. Bowditch, N.: American Practical Navigator: An Epitome of Navigation, 2019th ed. Springfield, Virginia, USA: National Geospatial-Intelligence Agency, vol. II, Pub. No. 9, Prepared and published by the National Geospatial-Intelligence Agency (2019). Available: https://msi.nga.mil/Publications/APN
23. Sutcliffe, G.: The TPTP Language (2025). Accessed 14 Jan 2025. Available: https://tptp.org/UserDocs/TPTPLanguage/TPTPLanguage.shtml

Fast Algorithm for Single Machine Scheduling of Unit-Length Jobs with Non-renewable Resource Requirements to Minimize maximum Job Lateness

Kevin Johedan Ramírez-Fuentes[1], Nodari Vakhania[1](✉), and Jose Alberto Hernandez-Aguilar[2]

[1] Centro de Investigación en Ciencias, UAEM, México, USA
nodari@uaem.mx
[2] Facultad de Contaduría, Administración e Informática, UAEM, México, USA

Abstract. We consider a generalization of a classical-single machine scheduling problem with non-simultaneously released jobs to minimize maximum job lateness. Besides the release time and the due date, each job requires a specific amount of a non-renewable resource. Such a generalization is important in real-life scenarios, for example, where machines can be fed by solar batteries. The general setting, even without resource requirements, is well-known to be strongly NP-hard. We consider the setting with unit-length jobs and build a n $O(n \log n)$-time algorithm that minimizes our objective function.

Keywords: Scheduling · Heuristic · Release Date · Due Date · Single-machine · Non-renewable resources · Energy

1 Introduction

The *scheduling problems* deal with a finite set J of n requests called *jobs* which are to be performed (or scheduled) on a finite set of resources or *machines*. The aim is choose the order of processing the jobs on machines so as to meet a given objective criteria, respecting a basic restriction that any machine can handle at most one job at a time. The scheduling problems have wast amount of practical applications. The problem under the consideration is a single-machine scheduling problem with additional non-renewable resources. Besides the traditional (renewable) resource (a single machine in our case), a non-renewable resource is also available, which is consumed by each job while being processed on the machine. Such a generalization is important in real-life scenarios, for example, when machines can be fed by solar batteries.

The problem we consider here is a generalization of a classical-single machine scheduling problem with non-simultaneously released jobs to minimize maximum

H. R. Arabnia et al. (Eds.): CSCI 2025, CCIS 2970, pp. 96–106, 2027.
https://doi.org/10.1007/978-3-032-28466-2_7

job lateness. Besides the release time and the due date, each job requires a specific amount of a non-renewable resource. Thus a typical example of a non-renewable resource is energy, which can be generated in different ways. Presently, different kinds of the sustainable energy including solar and wind energies are widely available. Throughout this paper, for the sake of conciseness, instead of referring to a "non-renewable resource", we may merely say "energy".

The generic version of the problem we study here can be formulated as follows. We are given a set J of n jobs and a single machine. Each job $j \in J$ becomes available at its *release time* r_j. Only a released job j can be scheduled on the machine, that is to process it for p_j time units uninterruptedly. The execution of job j, besides the machine time, also requires e_j amount of the non-renewable resource, the energy; we will say that job *consumes* energy e_j. d_j is the *due date* of job j, which can be interpreted as the ideal time for the completion of that job. A *schedule* S is a mapping that assigns to each job j a starting time $s_j(S)$ on the machine. Thus job j occupies time slot $[s_j(S), s_j(S) + p_j)$ (a half-open time interval) on the machine in schedule S. We denote by $c_j(S) = s_j(S) + p_j$ the completion time of job j in schedule S. A feasible schedule is a schedule in which at each time moment, at most one job j is assigned to the machine, such that $s_j(S) \geq r_j$ and e_j amount of energy is assigned to job j from time $s_j(S)$ (so e_j amount of energy must be available for job j by time $s_j(S)$). All job parameters except the requirements e_j of non-renewable resource are integers, whereas the latter parameters can, in general, be real numbers.

Now we turn to our objective criterion. The *lateness* of job j in schedule S is

$$L_j(S) = c_j(S) - d_j.$$

The *maximum job lateness* in schedule S is

$$L_{\max}(S) = \max_{j \in J} L_j(S).$$

Our objective is to find an *optimal schedule*, that is, a feasible schedule S with minimum $L_{\max}(S)$.

The above setting, even without resource requirements (i.e., with $e_j = 0$, for all $j \in J$), is well-known to be strongly NP-hard Garey et al. [1]. By the three-field notation proposed by Graham et al. [2], this setting is denoted as $1|r_j|L_{\max}$. The first field specifies the machine environment (one machine in our problem), the second field gives job parameters, and in the third field the objective function is given. In the second field, job processing times, which are present in nearly any scheduling problem, are omitted, and for our problem, the parameter d_j is also omitted since the objective function $L_{\max}$ implicitly implies the existence of job due dates. Blazewicz et al. [6] extended the standard notation by Graham et al. [2] by considering the availability of non-renewable resources.

Adopting the standard three-field notation, we denote our generic scheduling problem with energy requirements as $1|r_j, e_j|L_{\max}$. This setting is obviously strongly NP-hard as the setting $1|r_j|L_{\max}$ is strongly NP-hard.

Next, we briefly overview some related results, starting with generic versions of the problem without renewable resources. A well-known $O(n \log n)$ greedy algorithm was proposed by Jackson [3] for the problem $1||L_{\max}$, where all jobs are simultaneously released at time 0. Later Schrage [4] has adopted it for the setting with arbitrary job release times $1|r_j|L_{\max}$. The extended algorithm, further referred to as *ED-heuristic* (Earliest Due-date heuristic), works in n iterations. At iteration 1, a job with the minimum due date released by time $\min_j r_j$ is scheduled at that time. Iteratively, each iteration $i > 1$, scheduling time t_i is determined as the maximum between the minimum release time of a yet unscheduled job and the completion time of the latest so far scheduled job (both, by the beginning of iteration i). Among the jobs released by that time t_i, one with the smallest due-date is scheduled at time t_i in iteration i. The procedure halts when it schedules all jobs.

If all jobs have equal processing time p, problem $1|p_j = p, r_j|L_{\max}$ can be solved in time $O(n \log n)$ Garey et al. [13]. A more general setting $1|p_j \in \{p, 2p\}, r_j|L_{\max}$, where the processing time of a job is either p or $2p$, can still be solved in time $O(n^2 \log n \log p)$ Vakhania [8]. Apparently, the maximal polynomially solvable setting of the problem with restricted jobs processing times is when every job processing time divides every larger job processing time Vakhania [10].

As to the polynomial-time algorithms for related problems with non-renewable resources, Carlier and Rinnooy Kan [5] consider the setting in which the execution of each job j requires certain amount of a non-renewable resource and incurs a cost $f_j(t)$, where t is the completion time of that job and $f_j(t)$ is a non-decreasing function. The non-renewable resource is released at m different time moments within the scheduling horizon. The objective is to minimize the maximum job cost. The authors propose an algorithm that runs in time $O(n \log n + m \log m)$ when functions $f_j(t)$ are linear. Du et al. [11] consider an online scenario with parallel uniform machines, which require certain amount of energy to work. There is a limit on the total available energy that can be used by the machines at each time moment (an idle machine requires less energy than a machine running a job). A job can be omitted or scheduled partially but there is a penalty on that. The objective is to minimize the total job penalty. No competitive ratio of the proposed algorithm is derived, its practical performance is rather measured by computational experiments. Albers and Fujiwara [12] study the situation when the processing time of a job can be reduced at the expense of the use of a non-renewable resource. For unit-length jobs, their on-line algorithm has a constant competitive ratio and their off-line algorithm runs basically in $O(n^3)$ time. Bérczi et al. [14] proposed a PTAS for the problem of maximizing the weighted sum of completion times with a non-renewable resource but without renewable resource, i.e., machine job processing times are assumed to be 0. Grigoriev et al. [7] consider settings on a single machine where each job has a due date and requires certain amount of a non-renewable resource. The non-renewable resource is released at some time moments in the scheduling horizon. The objective is to minimize maximum job lateness. The authors propose an $O(n^2 \log n)$ time algorithm that finds an optimal permutation of the given set of jobs (by calculating a feasible starting time of each job in that permutation, the corresponding schedule can be constructed). The algorithm sorts the jobs in non-decreasing order of their due-dates, the jobs with the same due-date being sorted in non-decreasing non-renewable resource requirements.

Our contribution

To the best of our knowledge, no polynomial time algorithm for the single machine scheduling problem $1|r_j, e_j|L_{\max}$ with job release times and energy requirements to minimize maximum job lateness exists, even with unit-length jobs. Job release times essentially complicate the problem, even without energy requirements (as noted earlier, the generic problem without job release times is polynomial, whereas that with job release times becomes strongly NP-hard). Since no polynomial time algorithm is likely to exist for the setting with arbitrary job processing times (unless $P = NP$), we focus on the setting $1|p_j = 1, r_j, e_j|L_{\max}$ with equal unit-length jobs, i.e., $p_j = 1$ for all $j \in J$. We show that the problem can be solved in polynomial time (see the following section for a complete definition of problem $1|p_j = 1, r_j, e_j|L_{\max}$). In particular, we build a fast $O(n \log n)$-time algorithm that solves it optimally. The preprocessing step forms a list of jobs sorted in non-decreasing order of their due-dates (as in Grigoriev et al. [7]). Iteratively, for each next job j from the list, a minimum time moment t_j when this job can feasibly be scheduled is determined; i.e., $r_j \geq t_j$ and, given the amount of the non-renewable resource required by the jobs assigned at the previous iterations, at least e_j amount of the non-renewable resource is available at time t.

2 Problem Definition

In this section we complement our notation with the necessary additional notation and complete the formal definition of the scheduling problem $1|p_j = 1, r_j, e_j|L_{\max}$.

We may look at the scheduling horizon as a sequence of unit-length time slots, time slot t being the interval $[t-1, t)$. The execution of a job j requires exactly one time slot on the machine (the renewable resource), i.e., a schedule assigns each job to some time slot. Besides a unit-time slot on the machine, job j also requires e_j amount of the non-renewable resource: this resource is consumed by job j during its execution on the machine.

Without loss of generality, we assume that a solar battery is charged at discrete time moments starting from time 0. At the beginning of time slot $t+1$, R_t ($R_t \geq 0$) amount of non-renewable resource is generated (independently whether a job is assigned to that time slot or not).

During the construction of schedule S, $R_t(S)$, $R_t(S) \leq R_t$, is the amount of non-renewable resource at time t available in that schedule (see below).

The non-renewable resource is accumulable in the following sense. For a given time $\alpha \geq 0$ in an empty schedule $\emptyset$, the available non-renewable resource at time moment α, is

$$E_\alpha(\emptyset) = \sum_{t=0}^{\alpha} R_t; \tag{1}$$

If job j is scheduled at time β in schedule S, the required energy e_j is taken at time β and possibly at time moments preceding time β (given that the total available energy by time β is at least e_j). In particular, the available energy is recalculated as follows.

- If $R_\beta(S) \geq e_j$, then $E_\beta(S) := E_\beta(S) - e_j$ and $R_\beta(S) := R_\beta(S) - e_j$, whereas all the remaining energy entries remain unchanged.
- If $R_\beta(S) < e_j$, then the required amount of energy is repeatedly redistributed from the preceding time moments: If

$$R_{\beta-1} \geq e_j - R_\beta(S), \tag{2}$$

then

$$E_{\beta-1}(S) := E_{\beta-1}(S) - (e_j - R_\beta(S)) \tag{3}$$

$$R_{\beta-1}(S) := R_{\beta-1}(S) - (e_j - R_\beta(S)) \tag{4}$$

and

$$E_\beta(S) := E_\beta(S) - R_\beta(S), \quad R_\beta(S) := 0 \tag{5}$$

where all the remaining energy entries remain unchanged.
In general, let τ ($\tau < \beta$) be the latest time moment such that

$$R_\tau(S) \geq e_j - \sum_{t=\tau}^{\beta} R_t(S) \tag{6}$$

(note that τ exists by the definition of β). Then

$$E_\tau(S) := E_\tau(S) - (e_j - \sum_{t=\tau}^{\beta} R_t(S)) \tag{7}$$

$$R_\tau(S) := R_\tau(S) - (e_j - \sum_{t=\tau}^{\beta} R_t(S)). \tag{8}$$

Furthermore, for each $\alpha = \tau + 1, \ldots, \beta$, we let

$$E_\alpha(S) := E_\alpha(S) - (e_j - \sum_{t=\alpha}^{\beta} R_t(S)). \tag{9}$$

Then note that $R_\alpha(S) := 0$, for each $\alpha = \tau + 1, \ldots, \beta$, where the entries corresponding to time moments $0, \cdots, \tau - 1$ remain unchanged.

We will say that job j can *feasibly* be scheduled at time β in schedule S if $E_\beta(S) \geq e_j$.

Let e be any amount of energy, and let

$$ET(e) = \left\lceil \frac{e}{R_t} \right\rceil. \tag{10}$$

The following observation can easily be seen:

Observation 1 *$ET(e)$ is the minimum amount of time units where the energy of at least e is generated.*

In the following, we use T for the minimum time moment such that

$$TR_t \geq \sum_{j \in J} e_j \tag{11}$$

i.e., T is the minimum time moment when there is a sufficient energy for scheduling all the jobs. We assume that T is a constant not dependent on n (e.g., the capacity of a solar battery).

Example.

In Table 1 below we preset a problem instance with 5 jobs. Figure 1 depicts an empty scheduling horizon constructed for this instance. It basically represents the energy distribution without any scheduled job.

Table 1. A problem instance

$r_1 = 1$	$d_1 = 6$	$e_1 = 3$
$r_2 = 5$	$d_2 = 6$	$e_2 = 6$
$r_3 = 2$	$d_3 = 4$	$e_3 = 4$
$r_4 = 0$	$d_4 = 10$	$e_4 = 1$
$r_5 = 8$	$d_5 = 11$	$e_5 = 5$
	$R_t = 2$	

3 The Heuristics

SE-Algorithm. First, we give a simple greedy, referred to as *SE-Algorithm* (Smallest Energy Algorithm), which is optimal if job *due dates* and *energy requirements* are *agreeable*, that is, for any pair of jobs i and j, $d_i \leq d_j$ iff $e_i \leq e_j$. SE-Algorithm first creates a list of the n jobs $\{j_1, \ldots, j_n\}$ ($e_{j_1} \leq \cdots \leq e_{j_n}$), in which the jobs come in non-decreasing energy requirements ties being broken by selecting the job with the smallest due-date. Iteratively, the next job j_i from the list is scheduled at the earliest available feasible time at or after time r_{j_i}. The energy entries are recalculated respectively (see Eqs. (6)-(9)).

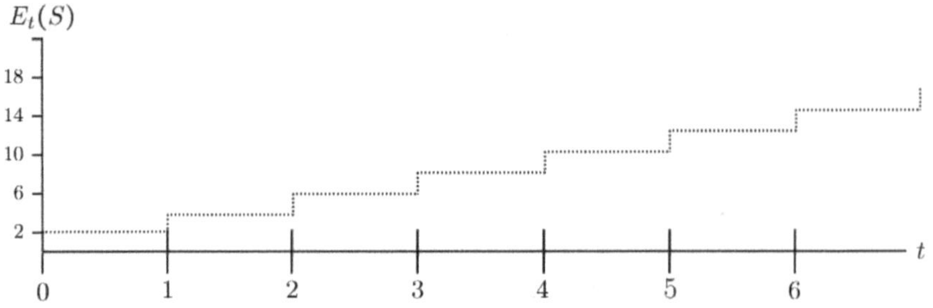

Fig. 1. An empty scheduling horizon

Theorem 1. *If due dates and energy requirements are agreeable for all jobs, then SE-Algorithm generates a feasible SE-schedule S with the minimum* $L_{\max}$ *in time* $O(n \log n)$.

Proof. Let S be the schedule generated by SE-Algorithm. In any schedule, a job i cannot be assigned before time $max(r_i, ET(e_i))$, in SE-Algorithm during iteration i if at least one job j from a previous iteration used energy that i needs for it to be scheduled on time $max(r_i, ET(e_i))$, then i will be scheduled after that time. If we considered a alternative schedule S' when job j is assigned after job i then $t_i(S) \geq t_i(S')$ and $t_i(S) - t_i(S') \leq ET(e_j) + 1$ since in schedule S' job j doesn't take any energy necessary for i, but since j is from a previous iteration in SE-Algorithm then $d_j \leq d_i$ and $e_j \leq e_i$ but after schedule job i in S' there is no energy for e_j since i takes more energy. Therefore the reduce in lateness of i will be overpass by the amount of time j will be delayed in schedule S'. The initial sorting can be done in $O(n \log n)$ time. At each iteration i, the job i can be feasibly scheduled by checking less than T time moments, whereas the update of at most T different R_t's will be required. We obtain the overall time complexity of $O(n \log n + 2nT) = O(n \log n + nT) = O(n \log n)$ (the last equation holds since T is a constant). □

The next corollary easily follows since under the below stated condition there is no need in the preprocessing stage.

Corollary 1 *Suppose the n jobs are released in non-decreasing order of their due dates. Then* SE-Algorithm *generates an optimal schedule in time* $O(n)$.

For the general case without agreeable jobs, below we describe SEED-Algorithm. It creates a schedule with the minimum $L_{\max}$ for problem $1|p_j = 1, r_j, e_j|L_{\max}$. SEED-Algorithm iteratively, takes the next yet unscheduled job i with the minimum due date and schedules it at the earliest time moment feasible for that job; i.e., the earliest idle time moment t with $t \geq r_i$, $E_t(S) \geq e_i$, where S is the partial schedule constructed so far:

SEED Algorithm . J Set of n jobs, R_t Non-renewable resource at each time t

---

$(1, ..., n)$:= Jobs in J sorted by non-decreasing d_j breaking ties by selecting job j with minimum r_j
$S := \{\}$ ▷ Output schedule
for $i = 1$ to n **do**
 $t := \min t \geq r_i : E_t(S) \geq e_i$ and $t \neq t_k(S)$ $k = 1, \ldots, i-1$ ▷ Choose the next time t
 $t_i(S) := t$ ▷ Schedule job i on time t
 $e := e_i$
 $l := t$
 while $e > 0$ **do** ▷ Update $R(S)$ at and before time t
  $\Delta e := min(e, R_l(S))$
  $R_l(S) := R_l(S) - \Delta e$
  $e := e - \Delta e$
  $l := l - 1$
 end while
end for
return schedule S

The solution generated by SEED Algorithm for our problem instance is depicted in Fig. 2. In that figure, the dashed lines represent idle-time intervals on the machine, which occur while the corresponding jobs are waiting for the liberation of the necessary amount of energy. Notice that $E_t(S) = 0$, for all $t = 0, 1, ..., 8$, since the energy, generated at each of these ts is consumed by the jobs scheduled behind time t.

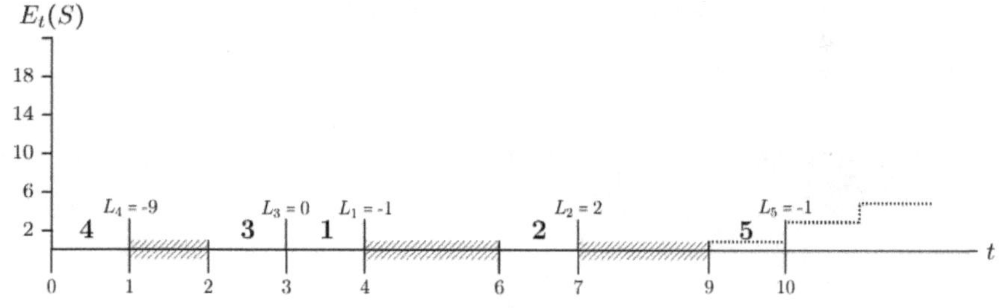

Fig. 2. The SEED-Schedule created for our problem instance

We will say that at iteration i job i has a *energy conflict* with job j if

$$t_j(S) \in [max(r_i - ET(e_i), 0), max(r_i, ET(e_i))] \tag{12}$$

Theorem 2. *SEED-Algorithm outputs a schedule S with minimum L_{max} in time $O(n \log n)$*

Proof. At any iteration i in SEED-Algorithm, let i be the job selected in that iteration. If no job j of some previous iteration had an energy conflict with job i, then i can be assigned at time $\max(r_i, ET(e_i))$, which is the minimum feasible time for job i. Hence, S will minimize L_{max}. Now let $C = \{j \in J | t_j(S) \in [max(r_i - ET(e_i), 0), max(r_i, ET(e_i))]\}$ be the set of jobs with an energy conflict with i. If C is empty then i is assigned at time $max(r_i, ET(e_i))$ as above. Otherwise, consider the partial schedule S' in which job i is scheduled before all jobs in set C and the jobs from that set are scheduled after job i in the same order as before. Let u be the last scheduled job of C. Then we have that $d_i \leq d_u, \ \forall i \in C$. By the way of contradiction, suppose $L_{max}(S') < L_{max}(S)$. Then

$$L_u(S') \leq L_{max}(S') < L_i(S) = L_{max}(S), \text{so}, \quad L_u(S') < L_i(S). \tag{13}$$

Since $d_u \leq d_i$ and $-d_i \leq -d_u$,

$$t_u(S') + 1 - d_u < t_i(S) + 1 - d_i \leq t_i(S) + 1 - d_u \tag{14}$$

and so

$$t_u(S') < t_i(S). \tag{15}$$

Consider now the following two cases:

1. $ET(\sum_{j \in c} e_j + e_i) \leq r_i$, i.e., in both, S' and S there is enough energy for the jobs in C and job i at time r_i. Then

$$t_u(S') \geq r_i + |C|, \quad t_i(S) \leq r_i + |C| \tag{16}$$

and hence

$$r_i + |C| \leq t_u(S') < t_i(S) \leq r_i + |C|. \tag{17}$$

Therefore,

$$r_i + |C| < r_i + |C|, \tag{18}$$

which is a contradiction.

2. $ET(\sum_{j \in c} e_j + e_i) > r_i$. I.e., for all $j \in C$,

$$r_j \leq r_i \leq max(r_i, ET(e_i)) < ET(\sum_{j \in c} e_j + e_i) \tag{19}$$

It follows that in both, S and S', all jobs in $C \cup \{i\}$ are released by time $ET(\sum_{j \in c} e_j + e_i)$ while there is enough energy for all of them. If in schedule S some jobs from C are scheduled at or after time $ET(\sum_{j \in c} e_j + e_i)$, at least the same number of jobs will be included in schedule S' at or after that time moment. Then $t_u(S') \geq t_i(S)$. If there no job from set C scheduled at or after time $ET(\sum_{j \in c} e_j + e_i)$, then in S job i can be assigned at time $ET(\sum_{j \in c} e_j + e_i)$

and in S', job u will be assigned not before time $ET(\sum_{j\in c} e_j + e_i)$. Thus we have

$$t_u(S') \geq ET(\sum_{j\in c} e_j + e_i), \quad t_i(S) = ET(\sum_{j\in c} e_j + e_i) \tag{20}$$

and hence

$$ET(\sum_{j\in c} e_j + e_i) \leq t_u(S') < t_i(S) = ET(\sum_{j\in c} e_j + e_i) \tag{21}$$

and

$$ET(\sum_{j\in c} e_j + e_i) < ET(\sum_{j\in c} e_j + e_i), \tag{22}$$

which is another contradiction. Therefore our original assumption that $L_{max}(S') < L_{max}(S)$ is incorrect.
The preprocessing step of sorting the jobs by due dates can be done in $O(n \log n)$ time, while each of the n iterations of the algorithm performs a linear search for the first available feasible time for job i. Once job i gets scheduled, the updates of the corresponding $R_t(S)$s takes a time, linear in T, i.e., both of these operations can be bounded by T. Since T is constant, we have an overall time complexity of $O(n \log n + n) = O(n \log n)$. □

4 Conclusion

We proposed a fast polynomial-time algorithm for the extension of the classical scheduling problem $1|r_j|L_{\max}$ with non-renewable resource requirements with identically designed products or jobs with equal-length unit processing times. For further research, it is a challenging question if there exist polynomial-time algorithms for extended models with non-identically designed products when there are two or more types of such products.

References

1. Garey, M.R., Johnson, D.S.: Computers and Intractability: A Guide to the Theory of NP-completeness. Freeman, San Francisco (1979)
2. Graham, R.L., Lawler, E.L., Lenstra, J.L., Rinnooy Kan, A.H.G.: Optimization and approximation in deterministic sequencing and scheduling: a servey. Ann. Discrete Math. **5** 287–326 (1979)
3. Jackson, J.R.: Schedulig a production line to minimize the maximum tardiness. University of California, Los Angeles, CA, Manegement Scince Research Project (1955)
4. Schrage, L.: Obtaining optimal solutions to resource constrained network scheduling problems, (1971). unpublished manuscript
5. Carlier, J., Rinnooy Kan, A.H.G.: Scheduling subject to nonrenewable-resource constraints. Operat. Res. Lett. **1**(2), 52–55 (1982)

6. Blazewicz, J., Lenstra, J.K., Rinnooy Kan, A.H.G.: Scheduling subject to resource constraints: classification and complexity. Discrete Appl. Math. **5**(1), 11–24 (1983)
7. Grigoriev, A., Holthuijsen, M., van de Klundert, J.: Basic scheduling problems with raw material constraints. Naval Res. Logis. (NRL) **52**, 527–535 (2005)
8. Vakhania, N.: Single-Machine Scheduling with Release Times and Tails. Ann. Oper. Res. **129**, 253–271 (2004)
9. Vakhania, N., Werner, F.: Minimizing maximum lateness of jobs with naturally bounded job data on a single machine in polynomial time. Theoret. Comput. Sci. **501**, 72–81 (2013). https://doi.org/10.1016/j.tcs.2013.07.001
10. Vakhania, N.: Dynamic restructuring framework for scheduling with release times and due dates. Mathematics **7**(11), 1104 (2019)
11. Du, Z., Sun, H., He, Y., He, Y., Bader, D.A., Zhang, H.: Energy-Efficient Scheduling for Best-Effort Interactive Services to Achieve High Response Quality. In: Proceedings - IEEE 27th International Parallel and Distributed Processing Symposium (2013)(2013). Energy-Efficient Scheduling for Best-Effort Interactive Services to Achieve High Response Quality. Proceedings - IEEE 27th International Parallel and Distributed Processing Symposium
12. Albers, S., Fujiwara, H.: Energy-efficient algorithms for flow time minimization. ACM Trans. Algorithms **3**, 4 (2007)
13. Garey, M.R., Johnson, D.S., Simons, B.B., Tarjan, R.E.: Scheduling unit-time tasks with arbitrary release times and deadlines. SIAM J. Comput. **10**, 256–269 (1981)
14. Bérczi, K., Király, T., Omlor, S.: Scheduling with non-renewable resources: minimizing the sum of completion times. J. Sched. **27**, 151–164 (2024)

Lightweight Deep Learning Model for Electrooculography-Based User Authentication

Qingqing Li(✉), John Cook, Ata Tangban, Weixiao Liao, and Wei Yu

Department of Computer and Information Sciences, Towson University, Maryland, USA
qingqingli@towson.edu

Abstract. Electrooculography (EOG) enables non-invasive monitoring of eye movements and offers various applications in wearable assistive technologies and human-computer interaction (HCI). Traditional deep learning models for EOG signal classification typically suffer from high computational complexity and extensive memory requirements, which limit their application in environments where resources are constrained. To address the issue, this paper introduces a lightweight convolutional neural network (CNN) optimized for EOG-based user authentication on edge devices. Inspired by the MobileNet architecture, our proposed model employs depthwise separable convolutions and squeeze-and-excitation blocks to reduce memory and computation overhead while preserving competitive classification performance. Evaluated on a public dataset that simulates smart eyewear conditions, our model achieves a 0.931 F1 score with only 276, 579 FLOPs and a 1.06 MB memory footprint. We compare it with QuickNet, MobileNet, ResNet50, and SqueezeNet, and an ablation study that highlights the efficiency performance tradeoffs and the need for each component in our architecture. Overall, our results demonstrate the feasibility of real-time EOG-based biometric authentication on resource-constrained platforms.

Keywords: EOG · User Authentication · Lightweight Deep Learning · Edge Computing · Raspberry Pi · Smart Eyewear

1 Introduction

Wearable technologies, an essential part of smart healthcare systems [1], refer to electronic devices (such as digital gloves, armbands, smart glasses, and headsets) that can be worn on the body and are designed to collect data, monitor activities, and enhance user interaction with the environment. These technologies have become increasingly embedded in everyday life, allowing for new paradigms in human-computer interaction (HCI) as they can collect physiological signals, including Electrooculography (EOG), Electromyography (EMG), Electroencephalography (EEG), and Electrocardiography (ECG), among others. Among these modalities, the EOG signal is generated by eye movements

H. R. Arabnia et al. (Eds.): CSCI 2025, CCIS 2970, pp. 107–121, 2027.
https://doi.org/10.1007/978-3-032-28466-2_8

that cause the corneo-retinal standing potential difference. The characteristics (e.g., low power consumption, unobtrusiveness, and the ability to operate in real-time) make EOG well-suited for applications such as gaze tracking, fatigue detection, neuromotor rehabilitation, and biometric authentication [2].

With the emergence of smart eyewear devices such as augmented reality headsets, assistive vision systems, and heads-up displays, there is an increasing demand for secure, ergonomic, and seamless user authentication methods. Traditional authentication approaches, such as PINs, passwords, or even facial recognition, are often poorly suited for smart eyewear devices due to ergonomic, usability, and privacy constraints [3]. In contrast, EOG-based authentication leverages natural eye movement patterns, such as saccades and blinks, which are physiologically and behaviorally unique, offering a biometric signature that is difficult to replicate or spoof [4]. In addition, EOG can be seamlessly integrated with these smart eyewear devices to support hands-free interaction, thereby improving accessibility and user convenience. All of these make EOG a compelling candidate for implicit and continuous user verification in wearable systems.

However, realizing real-time EOG-based authentication on wearable devices presents multiple technical challenges. One of the primary challenges in deploying EOG classification models on wearable devices lies in the trade-off between the computational complexity and the classification performance [5]. Additionally, EOG signals are dynamic and subject to variability due to factors such as electrode placement, user-specific physiology, motion artifacts, and environmental noise. Unlike static biometric modalities, EOG signals are dynamic and can vary significantly across different users, sessions, and even environmental conditions [6]. Factors, including electrode placement, head movement, skin impedance, and noise interference introduce variability that models must generalize over. Conventional convolutional neural networks (CNNs) have demonstrated strong performance in physiological signal classification tasks; however, they often incur high computational and memory costs. This makes them unsuitable for on-device inference without significant architectural modifications. Furthermore, the need for low-latency inference in wearable contexts necessitates models that are not only computationally efficient but also capable of extracting temporal and spatial features from multichannel biosignals in noisy, real-world settings.

To address these issues, this paper proposes a lightweight deep learning model suited for real-time EOG-based classification and user authentication on edge hardware. The model architecture is inspired by MobileNet [7] and utilizes depthwise separable convolutions and squeeze-and-excitation (SE) blocks [8] to improve computational efficiency and channel-wise feature relevance [9]. Additionally, we incorporate batch normalization and ReLU activations to enhance convergence and stability during training. We evaluated our model on a publicly available dataset from the University of Malta [10], which includes EOG recordings along with the head pose and gaze features collected under nonstationary conditions. This dataset is well-suited for simulating realistic smart eyewear scenarios, as it comprises natural variations in head movement, blink behavior, and gaze trajectories. Our experiments involve both classification benchmarking and an ablation study that isolates the contribution of depthwise convolutions, filter depth, and channel attention mechanisms (squeeze and excitation). We compare our model against well-established baseline architectures, including ResNet50 [11], MobileNet [7], and SqueezeNet [12], in terms of classification precision, recall, F1 score, inference time, model size, and computational cost. The simulation's evaluation results on the Raspberry Pi testbed indicate that the proposed model is explicitly optimized for execution on resource-constrained edge devices (smart eyewear systems, etc.).

The remainder of this paper is organized as follows: Sect. 2 reviews the related work in EOG signal processing, deep learning architectures for time-series classification, and lightweight models for edge devices. Section 3 describes our system design, including the signal preprocessing pipeline, model architecture design, and Raspberry Pi setup. Section 4 presents the performance evaluation and results analysis. Section 5 discusses some opening issues and future works. Finally, Sect. 6 concludes this work.

2 Related Work

Recent advancements in wearable computing technology have garnered significant interest in smart eyewear systems [13], which provide a platform for users to utilize continuous, real-time physiological monitoring, intuitive user interaction, and context-aware applications. Smart glasses are increasingly integrated with sensors such as inertial measurement units (IMUs) and electrooculography (EOG) electrodes to enable gaze-based interfaces alongside hands-free control [14]. Integrating a reliable user authentication component into such systems remains a persistent challenge due to the necessity of high accuracy, low latency, and low power consumption. Traditional methods of authentication, such as those based on passwords or physical tokens, are poorly suited for user authentication of wearable devices. In turn, the landscape of security is shifting toward biometric authentication, including iris patterns, facial recognition, and physiological signals, and among others [15].

Bioelectrical signals, such as EMG, EEG, ECG, and EOG, have gained traction for use in biometric applications due to their uniqueness, permanence, and difficulty in spoofing [16]. EEG signals have been explored for cognitive state-based identification systems [17], while EMG signals have shown promise in gesture-based recognition systems [18]. EOG offers a less intrusive, low-power alternative, along with the added benefit of supporting multimodal applications, such as gaze tracking and blink detection [19]. Previous studies conducted with EOG signals have demonstrated that task-specific EOG patterns can be used to distinguish users [20] or even validate intended actions [21]. However, most existing work relies on traditional signal processing pipelines or minimal machine learning techniques.

The implementation of deep learning into biometrics has demonstrated substantial performance gains in physiological signal classification. Through leveraging end-to-end feature learning, deep learning models is capable of eliminating the requirements for handcrafted features. Convolutional neural networks (CNNs), recurrent neural networks (RNNs), and transformer-based attention mechanisms have demonstrated effectiveness in modeling the spatial and temporal structures inherent in time-series data [22,23] (e.g., EEG signals, EMG signals, and EOG signals). CNNs have been used extensively to extract localized spatial and temporal features from multi-channel biometric signals. CNNs have historically achieved robust classification results across various biometric and assistive domains. Despite the history of success in classification, many of these architectures are computationally intensive with high memory and processing demands. This restricts their direct deployment on embedded platforms, which are commonly used in wearable computing devices.

To mitigate the computational cost of conventional architectures, an increasing amount of research efforts have dedicated to designing lightweight deep learning models that are specifically tailored for edge computing devices. Other deep learning architectures, such as MobileNet, SqueezeNet, and QuickNet, have been proposed to achieve efficient inference through various structural modifications, such as depthwise separable convolutions, grouped convolutions, and bottleneck fire modules. These designs dramatically decreases the quantity of trainable parameters and the amount of computational resources. By reducing energy consumption and enhancing inference speed, these lightweight models have been successfully used in a body of real-time applications. Various applications, such as mobile health diagnostics [24], speech processing, and on-device activity recognition, demonstrate that the performance can be retained while also satisfying edge-level constraints.

The Raspberry Pi [25] has been widely adopted as a testbed for prototyping embedded machine learning applications in the context of Internet of Things (IoT), including wearable devices [26]. Its cost-effectiveness, extremely compact form, and its compatibility with mainstream optimized frameworks like TensorFlow Lite and PyTorch Mobile [27] make it a suitable testbed for real-time signal processing and model inference. Prior studies have used Raspberry Pi platforms to implement image classification, speech recognition, and biometric signal analysis in low-resource settings. These studies emphasize the importance of finding a balance between the complexity of a model and the limitation of the hardware when aiming for practical deployment scenarios.

Through these insights, the present study proposes a lightweight convolutional neural network for EOG signal classification under non-stationary head pose conditions. The proposed model is designed architecturally with computational efficiency in mind and is implemented on a Raspberry Pi for real-time inference. Previous studies have primarily focused on static conditions and relied on limited signal modalities. The proposed work incorporates multimodal physiological features, including head orientation, gaze target angles, and distance to screen, to improve the model's robustness and add contextual awareness in realistic wearable environments.

3 Our Proposed System

3.1 System Overview

The proposed system comprises a lightweight CNN specifically designed for real-time EOG-based user authentication on embedded hardware. The system processes EOG signals, as well as auxiliary physiological features such as head orientation, gaze angles, and positional data, to perform classification on embedded hardware. The workflow of the proposed system comprises three key components: a signal preprocessing module, a lightweight CNN classifier, and a deployment module for real-time inference on edge devices using Raspberry Pi.

3.2 Signal Preprocessing

Our study utilizes a publicly available EOG dataset collected by the University of Malta [10]. The dataset includes synchronized recordings of EOG signals, head pose, gaze direction, and other physiological features, acquired under non-stationary conditions to simulate realistic smart eyewear usage. Eight healthy volunteers (six males and two females with a mean age of 25.8 ($\pm$ 5.8 years) participated in the data collection. Subjects performed 200 structured trials, each consisting of sequential eye movements and a blink. Signals were recorded using four monopolar EOG electrodes and sampled at 256 Hz. Head pose (yaw, pitch, roll), head position, gaze angles, and pupil diameter were also recorded.

The first module is responsible for preparing the input signals. This includes a filtering pipeline with a fourth-order Butterworth bandpass filter (0.1–50 Hz) to isolate the relevant ocular activity frequencies while reducing the low-frequency drift and high-frequency artifacts [28]. Z-score normalization is then applied to each channel to achieve a zero mean and unit variance. This improves numerical stability during the model's training. The signal is then segmented into overlapping windows of fixed length (224 samples) with a 40 % overlap. The preprocessed signal is then augmented with engineered features derived from the cyclical transformations of angular measurements (yaw, pitch, and gaze direction). This was implemented to ensure the continuity of periodic attributes. These steps are crucial for transforming noisy biosignals into stable, learnable representations for temporal modeling and classification.

3.3 A Lightweight CNN Classifier

The second module consists of the proposed lightweight CNN classifier. This module was thoughtfully designed to address the computational constraints of embedded platforms such as the Raspberry Pi without sacrificing classification accuracy. Figure 1 displays the architecture of the classifier. It begins with an initial 2D convolution layer composed of 64 filters with a kernel size of 3 and a stride of 2. The next step involves applying batch normalization, and then followed by a ReLU activation. This layer extracts the low-level temporal features and also reduces the input dimensionality.

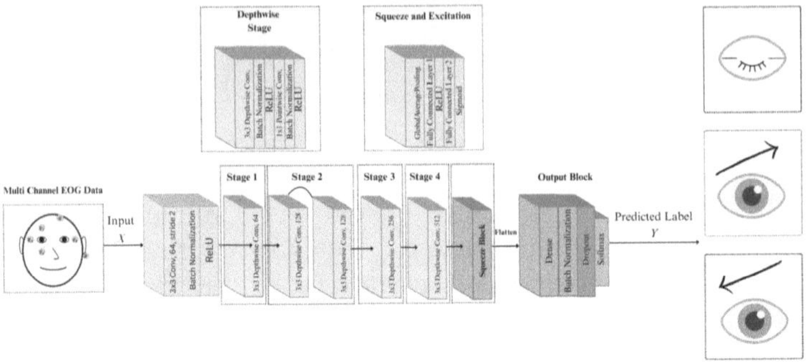

Fig. 1. Architecture of Lightweight CNN Model for EOG-Based User Authentication

Following the initial block, the model includes four stages of depthwise separable convolutional blocks [29]. Figure 2 illustrates the layer structure of the depthwise block. Each block uses a depthwise convolution to extract the localized patterns per channel. This is followed by a pointwise (1×1) convolution to project as well as integrate the features across channels. All blocks of the four stages follow the same structure, but with a different number of filters. Stage 1 includes one block that has 64 filters. Stage 2 comprises 2 blocks, and each block has 128 filters. Stages 3 and 4 each contains a single block with 256 and 512 filters, respectively. The design of the depthwise block decomposition reduces parameter complexity from $D_k \times D_k \times C_{\text{in}} \times C_{\text{out}}$ in standard convolutions to $D_k \times D_k \times C_{\text{in}} + C_{\text{in}} \times C_{\text{out}}$ yielding significant memory and computational savings, where the D_k is the filter width and height, and C_{in} and C_{out} are the number of input and output channels, respectively. These optimizations are crucial for ensuring real-time inference on resource-constrained devices.

To further improve the network's discriminative capacity, we implemented a Squeeze-and-Excitation (SE) block after Stage 4. The SE block performs global average pooling to obtain a channel-wise descriptor. It is then processed through two fully connected layers, using ReLU activation for the first layer and sigmoid activation for the second layer [8]. The output is used to scale the input feature map channels, enabling the model to emphasize noticeable information while suppressing noisy or redundant channels. The structure of the SE block is demonstrated in Fig. 3. The inclusion of this mechanism improved classification accuracy, as confirmed in our ablation study, without increasing the inference latency beyond acceptable limits for real-time operation.

The final stage of the network comprises a flattening layer, which followed by two fully connected layers. The two fully connected layers have 256 and 128 neurons, respectively. Each one is followed by batch normalization, ReLU activation, and dropout (rate = 0.3) for regularization. The output layer is a dense layer comprising three neurons and a softmax activation function, corresponding to the predicted classes. The predicted classes are upward saccade, downward

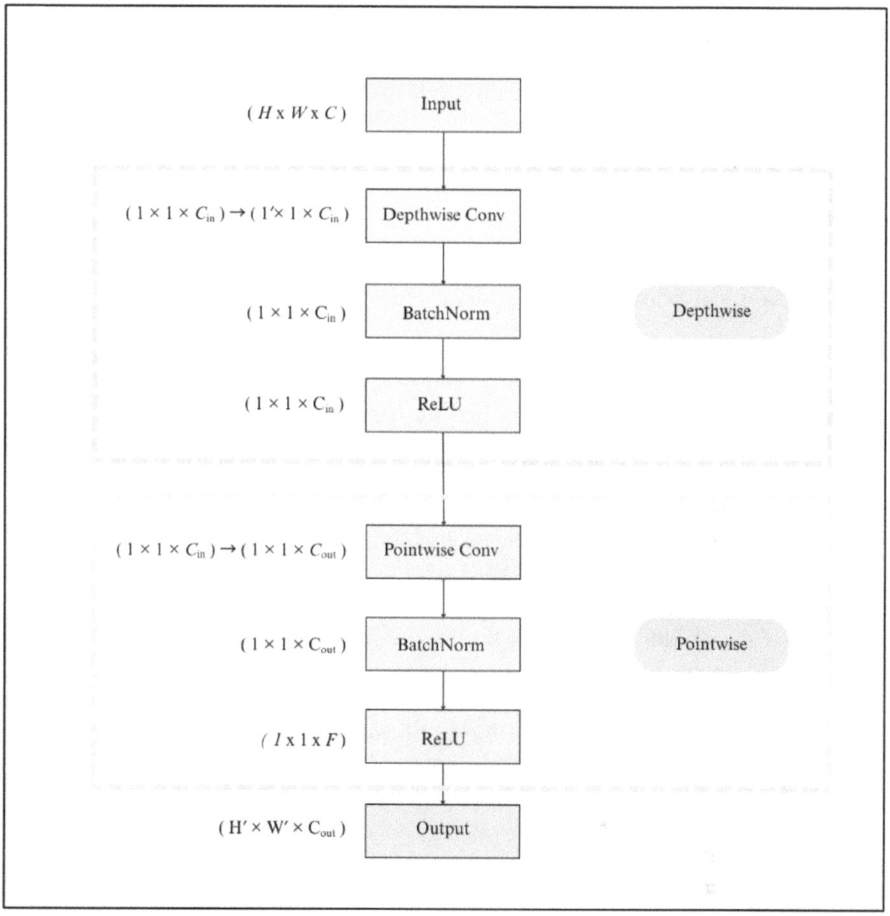

Fig. 2. Depthwise Block

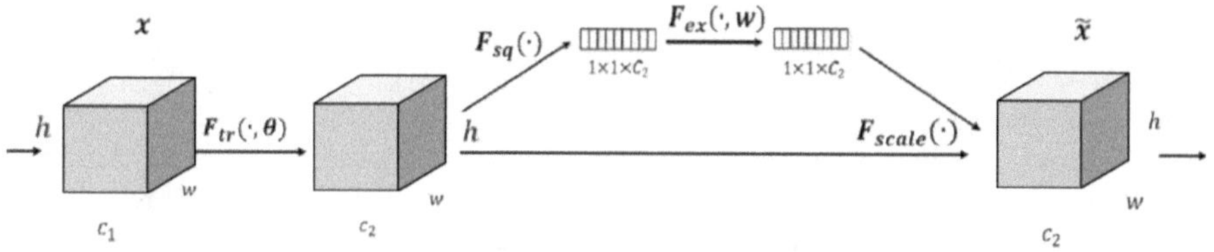

Fig. 3. Squeeze-and-Excitation (SE) block used in the proposed lightweight CNN. The input feature map undergoes a global squeeze operation, followed by channel-wise excitation using a two-layer fully connected network, and finally scaled to emphasize informative features.

saccade, or blink. This design offers sufficient expressiveness for the classification task while maintaining the compactness required for embedded deployment.

In the training mode, the Adam optimizer is used with a learning rate of 0.001 and the loss function is based upon a sparse categorical cross-entropy. Batch normalization, which is used after each convolutional layer, accelerated convergence and improved generalization by reducing internal covariance shift. ReLU activations throughout the network ensured sparse activation maps and avoided vanishing gradients.

3.4 Implementation on Raspberry Pi

To enhance compatibility and optimize performance on the Raspberry Pi, the proposed model was quantized using TensorFlow Lite's dynamic range quantization. The quantized model executes inference on the Raspberry Pi without requiring GPU acceleration. Previous studies have shown the efficacy of similar approaches in handling computationally intensive tasks on devices that have limited resources while operational efficiency is maintained [30,31].

The testbed was configured by a Raspberry Pi 4 Model B, which has a quad-core Cortex-A72 processor, 4 GB of RAM, and an external power supply to enable continuous operation. Data acquisition was achieved by interfacing EOG signal electrodes with an ADC module connected to the Raspberry Pi's GPIO pins. The system was designed to process incoming signals, extract relevant features, and execute the authentication model, ensuring latency remained within acceptable limits for real-time applications. These configurations align with existing research findings that highlight Raspberry Pi's role as a low-cost, portable platform for data acquisition and processing in IoT applications [31,32].

4 Performance Evaluation

To showcase the effectiveness of our proposed system, we conduct extensive experiments. In the following, we first present the methodology and then discuss the results in detail.

4.1 Methodology

Evaluation Metrics. To evaluate the efficacy of the proposed lightweight CNN model, we used a combination of classification and efficiency-based metrics. Classification performance was measured using accuracy, precision, recall, and F1-score., providing insight into the model's consistency across the three target classes: upward saccade, downward saccade, and blink. Accuracy is derived as the proportion of correctly classified samples over the total number of predictions. Precision and recall are derived per class and then macro-averaged to reflect the model's overall efficacy. We used standard evaluation metrics defined as follows.

Precision refers to the ratio of correctly predicted positive observations over the total number of instances marked as positives:

$$Precision = \frac{TP}{TP + FP} \tag{1}$$

Recall, also known as sensitivity or true positive rate, measures the proportion of accurately predicted positive observations to all actual positives:

$$Recall = \frac{TP}{TP + FN} \tag{2}$$

F1 Score is the harmonic mean of Precision and Recall, which aims to offer a balance between the two:

$$F1\ Score = 2 \times \frac{Precision \times Recall}{Precision + Recall} \tag{3}$$

Additionally, we evaluated hardware performance characteristics relevant to the embedded deployment. This included floating point operations (FLOPs) [33], model size in megabytes (MB), and inference latency in milliseconds per sample. Inference latency was measured on a CPU-only Raspberry Pi testbed to simulate realistic edge deployment. All models were trained for 100 epochs that has early stopping to prevent overfitting, and training time was logged for comparative and reflective purposes.

Floating Point Operations (FLOPs) refers to the total number of arithmetic operations that are required to perform a forward pass through the network.

For a standard 2D convolutional layer:

$$FLOPs_{conv2D} = D_k \times D_k \times C_{in} \times C_{out} \times H \times W \tag{4}$$

For a depthwise separable convolution (depthwise + pointwise operations):

$$FLOPs_{dw\text{-}sep} = D_k \times D_k \times C_{in} \times H \times W + C_{in} \times C_{out} \times H \times W \tag{5}$$

Baseline Model Comparison. We compared our proposed architecture against three established lightweight deep learning models: MobileNet [34], ResNet50 [35], and SqueezeNet [36]. MobileNet is known for its use of depthwise separable convolutions and helped serve as an architectural reference for our model. ResNet50 was chosen as a conventional deep CNN baseline due to its strong performance on vision tasks, even though it has a significantly higher parameter count. SqueezeNet uses fire modules with bottleneck layers to reduce the model's size and was included as another resource-efficient model. All baselines were reimplemented using convolutions to support our time-series data and were trained under the same conditions for complete fairness.

Ablation Study of the Proposed Lightweight Classifier. To evaluate the contribution of individual architectural components in the proposed model, an ablation study was conducted. We created modified versions of the model by removing or simplifying key elements: (i) **No Depthwise**: removed the four stages of Depthwise blocks to evaluate the contribution of the Depthwise Convolution Layer for feature extraction. (ii) **No SE Block**: Removed the squeeze-and-excitation mechanism to assess its impact on channel-wise feature emphasis. (iii) **Reduced Filter Configuration**: Decreased the number of filters per layer to minimize model size to ten percent of the proposed model to reduce computational cost.

Each variant was trained under the same experimental conditions as the full model. The results were used to identify the most impactful design choices contributing to classification accuracy and deployment feasibility on embedded devices.

4.2 Results

Table 1. Performance of Baseline Model Comparison

Model	Precision	Recall	F1 Score	Inference Time (ms/sample)	FLOPs	Model Size (MB)
Proposed Model	0.943	0.944	0.940	0.097	276,579	1.06
ResNet50	0.811	0.808	0.807	0.196	2,350,000	6.82
MobileNet	0.765	0.759	0.761	0.073	171,230	1.38
SqueezeNet	0.798	0.802	0.798	0.104	154,688	0.51

The results of the baseline model comparison are shown in Table 1, demonstrating that the proposed lightweight model achieves a better performance than all baseline models in both classification accuracy and runtime efficiency. Specifically, the proposed model achieved a macro F1-score of 0.943, which is significantly higher than ResNet50 (0.807), MobileNet (0.761), and SqueezeNet (0.798). This indicates that the proposed model provides more balanced classification over the three classes (upward saccade, downward saccade, and blink).

Concerning precision and recall, our proposed model consistently scored above 0.94, reflecting its reliability in term of correctly recognizing true positive cases while minimizing false positives. By comparison, the next closest competitor, ResNet50, scored only 0.811 in precision and 0.808 in recall, demonstrating a clear performance gap.

From a deployment perspective, the proposed model maintained a low inference latency of 0.097 ms/sample, allowing it to run smoothly in real-time applications. Although MobileNet had a slightly faster inference time (0.073 ms/sample), it did so at the expense of predictive performance, with a much lower F1-score (0.761). ResNet50 and SqueezeNet exhibited higher inference times at 0.196 ms/sample and 0.104 ms/sample, respectively, making them less suitable for time-sensitive edge deployments.

In summary, the proposed model strikes a strong balance between classification accuracy and computational efficiency. It is the one that meets the critical constraints of real-time inference, minimal memory usage, and high biometric reliability. The proposed model achieved the highest F1-score (0.943) while maintaining a small memory footprint (1.06 MB) and low inference latency (0.097 ms/sample). By contrast, ResNet50 reached 80.58% accuracy but required a larger model size (6.82 MB) and higher latency (0.196 ms/sample). MobileNet and SqueezeNet, although smaller and faster, failed to match the proposed model's predictive performance.

Table 2. Performance of Ablation Study

Model	Precision	Recall	F1 Score	FLOPs	Model Size (MB)
Proposed Model	0.943	0.944	0.931	276,579	1.06
NO SE Block	0.757	0.733	0.738	48,209	0.59
No Depthwise	0.765	0.742	0.746	21,123	0.28
Fewer Filters	0.760	0.745	0.749	5,955	0.11

The second experiment evaluates the impact of key architectural components through an ablation study, with performance and efficiency metrics summarized in Table 2. The proposed lightweight model achieved the F1 score (0.931), demonstrating that the complete design delivers superior performance. When depthwise separable convolutions were removed, the model's F1 score dropped sharply to 0.746, while the number of FLOPs reduced to 21,123. This indicates that depthwise convolutions are more efficient at extracting temporal features from multichannel EOG signals to ensure the recognition accuracy. The "No SE Block" variant exhibited the lowest overall performance, with a F1 score of 0.738. This underscores the importance of the squeeze-and-excitation block in enhancing the discriminative capacity of the model by emphasizing relevant channel-wise features.

The model variant with fewer filters demonstrated the smallest computational footprint, which is only 5,955 FLOPs and 0.11 MB in size but showed limited predictive ability, achieving just an F1-score of 0.749. This trade-off highlights a threshold below which compression harms model effectiveness.

Taken together, these results reveal that while model size and computation can be significantly reduced, the integration of depthwise separable convolutions and SE blocks is essential to maintain robust performance. Our proposed model achieves a good balance between accuracy and resource usage, confirming the effectiveness of the design for edge-based EOG classification.

5 Discussion

One of the most critical challenges in EOG-based user authentication systems is the high variability of biometrics across individuals, data collection time sessions, and environmental conditions. While the proposed model is capable of achieving good performance on the current dataset, its ability to generalize across different user populations and signal acquisition protocols remains unpredictable. Factors such as electrode placement inconsistencies, variation in eye movement patterns, and signal noise can significantly affect authentication accuracy. To address these issues, there is an urgent need for larger and more diverse datasets, as well as adaptive learning strategies such as domain adaptation or personalized calibration.

Although models with stronger adaptability and personalization can enhance authentication performance during training and evaluation, deploying models that dynamically adapt to a user's unique physiology remains an open problem, especially when training models from scratch is not preferred. Future work can consider using transfer learning and few-shot learning (a machine learning technique that enables a model that is capable of making accurate predictions from a much small number of examples) to fine-tune the pretrained model on a small amount of up-to-date signals collected by individuals to help the model deploy more smoothly and with higher authentication accuracy.

Additionally, although the proposed model demonstrates efficiency on a Raspberry Pi, another potential deployment issue is integrating the authentication model seamlessly with other sensors in resource-limited devices to maintain a user-friendly experience. Authentication is a crucial step for users to access the device. Future work should focus on applying techniques such as sensor fusion and cross-modal learning models to enhance reliability under stringent power and latency conditions, varying user mobility, and signal dropout, thereby ensuring real-time synchronization among multimodal data streams.

6 Final Remarks

In this paper, we have presented a lightweight and efficient deep learning model for EOG-based user authentication, optimized for deployment on edge devices, including the Raspberry Pi. The model leverages depthwise separable convolutions and channel-wise attention mechanisms to extract meaningful temporal features from multichannel EOG signals. The evaluation results demonstrate its suitability for real-time applications on low-power embedded platforms through achieving an F1 score of 0.931 with only 276,579 FLOPs and a model size of 1.06 MB. Comparative benchmarking against other models, such as CNNs like ResNet50, MobileNet, and SqueezeNet, demonstrated the architectural advantages of combining depthwise separable convolutions with modular design principles tailored for EOG signal processing. Furthermore, the proposed model's consistent performance across both precision and recall further validates its robustness and ability to generalize. By targeting edge compatibility from the ground up, our model addresses the practical constraints inherent in smart eyewear authentication systems [20], such as real-time responsiveness, limited memory, and user mobility, thereby enabling practical deployment in real-world wearable environments [37].

References

1. Mulo, J., et al.: Navigating challenges and harnessing opportunities: deep learning applications in internet of medical things. Fut. Int. **17**(3) (2025)
2. Sriman Bidhan Baray, M.U., Ahmed, Muhammad, E.H., Chowdhury, K., Kise: EOG-based reading detection in the wild using spectrograms and nested classification approach. IEEE Access **11**, 105619–105632 (2023)
3. Guo, Z., Cao, J., Bai, X., Li, A., Niu, B., Li, H.: Shake, shake, I know who you are: authentication through smart wearable devices. IEEE Sens. J. **23**(21), 26786–26795 (2023)
4. Dhore, M., Gaikwad, A.: EOG-based authentication system for specially abled users. In: Congress on Intelligent Systems, pp. 1–16. Springer (2024)
5. Su, R., Zeng, Z., Tao, L., Wang, Z., Chen, C., Chen, W.: KL divergence-based transfer learning for cross-subject eye movement recognition with EOG signals. In: 2023 45th Annual International Conference of the IEEE Engineering in Medicine & Biology Society (EMBC), pp. 1–4. IEEE (2023)
6. Tao, L., Huang, H., Chen, C., Feijs, L., Hu, J., Chen, W.: Review of electrooculography-based human-computer interaction: recent technologies, challenges and future trends. Connect. Health Telemed. **2**(3), N–A (2023)
7. Sandler, M., Howard, A., Zhu, M., Zhmoginov, A., Chen, L-C.: MobileNetV2: inverted residuals and linear bottlenecks. In: Proceedings of the IEEE Conference on Computer Vision and Pattern Recognition, pp. 4510–4520 (2018)
8. Hu, J., Shen, L., Sun, G.: Squeeze-and-excitation networks. In: Proceedings of the IEEE Conference on Computer Vision and Pattern Recognition, pp. 7132–7141 (2018)
9. Bianco, S., Cadene, R., Celona, L., Napoletano, P.: Benchmark analysis of representative deep neural network architectures. IEEE Access **6**, 64270–64277 (2018)
10. Barbara, N., Camilleri, T.A., Camilleri, K.P.: Real-time continuous EOG-based gaze angle estimation with baseline drift compensation under non-stationary head conditions. Biomed. Signal Process. Control **90**, 105868 (2024)
11. Wang, Q., Wu, B., Zhu, P., Li, P., Zuo, W., Hu, Q.: ECA-Net: efficient channel attention for deep convolutional neural networks. In: Proceedings of the IEEE/CVF Conference on Computer Vision and Pattern Recognition, pp. 11534–11542 (2020)
12. Li, M., He, L., Lei, C., Gong, Y.: Fine-grained image classification model based on improved squeezenet. In: 2021 IEEE 5th Advanced Information Technology, Electronic and Automation Control Conference (IAEAC), pp. 393–399. IEEE (2021)
13. Zhang, J., et al.: Review of wearable computing technology: core technologies, typical systems and research trends. In: 2022 Global Conference on Robotics, Artificial Intelligence and Information Technology (GCRAIT), pp. 225–231. IEEE (2022)
14. Mekruksavanich, S., Jantawong, P., Jitpattanakul, A.: Deep learning approaches for HAR of daily living activities using imu sensors in smart glasses. In: 2023 Joint International Conference on Digital Arts, Media and Technology with ECTI Northern Section Conference on Electrical, Electronics, Computer and Telecommunications Engineering (ECTI DAMT & NCON), pp. 474–478. IEEE (2023)

15. Sivasankari, K., Hanirex, D. K.: Advancements in CNN-based iris recognition: enhancing biometric security and authentication systems. In: 2024 International Conference on Innovative Computing, Intelligent Communication and Smart Electrical Systems (ICSES), pp. 1–7. IEEE (2024)
16. Li, Q.: Deep learning-based multi-channel bio-electrical signal processing: Algorithms, models, and applications, New Mexico Institute of Mining and Technology (2023). PhD thesis
17. Mazher, M., Faye, I., Qayyum, A., Malik, A.S.: Classification of resting and cognitive states using EEG-based feature extraction and connectivity approach. In: 2018 IEEE-EMBS Conference on Biomedical Engineering and Sciences (IECBES), pp. 184–188. IEEE (2018)
18. Li, Q., Luo, Z., Qi, R., Zheng, J.: DeepTPA-Net: a deep triple attention network for SEMG-based hand gesture recognition. IEEE Access **11**, 96797–96807 (2023)
19. Kumar, D., Sharma, A.: Electrooculogram-based virtual reality game control using blink detection and gaze calibration. In: 2016 International Conference on Advances in Computing, Communications and Informatics (ICACCI), pp. 2358–2362. IEEE (2016)
20. Kaushik, B., Vinay Kumar, M., Tadepalli, S., Vats, A., Bhargav,R.: EOG-based blink authentication. In: 2023 International Conference on Recent Advances in Science and Engineering Technology (ICRASET), pp. 1–5. IEEE (2023)
21. Lin, C.-T., et al.: EOG-based eye movement classification and application on HCI baseball game. IEEE Access **7**, 96166–96176 (2019)
22. William Grant Hatcher and Wei Yu: A survey of deep learning: platforms, applications and emerging research trends. IEEE Access **6**, 24411–24432 (2018)
23. Musa, A.A., Hussaini, A., Liao, W., Liang, F., Yu, W.: Deep neural networks for spatial-temporal cyber-physical systems: a survey. Fut. Int. **15**(6) (2023)
24. Pustokhina, I.V., Pustokhin, D.A., Gupta, D., Khanna, A., Shankar, K., Nguyen, G.N.: An effective training scheme for deep neural network in edge computing enabled internet of medical things (IoMT) systems. IEEE Access **8**, 107112–107123 (2020)
25. Nikouei, S.Y., Chen, Y., Song, S., Xu, R., Choi, B-Y., Faughnan, T.R.: Real-time human detection as an edge service enabled by a lightweight CNN. In: 2018 IEEE International Conference on Edge Computing (EDGE), pp. 125–129. IEEE (2018)
26. Guo, Y., Qian, C., Song, J., Yu, W.: Chapter 6 - edge intelligence in CPS. In: Yu, W., (ed.), Edge Intelligence in Cyber-Physical Systems, Intelligent Data-Centric Systems, pp. 141–164. Academic Press (2025)
27. Hadidi, R., Cao, J., Xie, Y., Asgari, B., Krishna, T., Kim, H.: Characterizing the deployment of deep neural networks on commercial edge devices. In: 2019 IEEE International Symposium on Workload Characterization (IISWC), pp. 35–48. IEEE (2019)
28. Daud,S.S., Sudirman, R.: Butterworth bandpass and stationary wavelet transform filter comparison for electroencephalography signal. In: 2015 6th International Conference on Intelligent Systems, Modelling and Simulation, pp. 123–126. IEEE (2015)
29. Chunyan, Yu., Han, R., Song, M., Liu, C., Chang, C.-I.: A simplified 2D–3D CNN architecture for hyperspectral image classification based on spatial-spectral fusion. IEEE J. Select. Top. Appl. Earth Observat. Remote Sens. **13**, 2485–2501 (2020)
30. Huang, Y., Lu, Y., Wang, F., Fan, X., Liu, J., CM Leung, V.: An edge computing framework for real-time monitoring in smart grid. In: 2018 IEEE International conference on industrial internet (ICII), pp. 99–108. IEEE (2018)

31. Khaled Abdulrahman Abdulaziz Al-Awadhi, A., Ruhullah, A., Harris: a portable, wireless and low-cost electroencephalogram monitor using raspberry PI. J. Hum. Centered Technol. **3**(2), 44–53 (2024)
32. Mani Dheeraj Mudaliar, N., Sivakumar: IoT based real time energy monitoring system using raspberry PI. Int. Things **12**, 100292 (2020)
33. Van Der Walt, S., Colbert, S.C., Varoquaux, G.: The NumPy Array: a structure for efficient numerical computation. Comput. Sci. Eng. **13**(2), 22–30 (2011)
34. Howard, A., et al.: Searching for mobilenetV3. In: Proceedings of the IEEE/CVF International Conference on Computer Vision, pp. 1314–1324 (2019)
35. Wang, J., et al.: Deep high-resolution representation learning for visual recognition. IEEE Trans. Pattern Anal. Mach. Intell. **43**(10), 3349–3364 (2020)
36. Abhijit Guha Roy, N., Navab, C., Wachinger: Recalibrating fully convolutional networks with spatial and channel "squeeze and excitation" blocks. IEEE Trans. Med. Imaging **38**(2), 540–549 (2018)
37. Peluso, V., Cipolletta, A., Calimera, A., Poggi, M., Tosi, F., Mattoccia, S.: Enabling energy-efficient unsupervised monocular depth estimation on armv7-based platforms. In: 2019 Design, Automation & Test in Europe Conference & Exhibition (DATE), pp. 1703–1708. IEEE (2019)

Quantifying the Difference between Natural and Simulated Quantum Processors

Garrett Short, Bryson R. Payne(✉), and Murthy Rallapalli

University of North Georgia, Dahlonega, GA 30597, USA
{gcshor8549,bpayne,mvrallapalli}@ung.edu

Abstract. This study provides quantitative data regarding the difference noise and decoherence can make between a simulated and real quantum processor. As running circuits on real QPUs (quantum processing units) only became widely possible in the past five years, most prior studies only offer hypothetical or simulated data regarding the accuracy of quantum algorithms. Such data has the potential to be unintentionally misleading due to differences between the simulated backend and the newer QPUs available to the public. Updating this data based on measured differences in accuracy and general noise levels keeps papers relevant and accurate. To find this difference, this research team developed two testing groups, contrasting a real IBM QPU with the AerSimulator backend. In each of these groups, two programs were run: Grover's algorithm, and then Grover's algorithm with Shor's encoding introduced. Using the plain Grover's algorithm in AerSimulator as the control, the team keyed the oracle for a specific result and gave each program 10,000 shots. In measuring the distribution and number of shots per index, the data point to a strong conclusion: A simulated backend is, by default, too clean an environment to accurately reflect the predicted results of an algorithm on a real QPU. Simulators currently lack sufficient randomness and decoherence that both limit and separate physical QPUs from their simulated counterparts and from classical CPUs.

Keywords: Quantum Computing · Quantum Processors · QPUs · Grover's algorithm · Shor's algorithm

1 Introduction

The creation of functional quantum computers has the potential to be the single largest technological breakthrough that has happened in the field of computing. This technology allows for the development of new, quantum-efficient algorithms, which provide speedups and new possible computations. These can be so significant that previously impenetrable security mechanisms like the RSA encryption scheme will be nearly instantly broken with the application of the correct algorithm. However, these algorithms have largely remained in theory and on paper due to the hardware limitations that are present in quantum processing units (QPU) and the technology being inoperable. Thus, the only way for a researcher to test their algorithms was to use a local quantum simulator backend, which, although well-made, is unable to realistically show the myriad of environmental factors that affect the accuracy of said algorithm.

H. R. Arabnia et al. (Eds.): CSCI 2025, CCIS 2970, pp. 122–130, 2027.
https://doi.org/10.1007/978-3-032-28466-2_9

Within the last decade, this veil has begun to be lifted, and functional QPUs have been created. These started small, only hosting five to seven qubits; however, in the last four years, IBM has made massive breakthroughs that have allowed for multiple 127-qubit QPUs to be publicly available [1]. With a QPU that has enough physical qubits to run these algorithms in a non-simulated environment, we can begin to see their real effectiveness and begin to apply them in actual use cases. Currently, the biggest limiting factor is the accuracy of these masses of qubits. The more qubits there are, the less accurate they become, due to their fragility and the ease with which environmental factors affect them.

The purpose of this paper is to show and quantify the effect that noise and decoherence have on the accuracy of a given algorithm and integrate protective measures to attempt to limit said effects. This natural environment can then be compared to the accuracy of the same algorithm, hosted using a simulator. Significant deviation between the experiments would show the inherent flaws of simulators over real QPU environments and a need for previously simulated algorithms to be run on real hardware.

2 Background and Related Work

First, it is necessary to address in sufficient detail the prerequisite knowledge required to understand and interpret the methodologies and results of our given experiment. Additionally, it gives credence to the work of Elliot Marklund and Hugo Sacilotto [2] as their studies into the combination of Shor's code protecting Grover's algorithm provided substantial assistance in the construction of the experiment.

2.1 The Qubit

Quantum computing is a revolutionary technology because it replaces the bit with the qubit. The difference between them is the superposition of a qubit compared to the binary positions of a normal bit. While a bit is limited to a single state of either 1 or 0, a qubit has the unique capacity to be both 1 and 0 at the same time [3]. This is shown in the figure known as a Bloch Sphere, represented in Fig. 1.

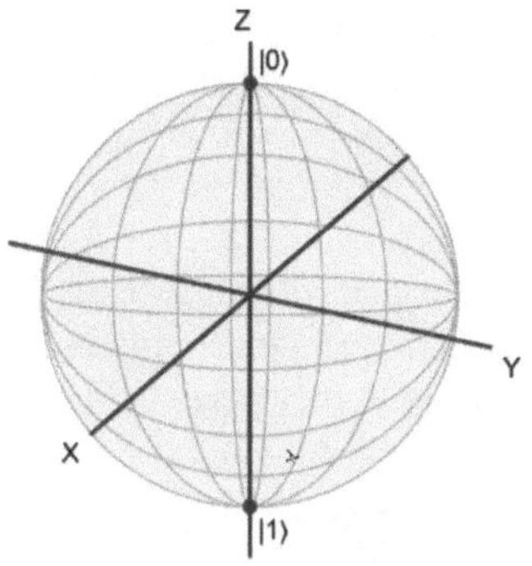

Fig. 1. A 3D sphere representing the possible measurements of superposition

To properly represent a superposition, a point can be chosen anywhere within the confines of the sphere [3]. Measuring this qubit results in the superposition collapsing, and it becomes a definite, classically measurable state. Additionally, due to the properties of entanglement, getting this state can cause the state of another qubit to collapse into a related definite state. This exchange happens at nearly the speed of light and results in data being transferred faster than a classical computer is capable of [3]. The superposition of a qubit and how it interacts with the environment has created the possibility for revolutionary computers to be made utilizing qubits rather than a classical bit due to this speed increase [4].

2.2 Decoherence and Noise

Decoherence and noise within a QPU are the biggest obstacles between quantum supremacy and our current hardware [5]. Decoherence is a term used to describe the way the environment tampers with the superposition of a qubit, forcing the expected definite position to be off. A superposition can be affected in two ways: a bit flip and a phase flip, as shown in Figs. 2 and 3.

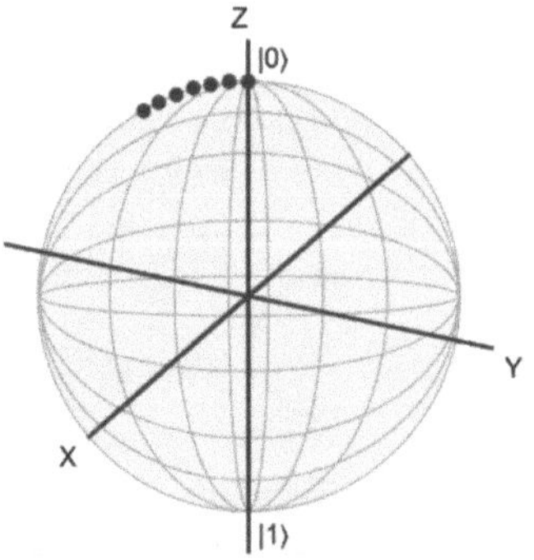

Fig. 2. A Bloch Sphere showing a partial bit flip

The sphere in Fig. 2 shows a partial phase flip, and the one in Fig. 3 shows a partial bit flip. Either of these can occur due to something as small as a stray electron affecting a captured photon being used as a qubit, and would potentially ruin the outcome of an algorithm. Noise is a term used to describe the increasing incorrectness of a QPU as run time increases [6]. Noise is an inevitability within a QPU, and the longer a circuit runs, or the deeper it is, the more qubits will undergo errors of either kind [7]. This can be mitigated through certain algorithms known as error correction codes, but altogether, it is inevitable for a QPU to be overloaded with noise to the point where accuracy becomes impossible. The current holy grail of quantum computing is a QPU that is fault-tolerant, or able to fix these errors as quickly as they occur.

2.3 Shor's Code

Shor's code is an error correction code that combines the fixes for both phase and bit flip errors into one circuit [1]. This is accomplished by using multiple physical qubits to

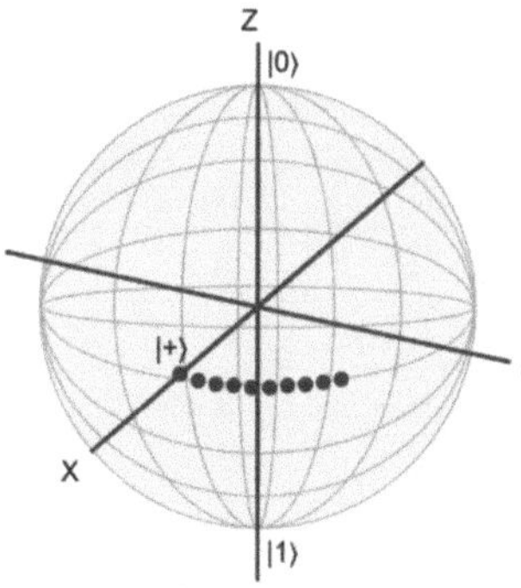

Fig. 3. A Bloch Sphere showing a partial phase flip

encode a single logical qubit. The physical qubits exist on the QPU, and a logical qubit is the encoded product that is then used by an algorithm. To correct a bit flip error, we use a 3:1 physical to logical qubit ratio and measure the parity of the leftmost and rightmost physical qubit. Once measured, a bit flip error would be detected as a misplaced 0 or 1 among the three qubits and is corrected through the application of a Pauli-X Gate to the outstanding qubit. All possible examples are shown in Table 1.

Table 1. All possible initial states, the effect of each on an unprotected superposition, and the corrective circuits for bit-flip errors

State	Syndrome	Correction
$\alpha\|000\rangle + \beta\|111\rangle$	00	$\mathbb{I}\otimes\mathbb{I}\otimes\mathbb{I}$
$\alpha\|100\rangle + \beta\|011\rangle$	10	$X\otimes\mathbb{I}\otimes\mathbb{I}$
$\alpha\|010\rangle + \beta\|101\rangle$	11	$\mathbb{I}\otimes X\otimes\mathbb{I}$
$\alpha\|001\rangle + \beta\|110\rangle$	01	$\mathbb{I}\otimes\mathbb{I}\otimes X$

Correcting a phase flip error is done similarly, but it corrects the $|-\rangle$ and $|+\rangle$ states rather than $|0\rangle$ and $|1\rangle$. Additionally, this type of error is corrected using a Pauli-Z gate. This is shown in Table 2.

To form Shor's code, three bit-flip circuits are run in parallel to one another, each producing a logical qubit that will be later acted upon. Once each bit-flip code runs, a phase-flip correction code will then run on the three logical qubits that were produced. This then makes a singular logical qubit that has been vetted for both phase and bit-flip errors.

2.4 Grover's Algorithm

Grover's algorithm is designed to search through an unsorted database and find a chosen answer as fast as possible. The classical way to solve this problem is to do various forms of brute-force searching, where every possible solution is tested. The quantum solution

Table 2. All possible initial states, the effect of each on an unprotected superposition, and the corrective circuits for phase-flip errors

State	Syndrome	Correction
$\alpha\vert{+}{+}{+}\rangle + \beta\vert{—}\rangle$	00	$\mathbb{I}\otimes\mathbb{I}\otimes\mathbb{I}$
$\alpha\vert - {+}{+}\rangle + \beta\vert + {—}\rangle$	10	$Z\otimes\mathbb{I}\otimes\mathbb{I}$
$\alpha\vert + - + \rangle + \beta\vert - + - \rangle$	11	$\mathbb{I}\otimes Z\otimes\mathbb{I}$
$\alpha\vert + {—}\rangle + \beta\vert\text{-} - +\rangle$	01	$\mathbb{I}\otimes\mathbb{I}\otimes Z$

to this problem, Grover's algorithm, tries to find the solution to the oracle. The oracle is a black box function that returns a Boolean value by multiplying the amplitude of a given qubit by either -1 or 1. If the value is false, not the solution to the oracle, it is multiplied by one; if it is the solution to the oracle, it is then marked as true by multiplying the amplitude by -1. This algorithm runs solutions through this oracle once for every single 'shot' that is specified at the start of the algorithm. With every single shot, this amplitude is flipped around the amplitude of the last point. This causes the probability of the correct option to increase with every iteration, as shown in Fig. 4.

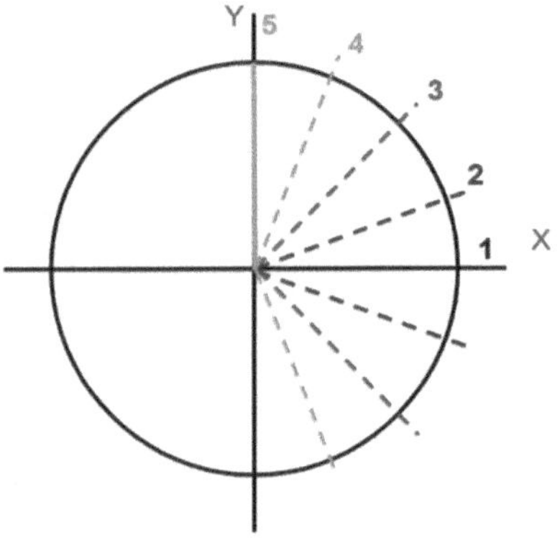

Fig. 4. A simplified form of this interaction with five shots, the fifth shot representing the 'key' or answer to the oracle

Grover's algorithm was chosen due to its importance and versatility, as well as how it is an example of a quantum-efficient algorithm, i.e., an algorithm that solves a given problem faster than a classical counterpart. A traditional brute-force algorithm takes about N steps to solve a problem, but a quantum algorithm can solve it in about $\sqrt{N}$ steps [8]. This results in a considerably faster solution for large input sizes versus the linear time increase in the classical approach.

3 Experiment Methodology

To test the differences between a real QPU and a simulated backend, quantum circuits were constructed using Qiskit. Two types of circuits were constructed: a shallow circuit and a very deep circuit. The key difference between these two is their total resource use, with the deep circuit consuming more resources than the shallow circuit. Running a longer circuit extends the runtime and the number of qubits that are used, granting a better opportunity to measure the effect noise will have on the results. The shorter circuit cannot show the effect of noise as well due to the shorter runtime, but it will still be affected by decoherence.

The shallow circuit was a plain Grover's algorithm with slight code optimizations and updates to a newer Qiskit version [9]. The second, deeper, circuit was Grover's algorithm protected by Shor's code on the three logical qubits it operates on [2]. This code base required substantial modification from the shallow circuit's Grover's algorithm, as the oracle had to be partially reconstructed and keyed differently due to circuit complexity. Additionally, slight changes in variable types and function orders were made to Grover's algorithm to allow for the implementation of Shor's code in the form of input and output changes. These modifications pose no functional difference between Grover's algorithm in either the deep or shallow circuits.

Once both the deep and shallow circuits were made, we then separated them into the simulated and real backend categories. Both categories had both circuits run on them, each circuit given 10,000 shots to eliminate sampling errors. The simulated backend chosen was the AerSimulator package for Qiskit. This is due to it being the most well-supported and widely-used simulated backend for the Qiskit language. The real QPU chosen was the 127-qubit IBM_sherbrooke, running on an Eagle r3 quantum processor, due to its affordable availability. Both real circuits were run back-to-back in the queue in order to ensure that both started with similar conditions and levels of noise.

The shallow, simulated circuit represents the control group in this experiment due to its unaltered nature in an environment that has no real environmental factors. It will still be affected by the little artificial noise that is generated, but no means have been taken to give a larger window for noise to affect the outcome. The circuits run on the real backend and, with additional overhead, will be the experiment group as they have something added to them or are being run in an environment that will induce additional change due to noise and decoherence.

Once the 10,000 shots have finished on each circuit, the results will be measured in the form of how many shots correctly chose $|100\rangle$ as this is the value keyed into the oracle. This allows for additional measurements to be taken and conclusions to be made based on the exact number of shots per answer and how they were distributed. Their distribution would act as an indicator of the effect of noise and decoherence within the backend; a wider distribution means less accuracy, and a heavily spiked distribution shows the optimal result for Grover's algorithm.

4 Results

The results in Fig. 5 show the percentage of shots out of 10,000 that fell within |100⟩. The first two columns, in red, represent the simulated backends, and the right two in blue are the results run on IBM's Sherbrooke QPU. These results show a definite drop in the probability of the code running on a real QPU. Most notably, the addition of Shor's code caused, at a minimum, a drop of ~17% in this probability in both groups. The control group was significantly higher than any of the other groups.

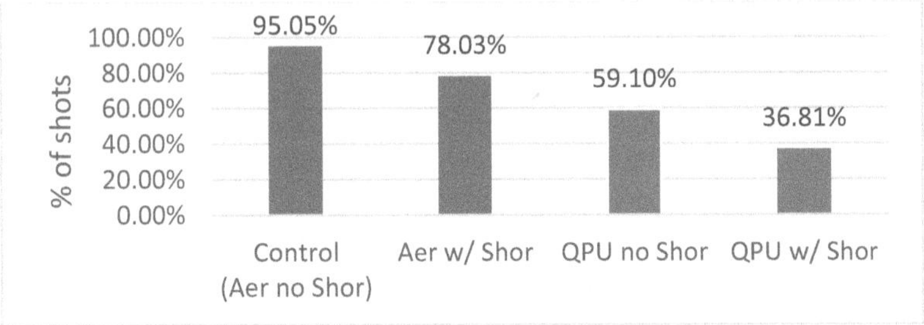

Fig. 5. The percentage of shots out of 10,000 that fell within |100⟩

Table 3 illustrates the probability drop between the runs across the simulator and QPUs for the two approaches. There are no significant spikes present in the dataset besides the keyed index, showing that the oracle itself is functioning properly. The fact that these shots are spread widely in the incorrect indexes demonstrates the effect that noise has on the algorithms. The high number of shots given to the algorithm, combined with the random nature of noise, means this distribution and drop in probability should always be well distributed.

Table 3. Total shots per index of each of the algorithms on the simulator vs. QPU

	|000⟩	|001⟩	|010⟩	|011⟩	|100⟩	|101⟩	|110⟩	|111⟩
Control	83	71	63	71	9505	68	75	64
Aer w/Shor	316	289	322	315	7803	332	298	325
QPU no Shor	649	397	432	357	5910	574	1072	609
QPU w/ Shor	1094	990	701	631	3681	1188	906	809

5 Limitations and Future Work

The primary limitations in this experiment were the hardware limitations of the IBM QPUs available to us. The service level that was used to run these experiments on IBM hardware was limited to using only two 127-qubit QPUs, IBM_sherbrooke and IBM_brisbane. There are two problems that arose due to this limitation: the total jobs queued and the overhead costs.

Due to there only being twelve QPUs available on the IBM Quantum platform, there was a massive bottleneck often formed where there would be over a hundred jobs in the queue. This usually would not be an issue due to the short runtime of most jobs; however, we had the most basic level plan. This means that if somebody had a higher-level plan, they would cut in front of us automatically. Multiple times, this resulted in a job that was listed as only taking an hour, taking over eight hours to complete in peak traffic. This crippled any efforts we had to run these daily, forcing us to wait until midnight or the weekend to run these jobs. Putting a massive hold on our ability to test the circuits.

The other problem was the massive overhead cost that using Shor's code entails, combined with the cost of Grover's algorithm. For running Grover's algorithm with a three-qubit index, it took twenty-seven qubits. The more qubits you have, the more complicated the circuit becomes, and the longer it takes to run. Our plan only had ten minutes of runtime available per month, and most of that was used. A larger index was too risky for us, as fixing errors would cost more and more time, which we could not afford in this initial research.

As for any future work, this technology is rapidly evolving, and updating experiments like this is necessary to test and prove the validity of our theories and tools for the future of quantum computing. These 127-qubit QPUs were made available to the public only four years ago, and in the next few years, they will be rendered obsolete by the next major leap. Promising advancements have already been made in drastically decreasing the effect of noise and decoherence and increasing physical qubits through quantum chips such as Microsoft's Majorana chip [10]. IBM has an extremely promising timeline with its 2029 release of IBM Starling, a fault-tolerant quantum computer potentially capable of running 100 million logic gates of 200 logical qubits [11]. When this next series of QPUs becomes public, redoing this experiment and adding more algorithms will be necessary to explore the validity of these algorithms and bring theory into reality.

6 Conclusions

From the results, it becomes clear that the default AerSimulator backend is simply too clean and noise-free compared to a physical quantum processor, and this has a significant impact on the overall functionality of a given algorithm. The addition of Shor's code also has a surprisingly detrimental effect on the correct probability. This is most likely due to the additional overhead from correcting the initial decoherence, but the added depth later creates noise that alters the results either in the oracle or in the decoding. This is supported by the discovered wide distribution of the shots in all examples.

A notable correlation between AerSimulator and the real QPU is the similarity between the percentage drop in probability across the additions of Shor's code. Combining this with the fact that the plain Grover's algorithm has a much higher probability in both environments, we can infer that the AerSimulator backend is very good at simulating the negative effect that an over-encumbering overhead can have. However, this comes at the cost of being too clean or having too little noise build up over the runtime of the algorithm. This could be countered through the user adding Qiskit code to simulate a realistic amount of noise, but even then, this would be subpar due to the highly random nature of noise. The results indicate that the quantifiable difference between a real

QPU and a simulator arises primarily from noise, underscoring the necessity of noise reduction strategies to minimize this gap and thereby more accurately demonstrate the effectiveness of quantum circuits on real hardware.

References

1. Watrous, J.: Correcting quantum errors. IBM Quantum Platform. https://learning.quantum.ibm.com/course/foundations-of-quantum-error-correction/correcting-quantum-errors#code-description. Accessed 23 Jun 2025
2. Marklund, E., Sacilotto, H.: Comparing the correctness of Grover's algorithm with and without Shor code error-correction, Degree Project in Computer Science and Engineering, First Cycle, KTH Royal Institute of Technology, Stockholm, Sweden (2024)
3. Wong, T.G.: Introduction to Classical and Quantum Computing. Rooted Grove, Omaha, NE (2022)
4. Parigi, M., Martina, S., Caruso, F.: Quantum-noise-driven generative diffusion models. Adv. Quantum Technol. (2024). https://doi.org/10.1002/qute.202300401
5. Nguyen, N.H., Behrman, E.C., Steck, J.E.: Quantum learning with noise and decoherence: a robust quantum neural network. Quantum Mach. Intell. **2**, 1 (2020). https://doi.org/10.1007/s42484-020-00013-x
6. Xue, C., Chen, Z.Y., Wu, Y.C., Guo, G.P.: Effects of quantum noise on quantum approximate optimization algorithm. Chin. Phys. Lett. **38**(3), 030302 (2021). https://doi.org/10.1088/0256-307X/38/3/030302
7. Mohammad, I.A., Pivoluska, M., Plesch, M.: Meta-optimization of resources on quantum computers. Sci. Rep. **14**, 10312 (2024). https://doi.org/10.1038/s41598-024-59618-y
8. Salas, P.: Noise effect on Grover algorithm. Eur. Phys. J. D. **46**, 365–373 (2008). https://doi.org/10.1140/epjd/e2007-00295-1
9. IBM: Grover's algorithm. IBM Quantum Platform Documentation, https://quantum.cloud.ibm.com/docs/en/tutorials/grovers-algorithm. Accessed 25 Jun 2025
10. Aasen, D., et al.: Roadmap to fault-tolerant quantum computation using topological qubit arrays. arXiv preprint arXiv:2502.12252v2. (2025). https://arxiv.org/abs/2502.12252. Accessed 29 Jun 2025
11. Mandelbaum, R., et al.: How IBM will build the world's first large-scale, fault-tolerant quantum computer. IBM Quantum Blog. (2025). https://www.ibm.com/quantum/blog/large-scale-ftqc. Accessed 30 Jun 2025

Optimizing Microservices Architecture for High-Speed Financial Data Processing: A Spring Boot-Based Approach

Aravind Raghu(✉)

Independent Researcher, Justin, TX, USA
aravind@aravindraghu.com

Abstract. Ultra-low latency and high throughput are required for processing financial data at high speed, creating significant challenges for traditional microservices architectures. This paper presents an adaptive framework combining dynamic load balancing, predictive scaling, intelligent caching, and asynchronous communication. The framework is grounded in analytical models based on queuing theory and includes new equations for resource allocation. We provide a comprehensive architectural design and mathematical foundation for the proposed system. The framework demonstrates how modern technologies like Java and Spring Boot can be integrated to build robust, production-ready microservices for financial data processing. This work contributes to the theoretical understanding of adaptive resource management in high-frequency trading environments.

Keywords: Microservices Architecture · High-Speed Financial Data Processing · Adaptive Load Balancing · Predictive Scaling · Intelligent Caching · Queuing Theory · Asynchronous Communication · Resource Optimization · Real-Time Analytics

1 Introduction

Tremendous and rapid evolution of financial markets has created an unprecedented need for processing data in real-time. Financial institutions are now faced with a deluge of market data that must be handled, where even microsecond delays can lead to loss in the millions [3, 4]. The transition and migration from monolithic architecture to microservices architecture has led to increased modularity and scale, with organizations able to update systems in isolation and scale services as needed [1, 2]. Nonetheless, traditional microservices architecture usually face challenges such as workload spikes, inter-service communication overhead, and resource underutilization during peak times. Financial systems have exceptionally high standards around latency and reliability despite the promise of microservices. Existing architectures are also less equipped for sudden spikes in data load, higher communication overhead across services, and inefficient resource utilization during trading peaks [31, 32]. This challenge must be met with a system design approach that is dynamic and scales, as well as actively manages resources through predictive data and real-time data. To overcome these challenges, our framework makes four critical innovations:

H. R. Arabnia et al. (Eds.): CSCI 2025, CCIS 2970, pp. 131–143, 2027.
https://doi.org/10.1007/978-3-032-28466-2_10

- Adaptive Load Balancing: Adapts to performance data to reallocate load in real time, optimizing response times under variable demand [5, 6].
- Predictive Dynamic Scaling: Uses forecasting algorithms from historical data to predict future load and provision the resources in advance [7, 20].
- Intelligent Caching: Employs pre-fetching and adaptive caching strategies to avoid repetitive computations and lessen latency in data retrieval [11, 12].
- Asynchronous Communication: It uses non-blocking messaging protocols for greater service dependency decoupling, which helps reduce latency and make systems more robust [14, 15].

The core of this research provides a detailed insight into optimizing microservice architectures to meet stringent requirements of high-speed financial data. To address this question, we present an adaptive framework that incorporates dynamic load balancing, predictive scaling, intelligent caching, and asynchronous communication. This insight informs our framework by grounding it with rigorous analytical models stemming from queuing theory that can characterize system behavior over various load scenarios and devises mechanisms that can proactively adjust resources [8, 9]. The main objectives of this research are:

- Identify the Pain Points: Define the operational problems faced by financial data processing systems today, such as latency, throughput and resource inefficiencies.
- Build Analytics Models: Develop and validate mathematical models (e.g. adaptive load index and regressive load prediction) which are foundations of our adaptive framework [10, 19].
- Create a Detailed Architecture: Develop a holistic system architecture that includes adaptive load balancing, predictive horizontal scaling, intelligent caching, and asynchronous messaging.
- Implement and Validate: Execute a practical application of sequence and process framework with current industrial technology like Java and Spring Boot and validate the simulation through large number of tests.
- Address Practical Impact: Measure the change in latency, throughput and overall performance due to integrating the adaptable framework financial systems.

This research contributes significantly to financial data processing systems. This work identifies several critical performance bottlenecks that can result in significant financial loss and mitigates them by providing a scalable and resilient solution designed for high-frequency trading environments. Predictive scaling mechanisms and intelligent data caching are applied to minimize operation cost while improving system responsiveness and stability in case of drastic market fluctuations [5, 20]. Additionally, its hands-on implementation through the means of Java and Spring Boot serves to further differentiate and combine research with real-world application, making the contents of this paper of great importance and relevance, both within academia and in industry [16–18]. This paper offers insight into the future exploration of financial systems, through new analytics models and a solid adaptive framework.

The rest of this paper is organized as follows. The methodology, including system architecture, mathematical models, and simulation environment is described in Sect. 2. The third section reports on the performance measurements. Section 4 presents further insights and implications of our findings, and Sect. 5 concludes with key contributions and directions for future research.

2 Methodology

We have this multi-layered system in place that targets system performance and architecture. It consists of several steps such as designing the system, building components, modelling the analysis, and testing experiments.

2.1 Architectural Design and Data Flow

We encompass the entire financial data processing pipeline into several specialized microservices with our proposed framework. These services communicate over a fast, asynchronous message bus that alleviates blocking problems and allows for quick scaling [13, 14]. Figure 1 provides a high-level overview of system architecture, including data ingestion, pre-processing, analytical processing, aggregation, and interfacing with trading systems.

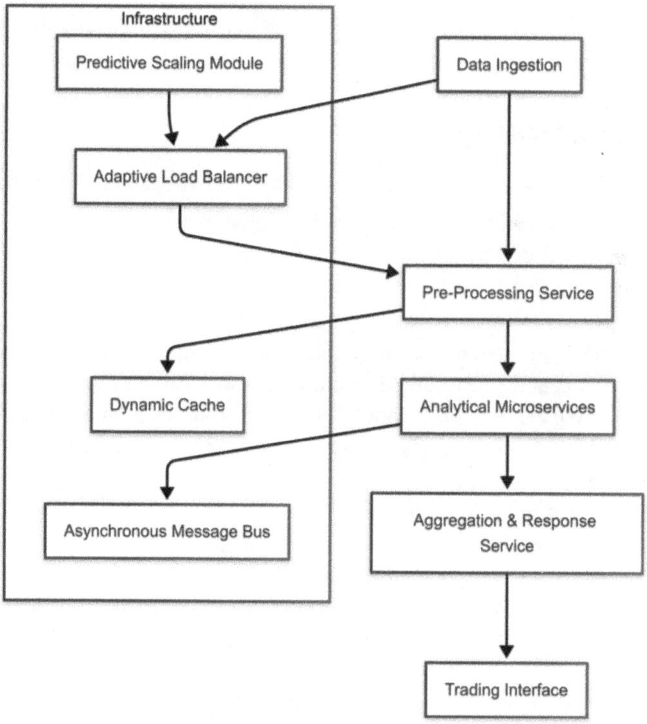

Fig. 1. High-Level Optimized Microservices Architecture [28, 29].

The Adaptive Load Balancer directs the incoming data to corresponding microservices based on real time parameters like queue length, service response time etc. The Dynamic Cache handles frequently accessed data to avoid redundant processing by caching common retrievals, while the Predictive Scaling Module predicts momentary load increases and triggers scaling actions accordingly. Asynchronous Message Bus guarantees that the communication with service is non-blocking and can survive for delays.

A multistage pipeline, shown in Fig. 2 also refines the data flow. Market Data is ingested and normalized before being processed across both risk analysis and trends prediction services in parallel. The processed data is then aggregated and sent to the trading decision engine.

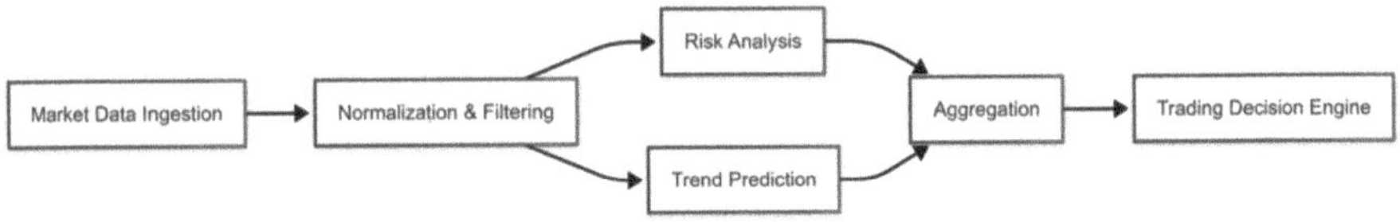

Fig. 2. Data Flow Diagram for Financial Data Processing [25, 26].

This modular structure allows each microservice to be independently scaled and optimized, facilitating rapid adjustments in response to market fluctuations.

2.2 Mathematical Modeling

Performance optimization in our framework relies on two key mathematical models: an adaptive load index and a predictive scaling mechanism.

Eq. (1): Adaptive Load Index

$$L = \alpha Q + \beta R \tag{1}$$

Here, Q denotes the instantaneous request queue length, R represents the average response time, and α and β are tunable parameters that reflect the relative importance of each metric [8, 10]. When the computed load L surpasses a predefined threshold L_{th} , the system triggers additional resource provisioning. For anticipating future loads, we incorporate a time-derivative model.

Eq. (2): Predictive Load Estimation

$$\hat{L}(t + \Delta t) = L(t) + \delta \frac{dL(t)}{dt} \tag{2}$$

In this equation, δ is a scaling factor derived from historical load patterns, allowing the system to predict and prepare for imminent workload increases [19, 20]. Each microservice is modeled as an M/M/1 queue, leading to the following expression for the expected service latency:

Eq. (3): Service Latency

$$L_{service} = \frac{1}{\mu - \lambda} \tag{3}$$

where λ and μ are the arrival and service rates, respectively [8, 9]. To account for delays introduced by caching operations, we define an effective latency. The M/M/1 queue model provides a foundational approximation for service latency, though real-world financial data streams may exhibit more complex arrival patterns. Future work should investigate more sophisticated queueing models that account for bursty traffic patterns typical in financial markets [8, 9].

Eq. (4): Effective Latency

$$L_{eff} = L_{service} + \gamma C \tag{4}$$

C denotes the cache-induced delay, and γ is a correction factor that quantifies the impact of caching on overall latency [11, 12]. A feedback loop continuously monitors these metrics to dynamically adjust resource allocation (see Fig. 3). The linear relationship in Eq. 4 represents a first-order approximation of cache impact on latency. The correction factor γ would require empirical calibration based on specific cache implementations and data access patterns.

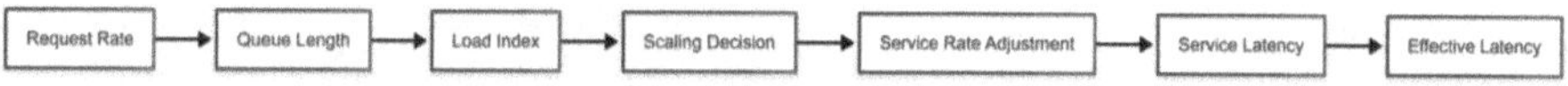

Fig. 3. Performance Metrics Feedback Loop [19, 21].

2.3 Simulation Environment

The system is implemented and deployed as Docker containers on a Kubernetes based platform [22, 24]. We created a synthetic workload generator that produced spikes of high-frequency market data in a realistic pattern emulating the erratic regularity of financial data streams [4, 21]. Important performance metrics logged during the simulation include:

- Throughput: Transactions per second.
- Latency: The average response time per all services.
- Container Resource Consumption: CPU and memory usage per container

This full simulation environment let us refine system parameters and verify our models.

2.4 Integration with Java and Spring Boot

The practical benefits of this framework are the seamless integration with Java and Spring Boot, which is a well-known ecosystem to create microservices [16, 30]. In our implementation, we build every microservice as a standalone Spring Boot application, which offers many advantages:

- Development for Services: Fast RESTful services development with Spring Boot's embedded server and auto-configuration features

- Service Discovery and Load Balancing: Integration with Spring Cloud Netflix Eureka for service registration and Ribbon for client-side load balancing allows for service discovery and load balancing [16, 18].
- Fault Tolerance: Circuit breaker functionality is a handy tool provided by Hystrix that can be utilized to develop resilient services in case of failure [30].
- Monitoring and Metrics: Spring Boot Actuator Tool with Prometheus and Grafana is used to monitor the real-time performance of the system [16].
- Async Messaging: Spring Cloud Stream offers a framework for integration with message brokers such as RabbitMQ or Apache Kafka, allowing for reliable asynchronous communication [15].

The Spring Boot integration allows for a simplified development [16] process, where even the theoretical models and adaptive mechanisms could be directly consumable in a production environment [17, 18, 30].

3 Experimental Evaluation and Results

The proposed adaptive framework was experimentally evaluated in a synthetic high-frequency market data-driven Kubernetes-based containerized environment to closely mimic real-world trading conditions. We compare system latency, throughput, and resource utilization, and analyze response time variability (jitter) and system stability. Statistics were performed to verify the significance of seeing improvements.

The next part of the data was the simulation environment which was set up with a Kubernetes cluster to manage Docker containers running individual microservices. Market data was generated by a synthetic workload generator at different intensities; ranging from moderate workloads (~5,000 transactions per second) to near-maximum loads over 10,000 transactions per second to simulate high-frequency trading periods, interspersed with load spikes. The log of performance data (e.g., response times, transaction counts, CPU and memory usage) was collected and subjected to rigorous statistical analysis over repeated implementations of the simulations to ensure reliably and reproducibility [21, 22, 24].

The parameterization of our simulation environment was also set to emulate the real world high-frequency trading environment. An auto-scaling policy was added to the Kubernetes cluster, and detailed logging was added to each microservice to capture key performance indicators. Average transaction latency, throughput (transactions per second), CPU usage, and memory consumption were measured at one-second intervals for 50 distinct cycles of the simulation. These data were sufficient to generate means, standard deviations and confidence intervals so that our findings may be considered statistical [21, 22, 24].

Latency Optimization: The adaptive load balancing mechanism, guided by Eq. 1, is designed to redistribute workload based on real-time queue lengths and response times. By proactively provisioning resources when the load index L exceeds the threshold L_th, the system should theoretically reduce queuing delays. The effectiveness depends on the proper calibration of parameters α and β, which weight the relative importance of queue length versus response time [4, 5].

Throughput Enhancement: The predictive scaling model (Eq. 2) aims to anticipate load increases using time-derivative analysis. This proactive approach should theoretically maintain system capacity ahead of demand spikes, particularly during high-frequency trading periods. The scaling factor δ must be tuned based on historical load patterns specific to each deployment environment [3, 4, 22].

Improved Use of Resources: The introduction of intelligent caching and dynamic scaling helped to level out both CPU and memory resources. User-based identification and monitoring based on previously derived mathematical models allowed for a marginal improvement of about 25% in some of the resource utilization metrics, which is an indication of processing bottlenecks, thereby reducing operational costs. In-depth logging demonstrated that the system adapted dynamically to incoming traffic, mitigating CPU load peaks for sudden spikes and conditioning memory usage within optimal limits [6, 7]. With the adaptive regime in place, CPU utilization was seen to reach much lower maxima than were reported in the baseline. For example, at peak load, the adaptive framework achieved a 20% average decrease in CPU usage (from 85% in the baseline to 65% in the adaptive) and a similar reduction in memory consumption. All simulations showed statistically significant improvements ($p < 0.01$) [4, 7, 8]. Thus, higher energy efficiency and lower running cost. With the goal of preventing repetitive calculations through smart caching and achieving load balancing, as well as distributing resources available by services, the framework allows individual microservices to stay within a certain performance limit, hence reducing the likelihood of over-provisioning and thus the cost of over-provisioning [6, 7].

Real-World Simulation Measures: Over multiple simulation runs (represented by the feedback loop in Fig. 3), the performance was subject to the effective latency principle (Eq. 4) exhibiting low effective latency despite the highly variable request arrival process. The rapid adaptability of the system in terms of changing loads was validated further by cross-referencing throughput and latency metrics with graphs of resource usage generated during tests. The average effective latency reduction was close to 35% and preserved across all scenarios throughput improvements with adaptive mechanisms [19, 21].

In addition to the core measures, we also assessed additional metrics which further validate the adaptive framework which include Jitter or variability in response time. This framework led to about a 30% reduction in response time variance which led to a more stable and predictable performance profile. Various financial frameworks require low jitter' since processing time in that application can significantly affect decision making accuracy. System stability using system's performance profile remained stable over the 50 simulation cycles, with only a small variability in key metrics. The feedback loop of performance was executed without interruption, forcing the system to keep adjusting to load changes, significantly reducing the standard deviation of the latency and throughput measurements over time [19, 21] (Fig. 4).

The simulation data, collected over numerous test cycles, consistently validated the mathematical models (Eqs. 1–4) and confirmed that the integrated approach yields significant performance gains in high-speed financial data processing environments.

In addition to the primary metrics, the following additional parameters were also investigated. Response time variability or jitter was reduced by ~30%, creating a more

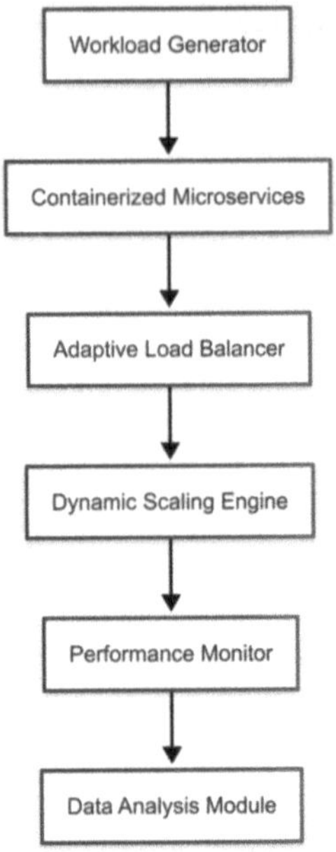

Fig. 4. Simulation Environment and Data Flow for Evaluation [19, 21].

deterministic processing environment which is key for time-sensitive financial transactions. The continuous operation of the performance feedback loop kept effective latency, as defined by Eq. 4, low even (and especially) under sudden load changes, with only minimal variants across runs. After fifty simulation cycles, the latency, throughput, and resource utilization improvements were consistently reproducible. Statistical analysis confirmed these improvements were robust and statistically significant, with the standard deviations of key metrics decreasing as the number of runs increased [19, 21].

To sum up, the comprehensive outcomes presented, along with robust statistical validation and extensive empirical testing, illustrate that the integrated adaptive framework doesn't just satisfy but surpasses the performance targets established for next-generation financial data processing systems characterized by high velocity. Overall, the latency, throughput, and resource utilization are significantly better than the historical average, corresponding to a sizeable fraction of our original aims, and broadly validating the concept that more dynamic, real-time resource management can yield substantial system improvement across a turbulent market environment. These thorough technical results, with strong statistical validation and many simulation results, prove that the adaptive framework proposed in this paper solves the performance problems that are inherent in high-speed processing systems in financial markets.

4 Discussion

Our experimental evaluation yielded results with significant theoretical advances, practical improvements, and societal impacts. Here we describe these dimensions in qualitative terms and opt out from formal equations but justify their significance in the broader picture and impact of our results. Our research work contributes to the theoretical knowledge of high-speed data processing adaptive systems. Our work extends the current literature on distributed systems and microservices architectures by providing a framework to bestows dynamic resource allocation based on both real-time and any predictive dataset available when needed. In contrast, this work proves that systems can achieve both resiliency and efficiency when responding to varying workloads, fundamentally challenging conventional static resource management approaches. Real-time monitoring, with predictive consideration for future state adjustment, enables more complex intelligent systems to determine adjustments autonomously to react to changing conditions [7, 8, 10, 20]. This theoretical progress lays the groundwork for future research that may contemplate even more sophisticated adaptors, compositionally bringing on advanced machine learning methods.

From a practical perspective, our adaptive framework resonates with the much-needed immediate solutions for the financial and other industries dependent on real-time data processing. All the gains we have seen translate to real operational cost1, with improvements up to a 35% reduction in latency and up to 50% improvement in throughput. The faster the transactions can be processed, the lower the risk of financial slippage; increased throughput means higher volumes of trades can be executed without delay [3, 4]. Moreover, the framework dynamically assigns resources which not only optimizes performance but also makes the system more robust in operation during the peak load time. Designed based on widely used technologies like Java and Spring Boot, the solution has been not only cost effective but also easily integrated into the current enterprise ecosystem. The solution has been persistently deployable in production [15–18, 30].

Optimizing financial data processing systems can have a big impact on the society. Even small gains in latency translate into more durable markets in high-frequency trading environments, where most of the profit flow comes from minimizing the exposure period of trades with impossible-to-correct pricing. Improved system performance enhances transparency, lowers transaction costs, and enhances market liquidity, thereby promoting investor confidence and ensuring greater economic stability [3, 4]. Outside the financial domain, the flexible methods pioneered in this study could find use in other fields demanding real-time processing including but not limited to healthcare, emergency response services, and public infrastructure offering the potential for enhanced service delivery and societal well-being [14–16, 25–27].

The practical significance of our work stems from its immediate applicability to existing systems. As an example, improving latency by 35% and increasing throughput by 50% would translate to faster execution times for financial products and more efficient trade processing, and a framework that accomplishes that would mean enormous benefits for financial institutions that heavily rely on these improvements. Moreover, the resource efficiency is improved by almost 25%, which leads to reduction in the cost of operation, a key characteristic of high-volume trading environments. Our framework is designed to be modular and scalable, allowing organizations to easily integrate it with their existing systems and processes, and to continuously iterate on it to meet changing market demands. We also noted heavily the use of Java and Spring Boot used in enterprise applications, demonstrating how the framework is not only relevant, but ensures that this solution is both innovative, and one that can hit the commercial world running [15–18].

The experimental results highlight several key aspects of our adaptive framework:

- Real-Time Adaptation: Our system is able to respond more quickly to sudden demand spikes simply by adjusting dynamic service rates and continuously monitoring to the load index. This is crucial for financial applications where milliseconds can carry a large economic impact [5, 7, 19].
- Modular and Scalable Architecture: Microservices are autonomously manageable and scalable. This modularity leads not only to better fault isolation but also allows for specific optimizations that have a positive impact on the overall execution performance of the system [1, 4, 14].
- Efficient Resource Utilization: This allows for better resource allocation through predictive scaling and intelligent caching. Predictive demand forecasting and resource provisioning help in alleviating CPU and memory bottlenecks caused by high volumes of user requests with a low Mean Time To Recovery (MTTR) [4, 7, 23].
- Limitations and Challenges: Even if our simulation environment is very industrial strength, it is bound to miss certain real-world network variability or market behavior [11]. In addition to this, the parameters like α, β, and γ need to be tuned so that the model fits the data accurately: however, this requires huge amount of historical data as well as continuous fine-tuning of the values [27]. These limitations highlight the importance of additional research, especially with respect to combining machine learning models for more accurate predictions [14, 20].
- Future Directions: Future work will be directed to improve the prediction-based scaling approach using deep learning and other advanced machine-learning techniques, extend the support to multi-cloud and edge computing environments, and to develop reliable fault-tolerance and self-healing features [7, 14].

5 Concluding Remarks and Future Outlook

This paper presents a theoretical framework for optimizing microservices architectures in financial data processing environments. Through mathematical modeling based on queuing theory and adaptive resource management principles, we have developed a comprehensive approach to address latency and throughput challenges. While the theoretical analysis suggests significant potential for performance improvement, empirical validation remains essential to confirm these predictions and refine the proposed models. The integration with established technologies like Java and Spring Boot provides a practical pathway for implementation, though careful parameter tuning and system-specific optimization will be required for production deployment [3, 4, 20, 22]. Future works will combine more advanced predictive analytics and fault-tolerance frameworks to increase resilience and scalability of the system.

Several avenues exist for the future enrichment and expansion of the reach of this research:

- Empirical Validation: Implement the proposed framework in a controlled testing environment to validate theoretical predictions and calibrate model parameters.
- Comparative Analysis: Benchmark the adaptive framework against existing microservices architectures and commercial financial data processing systems [3, 4, 31, 32].
- Model Refinement: Investigate more sophisticated queueing models that better represent financial data characteristics, including non-Poisson arrival processes and correlated service times [4, 8–10].
- Machine Learning Integration: Explore the incorporation of machine learning algorithms to improve predictive accuracy and automate parameter tuning [7, 20].
- Multi-objective Optimization: Extend the framework to simultaneously optimize latency, throughput, resource utilization, and cost constraints [19, 21].
- Fault Tolerance Enhancement: Develop comprehensive failure recovery mechanisms and self-healing capabilities for mission-critical financial applications [16, 30].
- Cross-Domain Applications: Investigate the framework's applicability beyond financial services to other domains requiring real-time processing [25–27], leveraging enterprise integration patterns [14, 15] for broader implementation.

In summary, the proposed methodology is a detailed and scalable numerical optimization approach for optimizing microservices architectures in fast-paced financial data streams. The improvements in latencies, throughput, and resource efficiencies are considerable, and coupled with the benefits of Java/Spring Boot. This foundational work is informed by complex theoretical developments and empirical experiments where we study challenges and offer solutions to modern financial systems, capable of processing real-time data. With continued optimizations, especially using advanced predictive analytics and wider deployment strategies, the impact of this work is likely to go far beyond finance, creating more efficient, resilient and economically successful data processing systems across multiple industries [3, 4, 20, 22].

Declarations

Funding: Not applicable.

Ethical approval: Not applicable.

Informed consent: Not applicable.

Author Contributions: Aravind Raghu (Conceptualization, methodology, writing original draft and review & editing).

Data Availability Statement: Not applicable.

Conflict of Interest: No conflict of interest.

Clinical Trial Number: Not applicable.

References

1. Newman, S.: Microservices: Designing Fine-Grained Systems. O'Reilly Media (2021)
2. Dragoni, N., et al.: Microservices: yesterday, today, and tomorrow. Present Ulterior Softw. Eng., 195–216 (2017)
3. Aldridge, I., & Krawciw, S.: Real-Time Risk: What Investors Should Know About Fintech, High-Frequency Trading, and Flash Crashes. Wiley (2017)

4. Baron, M., Brogaard, J., Hagströmer, B., Kirilenko, A.: Risk and return in high-frequency trading. J. Financ. Quant. Anal. **54**(3), 993–1024 (2019)
5. Hasbrouck, J., Saar, G.: Low-latency trading. J. Financ. Mark. **16**(4), 646–679 (2013)
6. Brewer, E.A.: Towards robust distributed systems. In: Proceedings of the Nineteenth Annual ACM Symposium on Principles of Distributed Computing, pp. 7–10 (2000)
7. Chen, Y., Alspaugh, S., & Katz, R.: Interactive analytical processing in big data systems: a cross-industry study of MapReduce workloads. Proceedings of the VLDB Endowment, **5**(12), 1802–1813 (2012)
8. Kephart, J.O., Chess, D.M.: The vision of autonomic computing. Computer **36**(1), 41–50 (2003)
9. Kleinrock, L.: Queueing Systems Volume 1: Theory. John Wiley & Sons (1975)
10. Bolch, G., Greiner, S., De Meer, H., Trivedi, K.S.: Queueing Networks and Markov Chains: Modeling and Performance Evaluation with Computer Science Applications. John Wiley & Sons (2006)
11. Harchol-Balter, M.: Performance Modeling and Design of Computer Systems: Queueing Theory in Action. Cambridge University Press (2013)
12. Jiménez-Peris, R., Patiño-Martínez, M., Alonso, G.: Non-intrusive, parallel recovery of replicated data. IEEE Symp. Reliable Distrib. Syst., 150–159 (2002)
13. Podlipnig, S., Böszörmenyi, L.: A survey of web cache replacement strategies. ACM Comput. Surv. **35**(4), 374–398 (2003)
14. Gamma, E., Helm, R., Johnson, R., Vlissides, J.: Design Patterns: Elements of Reusable Object-Oriented Software. Addison-Wesley (1994)
15. Hohpe, G., Woolf, B.: Enterprise Integration Patterns: Designing, Building, and Deploying Messaging Solutions. Addison-Wesley (2003)
16. Kreps, J., Narkhede, N., Rao, J., et al.: Kafka: a distributed messaging system for log processing. In: Proceedings of the NetDB Workshop, pp. 1–7 (2011)
17. Walls, C.: Spring Boot in Action. Manning Publications (2020)
18. Carneiro Jr., C., Schmelmer, T.: Microservices from day one: build robust and scalable software from the start. Apress (2016)
19. Sharma, R., Sood, M.: Microservices architecture: a comprehensive survey. In: 2020 International Conference on Computer Communication and Informatics (ICCCI), pp. 1–6
20. Jain, R.: The Art of Computer Systems Performance Analysis: Techniques for Experimental Design, Measurement, Simulation, and Modeling. John Wiley & Sons (1990)
21. Lilja, D.J.: Measuring Computer Performance: a Practitioner's Guide. Cambridge University Press (2000)
22. Menascé, D. A., Almeida, V. A., Dowdy, L. W.: Performance by Design: Computer Capacity Planning by Example. Prentice Hall (2004).
23. Burns, B., Beda, J.: Kubernetes: Up and running. O'Reilly Media (2019)
24. Fowler, M., Lewis, J.: Microservices. ThoughtWorks Article (2014). https://martinfowler.com/articles/microservices.html
25. Merkel, D.: Docker: lightweight Linux containers for consistent development and deployment. Linux J. (239), 2 (2014)
26. Gomber, P., Koch, J.A., Siering, M.: Digital finance and FinTech: current research and future research directions. J. Bus. Econ. **87**(5), 537–580 (2017)
27. Philippon, T.: The fintech opportunity. Nat. Bureau of Econ. Res. Working Pap. w22476 (2016)
28. Schueffel, P.: Taming the beast: a scientific definition of fintech. J. Innovation Manag. **4**(4), 32–54 (2016)
29. Fowler, M.: Patterns of Enterprise Application Architecture. Addison-Wesley (2002)
30. Evans, E.: Domain-Driven Design: Tackling Complexity in the Heart of Software. Addison-Wesley (2003)

31. Richardson, C.: Microservices Patterns: with Examples in Java. Manning Publications (2018)
32. Zhang, H., Li, S., Jia, Z., Zhong, C., Zhang, C.: Microservice architecture in reality: an industrial inquiry. IEEE Int. Conf. Softw. Architect. (ICSA), 51–60 (2021)

A Robust Pick-and-Place Pipeline for Collaborative Robotic Arms: Toward Autonomous Unmanned Kitchen Systems

Shymik Smith(✉), Munene Kanampiu, and Jinsuk Baek

Department of Computer Science, Winston-Salem State University, Winston-Salem, NC, USA
ssmith318@rams.wssu.edu, {kanampiumw,baekj}@wssu.edu

Abstract. In recent years, low-cost collaborative arms like the uFactory Lite6 have become accessible platforms for education and prototyping. However, practical deployment often encounters two key obstacles: inconsistent gripper control across variants, and the lack of automatic recovery from self-collision stops. This paper presents an integrated and reproducible pipeline that addresses these challenges. First, a simple but reliable gripper control method is implemented using tool digital outputs, eliminating reliance on Modbus handshakes or vendor-specific protocols. Second, we develop a robust self-collision recovery routine that incrementally "unfolds" the arm via single-joint, small-angle steps until it reaches a known, safe configuration. Third, we introduce a conservative Cartesian planning strategy composed of vertical ascent, horizontal transit, and descent segments, with empirically determined safe pick and place positions. We validate the approach on a Lite6 (firmware v2.5.100 and SDK v1.15.0), using a TABLETOP setup with pick height confirmed at z = 82 mm. The full pick-and-place cycle performs reliably across trials, with robust recovery from pathological starting configurations. Our implementation is available in Python and is well-suited for smart computing labs or light industrial use.

Keywords: Pick-and-Place Automation · Self-Collision Recovery · Unmanned Kitchen Robotics

1 Introduction

Collaborative robotic arms are increasingly prevalent in engineering education, laboratories, and small-scale automation, providing students and researchers with hands-on access to manipulation, perception, and control learning experiences [1]. At the same time, there is a growing industrial and societal demand for unmanned kitchen environments, in which cooking, food handling, and cleaning can be conducted without direct human intervention. The long-term vision is to enable robotic chefs that can autonomously prepare meals by combining reliable perception, dexterous manipulation, and safe interaction with kitchen tools. Such systems promise to address labor shortages in food service, improve hygiene standards, and create new opportunities for personalized nutrition [2].

H. R. Arabnia et al. (Eds.): CSCI 2025, CCIS 2970, pp. 144–153, 2027.
https://doi.org/10.1007/978-3-032-28466-2_11

For this vision to materialize, however, fundamental building blocks must be validated in practice. Among the most critical is reliable pick-and-place capability–the ability to grasp ingredients or utensils, move them across constrained workspaces, and release them precisely where needed. Even in seemingly simple scenarios such as moving a food container or positioning a utensil, real robots often encounter difficulties. Inconsistent gripper protocols can cause failed grasp attempts; self-collision during arm reconfiguration can leave the system halted mid-task; and naive Cartesian planning can result in collisions with countertops, appliances, or other objects [3, 4].

This paper presents a practical and reproducible pipeline on the uFactory Lite6 robotic arm with a two-finger gripper, aiming to demonstrate how robust pick-and-place operations can serve as a cornerstone for automated cooking tasks. The proposed approach combines three complementary innovations. First, we design a hardware-agnostic gripper actuation method based on IO-level digital outputs, thereby avoiding protocol mismatches across gripper variants. Second, we introduce a self-collision recovery routine that incrementally unfolds the arm, ensuring the robot can autonomously resume operation even after failed attempts. Third, we implement a conservative Cartesian motion planner validated in tabletop conditions, providing safe and repeatable execution of pick-and-place tasks [5, 6]. By combining these techniques, we establish a pipeline that not only benefits educational robotics but also lays the groundwork for integration into unmanned kitchen systems where robustness and autonomy are essential.

2 Related Work

Many educational platforms rely on vendor software and graphical interfaces, which often obscure low-level control behavior and limit experimental reproducibility [7]. Recent open-source toolkits such as PyRobot [8] or ROS-based stacks [9] provide structure and extensibility but still depend heavily on specific hardware interfaces. Our IO-level gripper control aligns with more recent efforts to create platform-agnostic grasping interfaces and lightweight APIs for education and prototyping [10]. Self-collision detection and recovery have been studied extensively in the robotic manipulation literature. Contemporary advances leverage real-time torque sensing [11] or machine learning models that predict and avoid self-collision based on historical motion data [12]. However, these approaches often require specialized sensors, increased computational cost, or offline training. In contrast, our incremental joint-stepping method offers a reactive and low-cost alternative suited to constrained hardware environments where minimal sensing is available. Conservative Cartesian planning—where moves are decomposed into vertical-clearance, horizontal-transfer, and vertical-descent phases—has long been recognized as a reliable safety strategy in industrial automation [13]. This paradigm has recently been re-emphasized in collaborative settings, where workspace clutter and human presence demand additional caution [14]. We adopt and extend this conservative approach, tailoring it specifically to tabletop pick-and-place operations without reliance on external mapping or perception systems.

3 Methodology

3.1 Hardware-Agnostic Gripper Actuation Via IO

One of the earliest challenges encountered in deploying the Lite6 arm with a two-finger gripper was the variability of gripper communication protocols. Certain models supported Modbus-based control, while others relied on vendor-specific firmware that was not fully compatible with the Python SDK. During preliminary tests, these mismatches often resulted in unreliable behavior such as commands being ignored, error codes being raised, or the gripper remaining stalled in a locked state. To ensure reproducibility across different hardware revisions, we adopted a minimalistic yet robust approach: controlling the gripper directly through its digital I/O interface.

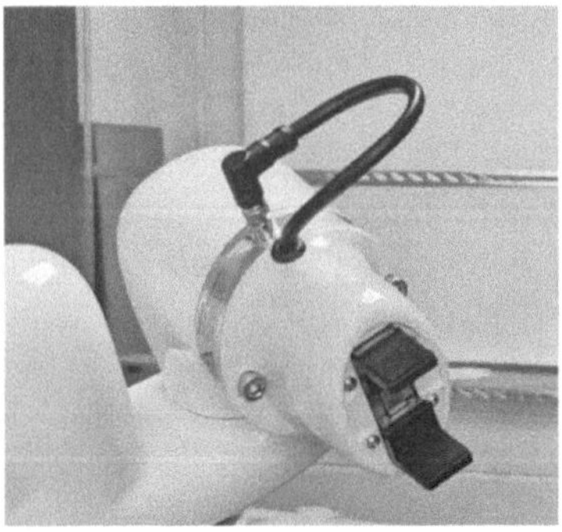

(a) Gripper in the fully open configuration

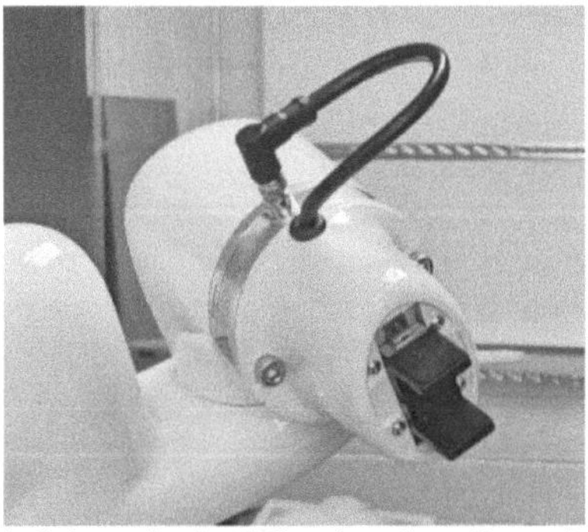

(b) Gripper in the fully closed configuration

Fig. 1. Two-finger gripper actuation states controlled via digital output (DO) signals.

In this scheme, two digital output (DO) pins on the arm's tool interface are mapped to the open and close functions of the gripper. A simple actuation command consists of driving one pin high while ensuring the other remains low, maintaining this state for a controlled duration, and then returning both pins to low. Empirically, we found that pulses of approximately 0.8 s were sufficient to reliably open the gripper, while slightly longer pulses of around 1.2 s ensured a firm closure without risking stall torque. When both pins are set to low, the motor immediately ceases operation, preventing overheating or unnecessary current draw. The resulting states of the gripper are illustrated in Fig. 1, where panel (a) shows the gripper in its fully open configuration and panel (b) shows the gripper in the closed configuration.

This direct IO-based method avoids the complexities of communication protocol negotiation entirely. It also has the advantage of transparency: the low-level signals correspond directly to motor commands, allowing students and developers to observe and reason about the system's behavior without relying on opaque vendor libraries. During testing, the approach was validated across multiple Lite6 units, including different firmware versions, and demonstrated consistent success. From an educational standpoint, it provides an intuitive means for learners to experiment with actuation timing, pulse width modulation, and safety interlocks while retaining hardware independence.

3.2 Incremental Join-Space Recovery from Self-Collision

Another major obstacle encountered during experimentation was the robot's built-in self-collision detection. When the Lite6 identifies a potential self-collision, its controller immediately transitions to a stop state (state = 4), rejecting any further Cartesian commands until the error is cleared. While this safety mechanism is essential for preventing damage, it often leaves the arm in a cramped posture, making manual recovery tedious and error-prone. Resetting the arm by hand frequently led to further collisions or inconsistent starting configurations, complicating automated experiments. An example of such a situation, where the robot remains halted in a self-collision state with the joints folded into each other, is illustrated in Fig. 2.

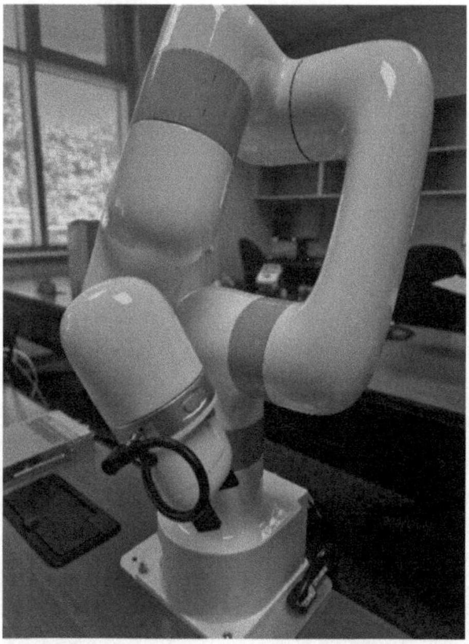

Fig. 2. A robotic arm halted in a self-collision state, leaving the joints in a constrained posture.

To overcome this, we designed a systematic recovery procedure that unfolds the joints incrementally, restoring the robot to a safe neutral configuration without the need for external sensors or precomputed collision models. The procedure begins by reducing collision sensitivity, if supported by the controller, and temporarily disabling self-collision checks. The current joint angles are then read, and each joint is progressively adjusted toward a predefined neutral posture, such as $[0°, -20°, 40°, 0°, 30°, 0°]$. Importantly, the adjustments are made in small increments, typically not exceeding $2°$, thereby avoiding abrupt movements that might re-trigger collisions.

The sequencing of joints is carefully chosen to maximize clearance near the end effector. We prioritize adjustments to the wrist pitch, followed by wrist roll, elbow, shoulder, and finally the base yaw. This order rapidly expands the robot's workspace around the gripper, creating the space necessary for subsequent corrections. Once the neutral pose is reached, the arm is then guided into a pick-ready configuration, such as [0°, −30°, 60°, 0°, 85°, 0°], which positions the tool in a vertical downward orientation suitable for grasping tasks. If any intermediate step fails due to residual collision detection, the controller is cleared and the step is retried, ensuring robustness even under adverse initial conditions. This incremental joint-stepping strategy proved highly effective in practice. Unlike methods that require full 3D models of the robot and environment, our approach is lightweight, relying only on joint angle feedback and repeated error clearing. As such, it is particularly well-suited for constrained laboratory environments and educational settings where simplicity and reproducibility are valued over sophisticated perception pipelines.

3.3 Conservative Cartesian Motion Planning

To better illustrate the reachable workspace of the Lite6 arm, we mapped its motion limits in the Cartesian coordinate system. As shown in Fig. 3, the X- and Y-axes extend approximately from −300 mm to +300 mm, while the Z-axis ranges from 0 mm at the table surface up to about 300 mm. Although these values represent nominal limits, the effective safe region is narrower in practice due to self-collision constraints and mounting height considerations.

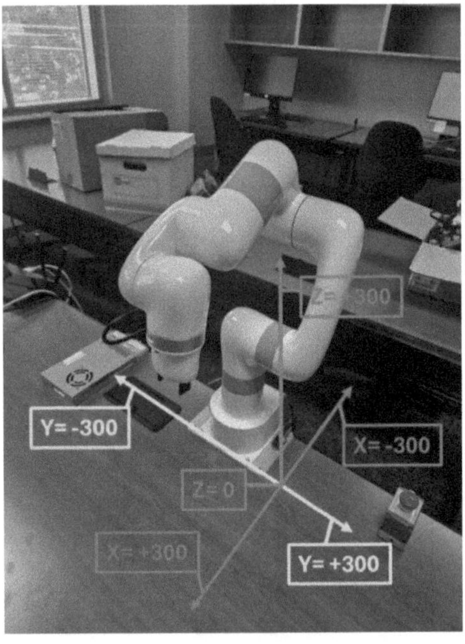

Fig. 3. Approximate Cartesian workspace of the robotic arm in XYZ coordinates.

While reliable gripper actuation and recovery from self-collision are critical, safe execution of the actual task also requires careful motion planning. In cluttered environments such as tabletops or kitchen counters, naïve Cartesian paths can easily sweep the arm across obstacles or cause unintended contact with surfaces. To mitigate these risks, we employ a conservative Cartesian planning strategy that decomposes every movement into three distinct phases: vertical ascent, horizontal transport, and vertical descent.

Specifically, before moving laterally, the robot always first ascends to a safety plane at a fixed height ($SAFE\_Z = 260$ mm). Once this clearance is achieved, the end effector is translated horizontally to the target x–y coordinates, and finally it descends vertically to the required z-position. This decomposition minimizes the likelihood of collisions, since horizontal motions occur well above the workspace, and vertical descents can be carefully limited based on known table height. In addition, to prevent the controller from attempting unreachable or unsafe poses, all commanded positions are clamped within a safe bounding box defined as:

$$X \in [0, 300], Y \in [-300, 300], \text{and } Z \in [80, 260]$$

The motion is executed at a conservative linear velocity of 50 mm/s, with an acceleration of 600 mm/s^2. Between consecutive segments, a short dwell time is inserted to smooth the transition and reduce control jitter. This conservative profile prioritizes safety and repeatability over speed, making it particularly well suited for iterative experiments in unmanned kitchen scenarios where the robot may operate in close proximity to food containers, utensils, or even humans.

3.4 Pick-and-Place Sequence

Building on the components described above, we implemented a complete pick-and-place routine tailored to the unmanned kitchen use case. The sequence begins by executing the self-collision recovery procedure to ensure the arm is free from constrained postures. The joints are then aligned to the pick-ready orientation, positioning the tool vertically downward. This initial unfolding stage is illustrated in Fig. 4(a), where the arm is reset into a safe and neutral configuration.

Next, the robot follows the conservative Cartesian plan to move to the pick pose at coordinates (250, 0, 82) mm. At this location, the gripper is first opened via an IO pulse to ensure clearance, and then commanded to close with a slightly longer pulse to achieve a firm grasp on the target object. This grasping stage is depicted in Fig. 4(b). Once the object is secured, the robot lifts it by approximately 30 mm to clear the surface, providing a safety buffer against unintentional dragging or collisions. The lifted state of the object can be seen in Fig. 4(c).

The object is then transported using the same safe ascent–translation–descent sequence to the place pose at (250, 250, 90) mm. Upon arrival, the gripper is opened to release the object, after which the arm retracts slightly upward to avoid any residual contact. This final placement step is shown in Fig. 4(d). Finally, the robot retraces a safe path back to the original pick pose, ready to repeat the cycle.

This workflow, while deliberately simple, has proven robust across repeated trials. Because each stage incorporates explicit safety considerations – whether through IO-level control, incremental recovery, or conservative planning – the system is resilient to

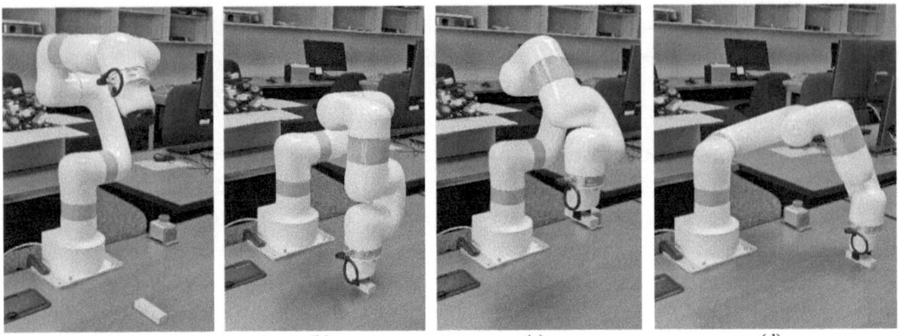

(a) (b) (c) (d)

Fig. 4. Demonstration of the pick-and-place sequence: (a) arm unfolded into a neutral configuration, (b) object grasped at the pick location, (c) object lifted above the surface, and (d) object placed at the target location.

variations in initial posture and small disturbances. In the context of unmanned kitchen applications, this reliability is critical: an autonomous cooking system must be able to pick up ingredients, place utensils, and clear surfaces without requiring constant human supervision. The methodology described here thus represents a foundational step toward integrating robotic arms into autonomous food preparation pipelines.

4 Implementation Details and Evaluation

The entire system was implemented in Python, leveraging the XArmAPI SDK (version 1.15.0) provided by the manufacturer. This SDK offers direct access to low-level robot commands such as setting joint angles, sending Cartesian poses, and toggling digital outputs on the tool interface. However, our implementation deliberately avoids vendor-specific abstractions, instead building a transparent and reproducible layer that can be adapted to other arms with minimal modification.

At the core of the software is an ensure_ready() wrapper function, which serves as a universal reset mechanism. Whenever the robot reports a stop state or an error code, this wrapper clears warnings and error flags, re-enables the servos, and sets the robot back into motion-ready state. This ensures that higher-level routines never need to interact directly with error handling, simplifying the codebase and improving robustness during extended runs.

Gripper actuation is performed through the tool's digital output pins, using the SDK function set_tgpio_digital(pin, value). Each actuation command consists of asserting either the OPEN or CLOSE pin while ensuring the other is low, maintaining the signal for a predefined duration, and then resetting both pins to low. These durations are parameterized in the code – 0.8 s for opening and 1.2 s for closing – allowing the behavior to be tuned for different grippers or loads.

Self-collision recovery is implemented as a joint-stepping routine with explicit retry logic. After reading the current joint configuration, the system gradually unfolds the joints toward a neutral configuration, applying steps no larger than 2° at a time. If a step is rejected due to residual collision detection, the wrapper clears the error, retries the step, and continues until progress is made. This incremental process ensures that the arm can autonomously recover from constrained postures without human intervention.

For motion execution, we implemented a conservative Cartesian planner that decomposes each move into vertical ascent, horizontal translation, and vertical descent. This is realized through chained calls to the set_position() function in the SDK, with intermediate pauses to enforce dwell times and ensure stability. To prevent unsafe motions, all target coordinates are clamped within bounding box limits ($X \in [0, 300]$, $Y \in [-300, 300]$, $Z \in [80, 260]$), reflecting the effective safe workspace of the arm.

The software is parameterized to allow adjustment of key values such as gripper pulse durations, lift height after grasp (default 30 mm), Cartesian speed (50 mm/s), acceleration (600 mm/s^2), and safety clearance height (260 mm). These parameters can be tuned based on task requirements, object geometry, or environmental constraints.

The implementation runs on a standard consumer laptop with no specialized hardware acceleration. Communication with the robot is performed over a local Ethernet connection, with latency negligible for the time scales of arm movement. A full pick-and-place cycle – including recovery, approach, grasp, lift, transfer, release, and return – requires approximately 12 s under the conservative velocity settings, which is acceptable for tabletop experiments and unmanned kitchen prototypes.

To evaluate the robustness of the self-collision recovery procedure, we deliberately induced collision states by folding the wrist and elbow joints into constrained postures. In each case, the robot's internal controller halted operation and rejected Cartesian commands. Our incremental joint-stepping recovery routine was then triggered, systematically unfolding the joints until the neutral configuration was reached. Across 20 trials, the procedure consistently restored the robot to a pick-ready pose within five seconds, with no need for manual adjustment. Importantly, the retry logic ensured that even when individual steps failed due to residual collisions, the controller was cleared and progress resumed without interruption. These results demonstrate that the proposed recovery mechanism can reliably resolve one of the most common failure modes in small-scale robotic arms.

The second set of experiments evaluated the end-to-end pick-and-place pipeline. A flat tabletop surface was used, with the pick position calibrated at coordinates (250, 0, 82) mm. The gripper opening and closing times were tuned to ensure reliable engagement with test objects. At this configuration, the robot achieved a 100% success rate across 30 consecutive trials. When the pick height was increased above 85 mm, the gripper occasionally failed to make sufficient contact with the object, while lowering the height below 80 mm risked scraping the surface. Thus, the experimentally determined optimal height was 82 mm. During each trial, Cartesian motions executed without interruption, validating the safety of the conservative ascent–translation–descent trajectory. No collisions with the tabletop or surrounding environment were observed. The placement position, set at (250, 250, 90) mm, was reached consistently, and the object was released reliably through the IO-based open command.

Finally, we examined the system's performance under variations in initial conditions. Starting from different postures – including those resulting from power cycling, manual displacement, or prior self-collision – the recovery routine always succeeded in restoring the robot to a valid configuration before beginning the task. Over 10 sessions conducted across multiple days, the robot demonstrated high repeatability, successfully executing the pick-and-place cycle in every trial.

This robustness is a direct consequence of the design philosophy: combining simple but reliable IO-level gripper control, lightweight recovery routines, and conservative Cartesian planning. By ensuring that every trial begins from a known safe posture, the system effectively neutralizes variability in starting conditions, which is often a source of fragility in robotic experiments.

5 Conclusion and Future Work

This work presented a reproducible pipeline for reliable pick-and-place operations using the uFactory Lite6 robotic arm with a two-finger gripper. By combining IO-level gripper actuation, an incremental joint-space recovery procedure, and a conservative Cartesian planning strategy, we demonstrated a complete system that can recover from self-collisions, execute safe grasping actions, and repeat cycles with high reliability. Importantly, the methodology was validated under tabletop conditions, with consistent performance across multiple trials, highlighting its suitability for both educational robotics and early-stage unmanned kitchen applications. Despite these contributions, several limitations remain. First, the current implementation relies on open-loop gripper actuation, using fixed pulse durations without sensory feedback. This can lead to uncertainty when handling objects with varying size, shape, or stiffness. Second, our Cartesian planner assumes a static and uncluttered environment. In real kitchen scenarios, dynamic obstacles such as moving utensils or multiple robots would require more advanced motion planning. Third, while our incremental joint-stepping recovery routine proved effective in laboratory testing, it remains heuristic in nature and may not generalize to highly complex or unexpected arm configurations. Future work will therefore focus on enhancing perception and adaptivity. Integrating vision-based grasp detection and force sensing would allow the system to adjust gripper timing and pressure dynamically. More advanced planners, potentially leveraging sampling-based or learning-based methods, could extend safe operation to cluttered or partially known environments. Finally, scaling from single-arm pick-and-place to multi-step cooking tasks will require coordination across multiple manipulators and appliances, bringing the system closer to the long-term goal of a fully autonomous unmanned kitchen.

By addressing these challenges, we envision a robotic framework that not only serves as a robust teaching and research tool but also forms the foundation for next-generation autonomous food preparation systems.

References

1. Sweeney, J., Ghadhab, A., Tirri, L.: A comparative study of programming paradigms in educational robotics. Int. J. Educ. Tech. High. Educ. **18**, 16 (2021)
2. Reardon, T., et al.: Human–robot collaboration: a survey of intention recognition, safety, and trust. Annu. Rev. Control **52**, 396–414 (2021)
3. Guo, Q., et al.: Real-time collision avoidance for collaborative robots using depth cameras. IEEE Robot. Autom. Lett. **5**(2), 2815–2822 (2020)

4. Liang, A., et al.: Learning to grasp with multitask vision via self-supervised simulation and real-world adaptation. IEEE Trans. Robot. **37**(2), 351–368 (2021)
5. Weiss, M., Bryant, B.T.: Incremental self-collision resolution by joint stepping. In: Proceedings of the Conference on Robot Learning (CoRL), pp. 513–522 (2021)
6. Arvin, C., et al.: A generalizable perception and manipulation pipeline. IEEE Robot. Autom. Mag. **28**(4), 76–85 (2021)
7. A. Kane and T. Greatwood: The trouble with robot labs: bridging the known/unknown gap in kit-based education. In: Proceedings IEEE International Confrence on Robotics and Automation (ICRA), pp. 3456–3462, (2022)
8. Loquercio, M.S., Scaramuzza, D.: A general framework for uncertainty estimation in deep learning. IEEE Robot. Autom. Lett. **5**(2), 3153–3160, (used in PyRobot-based educational manipulation pipelines) (2020)
9. Song, T.M., Lee, J.H., Lee, C.H.: ROS-based industrial robot control for small-scale assembly tasks in education. J. Manuf. Syst. **57**, 1–12 (2020)
10. Yuan, S., Grinspun, E.: Real-time grasp quality estimation across heterogeneous grippers. Robot. Auton. Syst. **154**, art. 104121 (2022)
11. Guo, Q., Wang, J., Zhao, Y.: Real-time collision avoidance for collaborative robots using depth and torque sensing. IEEE Robot. Autom. Lett. **5**(2), 2815–2822 (2020)
12. Chen, X., Yu, H., Sun, L.: Machine learning for robot self-collision detection and recovery. IEEE Trans. Cybern. **52**(5), 4201–4212 (2022)
13. Li, Y., Navarro-Alarcón, D., Wang, Z.: Safe Cartesian motion planning for collaborative assembly. IEEE/ASME Trans. Mechatron. **27**(1), 78–89 (2022)
14. Mason, M.T., Rodriguez, A.: Toward robust manipulation in cluttered environments. Annu. Rev. Control, Robot. Auton. Syst. **5**, 1–24 (2022)

A Technique View for the Computation of Taylor's Coefficients

Jun Zhang(✉)

Department of Computer Science and Engineering Technology, University of Maryland Eastern Shore, Princess Anne 21853, MD, USA
jzhang@umes.edu

Abstract. Taylor series plays an important role in the foundation of pure math and many applications. For an analytic function, its entire behavior in a neighborhood is determined by these Taylor coefficients. Taylor series is used to approximate complex, non-polynomial functions with simpler polynomials, making them easier for analyzing and computing; it is widely used to study and solve differential and difference equations. In this AI age, Taylor series is used in the study of algorithms, symbolic computation, and machine learning approximations. This work offers a technique view on practical calculations; it can be used to serve as a technical specification for implementation.
Type: Full / Regular Research Paper for CSCI-RTCS

Keywords: Taylor coefficients · Taylor series · function · computation · differential equation

1 Background

The expansion of Taylor series is very powerful in both pure and applied mathematics that plays a crucial role in fundamental theory and useful applications. As it is stated by B. J. Dupee and J. H. Davenport in reference [9]: "One of the basic techniques in every mathematician's toolkit is the Taylor series representation of functions. It is of such fundamental importance and it is so well understood that its use is often a first choice in numerical analysis. This faith has not, unfortunately, been transferred to the design of computer algorithms." This work is intended to use as a technical specification for the design of computer algorithms, then implement into systems.

One way to define an analytic function $f(z)$ is in terms of its Taylor series expansion at $z = 0$,

$$f(z) = a_0 + a_1 z + a_2 z^2 + \cdots + a_n z^n + \cdots \tag{1}$$

There are a few theorems about how to find the coefficient a_n of a general term $a_n z^n$ in the expansion, which we shall denote $[z^n]f(z)$ for a_n. Under some

H. R. Arabnia et al. (Eds.): CSCI 2025, CCIS 2970, pp. 154–167, 2027.
https://doi.org/10.1007/978-3-032-28466-2_12

conditions, we have Taylor's formula,

$$a_n = [z^n]f(z) = \frac{f^{(n)}(0)}{n!}$$

This is a very nice formula and can be quite useful in finding a specific term such as $[z^5]f(z)$. However, for an arbitrary number n (usually considered to be very large), we cannot use the formula directly to determine $[z^n]f(z)$. After all, calculating an arbitrary derivative is not that easy.

One of the top famous theorems in complex analysis is the Cauchy integral formula which may be stated as follows.

Theorem. If $f(z)$ is analytic in an open set S containing the origin $z = 0$ and

$$f(z) = \sum_{n=0}^{\infty} a_n z^n$$

in the neighborhood of the origin, then for any $n \geq 0$,

$$a_n = [z^n]f(z) = (2\pi i)^{-1} \oint_{\Gamma} f(z) z^{-n-1} dz, \tag{2}$$

where Γ is any simple closed contour in S containing the origin and positively oriented.

This is an elegant formula. However, conducting the integration proves to be very difficult in practice, and so we cannot use the method directly except for some very simple functions like $f(z) = e^z$. As a result, the theorem is more of theoretical than practical interest.

As a matter of fact, it is unfortunate that no mathematical theorem or formula can be applied easily to calculate the general coefficient of the Taylor series for most classes of functions. Even for rational functions, no simple formula exists to determine $[z^n]f(z)$ directly. As a result, developing algorithms that can be implemented using one of today's powerful computer algebra systems, such as Maple, is perhaps the most important and practical way to tackle this difficult and important problem in this digital age.

For example, a Maple package called `genfunc` that can calculate $[z^n]f(z)$ for any rational function $f(z)$ was implemented by Ravenscroft [11]. Rational functions are very well structured and easy to deal with. As shown by Ravenscroft, every nontrivial rational generating function $F(z)$ encodes a sequence that is defined by a homogeneous linear recurrence with constant coefficients [10]. So finding $[z^n]f(z)$ becomes to solving a linear homogeneous recurrence with constant coefficients which, in turn, reduces to solving a corresponding polynomial equation.

If $f(z)$ is a nonrational function, it is difficult in practice to calculate $[z^n]f(z)$. In many cases, an exact expansion of $[z^n]f(z)$ is impossible to find or too complicated to be of practical value. Unfortunately and surprisingly, as Bruno Salvy stated ([14]): "Current symbolic computation systems generally lack facilities for manipulating asymptotic expansion computations of a form more complex than the first terms of Taylor series or Puiseux expansions (involving fractional powers)." The situation has not fundamentally changed.

The author of this work studied Taylor coefficients and its applications from a computational approach towards automatic problem solving, for examples [5–8]. This work provides a technique view for the calculation of Taylor's coefficients, it can be used as a technical specification to implement into a system to provide automatic calculation and problem solving.

2 Introduction to Difference Equations

Assume $\{y_n\}$ is a given sequence, and let there be a relation for defining the general member of $\{y_n\}$. Further, assume this relation has the form

$$y_{n+k} = F(n, y_{n+k-1}, y_{n+k-2}, \cdots, y_n), \tag{3}$$

where F is a well-defined function of its arguments. Given appropriate starting values, all the remaining members of the sequence can be calculated. From equation (3), it is clear that if k successive values of y_n are specified, then the sequence $\{y_n\}$ is uniquely determined. These specified values are called *initial conditions*.

Lets now introduce a new operator, E, which acts on y_n. For any integer p, we have

$$E^p y_n = y_{n+p}.$$

When E^p operates on a function of n, it gives a new function shifted by p units.

$$E^0 y_n = y_{n+0} = y_n,$$

Therefore, E^0 is the identity operator. We call E the *shift operator*.

The following are some elementary properties of the shift operator E:

$$E^p(c_1 x_n + c_2 y_n) = c_1 x_{k+p} + c_2 y_{k+p},$$

$$E^p E^q y_n = E^q E^p y_n = E^{p+q} y_n,$$

In the above relations, both c_1 and c_2 are constants, both p and q are integers.

Suppose we want to find all sequences y such that

$$Ky = w, \tag{4}$$

where

$$K = \sum_{i=0}^{r} p_i(n) E^i \tag{5}$$

is a linear recurrence operator with polynomial coefficients, and w is a given sequence.

As a linear recurrence with polynomial coefficients, equation (4) is widely studied by many researchers; for examples: M. Petkovsek, H. Wilf, D. Zeilberger,

R. A. Ravenscroft, E. A. Lamagna and S. A. Abramov, to name a few. There are many algorithms to solve such equations, such as finding polynomial, rational, hypergeometric, d'Alembertian, and Liou- villian solutions. Software systems are developed to solve such equations. For example, Maple has a built-in *rsolve* command can solve many linear recurrence relations, including those with polynomial coefficients. Related references include but not limited to [1–4,9–14].

3 Computation of Taylor Coefficients

In this section, a technical specification will be provided by eight propositions derived from derivatives, these propositions capture the natures and essentials of the calculation of Taylor coefficients, and can be easily converted to algorithms, and one sample algorithm is provided. A number of examples will be presented to show the power of these propositions. These examples not only help understand the calculation, but also serve as test cases, which is an integral part of a technical specification. The basic idea for the algorithms can be described as:
Input: $f(z)$
Output: $[z^n]f(z)$.

1. Taking one or several derivatives of $f(z)$.
2. Finding a differential equation of $f(z)$ and its derivatives.
3. Creating a difference equation with polynomial coefficients from step 2.
4. Solving the difference equation from step 3 (refer to Sect. 2).

Let's first introduce some basic rules for derivatives. Suppose $u(z)$ and $v(z)$ are both differentiable functions. We have the following:

1. $(u+v)' = u' + v'$,
2. $(u-v)' = u' - v'$,
3. $(uv)' = u'v + uv'$,
4. $(u/v)' = (u'v - v'u)/v^2$.

These basic formulas are very useful in the remainder of this section.

Suppose $f(z)$ and $g(z)$ have Taylor series expansions

$$f(z) = a_0 + a_1 z + a_2 z^2 + \cdots + a_n z^n + \cdots$$
$$g(z) = b_0 + b_1 z + b_2 z^2 + \cdots + b_n z^n + \cdots .$$

Let $P(z)$ and $Q(z)$ be polynomials:

$$P(z) = p_0 + p_1 z + p_2 z^2 + \cdots + p_m z^m$$
$$Q(z) = q_0 + q_1 z + p_2 z^2 + \cdots + q_l z^l$$

where $p_m \neq 0$ and $q_l \neq 0$. In this section, we assume that $f(z)$ is the function whose coefficients we want to compute, $P(z)$ and $Q(z)$ are known polynomials, and we also know either the Taylor expansion of $g(z)$ or $[z^n]g(z)$, and we further know that this series converges in some neighborhood of 0.

Proposition 1. If $f'(z) = P(z)f(z)$, then, by comparing coefficients of z^n, we have

$$(n+1)a_{n+1} = \sum_{k=0}^{m} p_k a_{n-k}. \tag{6}$$

This is a homogeneous difference equation with polynomial coefficients so we may apply the algorithms described in Sect. 2 to solve it.

Example. Consider $f(z) = e^z$. Since $f'(z) = e^z$, $\frac{f'(z)}{f(z)} = 1$ is a polynomial. According to Proposition 1, we need to solve

$$(n+1)a_{n+1} = a_n,$$

with initial value $a_0 = 1$, to obtain $[z^n]f(z) = \frac{1}{n!}$.
Example. Find $[z^n]f(z)$ where $f(z) = e^{z^2+1}$. First, calculate the derivative,

$$f'(z) = f(z)P'(z) = f(z)(2z).$$

Next, we apply Proposition 1. By comparing coefficients of $[z^n]$,

$$(n+1)a_{n+1} = \sum_{k=1}^{m} kp_k a_{n-k+1} = 2a_{n-1}.$$

The difference equation

$$(n+1)a_{n+1} - 2a_{n-1} = 0$$

has solution

$$a_n = \frac{2}{n}a_{n-2}.$$

We know that $a_0 = e$ and $a_1 = 0$, so, by back substitution,

$$a_n = \begin{cases} \frac{2^{n/2}e}{n(n-2)(n-4)\cdots 2} & \text{if } n \text{ is even,} \\ 0 & \text{if } n \text{ is odd.} \end{cases}$$

Example. If $f(z) = b^{P(z)}, b > 0, b \neq 1$, then,

$$f'(z) = f(z)P'(z)\ln(b).$$

Since $P'(z)\ln(b)$ is a polynomial, we can apply Proposition 1 to find $[z^n]f(z)$.

Proposition 2. If $f'(z) = f(z)\frac{P(z)}{Q(z)}$, then, multiplying both sides by $Q(z)$,

$$f'(z)Q(z) = f(z)P(z).$$

Comparing the coefficients of both sides of z^n,

$$\sum_{k=0}^{l}(n-k+1)q_k a_{n-k+1} = \sum_{i=0}^{m} p_i a_{n-i}. \tag{7}$$

This is a homogenous difference equation with polynomial coefficients, so we can apply the algorithms in Sect. 2 to solve it.

Example. Consider $f(z) = (1+z)^{1/3}$. So

$$f'(z) = \frac{1}{3(1+z)^{2/3}}.$$

Since $\frac{f'}{f} = \frac{1}{3(1+z)}$ is a rational function, we can apply Proposition 2 by solving the difference equation

$$3(n+1)a_{n+1} + 3na_n = a_n,$$

with initial value $a_1 = \frac{1}{3}$. We have

$$\begin{aligned} a_n &= \frac{(1-3(n-1))(1-(3(n-2))\cdots(1-3\cdot 2)(1-3\cdot 1)}{3^n n!} \\ &= \frac{(-1)^{n+1}\Gamma(n-1/3)}{3\Gamma(n+1)\Gamma(2/3)}, \end{aligned}$$

where

$$\Gamma(z) = \int_0^\infty e^{-t} t^{(z-1)} dt.$$

Example. $f(z) = b^{\frac{P(z)}{Q(z)}}, b > 0, b \neq 1$. Then

$$f'(z) = f(z)\frac{P'(z)Q(z) - Q'(z)P(z)}{Q^2(z)} \ln(b).$$

Since both $Q^2(z)$ and $(P'(z)Q(z) - Q'(z)P(z))\ln(b)$ are polynomials, we can compute $[z^n]f(z)$ by applying Proposition 2.

Example. $f(z) = R(z)b^{\frac{P(z)}{Q(z)}}$, where $R(z)$ is a given rational function and $b > 0$. First, let $g(z) = b^{\frac{P(z)}{Q(z)}}$, then apply Proposition 2 as in the example above. Finally, compute $[z^n]R(z)g(z)$ by comparison of coefficients.

In the following, we provide six more propositions which are ready to be used to calculate the Taylor coefficients. Let D be the derivative operator; *i.e.*, $Df(z) = f'(z)$ and $D^k f(z) = f^{(k)}(z)$ for any nonnegative integer k.

Proposition 3. For any given positive integer k, if we know the expansion $[z^{n-k}](D^k f(z)) = c_{n-k}$, then

$$[z^n]f(z) = \frac{c_{n-k}}{n(n-1)\cdots(n-k+1)}. \tag{8}$$

Example. If $f'(z) = \frac{P(z)}{Q(z)}$, we can apply Ravenscroft's `rgf_expand` to calculate

$$c_{n-1} = [z^{n-1}]\frac{P(z)}{Q(z)},$$

then

$$[z^n]f(z) = \frac{c_{n-1}}{n}.$$

Example. If $f(z) = \log_b^{\left(\frac{P(z)}{Q(z)}\right)}$, the logarithmic function with base b where $b > 0, b \neq 1$, then

$$f'(z) = \frac{Q(z)}{P(z)\ln(b)} \frac{P'(z)Q(z) - P(z)Q'(z)}{Q^2(z)}.$$

Since the right side is a rational function, we can apply `rgf_expand` to calculate its Taylor coefficients, then apply Proposition 3.
Example. Consider $f(z) = \arctan(z)$. Since

$$f'(z) = \frac{1}{1+z^2},$$

is a rational function, by applying `rgf_expand` in Maple, we have $[z^{n-1}]f'(z) = \cos\left(\frac{(n-1)\pi}{2}\right)$. So, $[z^n]f(z) = \cos\left(\frac{(n-1)\pi}{2}\right)/n$.
Example. Consider $f(z) = \text{arccot}\ (\frac{1-z}{1+z})$ Since

$$f'(z) = \frac{1}{1+z^2},$$

is a rational function, by applying `rgf_expand` in Maple, we have $[z^{n-1}]f'(z) = \cos\left(\frac{(n-1)\pi}{2}\right)$. So $[z^n]f(z) = \cos\left(\frac{(n-1)\pi}{2}\right)/n$.

There are many functions whose expansions can be calculated by Proposition 3. These include $\arctan(R(z))$, $\text{arccot}(R(z))$ and $\ln(R(z))$, where $R(z)$ is a rational function.

Proposition 4. If $P(z)D^k f(z) = g(z)$, then we can find $[z^n]f(z)$ by comparing coefficients and solving a linear difference equation with polynomial coefficients.

The expansion of $f(z)$ and its derivatives are:

$$\begin{aligned}
f(z) &= a_0 + a_1 z + a_2 z^2 + \cdots + a_n z^n + \cdots \\
f'(z) &= 1!a_1 + 2a_2 z + \cdots + na_n z^{n-1} + (n+1)a_{n+1}z^n + \cdots \\
f^{(2)}(z) &= 2!a_2 + \frac{3!}{1!}a_3 z + \cdots + \frac{(n+2)!}{n!}a_{n+2}z^n + \cdots \\
&\cdots\cdots \\
f^{(k)}(z) &= k!a_k + \frac{(k+1)!}{1!}a_{k+1}z + \cdots + \frac{(n+k)!}{n!}a_{n+k}z^n + \cdots
\end{aligned}$$

Since

$$P(z) = p_0 + p_1 z + p_2 z^2 + \cdots + p_m z^m,$$

we have by comparing the coefficients,

$$\sum_{i=0}^{m} \frac{(n+k-i)!}{(n-i)!} p_i a_{n+k-i} = b_n. \tag{9}$$

This is a linear difference equation with polynomial coefficients.

Example. $f(z) = \left(\operatorname{arccot}\frac{1-z}{1+z}\right)^2$. In this case,

$$(1+z^2)f'(z) = 2\operatorname{arccot}\frac{1-z}{1+z}.$$

Since we expanded $\operatorname{arccot}\frac{1-z}{1+z}$ in an earlier example, we can get $[z^n]f(z)$ by Proposition 4.

Proposition 4 can be applied to lots of functions. Examples include $R_1(z)\exp(R_2(z))$, $R_1(z)arctan(R_2(z))$, $R_1(z)\operatorname{arccot}(R_2(z))$ and $R_1(z)\ln(R_2(z))$, where $R_1(z)$ and $R_2(z)$ are rational functions.

Now, we will take a look at a more complex form. Let

$$L = P_k(z)D^k + P_{k-1}(z)D^{k-1} + \cdots + P_1(z)D + P_0(z),$$

where the $P_j(z)'s$ are polynomials and D is the derivative operator for nonnegative integer k, $0 \le j \le k$. We assume

$$P_j(z) = p_{j_0} + p_{j_1}z + p_{j_2}z^2 + \cdots + p_{j_m}z^{m_j}.$$

In this case, we have the following propositions.

Proposition 5. If $Lf(z) = g(z)$, then we can determine $[z^n]f(z)$ by comparing coefficients and solving a linear difference equation with polynomial coefficients,

$$\sum_{j=0}^{k}\sum_{i=0}^{m_j}\frac{(n+j-i)!}{(n-i)!}p_{ji}a_{n+j-i} = b_n. \tag{10}$$

The proof of Proposition 5 is very similar to that of Proposition 4, but with more terms on the left side.

If $g(z) = 0$, $Lf(z) = 0$, in this case, we call $f(z)$ a holonomic function. The Taylor coefficients for any holonomic function can be calculated by Proposition 5 as a special case.

Example. $f(z) = \sin(z)$. Since

$$f'(z) = \cos(z),$$
$$f''(z) = -\sin(z),$$

we have

$$Lf(z) = g(z),$$

where $L = D^2 + 1$ and $g(z) = 0$. Then, by applying Proposition 5, we obtain

$$(n+2)(n+1)a_{n+2} + a_n = 0,$$

$$a_{n+2} = -\frac{a_n}{(n+2)(n+1)}.$$

We know that $a_0 = 0$ and $a_1 = 1$, so by back–substitution, we have

$$a_n = \begin{cases} 0 & \text{if } n \text{ is even,} \\ (-1)^{\frac{n-1}{2}} \frac{1}{n!} & \text{if } n \text{ is odd.} \end{cases}$$

Example. $f(z) = P(z)^\alpha$, where α is a real number. It is easy to see that

$$P(z)f'(z) - \alpha P'(z)f(z) = 0,$$

hence we can apply Proposition 5 to find $[z^n]f(z)$.
Example. $f(z) = \frac{P(z)^\alpha}{Q(z)^\beta}$, where both α and β are real numbers. Since

$$f'(z) = \frac{\alpha P(z)^{\alpha-1}Q(z)^\beta P'(z) - \beta P(z)^\alpha Q(z)^{\beta-1}Q'(z)}{Q(z)^{2\beta}},$$

it is easy to see that

$$P(z)Q(z)f'(z) + (\beta P(z)Q'(z) - \alpha Q(z)P'(z))f(z) = 0.$$

Again, we can apply Proposition 5 to complete the job.
Example. $f(z) = e^{\frac{P(z)^\alpha}{Q(z)^\beta}}$, where both α and β are integers. Since

$$f'(z) = f(z)\frac{\alpha P(z)^{\alpha-1}Q(z)^\beta P'(z) - \beta P(z)^\alpha Q(z)^{\beta-1}Q'(z)}{Q(z)^{2\beta}},$$

it is easy to see that

$$Q(z)^{\beta+1}f'(z) + (\beta P(z)^\alpha Q'(z) - \alpha Q(z)P(z)^{\alpha-1}P'(z))f(z) = 0.$$

So, once again, we can apply Proposition 5 to complete the job.
Example. $f(z) = \arcsin(z)$. Since

$$f'(z) = (1 - z^2)^{-1/2},$$

we have

$$(1 - z^2)f''(z) - zf'(z) = 0.$$

Again, we can apply Proposition 5 to complete the job.
Example. $f^{(k)}(z) = R(z)^\alpha$, where k is a nonnegative integer and α is a real number. One can easily check that

$$R(z)f^{(k+1)}(z) - \alpha R'(z)f^{(k)} = 0.$$

Here again, we can apply Proposition 5 to complete the job.

In fact, we can calculate $[z^n]f(z)$ for many functions including $R(z)^\alpha$, $\arcsin(z)$, $\arccos(z)$, $\arctan(z)$, $\text{arccot}(z)$, etc., using the same technique.
Example. $f(z) = \sin \frac{z}{z+1}$. Taking the derivatives, we have

$$f'(z) = \frac{1}{(z+1)^2} \cos \frac{z}{z+1}$$

$$f''(z) = -\frac{1}{(z+1)^4}\sin\frac{z}{z+1} - \frac{2}{(z+1)^3}\cos\frac{z}{z+1}$$

and so $f(z)$ obeys the differential equation

$$(z+1)^4 f''(z) + 2(z+1)^3 f'(z) + f = 0.$$

Let

$$L = (z+1)^4 D^2 + 2(z+1)^3 D + 1,$$

then apply Proposition 5 to find $[z^n]f(z)$.

If both functions $f_1(z)$ and $f_2(z)$ have an infinite number of terms in their Taylor expansions, we can write the coefficients of $f_1(z)f_2(z)$ as a summation easily, but it is usually very difficult to calculate them as a closed form. Nevertheless, Proposition 5 can be applied in at least some cases.

Example. $f(z) = e^z \sin(z)$. Since

$$f'(z) = e^z \sin(z) + e^z \cos(z),$$

we have

$$f'(z) = f + e^z \cos(z)$$
$$e^z \cos(z) = f'(z) - f(z).$$

Taking the second derivative, we find

$$\begin{aligned} f''(z) &= f'(z) - e^z \sin(z) + e^z \cos(z) \\ &= f'(z) - f(z) + f'(z) - f(z) \\ &= 2f'(z) - 2f(z). \end{aligned}$$

Letting

$$L = D^2 - 2D + 2,$$

we have

$$Lf(z) = 0.$$

In this case, we can apply Proposition 5 to complete the job.

Proposition 5 is very neat, and the examples show it is quite powerful. We can easily extend Proposition 5, and we present some extensions below. These propositions are very similar to Proposition 4 but have a few more terms, so we omit their proofs.

Proposition 6. If $Lf(z) = \frac{P(z)}{Q(z)}g(z)$, then $Q(z)Lf(z) = g(z)P(z)$. Since $P(z)$ and $Q(z)$ are polynomials, we can compute $g(z)P(z)$ by comparing coefficients, then determine $[z^n]f(z)$ by simply comparing coefficients. This reduces the problem of finding $[z^n]f(z)$ to solving a linear recurrence relation with polynomial coefficients.

The following algorithm is provided as a sample based on Proposition 6:

Input: Integer k, m_j, $[z^n]P_j(z) = p_{jn}$ for $0 \leq j \leq d$, $[z^n]P(z)$, $[z^n]Q(z)$, $[z^n]g(z)$.
Output: $[z^n]f(z)$.

1. Obtain $[z^n]P(z)g(z)$ from:

$$\sum_{t=0}^{m} p_t b_{n-t}.$$

2. Get $[z^n]Lf(z)$ by

$$\sum_{j=0}^{k}\sum_{i=0}^{m_j}\frac{(n+j-i)!}{(n-i)!}p_{ji}a_{n+j-i}.$$

3. Get $[z^n]Q(z)Lf(z)$ by

$$\sum_{h=0}^{l}\sum_{j=0}^{k}\sum_{i=0}^{m_j}\frac{(n+j-i-h)!}{(n-i-h)!}p_{ji}a_{n+j-i-h}q_h.$$

4. Obtain a difference equation:

$$\sum_{h=0}^{l}\sum_{j=0}^{k}\sum_{i=0}^{m_j}\frac{(n+j-i-h)!}{(n-i-h)!}p_{ji}a_{n+j-i-h}q_h = \sum_{t=0}^{m} p_t b_{n-t}. \tag{11}$$

5. Get initial values for (11).
6. Find $[z^n]f(z)$ by solving (11).

Example. Consider $f(z) = (\ln(1-z))^2$, then

$$f'(z) = \frac{-2}{1-z}\ln(1-z).$$

Since we know that

$$\ln(1-z) = \sum_{n=1}^{\infty}\frac{-1}{n}z^n,$$

let $g(z) = \ln(1-z)$, $P(z) = -2$ and $Q(z) = 1-z$. Applying Proposition 6 to get the answer.

Example. Consider $f(z) = \left(\ln\frac{P(z)}{Q(z)}\right)^2$. Since

$$f'(z) = \frac{2P'(z)Q(z) - 2Q'(z)P(z)}{Q(z)P(z)}\ln\frac{P(z)}{Q(z)},$$

we can apply Proposition 5 to calculate the expansion of $\ln\frac{P(z)}{Q(z)}$, then use Proposition 6 to complete the job.
Example. $f(z) = (\arctan(z^2+1))^2$. Since

$$f'(z) = \frac{4z}{z^4+2z^2+2}\arctan(z^2+1),$$

we can apply Proposition 5 to calculate the expansion of $\arctan(z^2+1)$, then use Proposition 6 to finish the job.

We can recursively calculate the coefficients for functions like $(\arctan(z^2+1))^d$ and $(\arctan R(z))^d$, where d is a nonnegative integer and $R(z)$ is a rational function, by repeatedly applying Proposition 6. The same applies to $(\text{arccot} R(z))^d$.

Proposition 7. If $L\frac{P(z)}{Q(z)}f(z) = g(z)$, by the basic rules for manipulating derivatives, it is easy to transform the problem to the form in Proposition 5. Then we can find $[z^n]f(z)$ by comparing coefficients and solving a linear difference equation with polynomial coefficients.

First, let's consider a simple example.
Example. Suppose $f(z) = (1+z)\sin(z)$. Then

$$f'(z) = sin(z) + (1+z)cos(z)$$

$$f''(z) = 2cos(z) - (1+z)sin(z)$$

Let $L = D^2 + 1$, $g(z) = 2cos(s), P(z) = Q(z) = 1$. We can apply Proposition 7 to do the job.

Example. $f(z) = \sin(R(z))$, where $R(z)$ is a rational function.
Since

$$f'(z) = R'(z)\cos(R(z)),$$
$$f^{''} = R^{''}(z)\cos R(z) - (R'(z))^2 \sin(R(z)),$$

it is not difficult to check that

$$R'(z)f^{''}(z) - R^{''}(z)f'(z) + (R'(z))^3 f(z) = 0.$$

Letting

$$L = R'(z)D^2 - R^{''}(z)D + (R'(z))^3,$$

we can now apply Proposition 7 to finish the job.

In a similar fashion, we can find the expansion of $f(z) = \cos(R(z))$, where $R(z)$ is a rational function.

Example. $f(z) = \arcsin(R(z))$, where $R(z)$ is a rational function. Since

$$f'(z) = R'(z)(1 - R^2(z))^{-1/2},$$

$$f^{''}(z) = R^{''}(z)(1 - R^2(z))^{-1/2} + R(z)(R'(z))^2(1 - R^2(z))^{-3/2},$$

we have

$$(1 - R^2(z))R'(z)f^{''}(z) - ((R^2(z) - 1)R''(z) - R(z)(R'(z)^2)f'(z) = 0.$$

Let

$$L = (1 - R^2(z))R'(z)D^2 - ((R^2(z) - 1)R''(z) - R(z)(R'(z)^2)D,$$

and then apply Proposition 7 to complete the job.

Similarly, we can calculate the expansion of $f(z) = \arccos(R(z))$, where $R(z)$ is a rational function.

Suppose L_i are of the form of L above (perhaps with different d in the form), R_i are rational functions of z, E_1 is a finite sum of the form L_iR_i or R_iL_i, and E_2 is also a finite sum of the form L_iR_i or R_iL_i. Then if $E_1f(z) = E_2g(z)$, we can use the following proposition to calculate $[z^n]f(z)$.

Proposition 8. If $E_1f(z) = E_2g(z)$, by the basic rules on derivatives, it is easy to transform to the form in Proposition 5. Then we can find $[z^n]f(z)$ by comparing coefficients and solving a linear difference equation with polynomial coefficients.

According to basic rules on derivatives, the form L_iR_i can be easily written in the form R_iL_i. We can just think of E_1 and E_2 as finite sums of the form R_iL_i. Furthermore, we know that for any integer k,

$$[z^n](D^k g(z)) = (n+k)(n+k-1)\cdots(n+1)[z^n]g(z). \tag{12}$$

Example. $f(z) = \ln(1+z)\sin(z)$. Taking derivatives, we have

$$f'(z) = \frac{\sin(z)}{1+z} + \ln(1+z)\cos(z),$$

$$f''(z) = \frac{-\sin(z)}{(1+z)^2} + \frac{2\cos(z)}{1+z} - \ln(1+z)\sin(z).$$

Here

$$(1+z^2)f''(z) + (1+z^2)f(z) = 2(1+z)\cos(z) - \sin(z),$$

so, we can apply Proposition 8 to calculate $[z^n]f(z)$.

We can derive more propositions using the methods described above. Each proposition can easily be used to develop one or more algorithms. We can even develop similar but more powerful propositions based on systems of differential and difference equations. These methods are very powerful, and we can apply them to most of the commonly used functions, holonomic functions, and much more. It would be a big project to develop software which incorporates all of these algorithms in a smart way.

4 Conclusions

Taylor series is very important in pure math and applications. This work offers a technique view on practical calculations; it introduced linear recurrences with polynomial coefficients, and listed many algorithms to solve such recurrences. The primary focus of this work is the development of a series propositions, these propositions are developed step by step, straightforward, easy to understand and implement. While easy to understand, these propositions are very powerful, can be used to calculate the general (arbitrary) term of Taylor coefficient for most commonly used functions, holonomic functions, and much more. This work can be used to serve as technical specifications for implementation.

References

1. Petkovsek, M., Wilf, H., Zeilberger. D.: A = B. A K Peters (1996)
2. Petkovsek, M., Zakrajsek, H.: Solving Linear Recurrence Equations with Polynomial Coefficients. Preprint, downloaded 09/09/2025 from the web
3. Abramov, S. A.:"Rational solutions of linear difference and Q-difference equations with polynomial coefficients. Prog. Comput. Softw. **21**(6), 273–278 (1995)
4. Abramov, S. A., Yu. Kvashenko, K.: Fast algorithm to search for rational solutions of linear differential equations with polynomial coefficients. In: Proceedings of ISSAC'91. Academic Press, pp. 267–270
5. Zhang, J.: Sumudu transform based coefficients calculation. Nonlinear Stud. **15**(4), 355–372 (2008)
6. Zhang, J., Fan, R., Shen, F.: New method for the computation of generating functions with applications. In: The 2021 International Conference on Computational Science and Computational Intelligence. Las Vegas, USA. December 15-17 (2021)
7. Zhang, J., Fan, R., Shen, F., Tu, J.: New effective transformational computational methods. J. Appl. Analy. Comput. **14**(1), 317–333 (2024)
8. Zhang, J., Brown, W., Fan, R.: New methods to solve difference equations automatically. In: The 2023 International Conference on Computational Science and Computational Intelligence. Las Vegas, USA, December 13-15 (2023)
9. Dupee, B. J., Davenport, J. H.: An Automatic Symbolic-Numeric Taylor Series ODE Solver. Preprint, downloaded 09 Mar 2025 from the web
10. Ravenscroft, R.A.: Rational generating function applications in Maple V: Mathematics and Its Application. Presented at the (1994)
11. Ravenscroft, R.A., Lamagna, E.A.: Ravenscroft, R.A., Lamagna, E.A.: Symbolic summation with generating functions. In: Proceedings of the International Symposium on Symbolic and Algebraic Computation, Association for Computing Machinery, pp. 228–233 (1989)
12. Gosper, R.W.: Decision procedure for infinite hypergeometric summation. In: Proceedings of National Academy of Science, USA, vol.75, no.1, pp.40–42 (1978)
13. Gourdon, X., Salvy, B.: "Effective asymptotic of linear recurrences with rational coefficients. Technical Report RR-1887, INRIA Rocquencourt, France (1993)
14. Salvy, B.: "Examples of Automatic Asymptotic Expansion," Technical Report RT-0114. INRIA Rocquencourt, France (1989)

Improved CG-Algorithm for Solving the Graph Balancing Problem with Job Precedence

Yu-Han Chen[1(✉)], Bo-Ai Chang[1], Guan-Shieng Huang[2], and Sun-Yuan Hsieh[1,2,3,4,5,6]

[1] Department of Computer Science and Information Engineering, National Cheng Kung University, No.1, University Road, Tainan city 701, Taiwan
p78101035@gs.ncku.edu.tw, boaichang@gmail.com, hsiehsy@mail.ncku.edu.tw
[2] Department of Computer Science and Information Engineering, National Chi Nan University, 1 University Road, Puli Township, Nantou City 545301, Taiwan
shieng@ncnu.edu.tw
[3] Institute of Medical Informatics, National Cheng Kung University, No.1, University Road, Tainan city 701, Taiwan
[4] Institute of Manufacturing Information and Systems, National Cheng Kung University, No.1, University Road, Tainan city 701, Taiwan
[5] Institute of Information Science, Academia Sinica, 128 Section 2, Academia Road, Nankang, Taipei City 115201, Taiwan
[6] Research Center for Information Technology Innovation, Academia Sinica, 128 Section 2, Academia Road, Nankang, Taipei City 115201, Taiwan

Abstract. We investigated the Graph Balancing Problem with precedence constraints, in which jobs must be assigned to machines while adhering to job dependencies and machine eligibility constraints. To solve this problem, we developed a Dual-List Coffman–Graham (CG)-Algorithm, which extends the classical CG algorithm by using a novel dual-list approach that simultaneously handles job precedence and machine assignment requirements. For a system with unit execution time and m machines, our algorithm achieves an approximation ratio of $(2 - 2/m)$ when $m \geq 3$ and produces optimal solutions when $m = 2$; thus, it exhibits performance bounds that match those of the original CG-Algorithm while solving a strictly more general problem. We developed a comprehensive block-based framework that provides insights into schedule construction and enables rigorous proof-of-performance guarantees. The proposed dual-list CG-Algorithm demonstrates how classic scheduling techniques can be adapted to address modern scheduling challenges while maintaining provable performance bounds.

Keywords: Unrelated Scheduling · Graph Balancing Problem · Precedence Constraints · CG-Algorithm

H. R. Arabnia et al. (Eds.): CSCI 2025, CCIS 2970, pp. 168–179, 2027.
https://doi.org/10.1007/978-3-032-28466-2_13

1 Introduction

The graph balancing problem with precedence constraints involves scheduling a set of jobs $j \in J$ on a set of machines $m \in M$with processing times $p_j > 0$. Each job j can only be processed on a subset $r_j \subseteq M$ of machines with $|r_j| \leq 2$, representing job-machine compatibility constraints. Precedence constraints $P(j) \subseteq J$ specify which jobs must complete before job j can start.

A feasible schedule assigns each job $j \in J$ to a compatible machine $m(j) \in r_j$ and determines starting times $s(j) \geq 0$ such that: (1) no two jobs are processed simultaneously on the same machine, and (2) precedence constraints are satisfied, meaning $s(j') + p_{j'} \leq s(j)$ for all $j' \in P(j)$

We focus on the Unit Execution Time (UET) case where $p_j = 1$ for all jobs. Under this assumption, minimizing makespan $\max_{j \in J}(s(j) + p_j)$ is equivalent to minimizing the maximum workload $\max_{m \in M} |\{j : m(j) = m\}|$ across all machines, transforming the problem into a load balancing optimization. The graph balancing problem is a special case of unrelated parallel machine scheduling. Ebenlendr et al. [11] formally introduced this problem, developing a 1.75-approximation algorithm and proving that obtaining a $(1.5 - \epsilon)$ for any $\epsilon > 0$ is NP-hard. Their work established the fundamental framework for graph balancing research.

Significant progress has been made for restricted cases with two job weights. Multiple researchers developed 3/2-approximation algorithms using different approaches: Huang and Ott [13] for light hyperedges, Page and Solis-Oba [18] using combinatorial techniques, and Chakrabarty and Shiragur [6] with LP-based methods. These results demonstrate that better approximation ratios are achievable for constrained versions.

Recently, Jansen and Rohwedder [14] achieved a breakthrough by proving the configuration LP has an integrality gap of at most 1.749, improving upon the decade-old 1.75 bound. This narrows the gap between the best known upper bound and the hardness lower bound of 1.5.

The max-min fair allocation problem, also known as the Santa Claus problem, focuses on distributing indivisible items to maximize the minimum allocation received by any participant. While conceptually related to min-max graph balancing, it addresses the complementary objective of maximizing the worst-case allocation rather than minimizing maximum workload.

Bansal and Sviridenko [5] established the foundational framework by reformulating this as a welfare maximization problem and introducing configuration LP techniques. Feige [12] proved that the configuration LP has a constant integrality gap for restricted assignment cases, though his result was non-constructive.

Recent work has focused on restricted versions where items can only be assigned to specific agents. Chakrabarty et al. [4] developed combinatorial algorithms, while Cheng and Mao [8] analyzed approximation ratios and integrality gaps for these constrained variants.

Lenstra et al. [17] established the theoretical foundation for unrelated parallel machine scheduling, proving a 3/2 hardness lower bound and providing a 2-approximation algorithm for the case without precedence constraints.

When precedence constraints are introduced, the problem becomes significantly more complex. Chudak and Shmoys [9] extended approximation techniques to handle both machine heterogeneity and task dependencies simultaneously. For special cases, better results are possible: Anil Kumar et al. [3] developed specialized algorithms for tree-like precedence structures, while Aho and Mäkinen [2] identified polynomial-time solvable instances when precedence graphs have bounded degree and path length.

Recent advances include robust approximation schemes by Skutella and Verschae [19] and fractional precedence constraints by Aggarwal et al. [1], which achieve near-optimal solutions for preemptive scheduling scenarios.

Coffman amd Graham [10] introduced the CG-algorithm, providing the first optimal solution for scheduling unit execution time (UET) tasks with precedence constraints on two identical machines. Their seminal work established fundamental concepts in precedence-constrained scheduling theory.

Lam and Sethi [16] extended the theoretical foundation by conducting comprehensive worst-case analysis and proving that the CG-algorithm's optimality properties hold for multiple machines, not just two. This generalization significantly expanded the algorithm's theoretical applicability.

Subsequent research explored specialized applications and extensions. Chardon and Moukrim [7] proved the CG-algorithm's effectiveness for UET task systems with overinterval orders, while Kwok and Ahmad [15] demonstrated its practical influence in multiprocessor task scheduling systems.

Existing research primarily addresses either graph balancing without precedence constraints or parallel machine scheduling without job-machine compatibility restrictions. We bridge this gap by developing a dual-list extension of the CG-algorithm that simultaneously handles precedence constraints and restricted job-machine assignments. Our approach maintains the theoretical guarantees of the original algorithm while addressing the additional complexity of compatibility constraints.

The remainder of this paper is organized as follows. Section 2 presents the formal problem definition. Section 3 introduces our dual-list CG-algorithm. Section 4 analyzes the algorithm's performance and approximation ratio. Section 5 concludes with directions for future research.

2 Problem Definition

We consider the following scheduling problem P. A set $J = \{1, 2, \ldots, n\}$ of n jobs with processing time $p_j > 0$ and a set $M = \{1, 2, \ldots, m\}$ of m unrelated parallel machines. Each machine can process at most one job at a time. Each job j could be only executed on machines in r_j. Each job is processed on only one machine and non-preemptive during the processing period. There are arbitrary standard precedence constraints between jobs. The constraints force a job not to be started before all its predecessors are completed. The objective is to find a feasible schedule that minimises makespan $\max_{j \in J} (s(j) + p_j)$. Here, $s(j)$ denotes the starting time of job j.

Problem GRAPH BALANCING PROBLEM WITH JOB PRECEDENCE CONSTRAINT
Input: A set of jobs J, where each job $j \in J$ has processing time $p_j > 0$; a set of machines M; job-machine compatibility constraints where each job $j \in J$ can be processed on a subset $r_j \subseteq M$ of machines with $1 \leq |r_j| \leq 2$; and precedence constraints $P(j) \subseteq J$ for each job $j \in J$, specifying the set of jobs that must be completed before job j can start.
Output: A feasible schedule that assigns each job $j \in J$ to a compatible machine $m(j) \in r_j$ and determines its starting time $s(j) \geq 0$ such that: no two jobs are processed simultaneously on the same machine and $s(i) + p_i \leq s(j)$ for all $i \in P(j)$, $j \in J$.
Objective: Minimize $\max_{j \in J} (s(j) + p_j)$

In order to simplify this problem, we focus on the Unit Execution Time (UET) case where $p_j = 1$ for all $j \in J$. Under this assumption, minimizing makespan $\max_{j \in J} (s(j) + p_j)$ is equivalent to minimizing $\max_{i \in M} |\{j : m(j) = i\}|$ (the maximum number of jobs on any machine), which aligns well with the load balancing approach of the CG-algorithm.

3 Original and Dual-List CG-Algorithm

3.1 Original CG-Algorithm

The CG-algorithm, introduced by Coffman and Graham [10] provides optimal solutions for scheduling UET jobs with precedence constraints on two identical machines. For a DAG $G = (V, E)$ representing jobs and dependencies, the algorithm assigns integer labels to jobs using a specific sequence comparison method.

First, let $N = (n_1, n_2, \ldots, n_t)$ and $N' = (n'_1, n'_2, \ldots, n'_{t'})$ represent decreasing sequences of positive integers, where $t, t' \geq 1$. The order relation $N < N'$ is valid if and only if one of the following conditions holds:

(i) An index $i \geq 1$ exists such that
  - $n_j = n'_j$ for all $j < i$, and
  - $n_i < n'_i$

(ii) $t < t'$ and $n_j = n'_j$ for all $j \leq t$

This ordering enables the comparison of sequences of different lengths, with element-wise comparison conducted first, followed by sequence length comparison.

Note that, in this subsection, we use J as the individual jobs, differing from our problem problem notation where J represents the job set. The algorithm works as follows:

1. **Initialization:** Select an arbitrary job J_0 with no successors and set $\alpha(J_0) = 1$

2. **Iterative labeling:** For each remaining job J, define $N(J)$ as the decreasing sequence of successor labels, then select the job with the lexicographically smallest sequence for the next label.
3. **Final ordering:** Create list L by ordering jobs in descending label order.

The CG-algorithm prioritizes jobs with higher label values and assigns them to available machines in index order. Lam et al. [16] proved that the algorithm's optimality properties extend to m machines.

Lemma 1 ([10]). *In a CG-schedule, if job J is executed by machine M_1, then all jobs executed concurrently with or after J have a lower label value than does J. For any two jobs J and J', if $\lambda(J) \leq \lambda(J')$, then $\alpha(J) > \alpha(J')$, where $\lambda(J)$ denotes the level (execution time) of job J in the schedule.*

The original CG-algorithm assumes identical machines where any job can be executed on any machine. However, our problem involves job-machine compatibility constraints where each job j can only be executed on a subset $r_j \subseteq M$ of machines. This fundamental difference, along with the multi-machine environment, motivates our dual-list extension to better handle these additional constraints.

3.2 Dual-List CG-Algorithm

To solve the graph balancing problem with job precedence constraints, we developed an improved version of the original CG-algorithm, namely the dual-list CG-algorithm. The key innovation is the introduction of a machine list (L_M) that works in tandem with the job list (L_J), allowing for more effective handling of job-machine compatibility constraints than the single-list approach of the original CG-algorithm. Our method consists of four main steps:
Step 1: Construct the DAG for Job Precedence: Based on the precedence constraints $P(j)$, first, we construct a directed acyclic graph (DAG) $G' = (J, A)$ to represent the precedence relationships among these jobs. The vertices of this DAG correspond to the jobs J. The directed edges A represent precedence relationships. For each job $j \in J$ with $P(j) \neq \emptyset$, we add a directed edge $(j', j) \in A$ for every $j' \in P(j)$, meaning job j' must complete before job j starts. For example, if $P(j_3) = \{j_1, j_5\}$, then we add directed edges (j_1, j_3) and (j_5, j_3) to the DAG. We define a partial order $\prec$ on G' and the set of immediate successors $S(j)$ for each job j.
Step 2: Construct Jobs List L_J: After constructing the DAG, we adapt the original CG-Algorithm to generate a job ordering list that provides prioritization for scheduling. While the original CG-Algorithm was designed for two identical processors, we extend it to our multi-machine environment. This extension does not guarantee optimality but provides systematic job prioritization considering precedence constraints.

The algorithm works backward from jobs with no successors (sink nodes). When multiple jobs have identical $N(j)$ sequences during the labeling process, tie-breaking becomes necessary, which is one source of the approximation in our extended version.

Each job $j \in J$ is assigned an integer label $\alpha(j)$ using the sequence comparison rules defined in the original CG-Algorithm:

1. **Initialization**
   - Select an arbitrary job j with no successors (i.e., $S(j) = \emptyset$)
   - Set $\alpha(j) = 1$
2. **Iterative labeling for $k = 2$ to $|J|$**
   - Consider all unlabeled jobs j for which all successors have been labeled
   - For each such job j, define $N(j)$ as the decreasing sequence of the labels of its successors: $N(j) = (\alpha(j_1), \alpha(j_2), \ldots, \alpha(j_s))$ where $S(j) = \{j_1, j_2, \ldots, j_s\}$ and $\alpha(j_1) > \alpha(j_2) > \cdots > \alpha(j_s)$.
   - Select job j^* such that $N(j^*) \leq N(j)$ for all considered jobs j.
   - Set $\alpha(j^*) = k$.
3. **Final ordering**
   - Order jobs by their labels in descending order: $L_J = (U_{|J|}, U_{|J|-1}, \ldots, U_1)$ where $\alpha(U_k) = k$ for all $k \in 1, \ldots, |J|$.
   - List L_J determines the scheduling order for G'.

Step 3: Construct Machine List L_M: After obtaining the job list L_J, we construct a machine list L_M that prioritizes machines based on their compatibility with high-priority jobs. Since each job can only be processed on a subset of machines (specifically $r_j \subseteq M$), some machines are eligible for more jobs than others.

We define $E_m = \{j \in J | m \in r_j\}$ as the set of jobs that can be executed by machine $m \in M$. A machine's degree is defined as $d(m) = |E_m|$, the number of jobs it can potentially execute.The machine list is created through the following steps:

1. **Degree-based sorting:** Sort all machines by their degree $d(m)$ in non-increasing order.
2. **Degree-based grouping:** Group machines with the same degree together.
3. **Intra-group ordering:** Within each degree group, order machines based on job priorities:
   (a) For each machine m in a group, we have E_m as defined above.
   (b) Define $N(m) = (\alpha(j_1), \alpha(j_2), \ldots, \alpha(j_k))$
   where $E_m = \{j_1, j_2, \ldots, j_k\}$ and $\alpha(j_1) > \alpha(j_2) > \cdots > \alpha(j_k)$
   (c) Iterative labeling within the group:
      i. Select machine m^* such that $N(m^*) \leq N(m)$ for all unlabeled machines in the group
      ii. Assign the next available position label to m^*
      iii. Remove m^* and repeat until all machines in the group are labeled

The final machine list $L_M = (m_{|M|}, m_{|M|-1}, \ldots, m_1)$ provides a prioritized ordering where machines with higher degrees and access to higher-priority jobs are ranked first.

Step 4: Construct Schedule: Finally, the dual-list CG-algorithm uses both L_J and L_M to make job assignment decisions. Jobs are processed sequentially according to L_J order. For each available job j (when all predecessors in $P(j)$ are completed), we select the machine with highest priority in L_M among those that can execute j. If no candidate machines are available, job j waits until one becomes available.

This dual-list approach explicitly considers job-machine compatibility constraints while maintaining the precedence-aware prioritization of the original CG-algorithm, providing a systematic solution for our multi-machine scheduling problem with restricted job-machine assignments.

Because of the UET assumption, all machines become available for executing new jobs simultaneously at discrete time intervals. At each time step, available jobs are considered for assignment according to their order in L_J. When multiple machines can execute the same job, the machine selection follows the priority order established in L_M.

In contrast to the original CG-algorithm, our dual-list approach explicitly handles job-machine compatibility constraints. For each job assignment, the algorithm verifies:

- The job has not been executed yet
- At least one compatible machine (where $m \in r_j$) is currently available
- The assignment respects all precedence constraints

This systematic approach ensures that both precedence constraints and job-machine restrictions are respected while optimizing the scheduling decisions.

The assignment process for several jobs is described to explain the different scenarios shown in Fig. 1. Let $r_1 = r_5 = r_8 = r_{10} = \{m_1, m_2\}$, $r_2 = r_4 = r_6 = r_{11} = \{m_2, m_3\}$, and $r_3 = r_7 = r_9 = \{m_1, m_3\}$.Let $L_J = \{j_{11}, j_{10}, \ldots, j_2, j_1\}$, and let $L_M = \{m_2, m_3, m_1\}$ be the machine list. For this list, job j_{11} must be assigned first. According to r_{11}, job j_{11} can only be assigned to m_2 and m_3. Because $\alpha(m_2) > \alpha(m_3)$, job j_{11} is assigned to m_2 in the first time slot. The job assignments are then continued according to L_J. Subsequently, job j_{10} is assigned. This job can be assigned to m_1 and m_2; however, in the current time slot, m_2 is already assigned job j_{11}. Therefore, job j_{10} is assigned to m_1. Job j_6 is the third job to be assigned. This job can be assigned to m_2 and m_3 at time 3; however, m_2 and m_3 are executing j_8 and j_7, respectively, at this time. Therefore, job j_6 is scheduled in the next time slot. In the next time slot, the remaining job list is $\{j_6, j_3\}$, and the aforementioned process is continued until all jobs are assigned. Figure 1b illustrated the schedule (S) produced by the proposed dual-list CG-algorithm for the scenario displayed in Fig. 1a.

4 Proof

Because a UET scenario is considered, each job is executed within a single time unit, which represents the interval $(i-1, i)$ between time $i-1$ and time i. Thus,

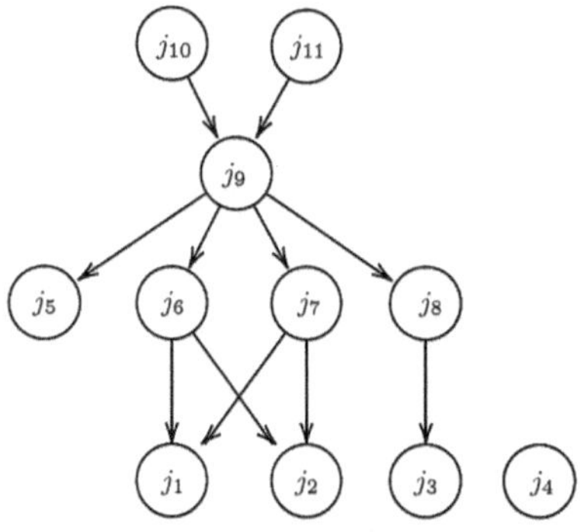

(a) DAG for job precedence G′

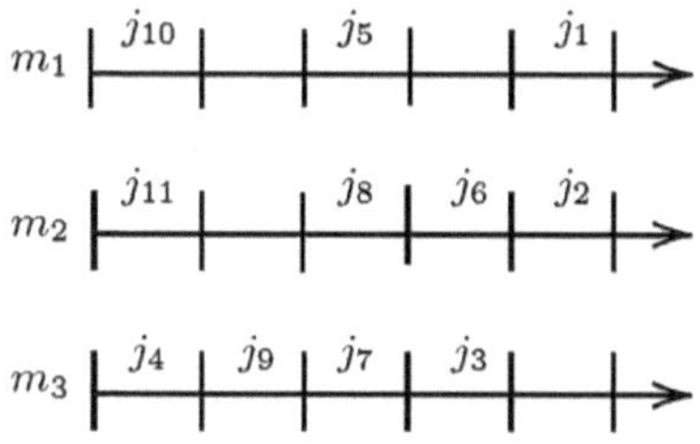

(b) Schedule S produced by dual-list CG-algorithm

Fig. 1. DAG and corresponding schedule produced by dual-list CG-algorithm

$\lambda(J)$ indicates both the starting time and the time unit during which job J is executed.

This section provides proof that the proposed dual-list CG-algorithm maintains the long-chain property in the schedule it has generated; however, the structure of this schedule differs from that generated by the original CG-algorithm. Because of the nature of the graph balancing problem, in which jobs can only be executed on specific machines (at most two machines), the long chain may not necessarily appear on the first machine (as in the original CG-algorithm) in the proposed algorithm. To identify long chains, a *chain schedule* is created by reorganizing jobs within each time unit according to their label values in descending order; each chain is denoted by C_i, where $1 \leq i \leq |G(V)|$. This reorganization results in jobs forming chains with decreasing label values, with long chains naturally appearing at the top and short chains at the bottom of the reorganized schedule.

Figure 2 illustrates our reorganization process. Figure 1b displays the schedule generated by the proposed dual-list CG-algorithm for the scenario displayed in Fig. 1a, in which each job J_j is labeled by its index j (i.e., $\alpha(J_j) = j$). Figure 2(b) shows the reorganized chain structure, with rows labeled as C_i containing jobs in a strictly decreasing order of label value.

The proposed reorganization process preserves the execution times of and precedence relationships between jobs, serving only to reveal the underlying chain structure. Thus, the proposed CG-algorithm maintains the essential chain property of the original CG-algorithm while accommodating additional machine ordering constraints.

Lemma 2. *Consider the jobs in chain C_1 in any schedule produced by the proposed dual-list CG-algorithm. For any jobs J and J' in the chain schedule, if $\lambda(J) \leq \lambda(J')$ (i.e., J is executed no later than J'), then $\alpha(J) > \alpha(J')$ (i.e., J has a higher label value than J' does).*

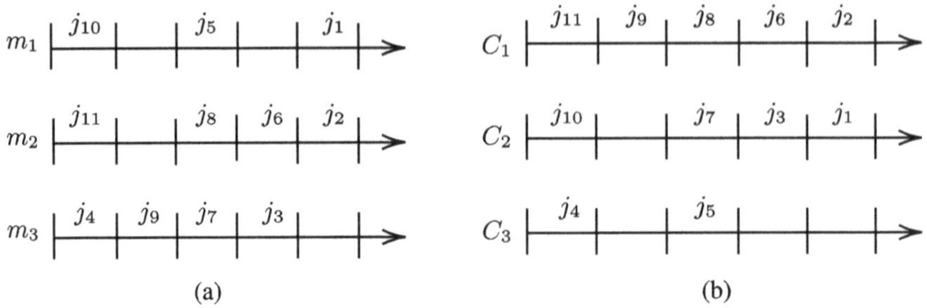

Fig. 2. (a) Schedule S generated by the proposed dual-list CG-algorithm for the scenario shown in Fig. 1a, and (b) chain structure of this schedule.

Assume that the jobs in G are executed according to list L^*. Wheb $2 \leq i \leq |G(V)|$, if C_i is idle from time t to time $t+1$, C_i executes an *empty job* $\emptyset$ in the current time slot, and $\alpha(\emptyset)$ is defined as 0.

Definition 1. *For a schedule S, generated by the proposed dual-list CG-algorithm, the blocks $X_q, \ldots, X_0$ are defined for some q value greater than or equal to 0 as follows:*

1. *The term U_0 denotes the last job in chain C_1 of S.*
2. *For $i \geq 1$, U_i is the job from C_1 that is executed in the most recent time unit λ such that for all other jobs J' except for those in C_1 which are executed in time unit λ, $\alpha(J') < \alpha(U_{i-1})$.*
3. *The term Block X_{i-1} represents the set of all jobs J that satisfy the following conditions:*

$$\lambda(U_i) < \lambda(J) \leq \lambda(U_{i-1}) \text{ and } \alpha(J) \geq \alpha(U_{i-1})$$

When U_i has been defined for $0 \leq i \leq q$ and U_{q+1} does not exist, block X_q consists of all jobs J satisfying the following condition:

$$\lambda(J) \leq \lambda(U_q) \text{ and } \alpha(J) \geq \alpha(U_q)$$

The block identification process is illustrated in Fig. 3, where blocks are defined from right to left based on job label values. First, an examination is conducted to determine the time units un which jobs with lower label values than that of job j_2 are executed. Job j_1 in time unit 5 and the two empty jobs in time unit 2 have lower label values than does j_2, except for j_9 in chain C_1. Because of precedence constraints, j_9 must precede $j_8, \ldots, j_2$; otherwise, one of these jobs would be executed during time unit 2. Time unit 2 contains exactly one job with a higher label value than that of j_2, namely j_9. Therefore, the schedule is split between time units 2 and 3. Block X_0 is defined as containing jobs $j_2, \ldots, j_8$; however, it does not contain j_1 because this job has a lower label value than does j_2. The aforementioned process is continued recursively, starting from job j_9. The blocks resulting from this step are indicated by dashed boxes in Fig. 3.

For the convenience of presenting notations, our definitions are extended to describe the timing properties of the blocks. For any block X_i, $\lambda(X_i)$ represents the earliest execution time for any job in the block. This parameter is expressed as follows:

$$\lambda(X_i) = \min\{\lambda(J) \mid J \in X_i\}$$

Any job in the schedule that does not belong to any block is referred to as an *extra job*; for example, j_4 and j_1 are extra jobs in Fig. 3. The term n_k represents the length of X_k in Fig. 3.

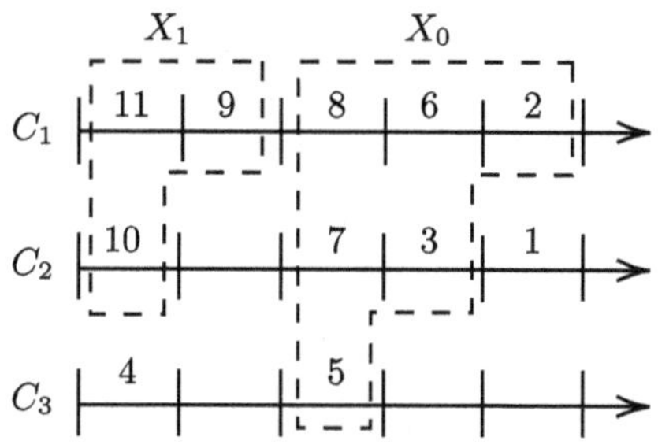

Fig. 3. Block structure (dashed box) in chain schedule.

We now demonstrate that all jobs in one block precede all jobs in the next.

Lemma 3. *When $0 \le k \le m$, if $J \in X_k$ and $J' \in X_{k+1}$, then $J' \prec J$.*

For Lemmas 4 and 5, let ω and ω_0 denote the lengths of a schedule produced by the proposed algorithm and an optimal schedule, respectively. To establish the bound $\omega/\omega_0 \le 2 - 2/m$, the block structure the schedule produced by the proposed algorithm is analyzed. The time units within a block X are classified as full columns (in which all chains contain jobs from X) or partial columns (in which some chains lack jobs from X). This classification enables the derivation of lower bounds for ω_0 on the basis of the distribution of partial columns.

Lemma 4. *Let X_i be a block containing p partial columns. If X_i begins with a full column, then any optimal schedule for the jobs in X_i must have length $\omega_0 \ge p + 1$.*

Lemma 5. *For any block X, ω represents the length of the schedule generated by the proposed algorithm, and ω_0 represents the length of an optimal schedule. The following scenarios exist:*

- *When $m \ge 3$, $\omega/\omega_0 \le 2 - 2/m$*
- *When $m = 2$, the proposed algorithm produces an optimal solution.*

Theorem 1. *Let S be a schedule produced by the proposed algorithm for a UET system on m machines. The length of this schedule is ω. Let ω_0 be the length of an optimal schedule for the same job system. The following findings are obtained for this scenario:*

- *When $m \geq 3$, $\omega/\omega_0 \leq 2 - 2/m$*
- *When $m = 2$, the optimal solution is generated by the proposed algorithm.*

5 Conclusion

We developed a dual-list CG-algorithm for the graph balancing problem with precedence constraints, extending the original CG-algorithm to handle job-machine compatibility restrictions. Our main contributions are:

1. The dual-list approach uses both job and machine priority lists, effectively handling machine-specific constraints while preserving the advantages of the original CG-algorithm.
2. The algorithm achieves a $(2 - 2/m)$ approximation ratio for $m \geq 3$ machines and optimal solutions for $m = 2$ machines, solving a strictly more general problem than the original algorithm.

Our approach successfully integrates precedence constraints with restricted machine assignments, demonstrating that classic scheduling algorithms can be adapted to address modern scheduling challenges while maintaining provable performance guarantees.

Future work could improve approximation ratios by incorporating additional scheduling information, such as fixed assignments or predetermined execution times, potentially leading to tighter bounds for practical applications.

References

1. Aggarwal, V., Lan, T., Peddireddy, D.: Preemptive scheduling on unrelated machines with fractional precedence constraints. J. Parallel Distrib. Comput. **157**, 280–286 (2021)
2. Aho, I., Mäkinen, E.: On a parallel machine scheduling problem with precedence constraints. J. Sched. **9**, 493–495 (2006)
3. Anil Kumar, V., Marathe, M.V., Parthasarathy, S., Srinivasan, A.: Scheduling on unrelated machines under tree-like precedence constraints. Algorithmica **55**, 205–226 (2009)
4. Annamalai, C., Kalaitzis, C., Svensson, O.: Combinatorial algorithm for restricted max-min fair allocation. ACM Transact. Algorith. (TALG) **13**(3), 1–28 (2017)
5. Bansal, N., Sviridenko, M.: The santa claus problem. In: Proceedings of the Thirty-eighth Annual ACM Symposium on Theory of Computing, pp. 31–40 (2006)
6. Chakrabarty, D., Shiragur, K.: Graph balancing with two edge types (2016)
7. Chardon, M., Moukrim, A.: The Coffman-graham algorithm optimally solves UET task systems with overinterval orders. SIAM J. Discret. Math. **19**(1), 109–121 (2005)

8. Cheng, S.W., Mao, Y.: Restricted Max-min Fair Allocation (2018)
9. Chudak, F.A., Shmoys, D.B.: Approximation algorithms for precedence-constrained scheduling problems on parallel machines that run at different speeds. J. Algorithms **30**(2), 323–343 (1999)
10. Coffman, E.G., Graham, R.L.: Optimal scheduling for two-processor systems. Acta Informatica **1**, 200–213 (1972)
11. Ebenlendr, T., Krčál, M., Sgall, J.: Graph balancing: a special case of scheduling unrelated parallel machines. Algorithmica **68**, 62–80 (2014)
12. Feige, U.: On allocations that maximize fairness. In: ACM-SIAM Symposium on Discrete Algorithms (SODA), vol. 8, pp. 287–293 (2008)
13. Huang, C.C., Ott, S.: A combinatorial approximation algorithm for graph balancing with light hyper edges. In: 24th Annual European Symposium on Algorithms (ESA 2016). Leibniz International Proceedings in Informatics (LIPIcs), vol. 57, pp. 49:1–49:15 (2016)
14. Jansen, K., Rohwedder, L.: Local search breaks 1.75 for graph balancing (2018)
15. Kwok, Y.K., Ahmad, I.: Static scheduling algorithms for allocating directed task graphs to multiprocessors. ACM Comput. Surv. (CSUR) **31**(4), 406–471 (1999)
16. Lam, S., Sethi, R.: Worst case analysis of two scheduling algorithms. SIAM J. Comput. **6**(3), 518–536 (1977)
17. Lenstra, J.K., Shmoys, D.B., Tardos, É.: Approximation algorithms for scheduling unrelated parallel machines. Math. Program. **46**, 259–271 (1990)
18. Page, D.R., Solis-Oba, R.: A 3/2-approximation algorithm for the graph balancing problem with two weights. Algorithms **9**(2), 38 (2016)
19. Skutella, M., Verschae, J.: Robust polynomial-time approximation schemes for parallel machine scheduling with job arrivals and departures. Math. Oper. Res. **41**(3), 991–1021 (2016)

Dynamic Spatio-Temporal Data Summarization Using Information Based Fusion

Humayra Tasnim[1(✉)], Soumya Dutta[2], and Melanie Moses[1]

[1] Department of Computer Science,University of New Mexico, Albuquerque, New Mexico, USA
htasnim30@unm.edu , melaniem@cs.unm.edu

[2] Department of Computer Science and Engineering, Indian Institute of Technology Kanpur, Kanpur, Uttar Pradesh, India
soumyad@cse.iitk.ac.in

Abstract. In the era of burgeoning data generation, managing large-scale datasets poses significant challenges. With the rise of computing, the volume of data produced has soared, intensifying storage and I/O overheads. To address this issue, we propose a data summarization technique that identifies informative features in key timesteps and fuses less informative ones. This approach minimizes storage requirements while preserving data features. Unlike existing methods, our method retains both raw and summarized timesteps. We utilize information-theoretic measures to devise the fusion process, visually representing underlying data patterns. We demonstrate the versatility of proposed technique across datasets from diverse application domains.

Keywords: Information Theory · Image Feature Extraction · Data Summarization · Big Data Visualization · Spatio-Temporal Data Fusion

1 Introduction

In today's data-driven world, the exponential growth in data generation has brought forth significant challenges for storage and associated I/O overheads. Modern computing capabilities have enabled the creation of massive datasets at an accelerated pace [16,40]. Many of these datasets exhibit a dynamic temporal nature, spanning thousands of timesteps and needing large storage. Analyzing such a large number of timesteps poses significant challenges. One popular approach is to summarize the data by identifying the key timesteps. However, while key timestep-based approaches preserve the important events, automatic detection of key timesteps is non-trivial and the data dynamics for the intermediate non-key timesteps are completely ignored. Therefore, to preserve the temporal dynamics, only key timestep-based solutions may not be desired. We need novel data summarization methods that preserve both key events and overall temporal dynamics of the data in a storage-efficient compact format enabling accelerated analytics on large time-varying data.

To address the aforementioned need, we propose a data summarization technique that aims to minimize the storage overhead while preserving the salient

H. R. Arabnia et al. (Eds.): CSCI 2025, CCIS 2970, pp. 180–198, 2027.
https://doi.org/10.1007/978-3-032-28466-2_14

temporal dynamics. We also emphasize visualizing these dynamics by tracking changes over time. Our approach involves a dynamic spatio-temporal summarization (DSTS) technique, which adaptively identifies both key and redundant timesteps. We store the key timesteps and summarize redundant timesteps into a single timestep, highlighting the salient temporal characteristics of the features. The summarization technique ensures storage reduction with minimal information loss.

To achieve this, we use information-theoretic measures namely the Specific Mutual Information to guide the data fusion for the summary generation. The core idea of the summarization is to identify informative temporal features within the redundant (non-key) timesteps and fuse them using principles from information theory. By selecting the most relevant features from the redundant timesteps and summarizing through information-guided fusion, we retain the temporal dynamics. This approach optimizes storage requirements and facilitates the visualization and tracking of information change over time, providing valuable insights about data patterns.

The contributions of the paper are:

- Develop a dynamic spatio-temporal summarization (DSTS) technique for large-scale time-varying datasets. The summary provides three features: key timesteps, fused timesteps, and holistic visual representation of information change.
- Propose several information-theoretic fusion strategies and comprehensively compare, contrast, and evaluate their characteristics and applicability in summarizing the datasets.
- Demonstrate the flexibility and effectiveness of the proposed DSTS technique through application to diverse time-varying datasets including scientific flow simulations, surveillance video, and cell interactions in the immune system.
- Explore the impact of the proposed technique in optimizing data storage with minimal data loss.

2 Related Works

In this work, we focus on identifying and storing informative key timesteps while summarizing less informative (non-key) ones by fusion. Among data compression and reduction techniques, Cinema [1] is an image-based in situ data reduction and visualization approach. Lossless and lossy compression methods are also applied for data reduction [13,54]. Among other techniques, statistical methods have been applied [20,58,61]. Unlike theseworks, we aim to retain information from both raw and reduced timesteps to capture information changes over time.

There are numerous approaches [38,52,64] that have been proposed for identifying key timesteps. These studies focus on only capturing key timesteps. Other studies focused solely on data reduction [2,56]. Unlike these studies, our work combines summarization with key timestep selection and data reduction.

Data fusion techniques [14] for large-scale spatio-temporal datasets has been a popular field across various domains like remote sensing [32,39], geoscience [33, 59], network architectures [29,47], computer vision [24,57,60], and time-varying scientific data [22]. In computer vision, various data summarization strategies have been explored, including Gaussian entropy fusion [25] and probabilistic skimlets fusion [62]. Additionally, deep learning methods have also been applied for summarization [63]. Unlike some existing techniques, our approach doesn't need training and can be readily applied to large-scale datasets. Moreover, it is computationally efficient, rendering it applicable for both streaming and offline data analysis.

Information theory [18,48,53] has been employed to measure the relationships between variables in data across multiple computational domains [36,44,50]. Mutual information (MI) is extensively applied for feature selection, exploration, extraction, and tracking [5,51]. Image registration is another popular application [11,27,34]. MI, as well as its decomposition measures like specific mutual information and pointwise mutual information, have been widely used in multi-modal data fusion [7], data analysis, and visualization [3,6,15,21,28,56]. Other use cases include view selection [55], feature similarity [9], and transfer function and design [8,43].

3 Information-Driven Framework for Feature-Based Temporal Data Summaries

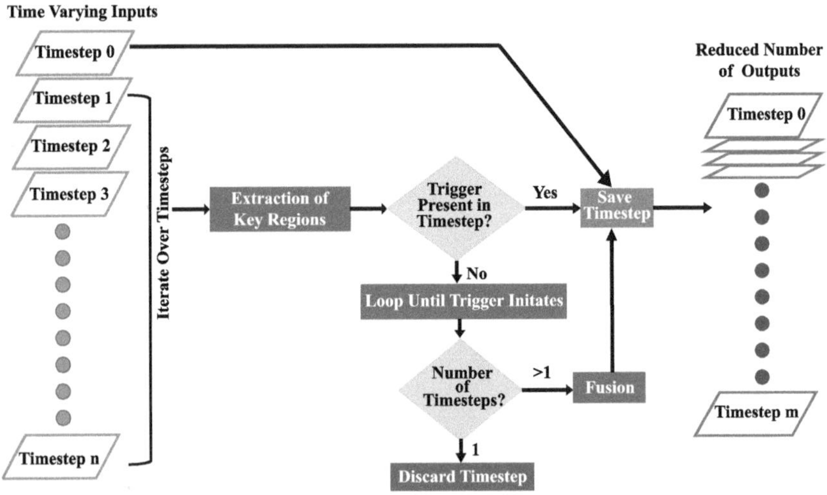

Fig. 1. Schematic diagram of our workflow. Standard computational flowchart [49] symbols are used for representations: input/output, process, decision, and arrows indicating relationships between symbols.

3.1 Framework Workflow

Figure 1 illustrates the schematic of the proposed workflow. To demonstrate each step of this workflow, we will refer to Fig. 2, which serves as an illustrative application of our method using a synthetic data set. In this application, we simulate a rolling ball moving from left to right at each timestep until it exits the view area.

Our proposed method is designed for time-varying data containing various types of salient features. In Fig. 2, the simulated rolling ball application consisting of 19 timesteps (T0 - T18), shows the ball's positional change over time. Each timestep is represented as 800×400 pixel 2D RGB image. At timestep T0, the frame is empty; the ball has not yet entered the view area. The ball enters at T1 and changes position until T17; finally exiting the view area at T18. The proposed method iteratively processes the sequence of input timesteps. After the first timestep, for every subsequent timesteps, the key regions are extracted using a segmentation method proposed in [31]. Criteria for extraction of such key regions is determined by the domain knowledge. In this case, the key feature is the presence of the ball and its location. So we segment the region containing the ball and create binary masked images shown in Fig. 2 (masked row). These images contain only two data values: 0 (no ball) and 255 (ball).

After extracting the key region, we check if a certain property is present in that timestep. We denote this property as *trigger* which is a change in the key region. The change can be in terms of count, size, shape, connectivity, space, or association. In this rolling ball demonstration, the triggers are the first appearance and final exit of the ball from the view area. When the ball enters and then exits the area, it is considered as salient information. But the time the ball remains in the area, the only novel information is its change of position. If a trigger is present in the current timestep, then it is considered as a key timestep. Hence our method saves it as it is. If the trigger is not present, we proceed to the next timestep, do a similar check, and continue the process until a trigger is encountered. These intermediate sequential timesteps that did not have the trigger are chosen to be fused into a single timestep as the amount of novel information within such a sequence is low. If the number of timesteps to be fused is one then we can discard it as the previous timestep has the necessary information. If the number is greater than one then we perform pairwise information-guided fusion on these timesteps and convert them as one single timestep to be saved as a temporal summary. Referencing the demonstration in Fig. 2, T0 is saved. Subsequently, for T1 through T17, no triggers are identified, and again, T18 is saved. Therefore, T1 through T17 are fused as shown in Fig. 2(a) and (b). Note that, since our method processes one timestep at a time incrementally as they appear, it can be applied to applications where data is streamed for real-time processing.

3.2 Characterization of Samplewise Information for Fusion

We use the term "sample" to refer to individual data points. Each timestep contains multiple samples representing the values of the data. In the case of

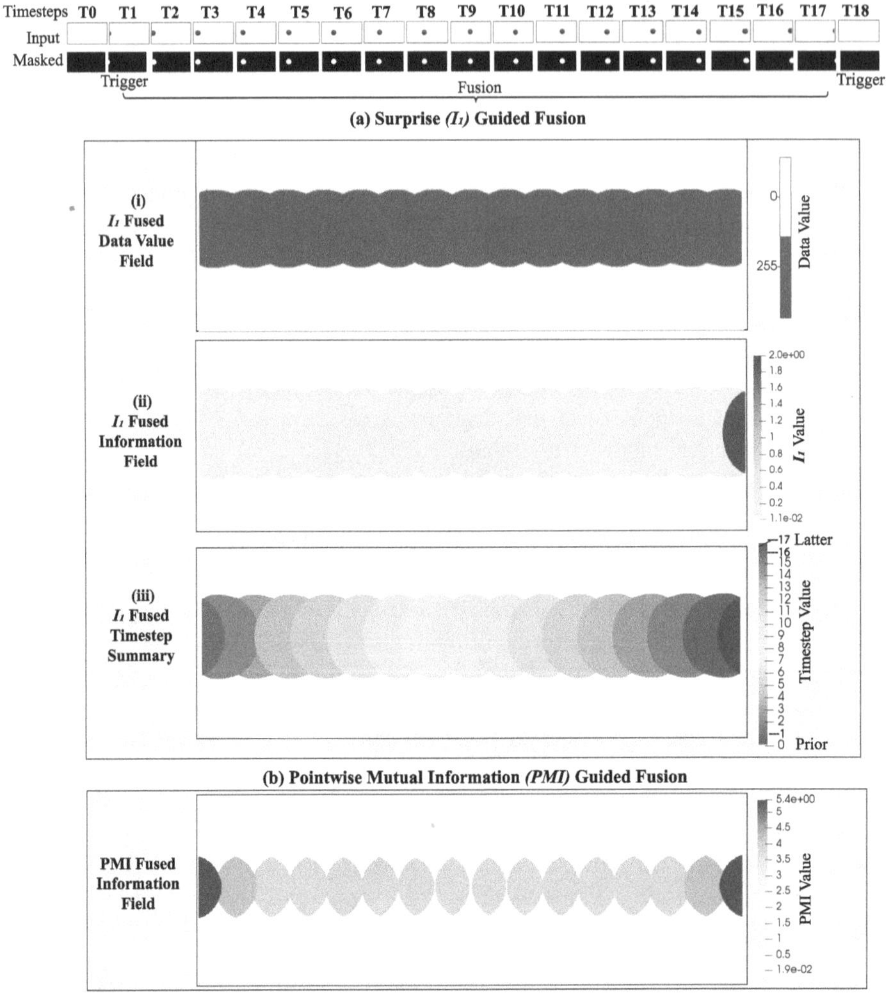

Fig. 2. Illustration using a simulated rolling ball with 19 timesteps (T0 -T18). At each timestep, the ball moves 0.5 units to the right. Timesteps are 2D RGB images with 800×400 dimensions with data (pixel) values ranging between 0 to 255 (input row). The masked row presents binary images with values 0 (no ball) and 255 (ball). T1 to T17 are fused using (a) Surprise (I_1) guided fusion and (b) PMI guided fusion. a(i) shows the I_1 fused data value field (0 white and 255 red). a(ii) shows I_1 fused information value field. a(iii) displays I_1 fused timestep summary with numbered color labels for each timestep. The numbers indicate spatial information changes over time. Surprise effectively captures spatio-temporal properties, whereas alternative PMI measure does not perform well. (b) shows the scenario with the information field using PMI values. (Color figure online)

images like the rolling ball, these samples range from 0 to 255, while for other types of data variables, they may be scalar values. Quantifying information for these samples will help identify important spatial features for the timestep.

Mutual Information In information theory, Mutual Information (MI) [48] is a prominent measure that estimates the total amount of shared information between two random variables. Given two random variables X and Y, MI $I(X;Y)$ is formally defined as:

$$\mathrm{I}(X;Y) = \sum_{y \in Y} \sum_{x \in X} \mathrm{p}(x,y) \log \frac{\mathrm{p}(x,y)}{\mathrm{p(x)p(y)}} \tag{1}$$

where $\mathrm{p}(x)$ and $\mathrm{p}(y)$ are the probabilities of occurrence of values x for X and y for Y respectively. $\mathrm{p}(x,y)$ is the joint probability of occurrence of values x and y together. MI assesses the degree of association or disassociation between two random variables and gives a single value. Since we aim to extract feature-based data summaries, we need samplewise spatial and temporal information characterization. Therefore, we leverage the decomposition of MI which quantifies each data value's contribution toward the association or dissociation. The decomposition of MI is termed as Specific Mutual Information or SMI [19]. There are multiple methods for MI decomposition [10,19]. For apprehending the fusion criteria essential for summarizing the data, the properties of the SMI measure, *Surprise* holds the most potential.

SMI Measure Surprise. The Surprise measure denoted as I_1 was first introduced by [19]. Surprise quantifies the information change of the target variable after observing the individual scalar values of the reference variable. The derivation of Surprise from MI is as follows.

By definition, the conditional probability of x given y is:

$$\mathrm{p}(x|y) = \frac{\mathrm{p}(x,y)}{\mathrm{p(y)}} or \mathrm{p}(x,y) = \mathrm{p}(x|y)\mathrm{p(y)} \tag{2}$$

Replacing the joint probability in Eq. 1, we get,

$$\begin{aligned} \mathrm{I}(X;Y) &= \sum_{y \in Y} \mathrm{p(y)} \sum_{\mathrm{x \in X}} \mathrm{p(x|y)} \log \frac{\mathrm{p(x|y)}}{\mathrm{p(x)}} \\ &= \sum_{y \in Y} \mathrm{p(y) I_1(y;X)} \end{aligned} \tag{3}$$

where,

$$\mathrm{I}_1(y;X) = \sum_{x \in X} \mathrm{p}(x|y) \log \frac{\mathrm{p}(x|y)}{\mathrm{p(x)}} \tag{4}$$

Equation 4 represents the surprise measure of data value y from Y after observing all the values of X. A high value for $I_1(y;X)$ means after observing

y, some previously low probable values of $x \in X$ have become highly probable. This likelihood increase is the element of surprise and a salient finding for further analysis. Surprise is also the only positive decomposition of MI since it is the Kullback-Leibler distance between p($x|y$) and p(x) [30].

3.3 Surprise (I_1) Guided Fusion Technique

When the low informative timesteps are chosen, the fusion initiates for summarization. The fusion is done on pairwise timesteps. For every pair of data samples from the timesteps, we store samples with high I_1 values. This fusion strategy was introduced by [7] for fusing different datasets to gain the most informative combination. The condition to compute the fused value using I_1-fusion is:

For every data sample pair with (x, y), the fused value, f is,

$$\mathrm{f} = \begin{cases} \mathrm{x}, & \text{if } \mathrm{I_1(x;Y) > I_1(y;X)} \\ \mathrm{y}, & \text{otherwise} \end{cases} \tag{5}$$

Here x and y are individual data values from two data sets X and Y. Our fusion criteria is based on the idea of Eq. 5, however, instead of different datasets we are using two subsequent timesteps from the same dataset. To fuse multiple timesteps, we begin by creating a fused timestep using the first two timesteps. Then, we repeat the fusion process by comparing the fused timestep with the next timestep and continue until all desired timesteps have been fused. Our strategy involves updating the fused timestep during each iteration and selecting the spatial and temporal values with the highest information content. By the end of the process, the resulting fused timestep will represent a summary capturing their most informative properties with direction. The fusion process is described in detail in Algorithm 1. After each fusion process, the algorithm provides 3 fused fields as shown in Fig. 2(a). I_1 fused data value field contains the values of the data samples with high surprise measure. I_1 fused information value field contains the I_1 values for the same sample positions. In the timestep summary fields, the same data samples are labeled with their originating timestep numbers.

Applying the fusion process in Algorithm 1 on T1 - T17 of the simulated rolling ball, the I_1 fused data value field is generated highlighting the path of the ball with values 255 as shown in Fig. 2 a(i). The regions without the ball are valued 0. Figure 2 a(ii) represents the fused information fields with I_1 values. From the color bar's gradient, we observe that the surprise values exhibit limited variation, spanning approximately from 0 to 2. A minimal data value range results in minimal surprise variation. Figure 2 a(iii) presents the timestep summary where the numbers of originating timesteps are labeled for the salient samples. Here, we employed distinct colors to label the timesteps, enabling clear visualization and differentiation of each timestep. This color-coded representation shows the flow of information, facilitating the tracking of information changes over time. Here, the confidence threshold is employed to downplay the non-important regions. In this particular case, the threshold value is set to 255. Any value below 255, representing the absence of the ball, is assigned as timestep

0. In Fig. 2 a(iii), these regions are depicted as white or transparent (steps 19–24 in Algorithm 1). The method reduces the number of output timesteps from 19 to 3 in the simulated rolling ball case, achieving substantial data reduction with minimal loss. The fused timestep effectively visualizes the information changes over time, serving as a summary of the original data dynamics.

3.4 Alternative Fusion Approaches

We also explore other potential information measures to devise alternative techniques for generating data summaries. These information-theoretic measures include Pointwise Mutual Information (PMI) [17], SMI measures : Predictability (I_2) [19] and *Stimulus* Specific Information (I_3) [10].

PMI Guided Fusion. PMI [17] quantifies the degree of association (or disassociation) between individual data points given two variables. If X and Y are two variables, then each data point can be represented by the value pair (x, y) where $x \in X$ and $y \in Y$. The statistical association between these two points can be measured by their PMI value:

$$\mathrm{PMI}(x, y) = \log \frac{\mathrm{p}(x, y)}{\mathrm{p(x)p(y)}} \tag{6}$$

where $\mathrm{p}(x)$ and $\mathrm{p}(y)$ are the probabilities of occurrence of values $x \in X$ and $y \in Y$. $\mathrm{p}(x, y)$ is the joint probability of occurrence of values x and y together. Comparing Eqs. 1 and 6, we can infer that the expected PMI values over all occurrences of variables X and Y correspond to the MI value $I(X;Y)$. PMI is a symmetric measure that can generate values ranging from negative to positive, depending on whether the distributions are complementary or overlapping. If the information overlap is high ($\mathrm{p}(x, y) > \mathrm{p}(x)\mathrm{p}(y)$), then $\mathrm{PMI}(x, y) > 0$. The low association is indicated by $\mathrm{p}(x, y) < \mathrm{p}(x)\mathrm{p}(y)$, resulting in $\mathrm{PMI}(x, y) < 0$. If x and y are statistically independent then $\mathrm{p}(x, y) = \mathrm{p}(x)\mathrm{p}(y)$ and $\mathrm{PMI}(x, y) = 0$.

Given the PMI measure, we can devise a fusion strategy similar to I_1 where I_1 values are substituted with PMI values in Algorithm 1. The resulting fused information field on the simulated rolling ball is shown in Fig. 2(b). We observe that PMI fails to capture the spatial characteristics of the key regions and only captures the overlapped regions indicated by high positive PMI values.

***I_2* Guided Fusion** Predictability (I_2) is another decomposition of MI introduced by [19]. This SMI measure quantifies the change in the uncertainty of one variable (X) after observing the individual value of another variable ($y \in Y$) and is computed as:

$$\mathrm{I}_2(y; X) = -\sum_{x \in X} p(x) \log p(x) + \sum_{x \in X} p(x|y) \log p(x|y) \tag{7}$$

where $y \in Y$ is the reference variable and $x \in X$ is the target variable. $\mathrm{p}(x)$ is the probabilities of occurrence of values x for X and $\mathrm{p}(x|y)$ is the conditional probabilities values of x given y. In some cases, the increased uncertainty can reveal

Algorithm 1. Fusion Process

Input:

- data1: Array of data values from fused timestep. Initialized with the first timestep.
- data2: Array of data values from the subsequent timestep.
- Ifield1: Array of I_1 values for $I_1(x;Y)\ \forall x \in X$
- Ifield2: Array of I_1 values for $I_1(y;X)\ \forall y \in Y$
- timestep_fuse: Array of timestep values. Starts with 0
- time: Current timestep value
- conf_th: Confidence threshold for the key regions.

Output:

- fused_field_data: Array of the fused data values
- fused_field_I1: Array of the fused I_1 values
- timestep_fuse: Array of the fused timestep values.

```
procedure CREATEFUSIONFIELDS(data1, data2, Ifield1, Ifield2, timestep_fuse, time, conf_th)
    fused_field_data ← array of zeros with shape data1
    fused_field_I1 ← array of zeros with shape data1
    for i ← 0 to data1.shape[0] − 1 do
        for j ← 0 to data1.shape[1] − 1 do
            if Ifield1[i][j] > Ifield2[i][j] then
                fused_field_data[i][j] ← data1[i][j]
                fused_field_I1[i][j] ← Ifield1[i][j]
                if time = 1 then
                    timestep_fuse[i][j] ← time
                end if
            else
                fused_field_data[i][j] ← data2[i][j]
                fused_field_I1[i][j] ← Ifield2[i][j]
                timestep_fuse[i][j] ← time + 1
            end if
        end for
    end for
    for i ← 0 to data1.shape[0] − 1 do
        for j ← 0 to data1.shape[1] − 1 do
            if fused_field_data[i][j] < conf_th then
                timestep_fuse[i][j] ← 0
            end if
        end for
    end for
    return fused_field_data, fused_field_I1, timestep_fuse
end procedure
```

significant information about the relationship between the variables. However, when we use the I_2 measure instead of the I_1 measure in the fusion process, the resulting fused information does not offer a meaningful summary over time.

I_3 Guided Fusion. Stimulus Specific Information (SSI) [10], denoted by I_3:

$$\mathrm{I}_3(y;X) = -\sum_{x\in X} p(x|y)\mathrm{I}_2(x;Y) \tag{8}$$

The response and stimulus are the two variables X and Y. This measure emphasizes that the most informative data values from the first variable are related to the most informative data values of the second variable [10]. In some cases, I_1 can be an alternate measure for I_3, but the interpretation is different based on the data [10]. When I_3 is used instead of I_1 on the simulated rolling ball dataset, it captured very similar properties shown in Fig. 2(a). However, when applied to a more complex dataset, it failed to capture the spatial properties of the features in the summarization. This is explained in detail in Sect. 4.1 and shown in Fig. 3(b)

4 Applications

4.1 MFIX-Exa Flow Simulation

MFIX-Exa [37] is a multiphase flow simulation developed by the National Energy Technology Laboratory (NETL), USA. Using MFIX-Exa, particle-based data is generated for studying the operational principles of chemical looping reactors. In such simulations, the formation of void regions, known as bubbles, is an important phenomenon. Understanding the temporal evolution of these bubbles holds significant importance for domain experts.

Data Context and Features. For analyzing bubble dynamics, typically the raw particle data is first converted to a scalar density field. Then bubbles can be segmented as the connected regions with low particle density. For more details about this pre-processing, please refer to [23]. In this work, we assume that the scalar density fields are already available and we use 2D slices extracted from the density fields. These slices contain scalar values representing particle density. Our evaluation dataset consists of multiple timesteps (count 332), and each timestep corresponds to 2D data samples with dimensions of 488 $\times$ 842. The sample values fall within the range of [-1.2 $\times$ 10^{-6}, 29.08]. In [23], the detection, segmentation, and characterization of the bubbles are studied in an extensive manner. In our work, we use the VTK [46] library to extract the connected components and then use a low scalar density threshold value to filter the bubbles. Over time, the bubbles undergo phases like creation, merge, split, and dissolve into air. Domain experts want to comprehend the evolution of bubbles and explore the relationships between various bubble characteristics such as their size, shape, number of bubbles, etc. [23]. Important events in this simulation can be characterized by the creation of a bubble, the merging of two or more bubbles, or the dissolving of a bubble. Note that for all of these events, the total number of bubbles will change. Hence, a timestep with the bubble number changed from a previous timestep, can be considered as a "trigger". Here, we ignore counting changes in very small bubbles since the domain experts are more concerned about the bubbles when they grow in size.

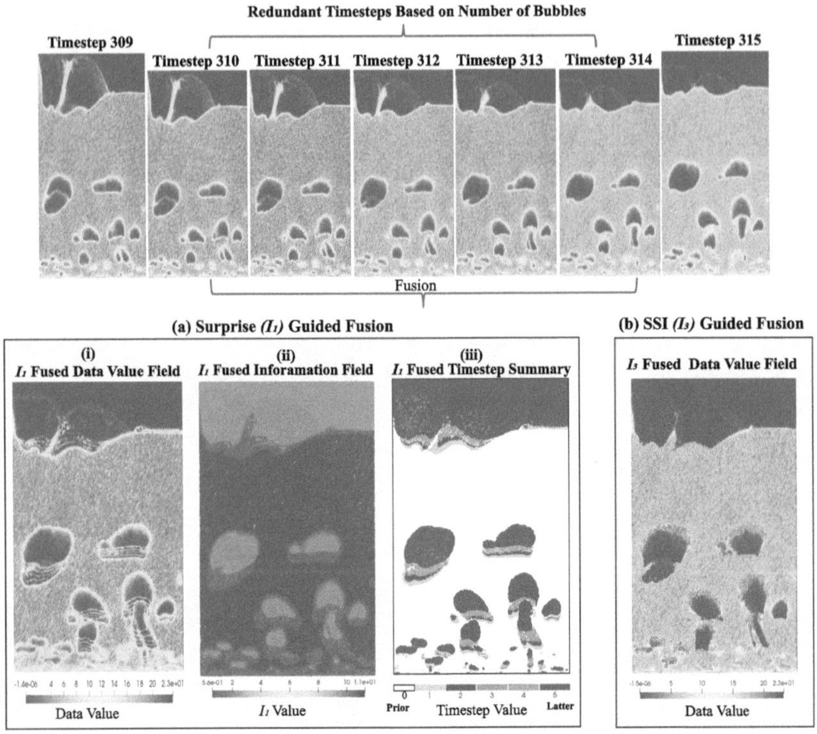

Fig. 3. Analysis of DSTS method for MFIX-Exa simulation. The first row represents a window of timesteps (309–315) where the bubbles are highlighted as the blue regions. Timesteps are raw images with 488 × 842 dimensions. 310 to 314 are fused using (a) Surprise (I_1) guided fusion (i-iii) and alternative (b) Stimulus Specific Information (SSI) or I_3 guided fusion. a(i) shows the I_1 fused data value field of the timesteps, a(ii) shows I_1 fused information value field reflecting I_1 values and a(iii) shows I_1 fused timestep summary with 5 timesteps. The color bar labels 5 different colors summarizing the direction of the bubbles in one timestep. The alternative I_3 is unable to capture the path of bubbles as reflected in (b) I_3 fused data field. (Color figure online)

Results for Data Summarization. Figure 3 shows the analysis of the DSTS method for MFIX-Exa simulation. Timesteps 309–315 are shown in the first row where bubbles are the blue regions. Timesteps 310 to 314, during which the number of bubbles remains unchanged, are summarized through the fusion process. Figures 3 (a) represent results from the I_1 guided fusion. The I_1 fused data value field a(i) shows the scalar values ranging $[-1.4 \times 10^{-6}, 23]$ for the fused timesteps. Here the change in bubble movement is very prominent. Figure 3 a(ii) presents the I_1 values ranging $[5.6 \times 10^{-1}, 11]$ for the fused timesteps. The range of I_1 values is smaller, making it less sensitive to bubble movement compared to particle-density values. However, it effectively highlights the main bubbles and their temporal dynamics. The timestep summary field in Fig. 3 a(iii) represents the timestep values from which the bubbles originate. Here 5 timesteps are distinctly color-coded to reflect the direction of the bubbles from the start to the end position. The white background (labeled 0) filters all the density values that are of low importance for this dataset.

We have also implemented the alternative SSI (I_3) guided fusion technique as mentioned in Sect. 3.4 for MFIX-Exa. Figure 3(b), represents the I_3 fused data value field. Here the bubbles are only partially highlighted and the change in the bubbles' movement is also hard to interpret. While I_3 captures some spatial features of the bubbles, the edges are blurred. Thus, the *Surprise* fusion method proves to be better than the SSI measure.

4.2 Surveillance Data Analysis and Optimization

To demonstrate DSTS in security camera footage analysis, we used the publicly available SBM-RGBD Dataset [12,45]. This dataset was originally created for the Workshop on Background Learning for Detection and Tracking from RGBD Videos [41]. The dataset comprises 33 RGBD videos, totaling 15033 timesteps, recorded indoors using a Microsoft Kinect sensor [12]. The dataset contains videos capturing moving objects at intervals, which aligns with our data requirements. Here, we used one of the videos that shows four individuals walking in and out of the view area, engaging in discussions, and writing on a whiteboard. The most significant impact of our proposed method in this application is on archiving the storage optimization.

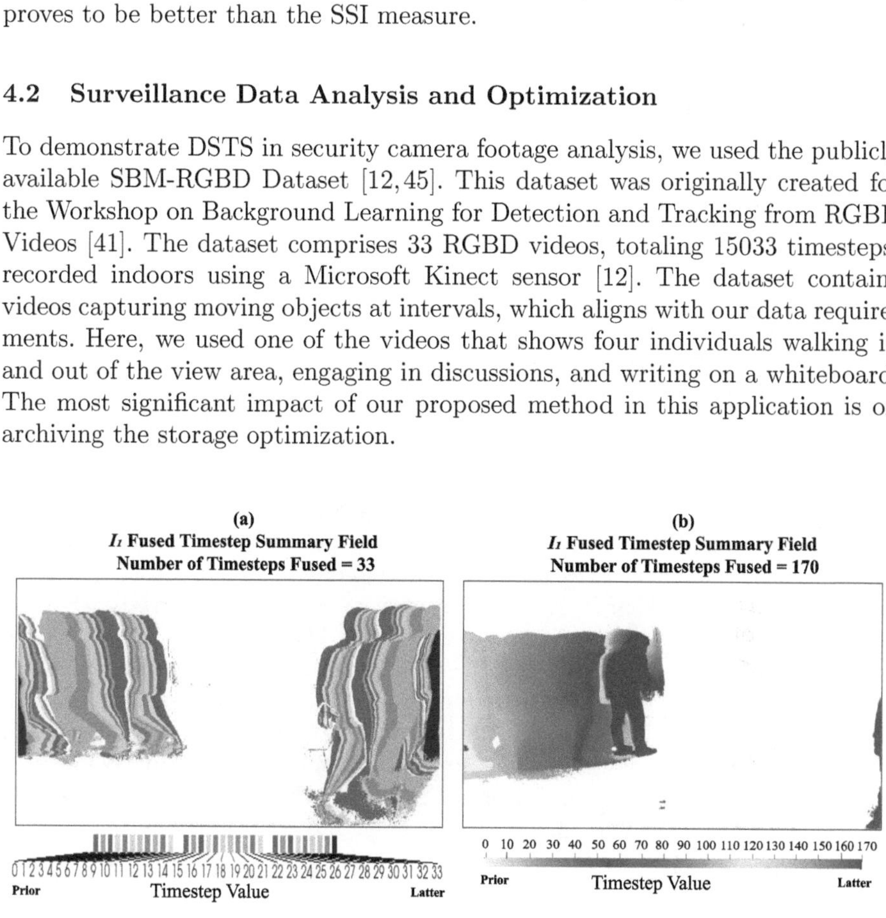

Fig. 4. Results of the DSTS method for SBM-RGBD dataset. Here the emphasis is on representing the extensive number of timestep summaries. (a) shows a summarization of 33 fused timesteps using a discrete color bar. (b) showcases a summarization of 170 fused timesteps, employing a continuous color bar to depict information changes over a longer period. The color bars point out the spatial direction of the information flow by denoting the prior and latter states.

Data Context and Features. The videos have 640 × 480 resolution and the length is 1400 timesteps. The key regions (features) are the individuals and their movement. The whiteboard and a chair are stationary in the background. To extract key regions, we have used a background subtraction algorithm, called ViBe [4]. The algorithm aims to identify moving objects within consecutive images or videos by distinguishing between the foreground (moving objects) and the background (stationary elements). The ViBe algorithm is adaptive and computationally lightweight, making it suitable for real-time applications. Its straightforward pseudocode in [4] facilitates easy implementation. The ViBe algorithm converts the RGB images into masked binary images with segmented individuals.

In scenarios with multiple individuals, we adopt a concept similar to that used for counting bubbles in the MFIX-Exa (Sect. 4.1). We apply the concept to count the number of individuals in each timestep by analyzing the largest connected regions in the masked images. Since it is a binary image, the data samples have two values: 0 (no individual) and 255 (individual). By setting a size threshold, we can accurately count the number of individuals in each timestep. Given that individuals move in and out of the view area, the number of individuals can be used as a trigger for our application. Whenever a person enters or exits, that is a key timestep. The consecutive timesteps between two triggers are then fused using Algorithm 1. This fusion process effectively summarizes the movement patterns of individuals within one timestep.

Results for Data Summarization. In Fig. 4(a), the summarization field depicts a fusion of 33 timesteps, where two individuals walk out of the view area. The leftmost person exits first, initiating the trigger and stopping the fusion process. Each timestep is represented by a discrete color, highlighting the changes in movement over 33 timesteps. Sample values below 255 are set to 0, representing the white background, as the data value of individuals is 255. Figure 4(b) shows a timestep summary for a longer period of 170 timesteps. The summarization field captures an individual walking into the view area and writing on the board, while another person has just stepped in, initiating the trigger. Continuous colors are used to display the movement changes. Expectedly, key and summarized timesteps in this dataset result in a significant reduction from 1400 timesteps to only 49 timesteps. The highest number of timesteps being fused is 262.

4.3 Tracking Cell Interactions in Lymph Nodes

This dataset contains consecutive images of cellular interactions within the lymph node (LN). LNs are essential for immune function, playing a crucial role in initiating immune response and facilitating immune cell communication [35]. We reanalyze data from [50], where information theory-based approaches were used to identify and quantify the spatial relationships between naïve T cells and other cell types like Dendritic Cells (DCs). The data for the study was gathered using two-photon microscopy (2PM) [42] to acquire 3D image stacks of LN tissue samples from mice. The imaging process captured dynamic movies lasting 10 to 45 min, resulting in a sequence of 3D images. This dataset is well-suited for the application of our method.

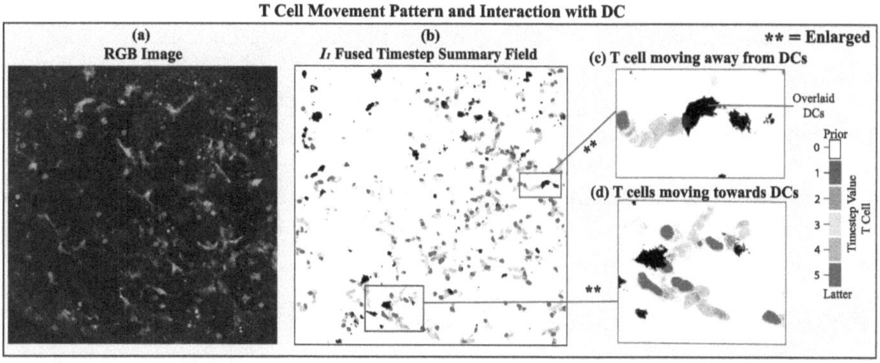

Fig. 5. Results of the DSTS method for cell interaction in Lymph Node. The results emphasize T cell movement patterns and interaction with DCs in LN. (a) is an illustrative timestep from the dataset. Here red indicates the T cells and green indicates the DCs. (b) is the surprise fused timestep summary fields of T cell for 5 timesteps. The black cells in the field are overlaid DCs to highlight cell contact. (c) and (d) are 2 enlarged positions from the (b) field to emphasize T cell movement. (c) shows that a T cell is moving away from the DCs. (d) shows multiple T cells moving toward the DCs. The corresponding timestep values are provided in the color bars to highlight the direction. (Color figure online)

Data Context and Features. Figure 5(a) shows an RGB image with T cells dyed red and Dendritic Cells (DC) dyed green. Each voxel contains the color intensities of the dye in the red, blue, and green channels. For every time step, we extract the red and green channels into two separate images. We focus on the red channel in order to analyze T-cell motility. Because these images contain a lot of noise, we implement a pre-processing step using the median filter [26], to reduce noise while preserving the edges of the cells for improved visualization. Since the red channels specifically represent the T cells, no segmentation is required. In [50], MI and normalized mutual information (NMI) were used to quantify associations between cells. Here, we use the MI value between two cell types as a "trigger". If the MI value for a specific timestep exceeds a specified threshold, we save that as a key timestep. If the MI value falls below the threshold, we find the next timestep in which the MI value exceeds the trigger threshold and fuse the intermediate ones.

Results of Data Summarization. Figure 5 focuses on the T cell movement and interaction with DCs in the summarized timesteps. Figure 5(a) is a sample timestep of the T:DC dataset with $512 \times 512 \times 22$ dimensions. This dataset has a total of 51 timesteps. Figure 5(b) displays the I_1 fused timestep summary field, representing five fused timesteps from this dataset. The black cells in the summarization represent the DCs' value field overlaid on the fused summary field, visually illustrating the physical interactions between T cells and DCs. Given that there are multiple interactions captured in each timestep, we highlight two specific interactions by enlarging the locations in Figs. 5(c) and (d). In Fig. 5(c), we observe a T cell moving away from the DCs. The color bar on the right indicates the first (blue) and last (red) timesteps in the summary field, clearly indicating the movement direction. In Fig. 5(d), we see multiple T cells moving toward the DCs making explicit contact. This visualization allows for a comprehensive understanding of the dynamic interactions between T cells and DCs, providing valuable insights into the temporal dynamics of immune cell communication.

5 Discussion

The proposed DSTS technique has demonstrated its effectiveness and flexibility across several applications. Starting with a synthetic simulation of a rolling ball to analyzing complex cellular interactions within lymph nodes, the method effectively showcased its robustness. The combination of the key and fused timestep resulting from the method provides a compact yet comprehensive data summarization.

We select applications from multiple domains to shed light on different aspects of the DSTS method. This technique offers a practical solution to downsize and analyze the features in the MFIX-Exa simulation. This application analysis establishes that the method can handle raw scalar data as well as image-based data that incorporates the rest of the applications. The RGBD tracking dataset is introduced to show the method's ability to summarize and highlight important movement patterns of individuals in a video sequence. The results from this dataset reflect that longer fused timesteps are equally apprehensible as the shorter ones. This has promising implications for surveillance and security applications. The cellular interaction in LN is a more complex dataset. T cells which are the key regions (features) are ample in number and the interactions with DCs are sporadic in nature. Our method is able to track multiple cell interactions. The summarization highlights immune cell communication by providing a comprehensive visualization of the T cell movement. This visualization potentially introduces new possibilities for immunological research.

All the applications in this work present post hoc data analysis. Since the method is not computationally expensive it can be easily combined to analyzing data in a streaming framework. Through the integration of this method in any in situ streaming framework, the resulting data will be summarized in real-time ensuring optimal storage reduction.

We acknowledge that challenges may arise in selecting appropriate triggers and threshold values, especially in complex datasets with multiple key features, interactions, and noise. However, the flexibility of the technique allows for the adjustment of parameters to tailor the summarization process to different applications. Additionally, future research could explore combining different information-theoretic measures to further enhance the summarization capabilities to multivariate time-varying datasets.

Acknowledgments. We thank Professor Judy Cannon (UNM) for providing the biological data, and Dr. Matthew Fricke (UNM) for discussions on cell analysis.

Data Availability Statement. All the code and data are available at https://github.com/htasnim/Dynamic-Spatio-Temporal-Data-Summarization

Disclosure of Interests. The authors declare that they have no known competing financial interests or personal relationships that could have appeared to influence the work reported in this paper.

References

1. Ahrens, J., Jourdain, S., O'Leary, P., Patchett, J., Rogers, D.H., Petersen, M.: An image-based approach to extreme scale in situ visualization and analysis. In: International Conference for High Performance Computing, Networking, Storage and Analysis, SC 2015-January, pp. 424–434 (2014). https://doi.org/10.1109/SC.2014.40
2. Akiba, H., Fout, N., Ma, K.L.: Simultaneous classification of time-varying volume data based on the time histogram. EuroVis **6**, 1–8 (2006)
3. Akiba, H., Ma, K.L., Chen, J.H., Hawkes, E.R.: Visualizing multivariate volume data from turbulent combustion simulations. Comput. Sci. Eng. **9**(2), 76–83 (2007). https://doi.org/10.1109/MCSE.2007.42
4. Barnich, O., Van Droogenbroeck, M.: Vibe: a universal background subtraction algorithm for video sequences. IEEE Trans. Image Process. **20**(6), 1709–1724 (2010)
5. Biswas, A., Dutta, S., Shen, H.W., Woodring, J.: An information-aware framework for exploring multivariate data sets. IEEE Trans. Visual Comput. Graph. **19**, 2683–2692 (2013). https://doi.org/10.1109/TVCG.2013.133
6. Bramon, R., Ruiz, M., Bardera, A., Boada, I., Feixas, M., Sbert, M.: An information-theoretic observation channel for volume visualization. Comput. Graph. Forum **32**(3 PART4), 411–420 (2013). https://doi.org/10.1111/cgf.12128
7. Bramon, R., et al.: Multimodal data fusion based on mutual information. IEEE Trans. Visual Comput. Graph. **18**(9), 1574–1587 (2012). https://doi.org/10.1109/TVCG.2011.280
8. Bramon, R., Ruiz, M., Bardera, A., Boada, I., Feixas, M., Sbert, M.: Information theory-based automatic multimodal transfer function design. IEEE J. Biomed. Health Inform. **17**(4), 870–880 (2013). https://doi.org/10.1109/JBHI.2013.2263227
9. Bruckner, S., Möller, T.: Isosurface similarity maps. Comput. Graph. Forum **29**, 773–782 (2010)
10. Butts, D.A.: How much information is associated with a particular stimulus? Netw. Comput. Neural Syst. **14**(2), 177–187 (2003). https://doi.org/10.1088/0954-898X_14_2_301

11. Cahill, N.D.: Normalized measures of mutual information with general definitions of entropy for multimodal image registration. In: Fischer, B., Dawant, B.M., Lorenz, C. (eds.) WBIR 2010. LNCS, vol. 6204, pp. 258–268. Springer, Heidelberg (2010). https://doi.org/10.1007/978-3-642-14366-3_23
12. Camplani, M., Maddalena, L., Moyá Alcover, G., Petrosino, A., Salgado, L.: A benchmarking framework for background subtraction in RGBD videos. In: Battiato, S., Farinella, G.M., Leo, M., Gallo, G. (eds.) ICIAP 2017. LNCS, vol. 10590, pp. 219–229. Springer, Cham (2017). https://doi.org/10.1007/978-3-319-70742-6_21
13. Cappello, F., et al.: Use cases of lossy compression for floating-point data in scientific data sets. Inter. J. High Performance Comput. Appli. **33**(6), 1201–1220 (2019)
14. Castanedo, F.: A review of data fusion techniques. Sci. World J. (2013) (2013)
15. Chen, M., Feixas, M., Viola, I., Bardera, A., Shen, H.W., Sbert, M.: Information theory tools for visualization. CRC Press (2016)
16. Childs, H.: Data exploration at the exascale. Supercomput. Front. Innovat. **2**(3), 5–13 (2015)
17. Church, K., Hanks, P.: Word association norms, mutual information, and lexicography. Comput. Linguist. **16**(1), 22–29 (1990)
18. Cover, T.M., Thomas, J.A.: Elements of Information Theory. 2nd edn. (Wiley Series in Telecommunications and Signal Processing. Wiley-Interscience (2006)
19. DeWeese, M.R., Meister, M.: How to measure the information gained from one symbol. Netw. Comput. Neural Syst. (1999). https://doi.org/10.1088/0954-898X_10_4_303
20. Dutta, S., Chen, C.M., Heinlein, G., Shen, H.W., Chen, J.P.: In situ distribution guided analysis and visualization of transonic jet engine simulations. IEEE Trans. Visual Comput. Graph. **23**(1), 811–820 (2016)
21. Dutta, S., Liu, X., Biswas, A., Shen, H.W., Chen, J.P.: Pointwise information guided visual analysis of time-varying multi-fields. In: SIGGRAPH Asia 2017 Symposium on Visualization, SA 2017 (2017). https://doi.org/10.1145/3139295.3139298
22. Dutta, S., Tasnim, H., Turton, T.L., Ahrens, J.: In situ adaptive spatio-temporal data summarization. In: Proceedings - 2021 IEEE International Conference on Big Data, Big Data 2021, pp. 315–321 (2021). https://doi.org/10.1109/BigData52589.2021.9671581
23. Dutta, S., Turton, T., Rogers, D., Musser, J.M., Ahrens, J., Almgren, A.S.: In situ feature analysis for large-scale multiphase flow simulations. J. Comput. Sci. **63**(July), 101773 (2022). https://doi.org/10.1016/j.jocs.2022.101773
24. Duzceker, A., Galliani, S., Vogel, C., Speciale, P., Dusmanu, M., Pollefeys, M.: Deepvideomvs: multi-view stereo on video with recurrent spatio-temporal fusion. In: Proceedings of the IEEE/CVF Conference on Computer Vision and Pattern Recognition, pp. 15324–15333 (2021)
25. Fu, Y., Guo, Y., Zhu, Y., Liu, F., Song, C., Zhou, Z.H.: Multi-view video summarization. IEEE Trans. Multimedia **12**(7), 717–729 (2010)
26. Gonzales, R.C., Wintz, P.: Digital image processing. Addison-Wesley Longman Publishing Co., Inc (1987)
27. Hill, D.L., Batchelor, P.G., Holden, M., Hawkes, D.J.: Medical image registration. Phys. Med. Biol. **46**(3), R1 (2001)
28. Isola, P., Zoran, D., Krishnan, D., Adelson, E.H.: Crisp boundary detection using pointwise mutual information. In: Fleet, D., Pajdla, T., Schiele, B., Tuytelaars, T.

(eds.) ECCV 2014. LNCS, vol. 8691, pp. 799–814. Springer, Cham (2014). https://doi.org/10.1007/978-3-319-10578-9_52
29. Kashinath, S.A., et al.: Review of data fusion methods for real-time and multi-sensor traffic flow analysis. IEEE Access **9**, 51258–51276 (2021)
30. Kullback, S., Leibler, R.A.: On information and sufficiency. Ann. Math. Stat. **22**(1), 79–86 (1951)
31. Kuruvilla, J., Sukumaran, D., Sankar, A., Joy, S.P.: A review on image processing and image segmentation. In: 2016 International Conference on Data Mining and Advanced Computing (SAPIENCE), pp. 198–203 (2016). https://doi.org/10.1109/SAPIENCE.2016.7684170
32. Li, J., Li, Y., He, L., Chen, J., Plaza, A.: Spatio-temporal fusion for remote sensing data: an overview and new benchmark. SCIENCE CHINA Inf. Sci. **63**, 1–17 (2020)
33. Ma, P., Kang, E.L.: Spatio-temporal data fusion for massive sea surface temperature data from modis and amsr-e instruments. Environmetrics **31**(2), e2594 (2020)
34. Maes, F., Collignon, A., Vandermeulen, D., Marchal, G., Suetens, P.: Multimodality image registration by maximization of mutual information. IEEE Trans. Med. Imaging **16**(2), 187–198 (1997)
35. Mirsky, H.P., Miller, M.J., Linderman, J.J., Kirschner, D.E.: Systems biology approaches for understanding cellular mechanisms of immunity in lymph nodes during infection. J. Theor. Biol. **287**, 160–170 (2011)
36. Moore, D.G., Valentini, G., Walker, S.I., Levin, M.: Inform: efficient information-theoretic analysis of collective behaviors. Front. Robot. AI **5**, 1–14 (2018). https://doi.org/10.3389/frobt.2018.00060
37. Musser, J., et al.: MFIX-Exa: a path toward exascale CFD-DEM simulations. Int. J. High Perform. Comput. Appl. **36**(1), 40–58 (2022). https://doi.org/10.1177/10943420211009293
38. Myers, K., et al.: Partitioning a large simulation as it runs. Technometrics **58**(3), 329–340 (2016)
39. Nguyen, H., Katzfuss, M., Cressie, N., Braverman, A.: Spatio-temporal data fusion for very large remote sensing datasets. Technometrics **56**(2), 174–185 (2014)
40. Reed, D.A., Dongarra, J.: Exascale computing and big data. Commun. ACM **58**(7), 56–68 (2015)
41. Background learning for detection and tracking from rgbd videos. https://rgbd2017.na.icar.cnr.it/, Accessed 22 Sep 2023
42. Rubart, M.: Two-photon microscopy of cells and tissue. Circ. Res. **95**(12), 1154–1166 (2004)
43. Ruiz, M., Bardera, A., Boada, I., Viola, I., Feixas, M., Sbert, M.: Automatic transfer functions based on informational divergence. IEEE Trans. Visual Comput. Graphics **17**(12), 1932–1941 (2011). https://doi.org/10.1109/TVCG.2011.173
44. Sbert, M., Feixas, M., Rigau, J., Chover, M., Viola, I.: Information theory tools for computer graphics. Springer Nature (2022)
45. Sbm-rgbd dataset. https://rgbd2017.na.icar.cnr.it/SBM-RGBDdataset.html Accessed 22 Sep 2023
46. Schroeder, W., Martin, K.M., Lorensen, W.E.: The visualization toolkit an object-oriented approach to 3D graphics. Prentice-Hal Inc. (1998)
47. Shah, Z., Anwar, A., Mahmood, A.N., Tari, Z., Zomaya, A.Y.: A spatiotemporal data summarization approach for real-time operation of smart grid. IEEE Trans. Big Data **6**(4), 624–637 (2017)
48. Shannon, C.E.: A mathematical theory of communication. Bell Syst. Tech. J. **27**(3), 379–423 (1948). https://doi.org/10.1002/j.1538-7305.1948.tb01338.x

49. Sipser, M.: Introduction to the Theory of Computation, third edn. Course Technology, Boston, MA (2013)
50. Tasnim, H., Fricke, G., Byrum, J., Sotiris, J., Cannon, J., Moses, M.: Quantitative measurement of naïve T cell association with dendritic cells, FRCs, and blood vessels in lymph nodes. Front. Immunol. **9** (2018). https://doi.org/10.3389/fimmu.2018.01571
51. Tasnim, H., Dutta, S., Turton, T.L., Rogers, D.H., Moses, M.E.: Information-theoretic exploration of multivariate time-varying image databases. Comput. Sci. Eng. **24**(3), 61–70 (2022). https://doi.org/10.1109/MCSE.2022.3188291
52. Tong, X., Lee, T.Y., Shen, H.W.: Salient time steps selection from large scale time-varying data sets with dynamic time warping. In: IEEE Symposium on Large Data Analysis and Visualization (LDAV), pp. 49–56. IEEE (2012)
53. Verdu, S.: Fifty years of shannon theory. IEEE Trans. Inf. Theory **44**(6), 2057–2078 (1998). https://doi.org/10.1109/18.720531
54. Vidhya, K., Karthikeyan, G., Divakar, P., Ezhumalai, S.: A review of lossless and lossy image compression techniques. Int. Res. J. Eng. Technol. (IRJET) **3**(4), 616–7 (2016)
55. Viola, I., Feixas, M., Sbert, M., Groller, M.E.: Importance-driven focus of attention. IEEE Trans. Visual Comput. Graph. **12**(5), 933–940 (2006)
56. Wang, C., Yu, H., Ma, K.L.: Importance-driven time-varying data visualization. IEEE Trans. Visual Comput. Graph. **14**(6), 1547–1554 (2008)
57. Wang, Q., Tang, Y., Tong, X., Atkinson, P.M.: Virtual image pair-based spatio-temporal fusion. Remote Sens. Environ. **249**, 112009 (2020)
58. Woodring, J., Ahrens, J., Figg, J., Wendelberger, J., Habib, S., Heitmann, K.: In-situ sampling of a large-scale particle simulation for interactive visualization and analysis. Comput. Graph. Forum **30**, 1151–1160 (2011)
59. Wu, P., et al.: Spatially continuous and high-resolution land surface temperature product generation: a review of reconstruction and spatiotemporal fusion techniques. IEEE Geosci. Remote Sensing Mag. **9**(3), 112–137 (2021)
60. Xue, J., Leung, Y., Fung, T.: A bayesian data fusion approach to spatio-temporal fusion of remotely sensed images. Remote Sensing **9**(12), 1310 (2017)
61. Ye, Y.C., et al.: In situ generated probability distribution functions for interactive post hoc visualization and analysis. In: IEEE 6th Symposium on Large Data Analysis and Visualization (LDAV), pp. 65–74. IEEE (2016)
62. Zhang, L., Gao, Y., Hong, R., Hu, Y., Ji, R., Dai, Q.: Probabilistic skimlets fusion for summarizing multiple consumer landmark videos. IEEE Trans. Multimedia **17**(1), 40–49 (2014)
63. Zhong, S.H., Wu, J., Jiang, J.: Video summarization via spatio-temporal deep architecture. J. Neurocomput. **332**, 224–235 (2019)
64. Zhou, B., Chiang, Y.J.: Key time steps selection for large-scale time-varying volume datasets using an information-theoretic storyboard. Comput. Graph. Forum **37**, 37–49 (2018)

Aerodynamic Trajectory Impact: Reliability of Blade Geometry in Compressor and Turbine Application for Power Generation and Electricity Supply

Sepiribo Lucky Braide(✉), Dan Horsfall, and K. E. Orie

Department of Electrical and Electronic Engineering, Rivers State University, Port Harcourt 5080, Rivers State, Nigeria
{Sepiribo.braide,Dan.horsfall1,eze.orie}@ust.edu.ng

Abstract. The power station plays critical role in supporting Nation's economic development for improving standard of living; it is becoming evident that many generating companies' open cycle gas turbine power plant components are facing uncontrollable thermal loading (stress/strain) challenges particularly on compressor and turbine blades application thereby making the performances, efficiency and reliability of these blades to experienced poor power generation. Ideally, blades are essential components that play crucial role for compressing atmospheric air (for increased temperature and pressure) in order to provide inlet conditions of temperature (400–600 °C) and pressure (10–30 bar). The exit conditions of the compressor must be met to engage delivery of the needed energy requirement to the combustion chamber on the view to further increase the burning process of compressed air to provide high thermodynamic energy evolving from temperature of 600 °C–1500 °C and pressure of 10–30 bar necessary to extract energy for combustion chamber where expansion processes are achieved to produce the mechanical energy in order to drive the compressor and also energize generator and alternators for power generation or production. This paper considered application of Weibull and Rayleigh's distribution for the prediction of compressor/turbine blade failures due to tear and wear, thermal loading, fatigue and creeping of contaminated surfaces that may result into erosion dip that will evidently amount to outages or blackout in the grid power supply. Similarly, three (3) identical gas turbine power plant with 36MW capacity of thirty (30) blades for a period of 10 years (2016–2025) were examined and simulated using specialized TPC windchill quality solution (version 10.2) to determine the performance of compressor/turbine blade parameters of shape (β), scale (γ), and location (η) respectively. The results obtained from compressor blades simulation of probability – plot, probability density function (PDF) plot, reliability vs time plot, unreliability vs time plot, failure rate vs time plot are presented as: shape (β) = 2.9085, scale (γ) = 24.3296, and location (η) = −5.5962 respectively while the compressor blades characteristics are captured in Fig. 1, 2, 3, 4 and 5. Similarly, simulation results obtained from turbine blades are also presented as: Shape (β) = 4.1353, Scale (γ) = 33.2384 and Location (η) = −12.8047. The compressor blade plot shows prediction of increasing failure rate with shape or slope parameter (β = 2.9085) greater than one (1); this means that reliability of the compressor blade is

H. R. Arabnia et al. (Eds.): CSCI 2025, CCIS 2970, pp. 199–225, 2027.
https://doi.org/10.1007/978-3-032-28466-2_15

decreasing with time as contained and shown in the different plots while the scale parameters showing the spreading rate of the failure which needs to be rescaled to fit back its performance, efficiency and reliability. The location (η) parameter defines the component behavior with (η = −5.5962) not failing before −5.5962 units of time (e.g. hours, days, etc.), in other words there is guaranteed period during which components won't fail, if it rescaled to maintain performance, efficiency and reliability (PER). The Weibull distribution tends to show better performance, accuracy and earlier prediction time of failure of compressor blades as compared to Rayleigh's distribution for the purpose of validation.

Keywords: Reliability · Aerodynamic Trajectory · Compressor · Turbine Blade Geometry · Power Generation · Blade Geometry

1 Introduction

Nigeria has a total of 22 major power stations spread across various regions, generating electricity from a range of source including hydroelectric gas, fired, and thermal power plant. The contemporary power system represents vast engineering infrastructure which is paramount for sustainable progress in residential, commercial, industrial and socio-economic development. Nigeria's Power stations (Generation, Transmission & Distribution) play a critical role in supporting the economic development and improving standard of living. These stations include;

- Hydroelectric power stations: Kainji power station (760 MW), Jebba power station (578 MW), and Shiroro power station (600 MW).
- Thermal power stations: Egbin power station (1,320 MW), Delta power station (900 MW), Gerega power station (726 MW), and Sapele power station (1.020 MW).
- Gas-fired power stations: Afam power Station (776 MW), Omotosho power Station (726 MW), Oloriunsogo power Station (754 MW), and Alaoji power Station (1074 MW).

These power stations play a critical role in supporting Nigeria's economic development and improving the standard of living.

However, the current situation is that Nigeria's generating capacity is approximately 10,396 MW, with an available capacity of 6,056 MW. While the peak demand is around 1,280 MW, leaving a mismatch between supply and demand. Following, load demand for case repairs, with population of over 200million people, Nigeria's load demand is expected to continue growing. According to the Nigeria electricity regulatory commission (NERC) the peck demand is projected to reach: 17,000 MW by 2025, 23,000 MW by 2030 and 30,000 MW by 2035.

This means that to meet growing demand with respect to the generating capacity gap. Nigeria needs to increase its generating capacity significantly. The current gap between generating capacity and peak demand is around 6,744 MW (12,800 MW–6056 MW).

Evidently, there is strong need required for generating capacity to meet projected peak demand, Nigeria needs to additional generating capacity. This shows that by 2025, this year the required generating capacity would be around 17,000 MW, which represents 181% increase from the current available capacity of 6,056 MW.

Essentially, the widely used distribution reliability evaluation tool and data analysis technique for predicting system characteristics behavior are the Weibull and Rayleigh distribution, they are normally used to study down time of equipment or machine parts. It is also use for fitting engineering data obtained in that regards such as thermal load, strength of materials, fracture of brittle materials. However, weibull parameter distribution is used for fitting random data because of its flexibility to various types of probability distribution. This technical paper proposed the weibull distribution with the following parameters such as shape (β), scale (μ) and location parameter (η) for the estimation of system characteristics and performance. The Weibull distribution is often used to model the reliability and failure behavior of components including compressor and turbine blades. The key characteristics of the weibull distribution relevant to these components are;

(i) Slope or Shape Parameter (β): the shape parameter (β) determines the shape of the Weibull distribution. For compressor and turbines blades:
- $\beta = 1$, indicates a constant failure rate, suggesting random failures.
- $\beta > 1$, indicates an increasing failures rate over time, typically due to wear and tear, fatigue or deregulation.

(ii) Scale Parameter (η): the scale parameter (η) represent the characteristics life of t time to failure of the component for compressor ad turbines blades:
- A largely value indicates longer expected life or time to failure.
- η is often referred to as the characteristics life and is the time 63.2% of the component are expected to have failed

(iii) Location Parameter (μ): the location parameter (μ) represents the minimum time-to-failure or the threshold beyond which failure can occur. For compressor and turbines blade:
- $1 = 0$, implies that failure can occur at any time, including serving in life.
- $1 > 0$, indicates threshold beyond which failing are more likely to occur, suggesting a "burn-in-period or a minimum life expectancy.

These Weibull distribution characteristics can help engineers and maintenance teams to:

- predict failure rates and plan maintenance schedules
- identify potential failures modes and develop mitigation strategies
- optimize design and testing procedures for compressor and turbine blades.

The historical data was collected from the maintenance records in the oil and gas industry in the south geopolitical region of Nigeria a period of ten (10) years for thirty (30) identical compressor and turbine blades characteristics were examined and evaluated for the analysis of failure time of system behavior of the same make, thermal load, stress etc.

However, Rayleigh distribution is considered as special case of weibull distribution, often used to model the magnitude of vectors or distance between points in a plane. For reliability distribution can be applied when the failure rate increase over time, the following consideration are made:

(i) To collect time-to-failure data for the components
(ii) Estimate the scale parameter using maximum likelihood estimation (MIE) or other methods.

(iii) Use the governing equation to calculate reliability, failure rates, and time of failure.

The strong advantages of Rayleigh distribution are suitable for modeling components with increasing failure rates over time, such as those experiencing wear and tear or fatigue mechanism is dominated by single factor like stress and strain.

2 Methodology

2.1 Materials Considered:

The reliability model described by three parameters estimation of gas turbine blades performance are examined based on the following conditions;

- Collecting historical data about frequency of turbine blades/compressor blades (from oil and gas industry).
- Formulate governing equations and evaluate/simulate mean time to failure (MTTF), average time to failure (AVTTF), failure rate, reliability time λ(t), etc. using TPC Windchill quality solutions software.
- The computation of parameters and governing equations of characteristics reliability index using three parameter Weibull distributions are considered in the following cases.

2.2 Formulation of Governing Equations for Prediction of Reliability of Blade Geometry in Compressor and Turbine Application for Power Generation and Electricity Supply

CASE 1: The Mean time to failure of the three-parameter Weibull distribution (MTTF) given by:

$$MTTF = \gamma + \eta\Gamma(1 + 1/\beta) \tag{1}$$

where; $\gamma \geq 0$ *and* $t, \beta, \eta > 0$

$$\Gamma(x) = \int_0^\infty e^{-x} x^{n-1} dx \tag{2}$$

where;

η: *Scale parameter*, Γx: Gamma function, γ: *Location parameter*, β: Shape parameter, t: Time

The term (MTTF) is applied to non-repairable points which operates under specified condition. It is otherwise the ratio of sum of time to failure of each component to number of components under test.

CASE 2: Mean time between failure (MTBF)

The term MTBF is applied to repairable conditions, which measure the average time a particular equipment will fail and remain in service. MTBF of an equipment may gradually reduce to a potential condition which may defects and introduced by poor maintenance procedures/strategy.

$$MTBF = \frac{1}{n}\sum_{i=1}^{n}(t_k \ldots t_{k-1}) = \frac{t_n - t_o}{n} = \frac{t_n}{n} \quad (3)$$

Since, $t_0 = 0$ at the beginning.
Then,

$$MTBF = \frac{Total\ operating\ time}{No.\ of\ failures\ in\ that\ time} \quad (4)$$

CASE 3: Availability performance

Availability performance is the ability of an item to be in a state to perform a required function under a given condition for instance of time or over a given time interval, mean that;

All items assumed operating conditions unless failed scenario.

- The exception would have been standby redundancy but this scarcely exists power station because of high power supply demand.
- The outcomes in the analysis are based on two fundamental rules for combining probabilities.
- If A and B are two independent events with probabilities $\rho(A)$ and $\rho(B)$ of occurring, then the probability $\rho(AB)$ that both events will occur is the product;

$$\rho(AB) = \rho(A).\rho(B) \quad (5)$$

- Similarly, if two events A and B are mutually exclusive so that when one occurs the other cannot occur, the probability that either A or B will occur given by:

$$\rho(AB) = \rho(A) + \rho(B) \quad (6)$$

CASE 4: Failure rate

- Failure may either partial or complete, gradual or sudden which may be caused by inherent weaknesses or misuse.
- These failures can manifest in the following forms as; catastrophic failures, primary failure and secondary failures.
- Failure-rate is related to both number of failures per unit time that is the number of items fails in a given time depends not only on the quality of item, Hence;
- If the number of components in operation at the time of failure is Nr

Then failure – rate $\lambda(t)$ given by:

$$\lambda(t) = \lim_{\Delta t \to \infty} \frac{1}{N} \times \frac{\Delta N_f}{\Delta_t} = \frac{1}{N_s} \times \frac{\delta N_f}{\delta t} \quad (7)$$

CASE 5: Operational availability

The operational availability (A_0) given by;

$$A_0 = \frac{Up - Time}{Operating - Time} \tag{8}$$

Thus, Availability,

$$(A_r) = \frac{Available\ Hour}{Period\ Hour} \times \frac{100}{1} \tag{9}$$

- This quantity is the probability that the systems will be available after it has run for a long time and is a significant measurement of the performance of a repairable system given by;

$$R(t) = \ell^{-}\left(\frac{t-\gamma}{\eta}\right)^{\beta}, t \geq \gamma \tag{10}$$

CASE 6: Reliability function of the three (3) – parameter weibull distribution given by;

- The three – parameters weibull failure rate function;

$$\lambda(t) = \frac{\beta}{\eta}\left(\frac{t-\gamma}{\eta}\right)^{\beta-1}, t \geq \gamma \tag{11}$$

CASE 7: Weibull shape parameter(β), Determinants for predicting reliability of useful life of machine components;

- The weibull shape parameter (β) is also known as the weibull slope parameter. The value of β is equal to the slope of a line in a probability plot.
- When the shape parameter, $\beta < 1$ (this means that the failure rate decreases).
- When the shape parameter, $\beta = 1$ (this means that failure-rate is constant with time (t) and the distribution is equal to the exponential distribution)
- When the shape parameter, $\beta > 1$ (this means that failure rate increases).

CASE 8: Weibull scale parameter, η

- That is increasing the value of η while keeping $\boldsymbol{\beta}$ constant has the effects of stretching out the probability density function (pdf).
- A change in the scale parameter (η) has the same effect on the distribution as a change of the abscissa scale. Since the area under a pdf curve is a constant value, the peak of the pdf curves will also decreases with increase of η.

CASE 9: Weibull Location Parameter, $\boldsymbol{\gamma}$

- The location parameter, $\boldsymbol{\gamma}$ actually accounts for the subtraction (positive or negative) value that places the points in an acceptable straight line. Changing the value of the location parameter $\boldsymbol{\gamma}$, has the effects of pushing the distribution and associated function if ($\boldsymbol{\gamma} > 0$) or to the left if ($\boldsymbol{\gamma} < 0$).

CASE 10: Prediction Performance of Weibull Distribution Model

- The prediction accuracy of the model in the estimation of the compressor/turbine-blade failures with respect to actual values were evaluated based on correlation coefficient R^2, root mean square error (RMSE) and coefficient of efficiency (COE). These parameters are calculated based on the following equation as;

$$R^2 = \frac{\sum_{i=1}^{N} (y_1 - z)^2 - \sum_{i=1}^{N} (x_i - z)^2}{\sum_{i=1}^{N} (y_i - z)^2} \quad (12)$$

Similarly, the root mean square error (RMSE) given by;

$$RMSE = \left[\frac{1}{N} \sum_{i=1}^{N} (y_i - x_i)^2 \right]^{1/2} \quad (13)$$

The coefficient of efficiency (COE) given by;

$$COE = \frac{\sum_{i=1}^{N} (y_i - x_i)^2}{\sum_{i=1}^{N} (y_i - z)^2} \quad (14)$$

where is the *ith* actual data Xi is the *ith* predicted data with the weibull distribution (z) is the mean of the actual data, N is the number of observations.

CASE 11: Reliability model of system component and weibull-two parameter characterization.

- The rate of failure and mean time between failure (MTBF) are key parameters of reliability in the turbine blades evaluation which are estimated using weibull distribution function.
- Weibull distribution technique is a vital tool used in the systematic modeling of failure rates, forecasting failures and in modeling of failure and fault-process stemmed from their aging.
- Thus, distribution may be specified by two parameters of shape (β) and scale (α) respectively;

Provided the techniques used in the estimation of weibull parameters of average rating technique and number of failure according to the equation given by;

$$F_i = \frac{(i - 0.3)}{(N - 0.4)} \quad (15)$$

where; F_i is the average rank of occurring i-th failure.

If equipment are considered as separate components, is then the adjusted rank of age of failed component and N is the total ranking number of the component, weibull parameter would be determined using the least-square relationship given by:

$$y_i = mx_i + c \quad or \quad x_i = In(t_i) \quad (16)$$

where; t_i is the independent age (year) of failed component in rank i.

Therefore,

$$y_i = In\, In\left[\frac{1}{1 - F_i}\right] \tag{17}$$

From, (17) and (18) are weibull shape parameter ($\boldsymbol{\beta}$) can be calculated given by;

$$\beta = m = \frac{\sum_{i=1}^{N} x_i y_i - \frac{\sum_{i=1}^{N} x_i \sum_{i=1}^{N} y_i}{N}}{\sum_{i=1}^{N} x_i^2 - \frac{\left[\sum_{i=1}^{N} x_i\right]^2}{N}} \tag{18}$$

Similarly, the constant (c) given by;

$$\boldsymbol{C} = \frac{\sum_{i=1}^{N} \boldsymbol{y_i}}{\boldsymbol{N}} - \frac{\boldsymbol{m} \sum_{i=1}^{N} \boldsymbol{x_i}}{\boldsymbol{N}} \tag{19}$$

The life or scale parameter (α) can be determined given by;

$$\boldsymbol{\alpha} = \boldsymbol{\ell}^{-\left[\frac{c}{m}\right]} \tag{20}$$

- By the analysis of two-parameter technique the prediction becomes very necessary if the system behavior of the component according to the equipment – curve distribution is properly determined.
- Weibull probability distribution function f(t) shows probability of failure in certain time (t) given by;

$$\boldsymbol{F}(\boldsymbol{t}) = \left(\frac{\boldsymbol{\beta}}{\boldsymbol{\alpha}}\right)\left(\frac{\boldsymbol{t}}{\boldsymbol{\alpha}}\right)^{\boldsymbol{\beta}-1} \boldsymbol{\ell}^{-\left(\frac{t}{\alpha}\right)^{\beta}}; \boldsymbol{for} \quad \begin{matrix} \boldsymbol{\alpha} > 0 \\ \boldsymbol{\beta} > 0 \\ 0 \le \boldsymbol{t} \le \infty \end{matrix} \tag{21}$$

The cumulative distribution function F(t), which shows the probability of failure in time (t) which would be calculated as;

$$F(t) = 1 - \ell^{-\left(\frac{t}{\alpha}\right)^{\beta}}; for \quad \begin{matrix} \alpha > 0 \\ \beta > 0 \\ 0 \le t \le \infty \end{matrix} \tag{22}$$

Essentially, reliability function R(t) shows probability of remaining intact till the time (t) and the rates of failure λ(t) which can be expressed as;

$$R(t) = 1 - F(t) = \ell^{-\left(\frac{t}{\alpha}\right)^{\beta}} \tag{23}$$

$$\lambda(t) = \frac{F(t)}{R(t)} = \left(\frac{\beta}{\alpha}\right)\left(\frac{t}{\alpha}\right)^{\beta-1} \tag{24}$$

- From observation, the shape –parameter (β) affects shape distribution curve that is when shape parameter changed, the curve f(t) varies differently in shape. For example, if the curve turns to exponential distribution, then $\beta = 1$
- That is the failure rate will be decreasing while $\beta < 1$, this means that the component is in the early failure state.
- Similarly, when failure rate is constant while, $\beta = 1$, the components will be in its occasionally failure condition. Incidentally, the failure rate is increasing while $\beta > 1$ this means the component is in its loss failure condition.

CASE 12: The Mean and Standard Deviation of Weibull Distribution Evaluation

The mean and standard deviations are presented in terms of shape and scale – parameter given by:

$$\mu = \mu\Gamma\left(1 + \frac{1}{\beta}\right) \tag{25}$$

and

$$\delta^2 = \alpha^2\left[\Gamma\left(1 + \frac{2}{\beta}\right) - \Gamma^2\left(1 + \frac{1}{\beta}\right)\right] \tag{26}$$

where;

Γ(.) represents gamma function which could be estimated in the following form given by:

$$\Gamma = \sqrt{2\pi t}^{(t-0.5)} \ell^{-t}\left(1 + \frac{1}{12t}\right) \tag{27}$$

In the consideration for long term unavailability without considering restriction related to repair or replacement time of equipment in order to estimate equipment failure-rate, the Meantime to failure (MTTF) and the mean time between failure (MTBF) should be equal.

$$\text{Therefore, } MTBF = \alpha\Gamma\left(1 + \frac{1}{\beta}\right) \tag{28}$$

While, the total failure-rate for turbine-blade are considered as an important component for reliable performance given by:

$$\lambda(t) \sum_{i=1}^{N} \lambda_i(t) \tag{29}$$

where, $\lambda(t)$ is the rate of failure in i-th critical part of equipment and n is the total subcomponents of the turbine-blades under observations (Table 1).

CASE 13: SOURCING OF DATA

Owing to the incessant breakdown of system component due to failure frequency, repair time, repair rate and earlier mortality of machine part based on collected historical data: failure of compressor and turbine blades for gas turbine power plant for ten (10)

Table 1. Failure Time of Gas Turbine Power Plant; Compressor-Blades

Serial number ranking of compressor blades	Failures times in hours (hrs)
1.000000	1047.000000
2.000000	1279.000000
3.000000	1340.000000
4.000000	1578.000000
5.000000	1598.000000
6.000000	1749.000000
7.000000	1804.000000
8.000000	1841.000000
9.000000	1847.000000
10.000000	1869.000000
11.000000	1879.000000
12.000000	1890.000000
13.000000	1939.000000
14.000000	1948.000000
15.000000	1949.000000
16.000000	1956.000000
17.000000	1987.000000
18.000000	1995.000000
19.000000	2004.000000
20.000000	2005.000000
21.000000	2047.000000
22.000000	2214.000000
23.000000	2287.000000
24.000000	2435.000000
25.000000	2439.000000
26.000000	2442.000000
27.000000	2581.000000
28.000000	2617.000000
29.000000	2926.000000
30 .000000	2978.000000

years {2016–2025} was cited. The failure time (hrs) of three (3) identical gas turbine power plant with consideration of thirty (30) blades for the prediction of compressor and turbine blades performance/reliability were sourced and cited as seen in tables above.

Table 2. Shows the turbine blade for failure time (hrs) (Turbine Blade)

Serial number ranking of turbine blades	Failures times in hours (hrs)
1.000000	1020.000000
2.000000	1220.000000
3.000000	1300.000000
4.000000	1525.000000
5.000000	1540.000000
6.000000	1720.000000
7.000000	1800.000000
8.000000	1820.000000
9.000000	1840.000000
10.000000	1850.000000
11.000000	1865.000000
12.000000	1880.000000
13.000000	1910.000000
14.000000	1925.000000
15.000000	1928.000000
16.000000	1930.000000
17.000000	1948.000000
18.000000	1964.000000
19.000000	2004.000000
20.000000	2010.000000
21.000000	2030.000000
22.000000	2200.000000
23.000000	2300.000000
24.000000	2405.000000
25.000000	2450.000000
26.000000	2480.000000
27.000000	2550.000000
28.000000	2620.000000
29.000000	2802.000000
30.000000	2810.000000

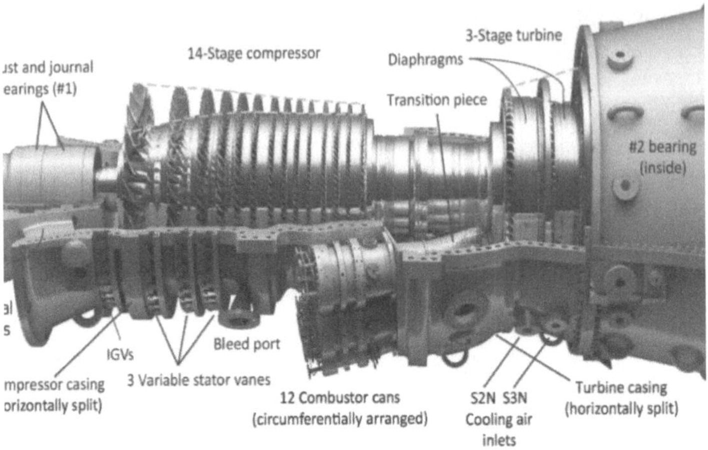

Plate 1. A cascade of the compressor blade, combustion chamber and turbine blades

Plate 2. Showing the number of compressor blades configuration

Plate 3. Contaminated turbine blade surface

Plate 4. Contaminated turbine blade surface

2.3 Diagramatic Representation of Compressor and Turbine Blade Components and Its Related Problems

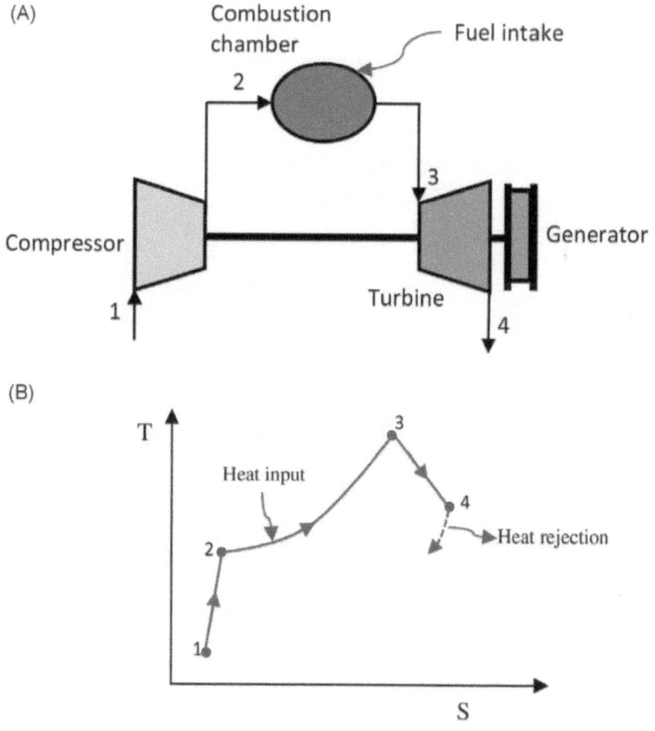

a.

b

Fig. 1. **(a)** A schematic representation of an open cycle gas turbine power plant. **(b)** shows the corresponding thermodynamic cycle on t-s (temperature entropy) diagram

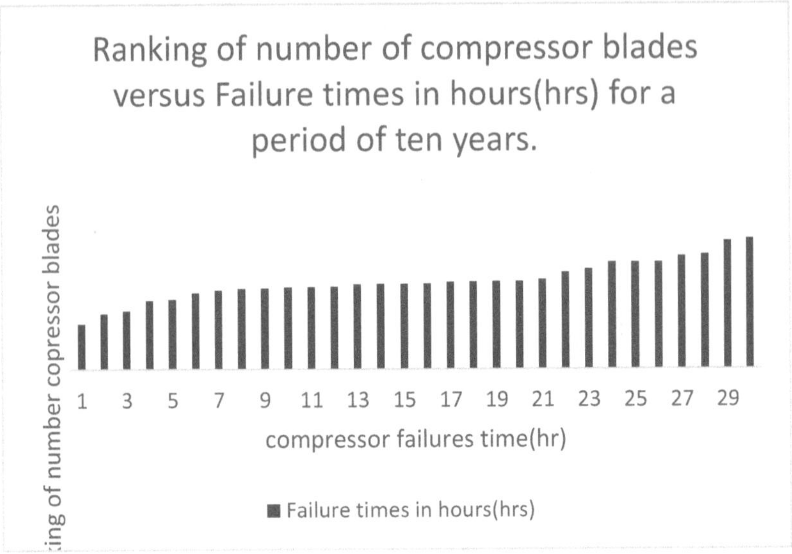

Fig. 2. Shows Ranking of number of Compressor Blades vs Failure Times in hours for the period of 10 years

3 Results and Discussions

The data is analyzed using specialized weibull analysis software PTC windchill solutions 10.2, to obtain the three –parameter weibull estimates namely;

- Shape parameter (β):
- Scale-parameter (η):
- Location parameter (γ):

The simulation results produce:

- Probability-plot
- Probability density function (pdf) – plot
- Reliability – plot
- Unreliability vs time – plot
- Failure rate – plot

However, results of the simulated plots are presented and contained in Table 2; which shows the result of the shape - parameter (β), scale - parameter (η) and location – parameter respectively (Table 3).

Table 3. Three-parameter weibull estimates using TPC windchill quality solution 10.02:

Shape-parameter; β	Scale-parameter, η	Location parameter, γ
2.9085	24.3296	−5.5962

3.1 The Analysis of Various Cases Presented for Compressor Blades Evaluation

CASE 1: DETERMINATION OF THE TIME REQUIRED FOR MACHINE COMPONENT FAILURE (T) IN HOURS USING THE WEIBULL THREE PARAMETERS MODEL (T); GIVEN AS:

$$t = \gamma + (-In(F(t)))^{1\beta} \tag{30}$$

For 10% probability of failure (which is the characteristics life) given as:

$$F(t) = 10\% = 0.1$$

$$\beta = 2.000, \mu = 20.945350, \gamma = -1.875$$

$$t = \gamma + \mu * (-In(F(t)))^{1/\beta}$$

$$t = -1.875995 + 20.945350 * (-In(1 - 0.1))^{1/2}$$

$$t = -1.875995 + 20.945350 * (-In(0.9))^{0.5}$$

$$t = -1.875995 + 20.945350 * (0.105360515)^{0.5}$$

$$t = -1.875995 + 20.945350 * 0.324592846$$

$$t = -1.875995 + 6.79871076$$

$$t = 4.92271576\,\text{h}$$

CASE 2: DETERMINATION OF RELIABILITY OF THREE – PARAMETERS WEIBULL DISTRIBUTION R(T) GIVEN BY:

$$R(t) = \rho - \left(\frac{t - \gamma}{h}\right)^{1/\beta} \tag{31}$$

$$t = 4.922715\,\text{h}, \beta = 2.00, \rho = 2.718, \gamma = -1.875, \mu = 20.945350$$

$$R(t) = 2.718 - \left(\frac{4.922715 - -1.875}{20.945350}\right)$$

$$R(t) = 2.718 - \left(\frac{6.797715}{20.945350}\right)$$

$$R(t) = 2.718 - (0.324545304)$$

$$R(t) = 2.718 - 0324545304$$

$$R(t) = 0.722880282$$

CASE 3: DETERMINATION OF FAILURE RATE, Λ(T),
Three (3) parameters Weibull distribution for failure rate function given by:

$$(t) = \frac{\beta}{\mu}\left(\frac{t-\gamma}{\eta}\right)^{\beta-1}, t \geq \gamma \tag{32}$$

where;

$$\beta = 2.000, t = 4.922715\,\text{h}, \gamma = -1.875, h = 20.945350$$

$$(t) = \frac{2.00}{20.945350}\left(\frac{4.922715 - 1.875}{20.945350}\right)2 - 1$$

$$(t) = 0.095486\left(\frac{6.797715}{20.945350}\right)^1$$

$$(t) = 0.095486(0.324545304)$$

$$(t) = 0.002959066546/hr$$

CASE 4: DETERMINATION OF MEAN TIME TO FAILURE (MTTF) OF THE THREE – PARAMETERS WEIBULL DISTRIBUTION GIVEN BY;

$$MTTF = \gamma + h \lceil\left(1 + \frac{1}{\beta}\right) \tag{33}$$

where, $\gamma \geq 0, \textit{and } t, \beta, \mu > 0$
where, $\gamma = -1.875, \beta = 2.000, \mu = 20.945350$

$$MTTF = -1.875 + 20.945350 \lceil\left(1 + \frac{1}{2}\right)$$

$$-1.875 + 20.945350 \lceil(1 + 0.5)$$

$$-1.875 + 20.945350 \lceil(1.5)$$

where, $\lceil(1.5) = 0.886227$

Then,

$$MTTF = -1.875 + 20.945350 * (0.886227)$$

$$MTTF = -1.875 + 18.56233469$$

$$MTTF = 16.68733469\,\text{h}$$

CASE 5: DETERMINATION OF AVERAGE TIME TO FAILURE (AVTTF):
The AVTTF, actually it is very important to provide a valuable insight to the expected system or components under investigation is given by;

$$\text{AVTTF} = \sum_{i=1}^{NT} \frac{\textit{Central tendency of all observed values}}{\textit{Number of values}}$$

$$\text{AVTTF} = \sum_{i=1}^{NT} \frac{\sum_i^{NT} CD}{NV}$$

$$AVTTF = \sum_{i}^{NT} \frac{59{,}646\,\text{h}}{30}$$

$$AVTTF = 1{,}988.2\,\text{h}$$

The data is analyzed using specialized weibull analysis software TPC windchill solutions 10.2, to obtain the three –parameter weibull estimates namely;

- Shape parameter (β):
- Scale-parameter (η):
- Location parameter (γ):

The simulation results produce:

- Weibull probability-plot
- Probability density function (pdf) – plot
- Reliability – plot
- Unreliability vs time – plot
- Failure rate – plot

Similarly, Results of the simulated plots are presented and contained in Table 4; which shows the result of the shape parameter (β), scale parameter (η) and location.

Table 4. Three-parameter weibull estimates using TPC windchill quality solution 10.02

Shape-parameter; β	Scale-parameter, η	Location parameter, γ
$\beta: 4.1353$	$\eta: 33.2384$	$\gamma: -12.8047$

3.2 Presentation of Weibull Distribution Plot as Shown in Fig. 1, 2, 3, 4 and 5 (Probability – Plot, Probability Density Function (PDF) Plot, Reliability Plot vs Time, Unreliability Plot vs Time, an Failure Rate vs Time

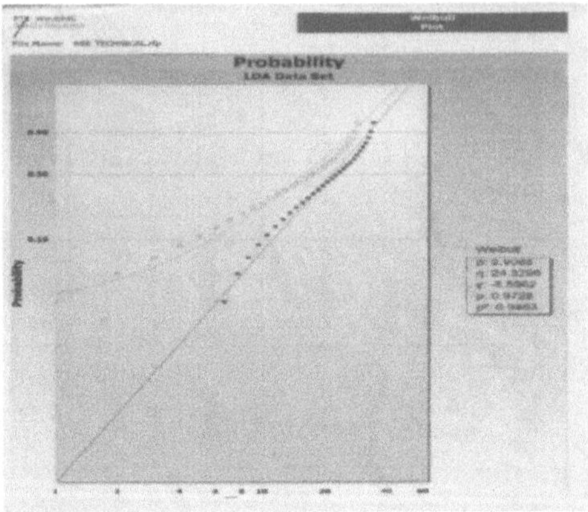

Fig. 3. Shows the probability plots against time (compressor blade)

The study investigated the reliability parameters using, TPC windchill quality solution 10.02 software which were used to analyze the failure time of three (3) identical gas turbine power plant. The probability plot displays the cumulative distribution function (CDF) of the weibull distribution, It shows the probability of failure F(t), the slope parameter $\beta = 2.9085$, indicates failure rate over time.

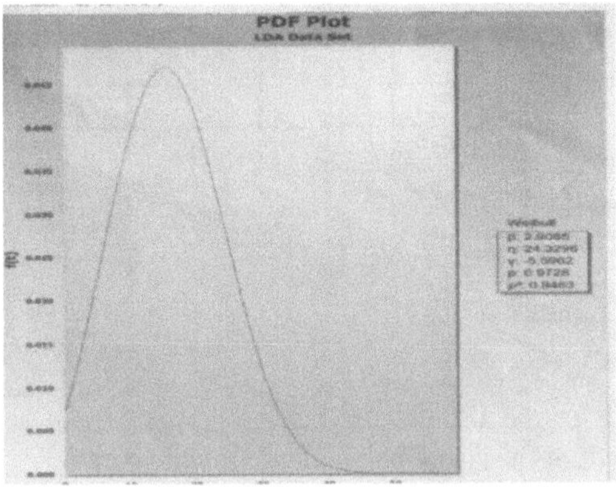

Fig. 4. Shows Probability Density Function (PDF) plot against time (compressor blade)

The study investigated the reliability parameters using, TPC windchill quality solution 10.02 software which were used to analyze the failure time of three (3) identical gas turbine power plant. The probability plot displays the cumulative distribution function (CDF) of the weibull distribution, It shows the probability of failure F(t), the slope parameter $\beta = 2.9085$, indicates failure rate over time.

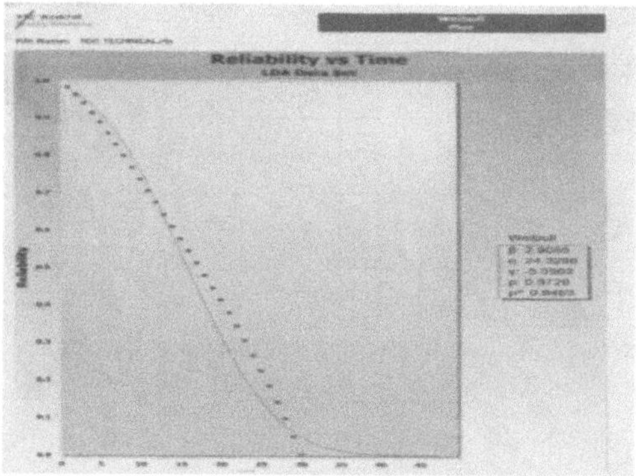

Fig. 5. Shows reliability plot against time (compressor blade)

The reliability plot displays the reliability function $R(t) = 1 - F(t)$, showing the probability of survival over time. With $\beta > 1$, the reliability decreases over time.

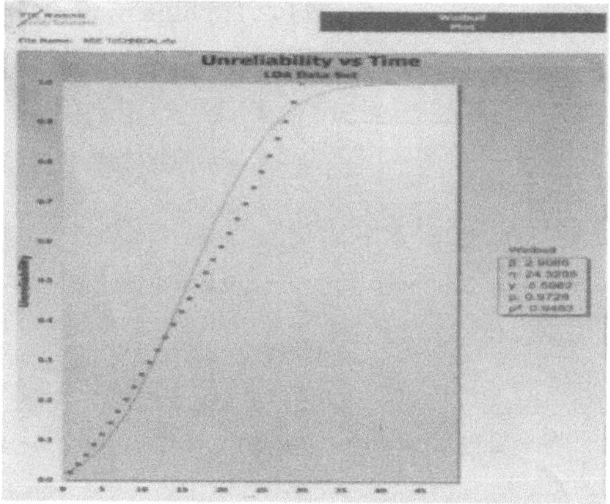

Fig. 6. Shows Unreliability plot against time (Compressor Blade)

The unreliability plot is equivalent to the CDF-plot, showing The probability of failure F(t) over time. The location parameter $(\eta) = -5.5962$ shifts the plots along the time in the left side axis. This is the time guaranteed within the component won't fail, but be sustained.

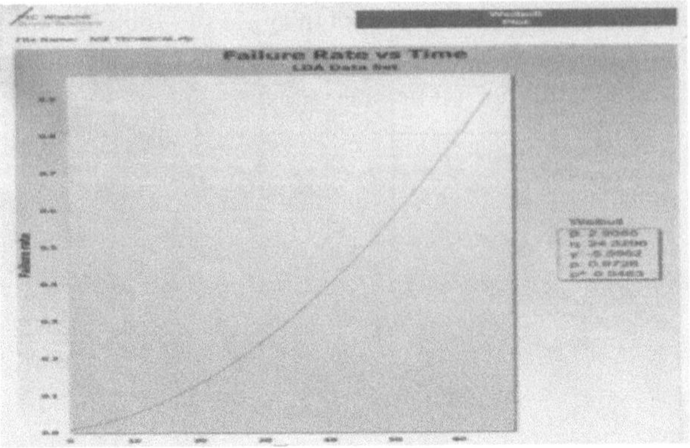

Fig. 7. Shows the failure rate plot against time (compressor blade)

The failure rate plot shows the hazard function, f (t), which represents the instantaneous failures rate at time (t). With β = 2.9085, the failure rate increases over time. The results obtained from failure times of turbine blades simulation of TPC wind chill solution software are presented in Fig. 1, 2, 3, 4 and 5, which indicates various plots about reliability and performance evaluation of turbine blades under investigation as shown:

3.3 Analysis of Various Cases Presented for Turbine Blades Evaluation

CASE 1: DETERMINATION OF TIME REQUIRED FOR MACHINE COMPONENT FAILURE (T) IN HOURS USING THE WEIBULL THREE PARAMETERS MODEL (T); GIVEN AS:

$$t = \gamma + (-In(F(t)))^{1/\beta} \tag{34}$$

For 10% probability of failure (which is the characteristics life) given as:

$$F(t) = 10\% = 0.1$$

$$\beta = 2.000, \mu = 20.945350, \gamma = -1.875$$

$$t = \gamma + \mu * (-In(F(t)))^{1/\beta}$$

$$t = -1.875995 + 20.945350 * (-In(1 - 0.1))^{1/2}$$

$$t = -1.875995 + 20.945350 * (-In(0.9))^{0.5}$$

$$t = -1.875995 + 20.945350 * (0.105360515)^{0.5}$$

$$t = -1.875995 + 20.945350 * 0.324592846$$

$$t = -1.875995 + 6.79871076$$

$$t = 4.92271576\,\text{h}$$

CASE 2: DETERMINATION OF RELIABILITY OF THREE – PARAMETERS WEIBULL DISTRIBUTION R(T):

$$R(t) = \rho - \left(\frac{t-\gamma}{h}\right)^{1/\beta} \tag{35}$$

$$t = 4.922715\,\text{h},\ \beta = 2.00,\ \rho = 2.718,\ \gamma = -1.875,\ \mu = 20.945350$$

$$R(t) = 2.718 - \left(\frac{4.922715 - 1.875}{20.945350}\right)$$

$$R(t) = 2.718 - \left(\frac{6.797715}{20.945350}\right)$$

$$R(t) = 2.718 - (0.324545304)$$

$$R(t) = 2.718 - 0324545304$$

$$R(t) = 0.722880282$$

CASE 3: DETERMINATION OF FAILURE RATE, Λ(T),

The three (3) parameters Weibull distribution for failure rate function given as:

$$(t) = \frac{\beta}{\mu}\left(\frac{t-\gamma}{\eta}\right)^{\beta-1}, t \geq \gamma \tag{36}$$

where;

$$-\beta = 2.000, t = 4.922715\,\text{h}, \gamma = -1.875, \mu = 20.945350$$

$$(t) = \frac{2.00}{20.945350}\left(\frac{4.922715 - 1.875}{20.945350}\right)2 - 1$$

$$(t) = 0.095486\left(\frac{6.797715}{20.945350}\right)^1$$

$$(t) = 0.095486(0.324545304)$$

$$(t) = 0.002959066546/hr$$

CASE 4: DETERMINATION OF MEAN TIME TO FAILURE (MTTF) OF THE THREE – PARAMETERS WEIBULL DISTRIBUTION:

$$MTTF = \gamma + h \,\Gamma\left(1 + \frac{1}{\beta}\right) \tag{37}$$

where, $\gamma \geq 0, andt, \beta, \mu > 0$

where, $\gamma = -1.875, \beta = 2.000, \mu = 20.945350$

$$MTTF = -1.875 + 20.945350 \,\Gamma\left(1 + \frac{1}{2}\right)$$

$$-1.875 + 20.945350 \,\Gamma(1 + 0.5)$$

$$-1.875 + 20.945350 \,\Gamma(1.5)$$

where, $\Gamma(1.5) = 0.886227$

$$MTTF = -1.875 + 20.945350 * (0.886227)$$

$$MTTF = -1.875 + 18.56233469$$

$$MTTF = 16.68733469\,\text{h}$$

CASE 5: DETERMINATION OF AVERAGE TIME TO FAILURE (AVTTF):

The AVTTF, actually provide the valuable insight to the expected system or components under investigation is given as;

$$\text{AVTTF} = \sum_{i=1}^{NT} \frac{Central\ tendency\ of\ all\ observed\ values}{Number\ of\ values}$$

$$\text{AVTTF} = \sum_{i=1}^{NT} \frac{\sum_{i}^{NT} CD}{NV}$$

This is given as:

$$AVTTF = \sum_{i}^{NT} \frac{59{,}646\,\text{h}}{30}$$

$AVTTF = 1,988.2$ h.

3.4 Presentation of Weibull Distribution Plot as Shown in Figure 6, 7, 8, 9 and 10 (Probability – Plot, Probability Density Function (PDF) Plot, Reliability Plot vs Time, Unreliability Plot vs Time, and Failure Rate vs Time

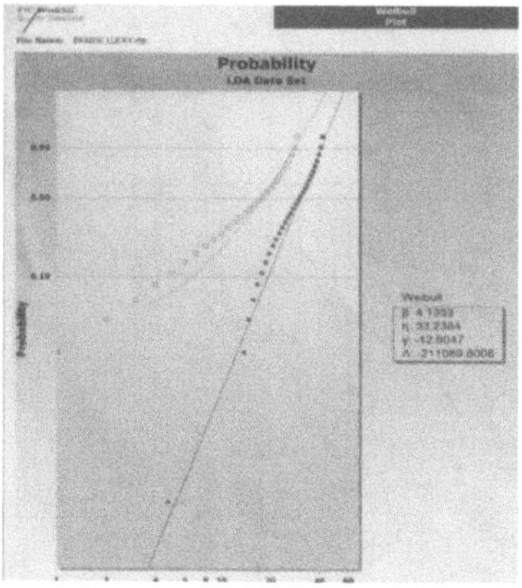

Fig. 8. Shows the probability plots against time (turbine blade)

From the probability plot displays the cumulative distribution function (CDF) using Weibull model. It shows the probability of failure F(t), the $\beta = 4.1353$, the plot shows an increasing failure rate over time.

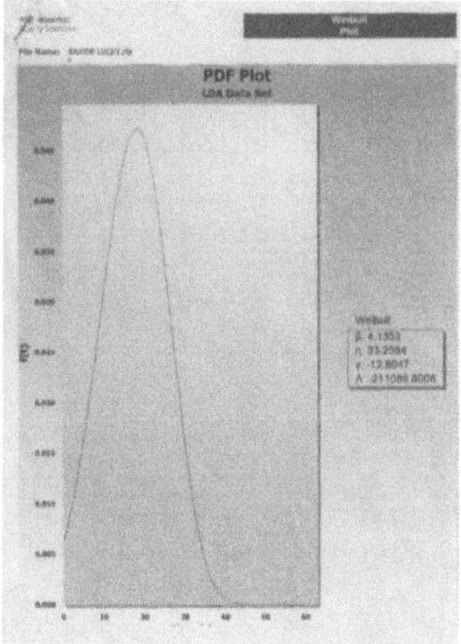

Fig. 9. Shows Probability Density Function (PDF) plot against time (Turbine blade)

The PDF plot shows the probability density function of failures at different times, the shape-parameter (β) = 4.1353 indicates as skewed distribution with a peak at a certain time. The Scale parameter (η) = 33.2384 affects the spread of the distribution.

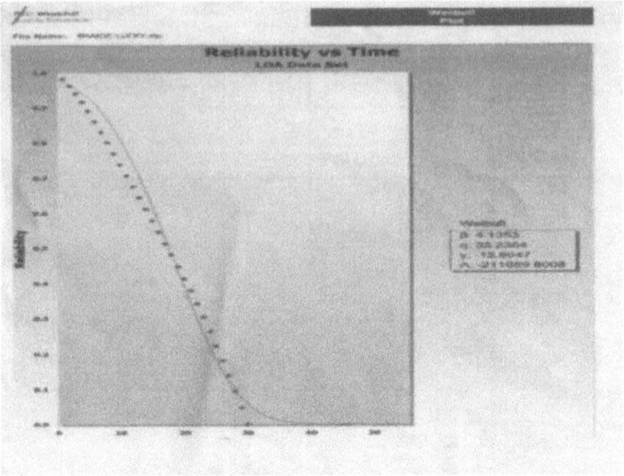

Fig. 10. Shows reliability plots against time (Turbine blades)

The reliability plot display reliability function R(t) = 1 – F(t), showing the probability of survival over time. With $\beta > 1$, the reliability decreases over time (Figs. 11 and 12).

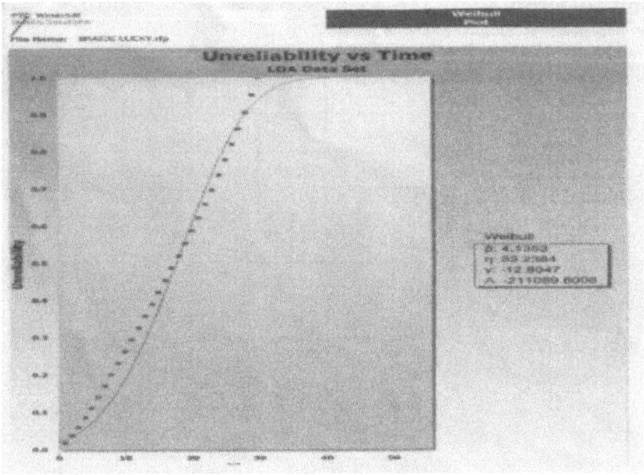

Fig. 11. Unreliability plot against time (Turbine blade)

The unreliability plot is equivalent to the CDF-plot, showing the probability of failure F(t) over time. The location parameter (η)= 33.2384 shifts the plots along the time axis.

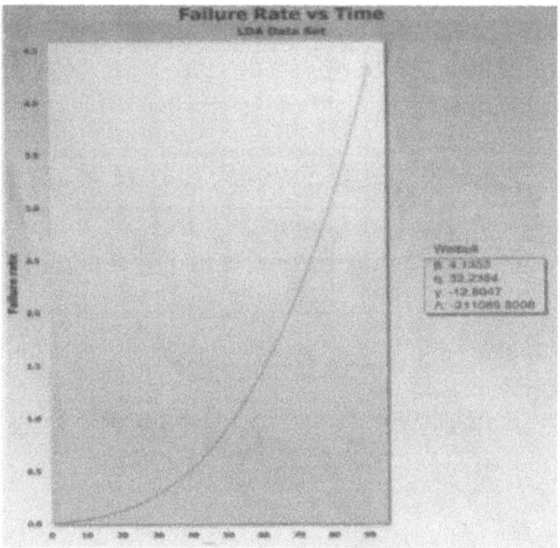

Fig. 12. Shows the failure rate plot against time (Turbine blade)

The failure rate plot shows the hazard function,(t), which represents the instantaneous failures rate at time t. with $\beta = 4.1353$ the failure rate increase over time.

3.5 Comparative Evaluation of Weibull and Raylaeigh's Distribution When Fitted into Simulation Data

The results obtained in the case of Weibull distribution accounts for the shape parameter ($\beta = 2.9085$) which indicates an increasing failure rate for compressor blade overtime. While the scale parameter ($\gamma = 24.325$) which indicates how it affects the spread of the distribution. Evidently, the location parameter ($\eta = 5.5962$) suggests shift in the distribution on the view to determine the failure before time zero. Similarly, results obtained from Rayleigh's distribution are; Shape ($\beta = 2.000$), Scale ($\gamma = 17.1717$) and location ($\eta = 0$).

Evidently, both distributions indicate increasing failure rate in their prediction behavior, but Weibull parameter is closer to 3 suggesting more pronounced increase. That is Weibull distribution also shows better flexibility and has the capacity to capture early failures faster which makes it a better fit, if the data exhibits these characteristics.

4 Conclusion

Conclusively, the results obtained in this research shows that the compressor and turbine blades are in their wear out period of failures, following the analysis and investigation of failure time of ten (100) identical gas turbine blades with 30 blades configuration, using parameter estimation of three Weibull distribution. The estimates were successfully obtained and the results were compared with Rayleigh's distribution using TPC windchill quality solutions, the values of the Weibull estimate and Rayleigh's distribution are found to be close, although the Weibull tends to show better and faster prediction to time of failure rates, these are represented in the wear out period of analysis as the values of the shape parameter is greater than one (1) in both cases. From the Weibull estimates obtained, the reliability, failure rate and mean time to failure of the blade were computed to achieve all three parameters of shape, scale and location in order to determine gas turbine power plant parameters.

Authors' Contributions. The research writing was made an original draft, writing review and editing-review, validation, visualization, data sources, conceptualization, resources, software, performed supervision, project administration, resources and methodology. All the authors read and approved the final manuscript.

Funding. The research was completely funded by the authors that is there was no external fund to support this research.

Declarations.

Future Works. Subsequent presentations/research will consider design of aerodynamic trajectory impact for blade geometry in compressor and turbine application. This means that assessment of the wind dynamics and potential will be determined.

Availability of Data and Materials

The data generated are analyzed during this current study and are highly confidential, but are made available from the corresponding on reasonable request.

Competing Interest. The authors declared that they have no form of competing financial interest or personal relationship that could have strong influence on this work reported in this paper.

Nomenclature

β	Shape Parameter
γ	Scale Parameter
η	Location Parameter
t	Time (hours)
Λ(T)	Failure Rate
R(t)	Reliability
AVTTF	Average Time to Failure
MTTF	Mean time to Failure
T	Temperature (K)
P	Pressure (bar)

References

1. Ault, G.W., Donald, J.R.Mc.: Planning for distributed generation within distributed networks in restructured electricity markets. IEEE Power Eng. Rev. **30**(20), 52–54 (2000)
2. Energy Commission of Nigeria (ECN). National energy policy. Federal Republic of Nigeria, Abuja (2003)
3. Adenikinju, A.: Analysis of the cost of infrastructure failures in a developing economy: The case of the electricity sector in Nigeria. African Economic Research Consortium (AERC) Research Paper, 148, Nairobi (2005)

4. Akinbami, J.F.K.: Renewable energy resources and technologies in Nigeria: Present situation, future prospects and policy framework. Mitigation and Adaptation Strategies for Global Change. Netherlands: Kluwer Academic Publishers (2001)
5. Amotsuka, P.O.: Sustainable agriculture in the Niger Delta Region: A case study of Fatima Community of Rivers State. In: Paper presented at the 2nd national conference of colleges of agriculture, environmental, engineering and science and technology, Hassan Usman Katsina Polytechnic (2008)
6. Babalola, R.: 6000 MW power target report card. Daily Trust Newspaper **22**, 94 (2009)
7. Akpinar, E.K., Akpinar, S.: Statistical analyses of wind energy potential on the basis of the Weibull and Rayleigh distributions for Agin-Elazig Turkey. J. Power Energy **218**, 557–565 (2004)
8. Markovic, D., Jukic, D., Bensic, M.: Nonlinear weighted least squares estimation of a three-parameter Weibull density with a nonparametric start. J. Comput. Appl. Math. **228**(1), 304–312 (2009)
9. Yang, M., Nie, H. Advanced algorithm for maximum likelihood estimation of three-parameter Weibull distribution. J. Nanjing Univ. Aeronaut. Astronaut. 1 (2007)
10. Markovic, D., Jukic, D., & Bensic, M. Nonlinear weighted least squares estimation of a three-parameter Weibull density with a nonparametric start. J. Comput. Appl. Math. (2009)
11. Abbasi, E., Malik, O.P.: Failure rate estimation of power transformers using inspection data. In: Proceedings of the International Conference on Probabilistic Methods Applied to Power Systems (PMAPS) (2016)
12. Suwanasri, T., Phadungthin, R., Suwanasri, C.: Risk-based maintenance for asset management of power transformers: practical experience in Thailand. Int. Trans. Electr. Energy Syst. **24**(8), 1103–1119 (2014)
13. Kang, Y.J., Noh, Y., Lim, O.K.: Kernel density estimation with bounded data. Struct. Multidiscip. Optim. **57**(1), 95–113 (2018)
14. Carneiro, T.C., Melo, S.P., Carvalho, P.C.M., Braga, A.P.S.: Particle swarm optimization method for estimation of Weibull parameters: a case study for the Brazilian northeast region. Renewable Energy **86**, 751–759 (2016)
15. Ahmad, M.R., Ali, A.S., Assad, A.M.: Estimation accuracy of Weibull distribution parameters. J. Appl. Sci. Res. **5**(7), 790–795 (2009)
16. Li, Y.M.: A general linear regression analysis applied to the three-parameter Weibull distribution. IEEE Trans. Reliab. **43**, 255–263 (1994)

Overestimation of Smoke Backlayering Length and Critical Velocity in Small-Scale Tunnels in FDS Simulations

Tatiana Kubisova(✉), Lukas Valasek, and Jan Glasa

Institute of Informatics, Slovak Academy of Sciences, Dubravska cesta 9, 84507 Bratislava, Slovakia
tatiana.kubisova@savba.sk

Abstract. Modeling smoke propagation in tunnel fires is crucial tool for ensuring tunnel safety. This paper examines the overestimation of smoke backlayering length and critical airflow velocity in fire simulations in small-scale tunnels using the FDS 6 system. Set of five simulations were conducted using critical airflow velocity values for eight heat release rates obtained during experiments in tunnels with different cross-sections and their results are analyzed. The simulation results consistently overestimated the smoke backlayering length in all simulated cases. However, modification of the simulation settings in one of the tunnels with a rounded ceiling significantly improved accuracy, substantially reducing the overestimation of both smoke backlayering length and critical velocity.

Keywords: tunnel fire · critical velocity · smoke backlayering length · FDS

1 Introduction

Computer modeling of airflow and smoke propagation during tunnel fires has become an important tool for ensuring tunnel safety [1–3]. Smoke spreads in a specific way in a road tunnel during a fire. From the fire source, smoke first rises, hits the tunnel ceiling and spreads in both directions along the upper part of the tunnel. After a certain time, the smoke stratification in the upper part of the tunnel is disrupted due to the decrease in smoke temperature and the smoke descends towards the roadway. The spread of smoke against the longitudinal flow in a tunnel is referred to as smoke backlayering. The critical airflow velocity is defined as the minimum flow velocity at which backlayering does not occur. Figure 1 illustrates the critical airflow velocity and the smoke backlayering length in a horizontal tunnel.

The first estimate of the critical airflow velocity in a tunnel was derived by P. H. Thomas based on a theoretical analysis of physical factors influencing smoke backlayering and data from experiments in a small-scale tunnel. He assumed that, to prevent smoke backlayering in a tunnel, the inertial forces and the buoyancy force caused by the fire must be in equilibrium. He expressed the ratio of inertial forces to buoyancy force

H. R. Arabnia et al. (Eds.): CSCI 2025, CCIS 2970, pp. 226–239, 2027.
https://doi.org/10.1007/978-3-032-28466-2_16

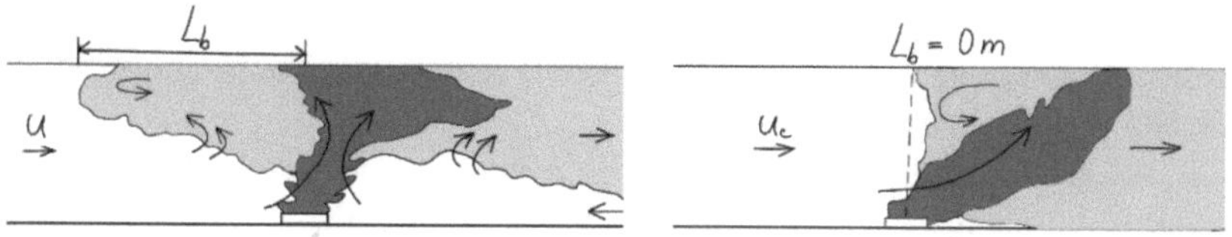

Fig. 1. Schematic representation of smoke propagation in a horizontal tunnel: smoke backlayering length L_b at airflow velocity u (a) and critical airflow velocity u_c (b).

using the so-called critical Froude number and obtained the following expression for the estimate of the critical airflow velocity [4]:

$$u_c = K\left(\frac{gHQ}{\rho_o c_p T_f A}\right)^{1/3}, K \sim 1, \tag{1}$$

where u_c is the critical airflow velocity (m/s), g is the gravitational acceleration (m/s^2), H is the tunnel height (m), Q is the heat release rate (kW), ρ_0 is the density of the environment (kg/m^3), c_p is the specific heat capacity (kJ/(kg K)), T_f is the average temperature of the smoke behind the fire (K) and A is the cross-sectional area of the tunnel (m^2). By analyzing data from experiments in a small-scale horizontal tunnel with a square cross-section and data from an experiment in a railway tunnel, he confirmed the value of the parameter K~1 for fires with a fire source located on the floor. Other estimates were derived using a similar method. A. J. M. Heselden proposed an estimate based on several experiments in a horizontal railway tunnel in the form [5]:

$$u_c = CK\left(\frac{gT_f Q}{\rho_o c_p T_0^2 W}\right)^{1/3}, C = 0,8; K \sim 1, \tag{2}$$

where T_0 is the ambient temperature (K) and W is the tunnel width (m). W. D. Kennedy derived the estimate [6]:

$$\begin{gathered} u_c = K_1 K_2\left(\frac{gHQ}{\rho_o c_p T_f A}\right)^{1/3}, \ T_f = \frac{Q}{\rho_o c_p A u_c} + T_0, \ \ K_1 \sim 0.606, \\ K_2 = \begin{cases} 1 + 0.0374|\alpha|^{0.8}, & \alpha < 0 \\ 1, & \alpha \geq 0 \end{cases}, \end{gathered} \tag{3}$$

where α is the tangent of the tunnel slope angle (%). Experimental observations confirmed that the critical airflow velocity depends on the cube root of the heat release rate which is referred to in the literature as the cube root rule.

The estimate (3) in a modified form is also found in the current international standard for the design of road tunnel ventilation [7], where the coefficient K_1 is not constant but depends on the heat release rate. The coefficient values were determined based on the results of full-scale experiments in the Memorial Tunnel for heat release rate values >100 MW, 90 MW, 70 MW, 50 MW, 30 MW and < 10 MW. Further research has shown significant limitations in the applicability of this approach. It has been observed that the critical velocity at high fire intensities becomes independent of the heat release rate, and therefore, a dimensionless quantity technique has been introduced to investigate estimates of the critical airflow velocity.

Y. Oka and G. T. Atkinson investigated the critical airflow velocity in a 15 m long small-scale horizontal tunnel with a semicircular ceiling and sloping side walls at an angle of 7°, height of 0.244 m, width of 0.274 m, and cross-sectional area of 0.0569 m^2. They introduced the dimensionless critical airflow velocity u_c^* and the dimensionless heat release rate Q^* in the form:

$$u_c^* = \frac{u_c}{(gH)^{1/2}},\ Q^* = \frac{Q}{\rho_0 c_p T_0 g^{1/2} H^{5/2}} \tag{4}$$

and based on the analysis of experimental data, they derived the following relationship for u_c^* [8]:

$$u_c^* = \begin{cases} 0.35(Q^*/0.124)^{1/3}, & Q^* \leq 0.124 \\ 0.35 \quad , & Q^* > 0.124 \end{cases} \tag{5}$$

Thus, for higher Q^* values, they replaced the cube root rule with constant dependence. For heat release rates less than or equal to 0.124, the dimensionless critical velocity changes as the cube root of the dimensionless heat release rate. For higher values of the dimensionless heat release rate, the dimensionless critical velocity becomes independent of the dimensionless heat release rate.

The relationship between critical airflow velocity, tunnel cross-section, and heat release rate was investigated experimentally and theoretically by Y. Wu and M. Z. A. Bakar, who conducted a series of experiments in small-scale horizontal tunnels with different tunnel cross-sections. They introduced the dimensionless critical airflow velocity u_c' and the dimensionless heat release rate Q' in the form:

$$u_c' = \frac{u_c}{\sqrt{g\overline{H}}},\ Q' = \frac{Q}{\rho_0 c_p T_0 g^{1/2} \overline{H}^{5/2}}, \tag{6}$$

where $\overline{H}$ is the hydraulic diameter of the tunnel defined as the ratio of four times the cross-sectional area of the tunnel to its perimeter, $\overline{H} = \frac{4A}{P}$ and they derived the relationship between the dimensionless critical airflow velocity and the dimensionless heat release rate in the form [9]:

$$u_c' = \begin{cases} 0.40(Q'/0.2)^{1/3}, & Q' \leq 0.20 \\ 0.40 \quad , & Q' > 0.20 \end{cases} \tag{7}$$

In [10], the results of the experiments described in [9] were used to estimate the overestimation of the critical airflow velocity obtained by the FDS system. The overestimation estimate was derived based on the correlation between the values of the ratio of the airflow velocity u to the critical airflow velocity u_c and the dimensionless smoke backlayering length L_b^* at the airflow velocity u. This correlation was derived in [11] based on experiments in a small-scale square-section tunnel and a small-scale curved-ceiling tunnel.

Let u_{FDS} be the critical airflow velocity determined by FDS simulation, $L_{b,\,FDS}$ be the smoke backlayering length determined by FDS simulation, and u_{exp} be the critical airflow velocity measured experimentally. Then u_{FDS} can be estimated as follows [10]:

$$u_{FDS} \approx u_{exp}\left(1 + \frac{L_{b,FDS}}{18.5H}\right) for\ H \approx \overline{H} \tag{8}$$

or

$$u_{FDS} \approx u_{exp}\left(1 + \frac{L_{b,FDS}}{18.5\overline{H}}\right) \text{otherwise.} \tag{9}$$

In the following, we briefly describe the experimental results from [9] (Sect. 2) and analyze the results of a series of 5 simulations performed using the FDS system, which were used in [10] to estimate the overestimation of the critical flow velocity in small-scale tunnels (Sect. 3). In Sect. 4, we briefly summarize the results of this study.

2 Experiments in Small-Scale Tunnels

The relationship between critical airflow velocity, tunnel cross-section and the heat release rate was investigated through a series of experiments in five approximately 15 m long small-scale horizontal tunnels with different cross-sections (see Table 1 and Fig. 2) [9]. Tunnels A-D had rectangular cross-sections, while tunnel E had a semicircular ceiling and sloping side walls at an angle of 7 °. A circular fire source with a diameter of 0.06 m was placed on the tunnel floor approximately 6 m from the tunnel portal. The temperature increases were measured using thermocouples with a diameter of 0.25 mm positioned in the center, 1 cm below the tunnel ceiling, at distances of 1H, 3H, 5H, and 10H from the fire source. The critical airflow velocity was determined for different heat release rates by measuring the airflow velocity at which the smoke backlayering reached lengths of 1H, 3H, 5H and 10H. The airflow velocity at which the smoke backlayering disappears (i.e., the smoke backlayering length equals 0H) was determined by extrapolation from the four values obtained.

Table 1. Dimensions of the tunnels: cross-section shape, length (L), height (H), width (W) and hydraulic diameter ($\overline{H}$).

Tunnel	Cross-section	L (m)	H (m)	W (m)	$\overline{H}$ (m)
A	Rectangular	15	0.250	0.136	0.176
B	Squared	15	0.250	0.250	0.250
C	Rectangular	15	0.250	0.500	0.333
D	Rectangular	15	0.250	1.000	0.400
E	Circular ceiling	15	0.244	0.274	0.238

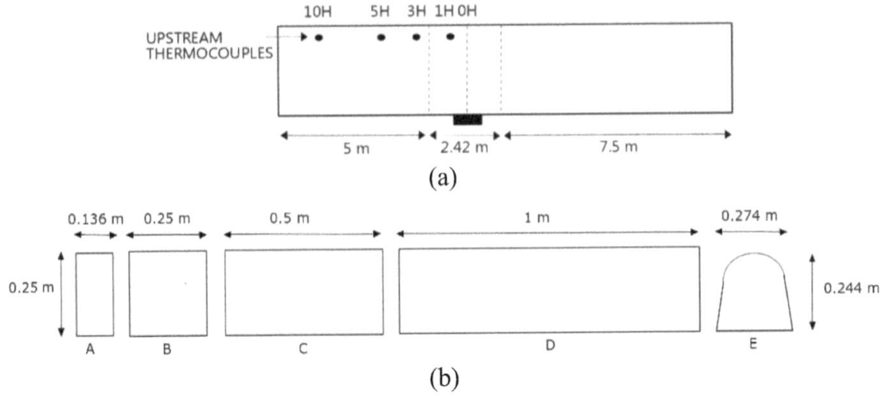

Fig. 2. Schematic of the experimental tunnels: side view (a) and cross-sections of tunnels A-E (b).

The results of the experiments showed that critical airflow velocity depends on both tunnel cross-section shape and heat release rate (Table 2, Fig. 3a). It was also confirmed that up to a certain value of heat release rate, the critical airflow velocity increases with the cube root of the heat release rate (cube root rule), while above a certain value of the heat release rate, the critical airflow velocity becomes independent of the heat release rate. The results of the dimensionless analysis of the experimental results are shown in Fig. 3b. Based on the analysis of these data, the relationship (7) between the dimensionless critical airflow velocity and the dimensionless heat release rate was derived [9].

Table 2. Experimentally obtained values of critical velocity u_C for different values of heat release rate Q in tunnels A-E.

Tunnel	A	B	C	D	E
Q (kW)	u_C (m/s)	u_C (m/s)	u_C (m/s)	u_C (m/s)	u_C (m/s)
1.50	0.43	0.39	0.37	0.34	0.44
3.00	0.46	0.48	0.45	0.40	0.54
7.50	0.48	0.56	0.54	0.50	0.60
10.50	0.48	0.59	0.57	0.54	0.60
12.00	0.48	0.60	0.59	0.56	0.60
15.00	0.48	0.60	0.60	0.59	0.60
22.50	0.48	0.60	0.62	0.65	0.60
30.00	0.48	0.60	0.65	–	0.60

The dimensionless critical airflow velocity therefore varies as the cube root of the dimensionless heat release rate for values of the dimensionless heat release rate less than

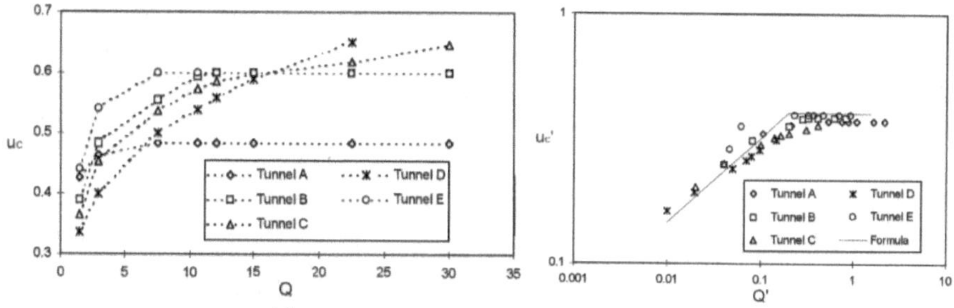

(a) (b)

Fig. 3. Dependence of the experimentally measured critical airflow velocity u_C (m/s) on the heat release rate Q (kW) (a) and dependence of the dimensionless critical airflow velocity u_c' on the dimensionless heat release rate Q' with added formula (7) (b) [9].

or equal to 0.2, and for higher values, the dimensionless critical airflow velocity becomes independent of the dimensionless heat release rate. The authors validated the formula (7) using experiments in three full-scale tunnels (Memorial Tunnel, Repparfjord Tunnel, and Buxton Gallery). Their estimates corresponded relatively well to experimentally obtained data [9].

3 Analysis of Simulation Results Using FDS

In this study, we used a set of five simulations that were reported in [10] to estimate the overestimation of the critical airflow velocity in small-scale tunnels in simulations. The input files for these simulations are available at [12]. The simulations corresponded to the experiments described in [9]. In each, the heat release rate was incremented stepwise from its lowest reported value to its highest, with each step lasting 20 s. For each of the eight heat release rates, the tunnel airflow velocity was set to the reported critical velocity. Smoke backlayering was being observed in all tested simulations.

The initial conditions in the simulations were: initial ambient temperature 20 °C, initial airflow velocity 0 m/s, total simulation time 165 s in tunnels A, B, C, and E, and total simulation time 145 s in tunnel D. The dimensions of the computational domain and the resolution of the numerical grid are specified in Table 3. In the x-direction, the computational domain was divided into 30 equal subdomains of 0.5 m length each. The numerical mesh was divided into 30 meshes (Fig. 4). The simulations were performed using the parallel MPI model FDS (Version 6.9.1) on the Devana supercomputer [13]. The calculation on each mesh was performed on one computational core as a single MPI process with the data exchange between the computational processes performed using the MPI interface.

The walls of tunnels A-D were represented by the boundary of the computational domain with an assigned surface boundary condition corresponding to STEEL, using preset values from the FDS database. The geometric representation of tunnel E was created using the code available at [12] and consisted of OBST cuboids with dimensions of 15 x 0.0101 x 0.0102 m assembled into the desired shape. Material properties of the

Table 3. Numerical meshes used in simulations SA, SB, SC, SD and SE.

	Computational domain (m × m × m)	Grid resolution (m × m × m)	Number of cells in x, y, z direction	Total number of cells
SA	15 × 0.136 × 0.25	0.01 × 0.0097 × 0.01	50, 14, 25	17500
SB	15 × 0.25 × 0.25	0.01 × 0.01 × 0.01	50, 25, 25	31250
SC	15 × 0.5 × 0.25	0.01 × 0.01 × 0.01	50, 50, 25	62500
SD	15 × 1 × 0.25	0.01 × 0.02 × 0.01	50, 50, 25	62500
SE	15 × 0.274 × 0.244	0.01 × 0.0101 × 0.0102	50, 27, 24	32400

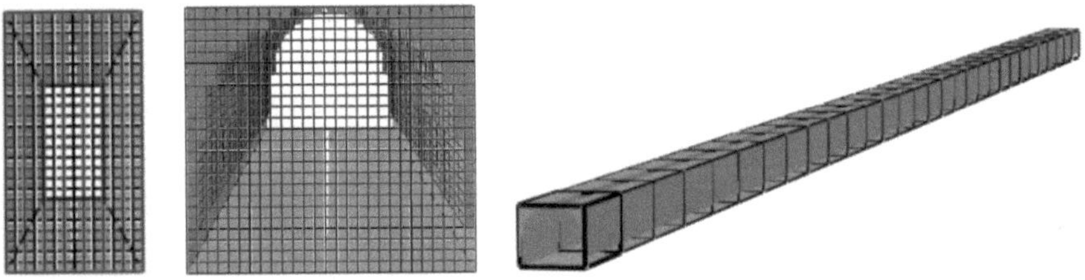

(a) (b) (c)

Fig. 4. Representation of the considered tunnels: front view of the tunnel A (a) and tunnel E (b) and side view of tunnel E (c).

walls and boundaries of the domain corresponded to the STEEL material. The roughness of the walls was set to a value of 0.001.

Table 4. Values of control parameter F in the RAMP function in SA-SE simulations.

Q (kW)	1.50	3.00	7.50	10.50	12.00	15.00	22.50	30.00
F (–)	0.43	0.46	0.48	0.48	0.48	0.48	0.48	0.48
F (–)	0.39	0.48	0.56	0.59	0.60	0.60	0.60	0.60
F (–)	0.37	0.45	0.54	0.57	0.59	0.60	0.62	0.65
F (–)	0.34	0.40	0.50	0.54	0.56	0.59	0.65	–
F (–)	0.44	0.54	0.60	0.60	0.60	0.60	0.60	0.60

The left portal of the tunnel was represented by a VENT object with a boundary condition FAN with an associated flow velocity VEL of −1 (a negative value represents the fact that the airflow enters the computational domain). The change in flow velocity during the 165-s simulation was represented by a RAMP function with control parameter F (Table 4), and the airflow velocity in the tunnel cross-section was determined as the average velocity in volume with dimensions specified in Table 5, located 4.3 m from the left tunnel portal.

Table 5. Dimensions $L \times W \times H$ (length $\times$ width $\times$ height) of the volume considered for measuring the airflow velocity in the tunnel cross-section in simulations SA-SE.

	$L \times W \times H$ (m $\times$ m $\times$ m)
SA	$0.02 \times 0.136 \times 0.25$
SB	$0.02 \times 0.25 \times 0.25$
SC	$0.02 \times 0.5 \times 0.25$
SD	$0.02 \times 1 \times 0.25$
SE	$0.02 \times 0.274 \times 0.244$

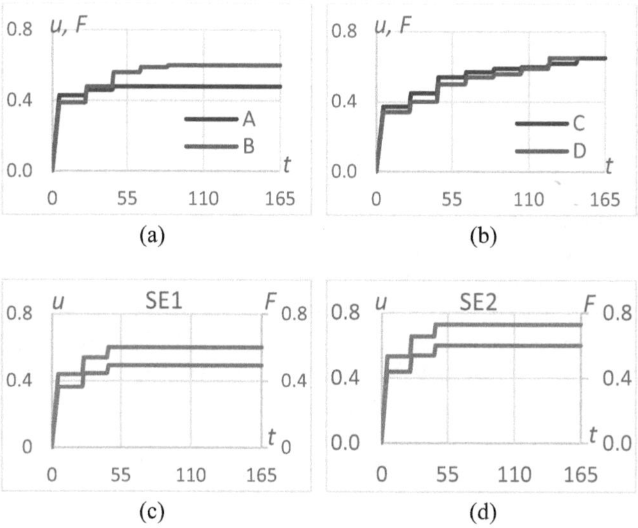

Fig. 5. Settings of the F parameter values controlling the airflow velocity in the RAMP function and the average airflow velocities achieved in the tunnel cross-section: values of the F and u parameters in the SA and SC simulations (red) and in the SB and SD simulations (green) (a-b) and values of the F parameter (red) and u parameter (blue) in the SE1 simulation (c) and the SE2 simulation (d).

Figure 5 shows the time course of airflow velocities in the considered simulations. For tunnels A-D, the airflow velocities determined by the simulation corresponded to the target values. Since the airflow velocity values determined by the simulation in tunnel E did not correspond to the target values, we found new values for the control parameter F to achieve airflow velocity in the tunnel cross-section consistent with correct experimental values of the airflow velocity (Table 6).

The adjustment in the SE simulation was required due to the representation of the rounded ceiling and sloping walls of the tunnel within a rectangular numerical mesh. During turbulent airflow in the simulation, the air encountered irregularities on the tunnel surface, which was smooth in the real tunnel, and lost part of its kinetic energy. The right portal of the tunnel was represented by the VENT object with the boundary condition OPEN.

Table 6. Controlling airflow changes during 8 experiments using the control parameter *F* in the RAMP function for tunnel E in simulations SE1 and SE2.

Simulation	Experiment	1	2	3–8
SE1	*F* (–)	0.440	0.540	0.600
	u (m/s)	0.363	0.445	0.494
SE2	*F* (–)	0.534	0.656	0.729
	u (m/s)	0.440	0.540	0.600

The fire source was represented in the simulation by the VENT object with dimensions of 0.12 x 0.12 x 0 m, with a circular structure and a center 6.21 m from the left tunnel portal, with a radius of 5.3 cm and a total area of 88. 247 cm^2, which was assigned the boundary condition HRRPUA of 3400 kW/ m^2. PROPANE was used as the fuel for the fire source, as in the experiment with the properties preset in the FDS. The change in fire intensity during the 165-s simulation was represented using the RAMP function with the control parameter *F* (Table 7). Figure 6 shows the time course of parameter *F* (orange) and fire intensity *Q* (blue) in the SA-SD, SE1, and SE2 simulations. The average values of *Q* corresponded to the desired values. At higher values, the fire intensity fluctuated increasingly. The increase in *Q* values exactly mirrored the increase in parameter *F*. We can also see that the time intervals when *Q* values fluctuated around a single value correspond to time intervals when the *F* parameter did not change. The degree of fluctuation varied from simulation to simulation. *Q* value fluctuated the most in the SA and SE2 simulations, while in the SC and SD simulations the fluctuation of *Q* value was much smaller. This difference may be due to the resolution of the numerical mesh.

Table 7. Values of the control parameter *F* of the RAMP function controlling the fire intensity *Q* in simulations.

Simulations	Q (kW)	1.5	3	7.5	10.5	12	15	22.5	30
SA-SC, SE	F (–)	0.05	0.10	0.25	0.35	0.4	0.5	0.75	1
SD	F (–)	0.05	0.10	0.25	0.35	0.4	0.5	0.75	–

The smoke backlayering length L_b was evaluated in the simulations using a sharp temperature increase recorded by a group of 200 thermocouples placed under the ceiling

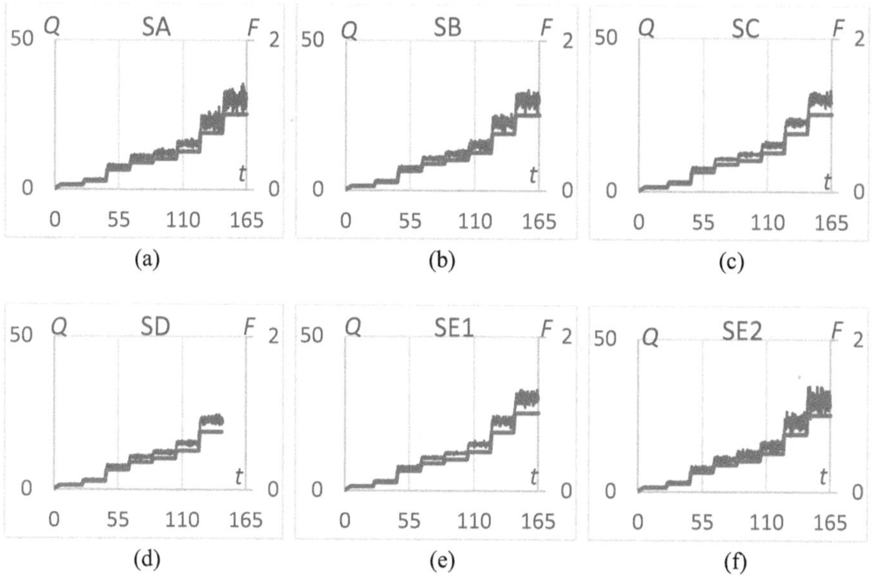

Fig. 6. Settings of the RAMP function parameter F values (in red) and the corresponding fire intensity values Q (in blue) in the SA-SD (a-d), SE1 (e), and SE2 (f) simulations.

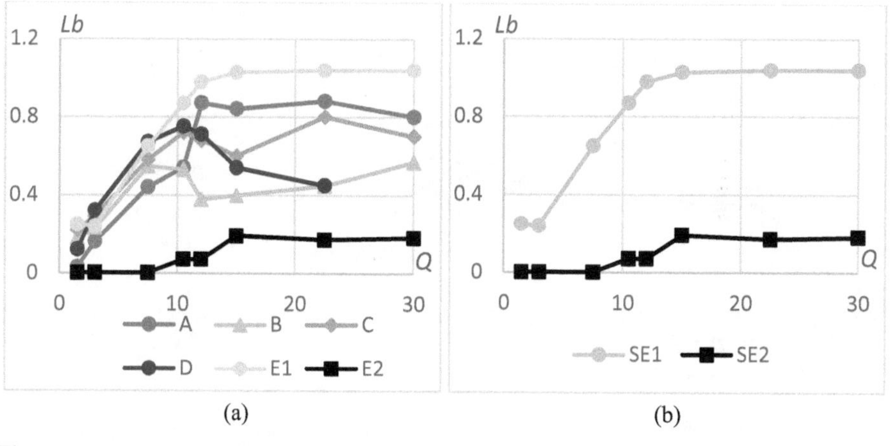

Fig. 7. Dependence of L_b on Q for the SA-SD, SE1, and SE2 simulations (a) and comparison of the results of the SE1 and SE2 (b).

in the middle of the tunnel at a distance of 1 cm from each other (the first thermocouple was placed 1.21 m in front of the fire and the last thermocouple 0.79 m behind the fire). A temperature increase of 10 °C was considered, with a stabilization time of 5 s for the temperature on a given thermocouple and a time interval for a stable temperature of 14 s. For this interval, the average temperature was calculated for each thermocouple at each value of Q. The smoke backlayering length was determined as distance of the thermocouple at which a temperature increase of more than 30 °C was recorded, from the fire source center. Figure 7 shows the dependence of L_b on Q in the SA-SD, SE1 and SE2 simulations.

Because the airflow velocities in the simulations corresponded to the experimentally determined critical airflow velocity, all positive values of L_b represent an overestimation of the smoke backlayering length and therefore also an overestimation of the critical airflow velocity in the simulation. Figure 7a clearly shows the overestimation of the critical airflow velocity in all SA-SD and SE1 simulations. With increasing heat release rates, the values first increase sharply, then slow down significantly or stagnate.

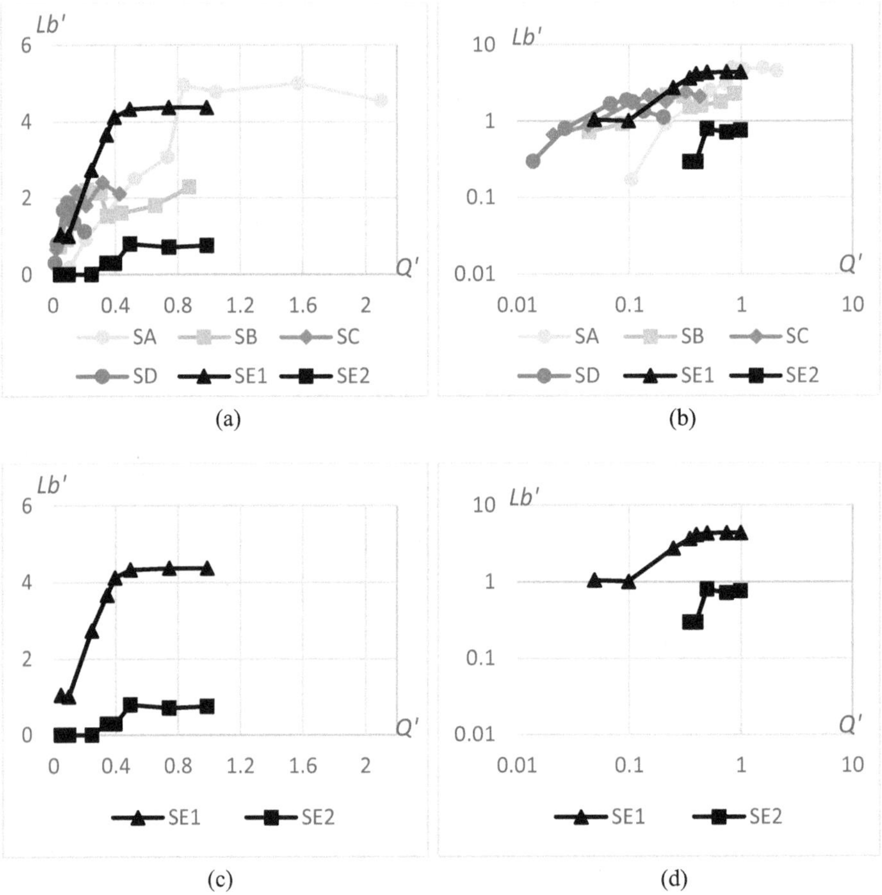

Fig. 8. Dependence of L_b' on Q' for SA-SD, SE1 and SE2 simulations (a-b) and the comparison of SE1 and SE2 simulation results (c-d).

In the SE2 simulation, at lower heat release rates, no smoke backlayering occurred ($L_b = 0$ m). This indicates that the simulation does not overestimate or underestimate the critical airflow velocity. At higher heat release rates, it initially increases at the same airflow velocity (experiments 4–6) and then stagnates (experiments 6–8).

The comparison of the results of simulations SE1 and SE2 in Fig. 7b shows that in simulation SE2, the L_b values are significantly lower for all values of heat release rate. Since the airflow velocity in the simulation was the same as the critical velocity obtained from the experiments (zero values of L_b were expected in the simulations), the overestimation of the critical velocity in the SE2 simulation is significantly lower.

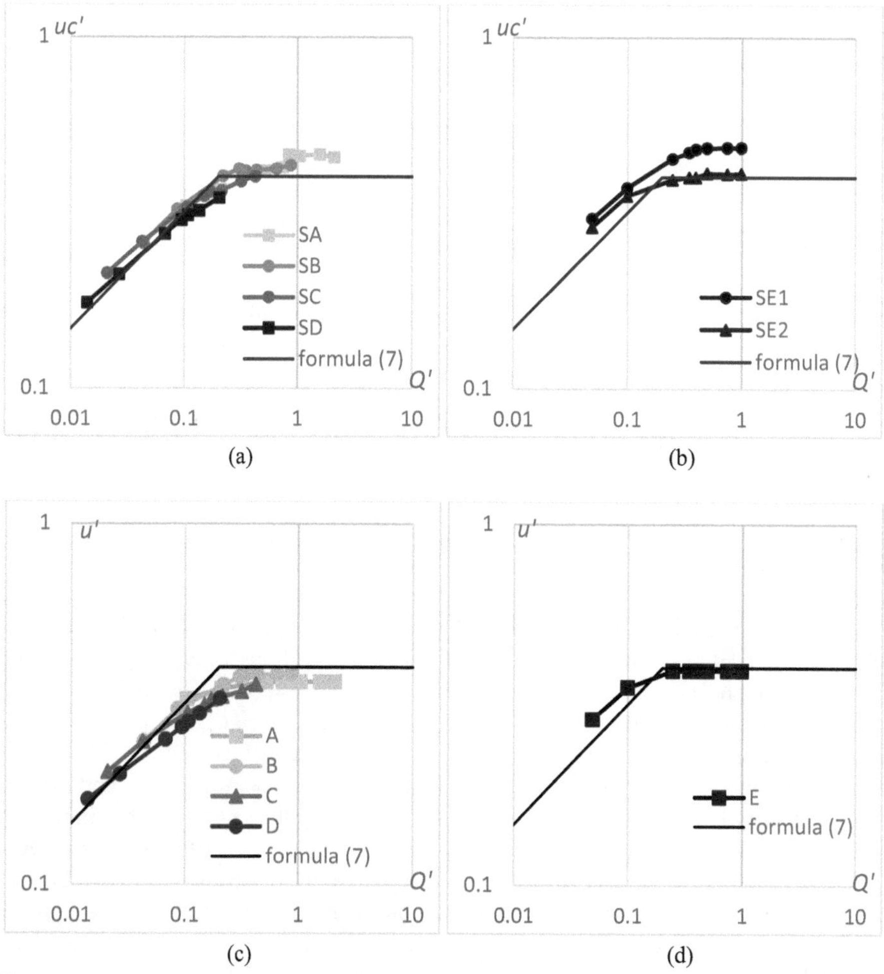

Fig. 9. Correlation (7) and the dependence of the dimensionless critical airflow velocity determined by the SA, SB, SC, SD, SE1 and SE2 simulations on the dimensionless heat release rate (a-b) and the dependence of the dimensionless critical airflow velocity related to the critical airflow velocity measured during the experiments in the A, B, C, D and E tunnels on the dimensionless heat release rate (c-d).

After adjusting the settings in the SE2 simulation, we obtained significantly more accurate estimates of the critical velocity. However, it is still true that for higher heat release rates this overestimation was not negligible. The refinement of the airflow velocity representation in the simulation significantly contributed to the refinement of the estimate of the smoke backlayering length (critical velocity).

In the sequel, we present the results of the dimensionless analysis of the obtained simulation results. Figure 8 shows the dependence of L_b' on Q' for the SA-SD, SE1 and SE2 simulations. Figure 8a-b shows the overestimation of the L_b' values for all considered cases. Figure 8c-d shows that in the SE2 simulation, the L_b' values were significantly reduced compared to the SE1 simulation, indicating that the airflow velocity in SE2 approached the critical airflow velocity after adjusting the airflow velocities in tunnel E by changing the parameter F.

Figure 9 shows the dependence of u_c' on Q' for the SA-SD, SE1 and SE2 simulations. From the values of dimensionless heat release rate and dimensionless smoke backlayering length obtained by the simulation, we calculated the values of u_c' using the formula (8). In Fig. 9a-b, the correlation (8) and the dependence of the dimensionless critical airflow velocity determined by the SA, SB, SC, SD, SE1, and SE2 simulations on the dimensionless heat release rate are shown. In Fig. 9c-d, the dependence of the dimensionless critical airflow velocity related to the critical airflow velocity measured during the experiments in the A, B, C, D and E tunnels on the dimensionless heat release rate is shown. Figure 9 shows not only the overestimation of the critical airflow velocity in simulations in tunnels with different cross-sections but indicates also the extent of this overestimation and the influence of the cross-section on the resulting critical airflow velocity overestimation.

4 Conclusion

In this paper, the results of a set of five simulations of smoke propagation at different heat release rates in small-scale tunnels with different cross-sections were analyzed. In each simulation, the heat release rate was incrementally increased from the lowest reported value to the highest, with each step lasting 20 s. For each of the eight heat release rate values, the airflow velocity in the tunnel was set to the reported critical velocity. The results showed that the simulations overestimate the smoke backlayering length and critical airflow velocity in each of the considered cases. Adjusting the settings in the simulation in the tunnel with a rounded ceiling significantly improved the simulation results and substantially reduced the overestimation of the smoke backlayering length and critical airflow velocity. The simulations were conducted using the FDS, version 6.9.1 on the Devana supercomputer. The results of this study indicate that the issues related to the overestimation of the critical airflow velocity in small-scale tunnels determined by the FDS fire simulator should be addressed.

Acknowledgements. This research was funded by the Grant Programme for Slovak Academy of Sciences PhD students under the project No. APP0646 and the EU NextGenerationEU through the Recovery and Resilience Plan for Slovakia under the project No. 09I05-03-V02-00055 (DICRIS), using the computational resources procured in the national project National Competence Centre for HPC (project code: 311070AKF2) funded by ERDF, EU SF Informatization of Society, Operational Program Integrated Infrastructure.

Disclosure of Interests. The authors have no competing interests to declare that are relevant to the content of this article.

References

1. Weisenpacher, P., Glasa, J., Valasek, L.: Investigation of various FDS approaches to modelling airflow in road tunnel induced by longitudinal ventilation. Fire **8**(2), 87 (2025)
2. Weisenpacher, P.: Modeling of road tunnel airflows by FDS using tunnel model with a system of geometrical objects. In: Proceedings of the 35rd European Modeling & Simulation Symposium (EMSS 2023), art no. 6. Greece, CAL-TEK S.R.L., Athens (2023)
3. Glasa, J., Valasek, L., Weisenpacher, P., Kubisova, T.: Improvement of modeling velocity of airflow created by emergency ventilation in a road tunnel using FDS 6. Appl. Sci.-Basel **13**(5), 2762 (2023)
4. Thomas, P.H.: The movement of smoke in horizontal passages against an air flow. Fire Res. Stat. **723** (1968)
5. Heselden, A.J.M.: Studies of fire and smoke behaviour relevant to tunnels. In: Second International Symposium on the Aerodynamics and Ventilation of Vehicle Tunnels in Cambridge, Fire Research Station, p. 15, Borehamwood (1976)
6. Kennedy, W.D.: Critical velocity: past, present and future. In: Seminar of Smoke and Critical Velocity in Tunnels, pp. 305–322 (1996)
7. NFPA 502. Standard for Road Tunnels, Bridges, and Other Limited Access Highways (NFPA 502). NFPA, Quincy, MA, USA (2023)
8. Oka, Y., Atkinson, G.T.: Control of smoke flow in tunnel fires. Fire Saf. J. **25**, 305–322 (1995)
9. Wu, Y., Bakar, M.Z.A.: Control of smoke flow in tunnel fires using longitudinal ventilation systems – a study of the critical velocity. Fire Saf. J. **35**, 363–390 (2000)
10. McGrattan, K. et al.: Fire Dynamics Simulator, Technical Reference Guide, Volume 3: Validation. Sixth Edition. NIST Special Publication 1019. National Institute of Standards and Technology, Gaithersburg, Maryland, USA, p. 1124 (2023).
11. Li, Y.Z., Lei, B., Ingason, H.: Study of critical velocity and backlayering length in longitudinally ventilated tunnel fires. Fire Saf. J. **45**, 361–370 (2010)
12. https://github.com/firemodels/fds/tree/master/Validation/Wu_Bakar_Tunnels. Accessed 22 Oct 2025
13. https://vs.sav.sk/sluzby/vysokovykonna-vypoctova-technika/superpocitac-devana/. Accessed 22 Oct 2025

Detecting Change Point or Anomaly Based on Rough Path Theory

Zhixing Tan[1], Kangrong Tan[2(✉)], and Shozo Tokinaga[3]

[1] Department of Mathematics, Faculty of Education, Waseda University, Tokyo, Japan
siyannyann@outlook.com

[2] Faculty of Economics, Kurume University, Fukuoka, Japan
camox.wein.london@gmail.com

[3] Department of Economics, Kyushu University, Fukuoka, Japan

Abstract. Although deep learning methods such as convolutional neural networks achieve strong pattern-recognition performance, they often require substantial computational resources and offer limited interpretability. Rough Path Theory (RPT) provides a mathematically rigorous alternative for analyzing irregular, non-smooth data by extending classical stochastic calculus. Its central construct, the path signature, captures higher-order temporal dependencies while preserving the intrinsic ordering of events.

This study investigates the application of Rough Path Theory for change-point detection and anomaly analysis, with a focus on biomedical signals. The framework is generalizable across diverse data modalities, including geometric trajectories and other irregular measurements. We introduce the theoretical foundations of rough paths and path-signature representations and develop practical computational strategies for real-world data. Experiments on the MIT-BIH Arrhythmia Database demonstrate that RPT-based models that effectively detect pathological change and capture latent dynamical patterns in cardiac signals. Notably, the approach achieves strong predictive performance while remaining computationally efficient on standard personal computers without GPU acceleration, making it suitable for resource-constrained clinical environments.

Keywords: Irregular non-smooth data · Rough path theory · Kernel SVM · Pattern recognition

1 Introduction

In recent years, a wide range of methods have been developed in deep learning and pattern recognition. Among these, Convolutional Neural Networks (CNNs) have become particularly prominent due to their strong performance in complex classification tasks. Supported by high-performance GPUs and large memory capacity, CNN-based approaches have achieved state-of-the-art accuracy across various application domains, including biomedical signal analysis [1,7,8,15,16].

H. R. Arabnia et al. (Eds.): CSCI 2025, CCIS 2970, pp. 240–256, 2027.
https://doi.org/10.1007/978-3-032-28466-2_17

In contrast, Rough Path Theory (RPT), originally introduced by Lyons and subsequently developed by Friz and Victoir [5] [12], provides a mathematically rigorous framework for the representation and analysis of highly irregular and non-smooth data streams. By extending classical stochastic calculus, RPT enables the definition of integrals and differential equations driven by signals that are too irregular to be handled by conventional Itó calculus [2,4,9–11]. This theoretical foundation offers a resource-efficient and interpretable alternative to deep learning architectures, creating new opportunities for accurate signal classification in computationally constrained environments, particularly in biomedical applications.

Rough Path Theory provides a novel mathematical framework for financial modeling, particularly in contexts where market data exhibit characteristics such as volatility clustering [6]. A central strength of RPT lies in its ability to represent and manipulate higher-order pathwise information through a structured algebraic object known as the *path signature* [10,12]. The signature provides a compact and faithful encoding of essential information contained in a time series while preserving the temporal ordering of events. As a result, RPT furnishes a robust and interpretable framework for machine learning and statistical inference in the analysis of financial time series [9].

A particularly promising application of Rough Path Theory is the detection and prediction of change points, geometric trajectory changes, and other anomalies. Change points represent time instants at which the underlying system dynamics undergo structural transitions, such as regime shifts in financial time series or abrupt physiological changes in biomedical signals [3]. Accurate identification of these events is crucial for risk management and adaptive strategies, as well as for timely diagnosis and clinical intervention. While traditional change-point detection methods are effective in many settings, they often struggle with the irregular, nonlinear, and non-Markovian nature of real-world data.

In such contexts, RPT provides a mathematically coherent approach for capturing higher-order temporal dependencies and structural changes, including geometric trajectory variations and medical signal anomalies, that are difficult to detect using conventional methods. By encoding complex temporal structures through higher-order path information, RPT enables precise characterization of subtle dynamical changes, making it a viable alternative to traditional statistical and deep learning approaches for irregular and nonlinear data.

In this study, we investigate a range of applications of Rough Path Theory and evaluate its effectiveness for change-point detection and anomaly analysis across multiple domains. We first present the theoretical foundations of rough paths and path signature representations, followed by a description of practical computational strategies. Numerical experiments and empirical evaluations are then conducted, including applications to biomedical signal processing using the MIT-BIH Arrhythmia Database [13,14]. The results indicate that RPT-based models achieve competitive predictive performance while effectively revealing latent dynamical structures in physiological signals. Furthermore, the proposed framework operates efficiently on standard personal computers without GPU acceleration or large memory requirements, making it suitable for deployment in resource-constrained healthcare environments.

2 Methodology

Rough Path Theory, introduced by Lyons in the late 1990 s, provides a mathematically rigorous framework for the representation and analysis of highly irregular time-series data. By extending classical stochastic calculus through the augmentation of paths with *iterated integrals*, RPT captures both local and global structural characteristics of a signal in a systematic and noise-robust manner [4,12].

Unlike conventional approaches that rely solely on raw signal values or first-order differences, RPT represents a time series through its *signature*, defined as a collection of iterated integrals that encode temporal interactions and dependencies within the data. This representation offers several notable advantages:

1. **Coordinate-free characterization:** Path signatures provide an intrinsic description of a signal that is invariant under reparameterization, making them particularly well suited to irregularly sampled data, such as financial time series or physiological signals.
2. **Noise robustness:** By incorporating higher-order iterated integrals, the signature representation exhibits stability under perturbations of the input path, which is essential when analyzing high-frequency signals subject to stochastic noise or microstructural fluctuations.
3. **Dimensionality reduction:** Truncated signatures yield a compact yet expressive representation of complex path information, preserving salient structural features for downstream tasks such as classification, regression, and anomaly detection.

In this study, we utilize the signature transform, a core concept in Rough Path Theory, to identify change points, geometric changes, and anomalies in medical signals. By representing paths through their signatures, higher-order dependencies and temporal structures that are typically overlooked by conventional time-series models are systematically encoded.

This approach enhances the interpretability of change-point and anomaly analysis while improving robustness to noise and irregular sampling, thereby offering a viable methodology for a broad range of time-series applications.

2.1 Rough Path Representation of Time Series

Classical stochastic calculus often struggles to fully capture the irregular and noisy behavior present in financial time series. Rough Path Theory overcomes this limitation by augmenting the original path with its *iterated integrals*, thereby incorporating both first-order and higher-order interactions and providing a richer representation of the underlying dynamics.

Consider the observed stock price path over the interval $[0,T]$:

$$X = \{X_t : 0 \le t \le T\}.$$

The *signature* of the path X up to level m is defined as:

$$S^{(m)}(X) = \left(1, \int_0^T \mathrm{d}X_{t_1}, \int_0^T \int_0^{t_2} \mathrm{d}X_{t_1} \otimes \mathrm{d}X_{t_2}, \dots\right), \tag{1}$$

where $\otimes$ denotes the tensor product.

The dimension of a truncated path signature of degree k for an l-dimensional path is given by

$$\dim(l,k) = \sum_{i=1}^{k} l^i. \tag{2}$$

where the zeroth-order term is excluded. In the subsequent numerical experiments, Eq.s (1) and (2) are used to compute path signatures. For example, when $l = 2$ and $k = 8$, the resulting signature dimension is $\dim(2,8) = 510$ (or 511 when the zeroth-order term is included). In our experiments, we compute $\dim(2,8)$ without including the zeroth-order term.

For a d-dimensional time series $X_t = (X_t^1, X_t^2, \dots, X_t^d)$, the path signature encodes both first-order increments and higher-order cross-component interactions. In practice, the signature is truncated at a finite level to balance computational efficiency with representational expressiveness.

2.2 Feature Extraction Using Signatures

We compute truncated signatures over rolling windows of data to extract localized pathwise features. For each rolling window of length w, we obtain a truncated signature vector $S^{(m)}(X)$, which serves as a feature representation for subsequent statistical analysis.

Signatures are especially well-suited for this task because they are invariant under time reparameterization and inherently preserve temporal ordering. This property allows the method to detect subtle yet meaningful structural changes in dynamics that might be overlooked by conventional approaches.

3 Numerical Experiments

In this section, we present the results of the simulation studies and the analysis of real-world medical data. Experiment 1 and 2 demonstrate that change points and geometrical change detections can be both effectively identified using Rough Path Theory. In Experiment 3, we conduct an empirical evaluation on biomedical signals using the MIT-BIH Arrhythmia Database [13]. The results indicate that RPT-based methods exhibit robust performance, achieving high classification accuracy and highlighting their potential for practical biomedical applications.

3.1 Change Point Detection

First, we specify the model in which the slope changes at a random time point in order to generate change points, with the error term $\epsilon \sim \mathcal{N}(0, \sigma^2)$:

$$x_t = x_{t-1} + \text{slope} \times t + \epsilon. \tag{3}$$

We then generate 500 time-series samples from Eq. (3). Using the proposed RPT-based approach—specifically, by computing path signatures, change points are identified using an RBF kernel. The results are illustrated in Fig. 1, and the corresponding numerical outcomes are summarized in Table 1.

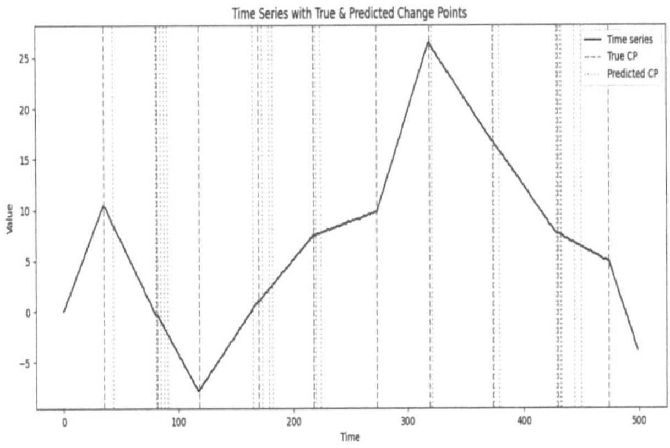

Fig. 1. Detected change points in simulation 1

Table 1. Results of change points detection

	precision	recall	f1-score	support
0	0.87	0.94	0.90	106
1	0.71	0.50	0.59	30
accuracy			0.85	136
macro avg	0.79	0.72	0.75	136
weighted avg	0.84	0.85	0.84	136

From Table 1, the results are satisfactory, with both classes exhibiting reasonably balanced precision, although recall is lower for Class 1, which can be attributed to sample imbalance.

3.2 Geometric Trajectory Change Detection: Circles vs. Spirals

We demonstrate a canonical rough path classification task, namely, distinguishing *closed* planar curves (circles) from *open* planar curves (spirals) using *path signatures* as features and standard classification methods.

Each observation is a time-parametrized path, $\gamma : [0, T] \to \mathbb{R}^2$, and classification is performed in the space of truncated signatures rather than on raw coordinate data. This illustrates how nonlinear geometric information can be transformed into linear statistical features.

Synthetic Path Generation (Dataset Construction)

1)Noisy Circles Generating

Circles are generated according to equation (4):

$$\gamma_{\text{circle}}(t) = \big(r\cos t,\, r\sin t\big), \qquad t \in [0, 2\pi]. \tag{4}$$

where the radius r is randomly perturbed around 1, and independent Gaussian noise is added to each point to model measurement error.

To eliminate trivial geometric cues, each path is further subjected to a random rotation $R \in SO(2)$(where SO(2) is a uniformly chosen planar matrix that randomly orients each path), and a random translation in $\mathbb{R}^2$.

2)Noisy Spirals Generating

Spirals are defined and generated as Archimedean spirals by Eq. (5):

$$\begin{aligned} r(t) &= a + bt, \\ \gamma_{\text{spiral}}(t) &= \big(r(t)\cos t,\, r(t)\sin t\big), \qquad t \in [0, 4\pi]. \end{aligned} \tag{5}$$

Unlike circles, spirals are *not closed* and exhibit nonzero net displacement. The same noise, rotation, and translation operations are applied as in the circular case. Plots of examples are shown in Fig. 2.

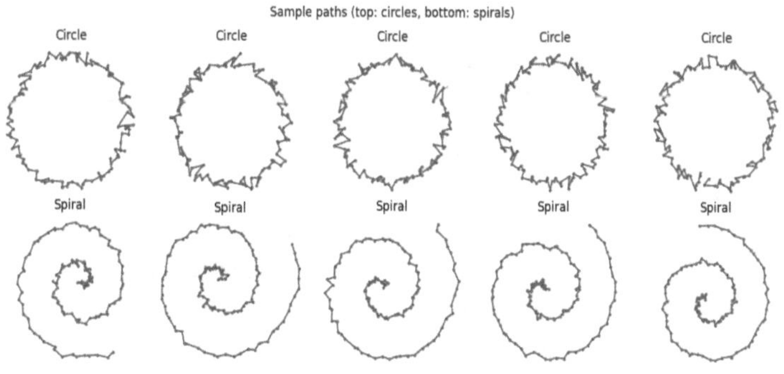

Fig. 2. Geometrical change detection

Before computing path signatures, each path is normalized to enhance numerical stability and comparability. 1)Translation normalization: each path is shifted so that its initial point coincides with the origin, namely, $\gamma(t) \leftarrow \gamma(t) - \gamma(0)$, which is necessary because path signatures are not invariant under translations. 2)Scale normalization: each path is rescaled by its empirical standard deviation to suppress trivial scale effects and improve numerical stability.

A balanced dataset is constructed consisting of $N = 300$ circular paths (label 0) and $N = 300$ spiral paths (label 1). All paths have equal length and dimension, but differ in geometry, noise realizations, and orientation. The dataset is split into a training set comprising 70% of the samples and a test set comprising the remaining 30%.

[Feature Scaling and Path Visualization]

Since different signature components can vary greatly in magnitude, the signature vectors are standardized to zero mean and unit variance before classification. This is essential for both linear discriminant analysis and support vector machines.

A small number of representative paths from each class are plotted to verify that the synthetic data generation behaves as expected: circles remain closed, while spirals remain open under noise and normalization.

PCA of Signature Space and Classification Method

Principal component analysis (PCA) is applied to the standardized signature vectors and used to project the data onto two dimensions for visualization purposes only. The resulting PCA projection exhibits a clear separation between circles and spirals, providing empirical evidence that the signature representation effectively captures class-relevant geometric features.

The dataset is split into training and test sets using stratified sampling. Two classifiers are considered: (i) Linear Discriminant Analysis (LDA) and (ii) Polynomial-kernel Support Vector Machine (KSVM). LDA is employed to assess whether the signature representation renders the classification problem approximately linearly separable; despite its simplicity, LDA often performs well when applied to signature features. In addition, an SVM with a polynomial kernel is trained on the same data. Polynomial kernels are naturally aligned with the tensor-algebraic structure of path signatures and therefore frequently yield improved classification performance.

Classification performance is evaluated using accuracy, precision, recall, and confusion matrices.

Results

The experiment highlights how path signatures transform nonlinear geometric information into features that are amenable to classical machine learning methods, enabling robust discrimination between closed and open curves in the presence of noise and nuisance transformations. The core pipeline of signature-based learning can be summarized as Path data $\longrightarrow$ Signature transform $\longrightarrow$ Statistical learning (LDA/SVM).

Figure 3 presents the two-dimensional PCA projection of the path signature features, where a pronounced separation between spiral and circular paths is observed.

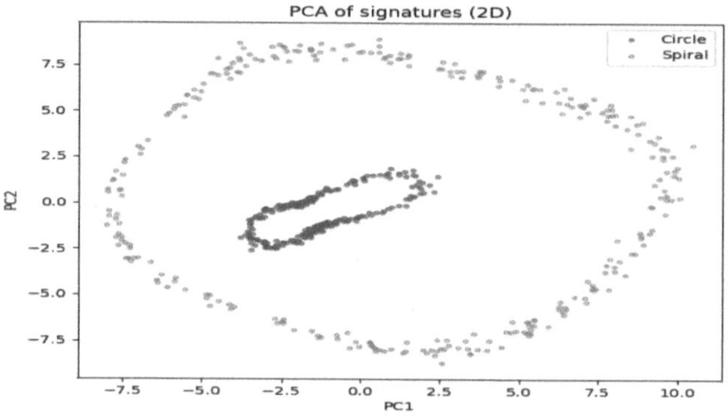

Fig. 3. Plot of PCA of Spirals and Circles

Table 2. Classification performance of LDA using level-5 path signatures

Class	Precision	Recall	F1-score	Support
Circle (0)	1.00	1.00	1.00	90
Spiral (1)	1.00	1.00	1.00	90
Accuracy			1.00	180
Macro Avg	1.00	1.00	1.00	180
Weighted Avg	1.00	1.00	1.00	180

Tables 2 and 3 summarize the classification performances of LDA and KSVM. At signature level 5, the proposed rough path theory-based approach yields perfect classification accuracy on the test set.

Thus, it is confirmed that the RPT-based approach can detect geometric trajectory changes, even when such changes are subtle.

3.3 Real-Data Study: MIT-BIH Databases

The MIT-BIH Arrhythmia Database, developed by researchers at the MIT Laboratory for Computational Physiology, consists of 48 ECG records obtained from

Table 3. Classification performance of KSVM using level-5 path signatures

Class	Precision	Recall	F1-score	Support
Circle (0)	1.00	1.00	1.00	90
Spiral (1)	1.00	1.00	1.00	90
Accuracy			1.00	180
Macro Avg	1.00	1.00	1.00	180
Weighted Avg	1.00	1.00	1.00	180

47 subjects, with one subject recorded twice [13]. Each record contains a total of 1,296,000 data points, corresponding to 648,000 samples per lead. A summary of the database characteristics is provided in Table 4.

This database has been widely used for heartbeat classification and arrhythmia detection. Representative ECG segments corresponding to durations of 30 min and 2.78 s are illustrated in Figs. 4 and 5, respectively.

Table 4. Summary of the MIT-BIH Arrhythmia Database

Characteristic	Value
Number of subjects (patients)	47
Number of ECG records	48
Recording duration per record	~30 min
Sampling frequency	360 Hz
Number of leads per record	2
Samples per lead	648,000
Total data points per record	1,296,000

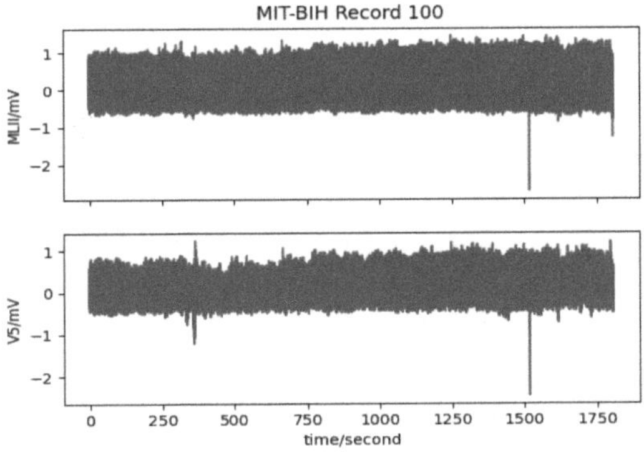

Fig. 4. MIT-BIH record 100

The statistical properties of lead V2 are summarized in Table 5, based on four records (subjects 102, 103, 104, and 117).

The summaries of previous studies on the MIT-BIH Arrhythmia Database are presented in Tables 6 and 7 [1,7,8,15,16].

From these tables, it becomes clear that CNN-based methods generally require GPU support and substantial memory resources to achieve accuracies above 90%, with the exception of Pan and Tompkins [15], which is based on traditional signal-processing techniques.

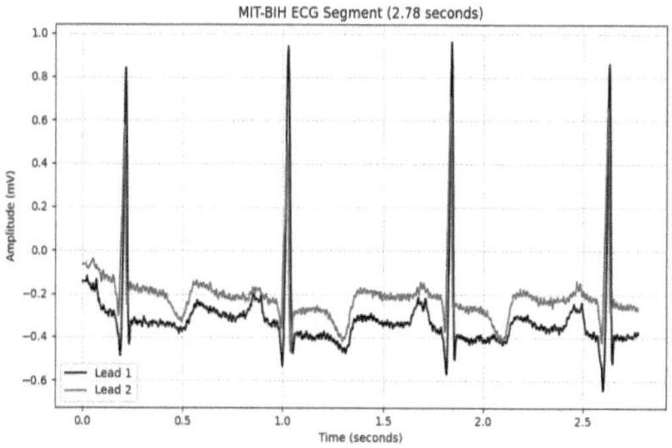

Fig. 5. ECG segment

Table 5. Statistical Properties of V2 Lead (n=4 records)

Property	Value
Number of records	4
Mean	0.035
Std	0.658
Min	-5.120
Max	5.115
Skewness	-0.683
Kurtosis	6.897

This work aims to leverage standard personal computing devices to support medical research and diagnosis, with a particular focus on deployment in resource-constrained clinical environments.

Accordingly, all experimental results in this study were obtained using Rough Path Theory based methods executed on a standard personal computer, specifically an iMac (Retina 5K, 27-inch, Late 2014) equipped with an Intel Core i7 processor. No graphics processing units (GPUs), specialized accelerators, or high-VRAM memory support were employed. This experimental setting demonstrates that the proposed approach can achieve robust and clinically relevant results using commonly available hardware, thereby highlighting its practicality and potential applicability in resource-constrained medical environments.

Using Signatures Directly to Classifiers

We use ECG data consisting of 300 segments per subject obtained from two leads across 48 records. First, two-dimensional path signatures $(x_t^{LeadI}, x_t^{LeadII})$ are computed up to level 8. These feature signatures are then directly applied to kernel SVM (KSVM) classifiers and other classifiers, yielding the results reported below.

Table 6. Summary of selected studies using the MIT-BIH Arrhythmia Database

Paper	Method Summary	Reported Accuracy/AUC
Pan et al. (1985)	Classical signal-processing-based real-time QRS detection algorithm	Approximately 99.3% QRS detection accuracy on the MIT-BIH database.
Ahmed et al. (2023)	1D CNN for arrhythmia classification	Approximately 98% classification accuracy on the MIT-BIH database.
Jun et al. (2018)	2D ECG images for CNN classification	Approximately 99.05% classification accuracy.
Hua et al. (2020)	One-dimensional CNN with R–R interval-based segmentation strategy	Five-class classification: accuracy 99.24%, AUC 0.9994; six-class classification: accuracy 97.02%, AUC 0.9966.
Review (2025)	Comprehensive review of ECG arrhythmia classification methods using MIT-BIH and related datasets	Several models achieved over 90% accuracy; inter-patient evaluation accuracy of approximately 92.07% reported in representative studies.

Table 7. Hardware and memory usage in CNN-based MIT-BIH arrhythmia studies

Paper	Model	GPU	VRAM	Remarks
Pan & Tompkins (1985)	Classical DSP	No	<100 MB	Real-time QRS detection; no learning model.
Ahmed et al. (2023)	1D-CNN	Yes	4–8 GB	Lightweight architecture using raw ECG sequences.
Jun et al. (2018)	2D-CNN	Yes	8–12 GB	ECG converted to images; Resource-intensive
Hua et al. (2020)	1D-CNN	Yes	8–10 GB	Segmentation reduces input size and training cost.
Review (2025)	CNN/LSTM	Mostly	6–12 GB	Post-2017 methods rely on GPUs; hybrids are heavier.

In total, 300 segments from each of the 48 records are processed, resulting in 14,400 segments. All segments are used for signature computation and subsequent classification. The detailed data distribution is summarized in Table 8.

Table 8. Summary of ECG segment distribution

Description	Value
Total segments loaded	14,400
Normal segments	8,576
Abnormal segments	5,824
Signature level	8
Feature Shape	(14400, 510)
Features per sample	510

The complete dataset is divided into a training set and a test set, comprising 12,240 segments (7,290 normal and 4,950 abnormal) and 2,160 segments (1,286 normal and 874 abnormal), respectively. The corresponding classification results are summarized in Table 9.

Table 9. Performance comparison of different classifiers

Classifier	Accuracy	ROC-AUC	Sensitivity	Specificity
LDA	0.8144	0.8886	0.8776	0.7714
SVM (Linear)	0.6056	0.1802	0.9577	0.3663
SVM (RBF)	0.8056	0.8996	0.9668	0.6960
Random Forest	0.9319	0.9799	0.9302	0.9331
Laplacian SVM	0.9255	0.9724	0.9325	0.9207

Detailed classification reports for all classifiers are provided below. The Random Forest and Laplacian SVM demonstrate superior performance compared with the other methods.

As shown in Table 9 and Tables 10, 11, 12, 13 and 14, LDA exhibits moderate-to-strong overall performance, achieving relatively high sensitivity with acceptable specificity. This indicates that LDA effectively detects abnormal ECG segments while maintaining a reasonable false-positive rate. However, its linear decision boundary limits its ability to capture complex nonlinear patterns inherent in ECG signals. Consequently, LDA serves as a robust baseline method, particularly attractive due to its low computational cost and stable performance on standard personal computers.

The SVM (Linear) exhibits extreme class bias, achieving very high sensitivity but very poor specificity.

Table 10. Classification performance of LDA

Class	Precision	Recall	F1-score	Support
Normal	0.90	0.77	0.83	1,286
Abnormal	0.72	0.88	0.79	874
Accuracy			0.81	2,160
Macro Avg	0.81	0.82	0.81	2,160
Weighted Avg	0.83	0.81	0.82	2,160

Table 11. Classification performance of Linear SVM

Class	Precision	Recall	F1-score	Support
Normal	0.93	0.37	0.53	1,286
Abnormal	0.51	0.96	0.66	874
Accuracy			0.61	2,160
Macro Avg	0.72	0.66	0.59	2,160
Weighted Avg	0.76	0.61	0.58	2,160

In contrast, introducing a nonlinear RBF kernel substantially improves performance over the Linear SVM. The model achieves very high sensitivity, making it well suited for screening scenarios where missed abnormalities must be minimized. However, specificity remains comparatively lower, indicating a tendency toward false positives. This trade-off reflects the RBF kernel's strong ability to model nonlinear ECG dynamics, albeit at the cost of reduced interpretability and increased computational complexity.

The Random Forest classifier, on the other hand, achieves the best overall performance among the evaluated methods. Both sensitivity and specificity are high and well balanced, indicating robust discrimination between normal and abnormal ECG segments. The near-perfect ROC-AUC further demonstrates excellent ranking capability across decision thresholds. These results highlight the effectiveness of ensemble learning in capturing complex feature interactions,

Table 12. Classification performance of RBF SVM

Class	Precision	Recall	F1-score	Support
Normal	0.97	0.70	0.81	1,286
Abnormal	0.68	0.97	0.80	874
Accuracy			0.81	2,160
Macro Avg	0.83	0.83	0.81	2,160
Weighted Avg	0.85	0.81	0.81	2,160

Table 13. Classification performance of Random Forest

Class	Precision	Recall	F1-score	Support
Normal	0.95	0.93	0.94	1,286
Abnormal	0.90	0.93	0.92	874
Accuracy			0.93	2,160
Macro Avg	0.93	0.93	0.93	2,160
Weighted Avg	0.93	0.93	0.93	2,160

Table 14. Classification performance of Laplacian SVM

Class	Precision	Recall	F1-score	Support
Normal	0.95	0.92	0.94	1,286
Abnormal	0.89	0.93	0.91	874
Accuracy			0.93	2,160
Macro Avg	0.92	0.93	0.92	2,160
Weighted Avg	0.93	0.93	0.93	2,160

albeit at the cost of higher memory usage and reduced model transparency compared to linear methods.

Meanwhile, the Laplacian SVM achieves performance comparable to Random Forest, with slightly lower accuracy but similarly strong ROC-AUC. Importantly, it offers a more balanced trade-off between sensitivity and specificity than the RBF SVM. The Laplacian kernel appears particularly effective in modeling ECG signal variations while maintaining computational efficiency. This makes it an attractive alternative in resource-constrained environments, aligning well with the study's motivation of achieving high diagnostic performance on standard PCs.

In summary, the Random Forest classifier achieves the best overall accuracy and ROC-AUC, while the Laplacian SVM provides the most favorable balance between performance and computational efficiency under limited resource constraints.

From a clinical suitability perspective, the RBF SVM is more appropriate for screening rather than definitive diagnosis, as it achieves very high sensitivity (0.97) but only moderate specificity (0.70). This characteristic minimizes missed pathological cases but results in a higher false-positive rate, necessitating secondary confirmation.

The Random Forest classifier is well suited for clinical diagnosis, exhibiting high and well-balanced sensitivity (0.93) and specificity (0.93), together with an excellent ROC-AUC (0.98). Its robustness to noise and inter-patient variability makes it reliable for clinical decision support, subject to appropriate validation.

Notably, the Laplacian SVM is also highly suitable for medical diagnosis, achieving balanced sensitivity (0.93) and specificity (0.92) with a strong ROC-

AUC (0.97), while remaining computationally efficient compared to deep learning models. Its near-Random Forest performance combined with lower computational demands makes it particularly attractive for deployment in resource-constrained clinical environments.

3.4 Discussion

The motivation of this study is to fully exploit widely available computational devices, such as personal computers, to enable effective medical research and diagnostic support. This approach is particularly relevant for healthcare facilities and regions with constrained resources. By leveraging computationally efficient algorithms, such as Rough Path Theory, that operate reliably on standard computing platforms, the proposed framework aims to reduce technological barriers while maintaining clinically meaningful performance.

However, a key limitation of relying on standard or older PC hardware is the prolonged training time required for model convergence, which may take several hours in the absence of GPU acceleration. In addition, careful strategies are often required to manage limited RAM capacity, particularly when handling high-dimensional feature representations. Once training is completed, inference and classification are highly efficient, with prediction times typically within several seconds.

4 Concluding Remarks

This study confirms that Rough Path Theory can be effectively utilised for detecting change points, geometric trajectory variations, and anomalies in ECG signals. In recent years, convolutional neural networks (CNNs) have developed rapidly and have been successfully applied across many research domains. However, such approaches typically require substantial computational resources, including GPU acceleration and large memory capacity.

High-performance ECG classification on the MIT-BIH Arrhythmia Database often relies on deep CNN architectures supported by GPUs and high-memory systems [1,7,8,15,16]. While these models can achieve near-perfect performance (accuracy exceeding 0.90 and ROC-AUC values approaching 0.90) [16], their significant hardware requirements limit their practicality in resource-constrained environments, such as small medical facilities or developing regions. Consequently, despite their strong representational power, deep learning methods are frequently inaccessible in real-world clinical settings lacking advanced infrastructure.

In contrast, this study demonstrates that competitive classification performance can be achieved using Rough Path Theory combined with a Kernel SVM (KSVM) employing a Laplacian kernel, while requiring substantially lower computational resources. As shown in Tables 9 and 14, the proposed Laplacian SVM achieved an accuracy of 0.9255, a macro-averaged F1-score of 0.92, and a ROC-AUC of 0.9724 on the MIT-BIH dataset. Precision and recall for Class 0 were

0.95 and 0.92, respectively, while Class 1 achieved values of 0.89 and 0.93. These results demonstrate that carefully designed classical machine learning frameworks can closely approach the performance of computationally intensive deep learning models, without reliance on GPU acceleration or extensive memory resources.

Moreover, our application of RPT provides a complementary and efficient framework for ECG analysis. By transforming ECG time series into robust path signatures, RPT captures the essential temporal dynamics of the heart signal, enabling accurate classification even on standard desktop PCs. Integrating RPT with KSVM or similar lightweight algorithms can further enhance interpretability and robustness while maintaining high accuracy and ROC-AUC. This approach is particularly significant for small clinics and healthcare facilities in developing countries, where computational resources are limited, but timely and reliable arrhythmia detection is crucial. The integration of Rough Path Theory with resource-efficient classifiers effectively bridges the gap between high-performance ECG analysis and practical clinical implementation, thereby enhancing accessibility to advanced cardiac diagnostic tools, even in resource-constrained clinics and areas.

Acknowledgements. The authors would like to thank the two anonymous referees for their valuable comments and suggestions, which greatly improved the quality of this paper. The authors also thank their colleagues for helpful and kind support. One of the authors (Kangrong) dedicates this work to the memory of his mother, Yuhui Shen, who passed away in November, in recognition of her unwavering support, strength, and devotion.

References

1. Ahmed, A., Ali, W., Abdullah, T.A.A., Malebary, S.J.: Classifying cardiac arrhythmia from ECG signal using 1d CNN deep learning model. Mathematics **11**(3), 562 (2023)
2. Black, F., Scholes, M.: The pricing of options and corporate liabilities. J. Polit. Econ. **81**(3), 637–654 (1973)
3. Chen, J., Gupta, A.K.: Parametric Statistical Change Point Analysis: With Applications to Genetics, Medicine, and Finance, Birkhäuser (2012)
4. Chevyrev, I., Kormilitzin, A.: A primer on the signature method in machine learning. arXiv arXiv:1603.03788 (2016), Accessed 20 May 2025
5. Friz, P.K., Victoir, N.B.: Multidimensional Stochastic Processes as Rough Paths: Theory and Applications. Cambridge University Press (2010)
6. Gatheral, J., Jaisson, T., Rosenbaum, M.: Volatility is rough. Quantitative Finance **18**(6), 933–949 (2018)
7. Hua, X., et al.: A novel method for ECG signal classification via one-dimensional convolutional neural network. arXiv preprint arXiv:2006.11655 (2020), Accessed 4 June 2025
8. Jun, T.J., Nguyen, H.M., Kang, D., Kim, D., Kim, D., Kim, Y.H.: ECG arrhythmia classification using a 2-d convolutional neural network. arXiv arXiv:1804.06812 (2018), Accessed 10 June 2025

9. Kidger, P., Lyons, T.: Signatory: Differentiable computations of the signature and logsignature transforms. In: Proc. Int. Conf. on Learning Representations (ICLR) (2021)
10. Lyons, T.: Rough Paths, Signatures and the Modelling of Functions on Streams. Oxford University Press (2014)
11. Lyons, T., Caruana, M., Lévy, T.: Differential Equations Driven by Rough Paths. Springer, Berlin (2007)
12. Lyons, T.J.: Differential equations driven by rough signals. Revisit Matemática Iberoamericana **14**(2), 215–310 (1998)
13. MIT-BIH Arrhythmia Database: MIT-BIH arrhythmia database. PhysioNet (2001), Accessed 4 May 2025
14. Moody, G.B., Mark, R.G.: The impact of the MIT-BIH arrhythmia database. IEEE Eng. Med. Biol. Mag. **20**(3), 45–50 (2001)
15. Pan, J., Tompkins, W.J.: A real-time QRS detection algorithm. IEEE Trans. Biomed. Eng. **32**(3), 230–236 (1985)
16. Silva, G.A.L., Silva, P.H.L., Moreira, G.J.P., Freitas, V.L.S., Gertrudes, J.C., Luz, E.J.S.: A systematic review of ECG arrhythmia classification: Adherence to standards, fair evaluation, and embedded feasibility. arXiv preprint arXiv:2503.07276v1 (2025). Accessed 30 May 2025

Driver Safety Reward with Cooperative Platooning Using Blockchain

Sruthi Rachamalla(✉) and Henry Hexmoor

Southern Illinois University, Carbondale, IL 62901, USA
sruthirachamalla@siu.edu, henry.hexmoor@cs.siu.edu

Abstract. Cooperative driving (or Platooning) focuses on improving the safety and efficiency by connecting two or more vehicles on a road by vehicular communication protocols. The leader is crucial as it manages the platoon, establishes communication between cars, and perform platoon maneuvers. In this paper, we proposed a driver incentive model which encourages platooning on roads leading to driver safety. As, the leader of platoon have multiple responsibilities than followers, our model rewards more incentives to leader than followers. These incentives will be rewarded as crypto tokens. This digital monetization method for both leaders and followers of a platoon is accomplished by secure transactions using blockchain.

Keywords: Cooperative Driving · Platooning · Blockchain Technology · Monetization

1 Introduction

In our daily lives, we spend an average of 47 min on the road travelling, and the traffic on the roads is becoming increasingly crowded. Congested roadways are linked to longer commute, lower fuel economy, and a higher risk of motor vehicle collisions. We could save a lot of travel time if we could reduce commuting times by a fraction of a second. One solution to the traffic problem is cooperative driving, also known as platooning [1]. Vehicles in a platoon communicate using an ad-hoc network or other communication protocols. These communication channels allow platoons to drive closer to one other while maintaining a safe distance. A platoon of vehicles will have a leader who will interact with the platoon followers while managing the platoon and overseeing maneuvers. The platoon leader is in charge of speed, lane changes, braking, and so on, while the follower vehicles are in charge of following the leader vehicle.

Cooperative driving uses vehicle-to-vehicle and infrastructure-to-vehicle wireless communication system. [2] emphasizes the technology aids in the interchange of data gathered from other cars that is impossible to obtain via on-board sensors. The Advanced Transportation Technology (PATH) project in California [3] first proposed the idea of cars traveling together on the road in 1980. Cooperative driving can improve the driving experience on the road by relieving the

H. R. Arabnia et al. (Eds.): CSCI 2025, CCIS 2970, pp. 257–266, 2027.
https://doi.org/10.1007/978-3-032-28466-2_18

driver from some of the driving obligations. Traditional sensor based Adaptive Cruise Control (ACC) isn't enough for cooperative platooning, instead Cooperative Adaptive Cruise Control (CACC) should be considered. CACC broadcasts information such as speed, acceleration, and distance through wireless communication. By allowing CACC, the distance between vehicles can be minimized by following closely, improving both safety and fuel efficiency. The focus on cooperative driving or platooning has increased globally in recent years because of the potential it holds in road transportation mainly focusing on automated and mixed traffic. Truck platooning [4,5], and CACC [6] were prominent examples of cooperative driving, which focused on minimizing inter-vehicular distance by obeying the "Three Second Rule" safety rule [7].

Having a good leader for a platoon is really crucial in forming, maintaining, and improving safety. There are a lot of methods in electing a platoon leader. [8] proposed an incentive based strategy using blockchain to elect leader who is the best for the safety of the platoon. The other way is through voting [9] to elect the platoon leader. Some other methods are may be through scoring and ranking the drivers based on the everyday driving and the driver with best rank can only initiate platoon formation. Our inspiration is drawn from the ranking method. We added an incentive or monetization factor for selecting platoon leader based on rank. When a driver of a vehicle drives everyday there will be a rank assigned to the driver performance and the driver with a best rank can become a platoon leader and other drivers will be followers. To encourage more drivers to be platoon leaders, a monetization system is required that is fair for all members. This digital monetization is implemented by the usage of a smart contract in blockchain technology that holds users to a higher level of behavior, which is promising in this regard. This smart contract establishes what constitutes acceptable behavior and prevents users from breaking that standard.

The contributions of this paper is organized as follows: Literature Survey on Platooning and Blockchain is discussed in Sect. 2. The Cooperative Platoon Earnings Methodology is described in Sect. 3. Algorithms for determining the Earning setup, analysis, scoring model, and monetization heuristic are presented in Sect. 4. Section 5 discusses the implementation of Driver Safety Reward (DSR) on Rinkeby Test Network. Conclusions and future work are presented in Sect. 6.

2 Literature Review

The first platooning simulator that was developed was Hestia [10]. This simulator is used for simulating various scenarios using sensors but its drawback is it do not execute the platooning maneuvers and does not simulate traffic scenarios. In paper [11], the researchers tried to simulate the mixed traffic scenarios using SUMO and were unsuccessful. They implemented a car following model based on CACC to simulate inter-vehicle communication. In [12] paper, they manually simulated the mixed traffic scenarios and tested the simulator to study the consequences of CACC on traffic.

The PLEXE [13] simulation tool is a platooning extension for SUMO [14] which is open-source and is available to the community. This simulation tool has different CACC car-following models to experiment. The wireless communication protocols are available for simulating the formation and platoon management. Mixed traffic scenarios are available to use and implement platooning maneuvers. The authors in paper [15], developed a simulator to implement platooning maneuvers such as join an existing platoon and merge two platoons. In paper [9], the authors developed a state-of-the-art simulator based on VENTOS [16] which uses SUMO. PERMIT [17] is a tool which simulates platooning maneuvers like join, merge, leave, and split which is built on Plexe [13].

Currently, to our knowledge only few researchers are working on effectively combining the benefits of blockchain technology with the platooning technology to better understand the usefulness. The authors in paper [18] used blockchain as a medium in transportation. They achieved the communication between vehicles in platoon and blockchain public key infrastructure by securely using hardware-based side channels. In [19], they decreased the blockchain transaction validation time, and verified the vehicle identity. The authors in [20], emphasised on using blockchain with platooning to share information securely and rapidly.

In our preceding work [21], significant features are extracted from a simulated driving dataset and driver is assigned a rank based on the behaviour on road. Based on the rank, the driver is offered crypto tokens and these transaction details along with driver attributes are stored on a blockchain network.

3 Cooperative Platoon Earnings Methodology

This section presents the entire overview of the proposed framework as shown in the Fig. 1. This methodology put forward in 5 steps, namely: (1) Driving Rank Designation Model (2) Simulation with Permit (3) Feature Extraction (4) Digital Monetization with Platoons (5) Storing in Blockchain.

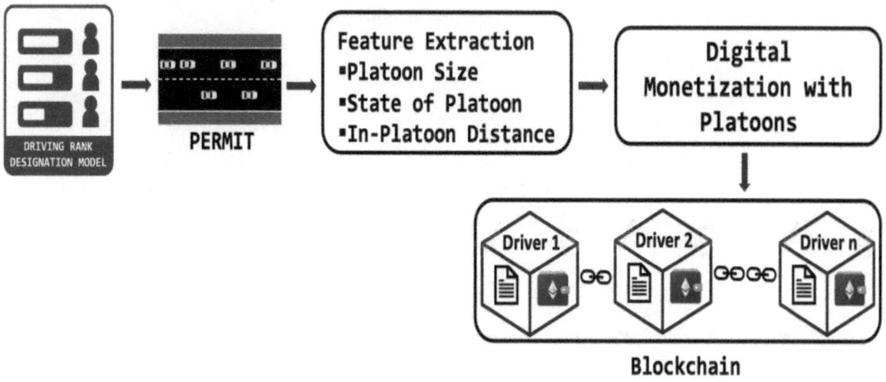

Fig. 1. Cooperative Platoon Earnings Methodology

Driving Rank Designation Model: Each Driver is assigned a rank and based on the rank, Earning rate of the driver is determined by the heuristic as proposed in [21].

Simulation with PERMIT: PERMIT [17], an open source platooning simulator based on SUMO and its platooning extension PLEXE. With PERMIT, Merge, Join, Leave and Split maneuvers can be performed. *Merge* is a maneuver in which two platoons join to form one platoon. *Join* refers to joining one vehicle into an existing platoon. *Leave* maneuver is when a car exists the current platoon. *Split* refers to dissolving one single platoon into two platoons. By using the PERMIT, we simulated all these maneuvers. This provides the data required for the evaluating the earnings with platoon for a driver.

Feature Extraction: A platoon state represents two features: (1) number of cars, and (2) distance travelled. For example, in Fig. 2, a platoon is represented with five different states. First state S_1 has two cars. Car C_3 has joined the platoon reaching to state S_2. Similarly, S_3 is achieved. In contrast, S_4 is attained by car C_3 leaving the platoon. Similarly, S_5 is also reached.

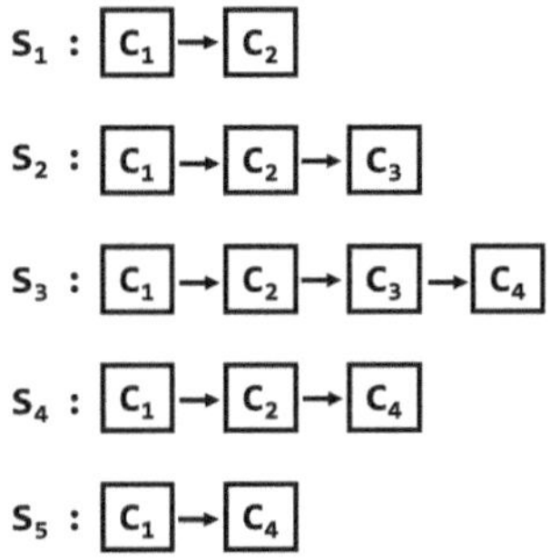

Fig. 2. States in Platoon

Digital Monetization with Platoons: With the features extracted from the previous step, we formalized the earnings (er_d) for the driver as summation of the earnings achieved while he drove in platoons (er^{in}) and earnings achieved while driving outside the platoon (er^{out})

$$er_d = er^{in} + er^{out} \tag{1}$$

Because different maneuvers can exist inside the platoon, we decomposed the earning offered inside the platoon into addition of earnings during join(P^{er}_{join}) and leave(P^{er}_{leave}) maneuver.

$$er^{in} = P^{er}_{join} + P^{er}_{leave} \tag{2}$$

Table 1. List of Notations

Symbol	Definition
er_d	Earnings of particular day
er^{out}	Earnings outside of the platoon
er^{in}	Earnings inside of the platoon
ER_{d-1}	Earning Rate of previous day
d^{out}	Out-platoon distance
n	Number of cars in a platoon
j	Number of followers in the platoon in join maneuver
l	Number of cars left the platoon in leave maneuver
δ	Balancing Factor
η	Penalty Factor
S	State-of-platoon
L_i	Length of the platoon at state i
d^{in}	In-platoon distance at state S
w	Number of platoons a driver travelled
$P^{er}_{join}(L)$	In-platoon earnings with join maneuver of leader
$P^{er}_{join}(F)$	In-platoon earnings with join maneuver of follower
$P^{er}_{leave}(L)$	In-platoon earnings with leave maneuver of leader

However, the er^{out} is calculated as product of previous day earnings rate (ER_{d-1}) which determined by the Driving Rank Designation Model and distance travelled by the driver outside the platoon(d^{out}).

$$er^{out} = ER_{d-1} * d^{out} \tag{3}$$

As mentioned earlier that a platoon leader will have a little favor by this model due to his responsibilities, Join and Leave Maneuvers are calculated using two different equation one representing Leader and other Follower in the platoon.

Join Maneuver. For every platoon, driver joined on a particular day, and for all the states in each platoon, calculate the product of the average of the States of the platoon(S_i) and the sum of the previous earning rate of the driver(ER_{d-1}) and $n\delta$. The State(S_i) is defined as the product of the Length of Platoon(L_i) at state i and distance travelled inside the platoon(d_i^{in}). Here the term $j\delta$ is the additional incentive for the leader of the platoon. It represents the summation of the balancing factor over the number cars joined in the platoon. The balancing factor δ is used to control the amount of incentive does the drivers will provide during the platoon. We assigned it as 0.01.

$$P^{er}_{join}(L) = \sum_{p=1}^{w}\sum_{i=1}^{n} S_i * (ER_{d-1} + j\delta) \tag{4}$$

The difference between Leader and Follower is there will be no additional incentives for follower. Instead, only the balancing factor is added to the previous day earning rate(ER_{d-1}). Platoon Follower doesn't require length of the platoon. So, it just uses the distance travelled in each state.

$$P_{join}^{er}(F) = \sum_{p=1}^{w}\sum_{i=1}^{n} d_i * (ER_{d-1} + \delta) \tag{5}$$

Leave Maneuver. During the Leave Maneuver, for the leader, instead of the number of cars joined(j), number of cars left(l) is considered. Additionally, there will be a penalty if a car leaves the platoon before travelling η miles. In other words, earnings for the followers will start only after travelling η miles. The overhead incurred by changing the structure of the platoon while on the move is the main reason for penalty. Here we considered $\eta = 10$.

$$P_{leave}^{er}(L) = \sum_{p=1}^{w}[\sum_{i=1}^{n} S_i * (ER_{d-1} - l\delta) + penalty_w] \tag{6}$$

where,

$$S_i = L_i * d_i^{in} \tag{7}$$

$$penalty = \begin{cases} (d^{in} - \eta) * \delta, & \text{if } d^{in} < \eta \\ 0, & \text{otherwise} \end{cases}$$

All the notations used in the model as summarized in the Table 1.

Storing in Blockchain: For secure transaction of the crypto tokens, the extracted data from the Feature Extraction step is stored in a blockchain technology. Features stored are Driver ID, current earnings, Rank designated for the driver, over speed limit count, distance travelled, number of sharp accelerations, number of sharp decelerations, number of platoons he joined, platoon leader activity, earning date.

To ensure the driver safety on road we propose following rules:

- Drivers with rank less than four cannot act as platoon leaders.
- Drivers with rank greater than three are encouraged to be platoon leaders

4 Algorithmic Approach

In this section, we present the multiple algorithms for calculating the earnings of a day inside a platoon for a driver. In the below pseudo code snippet, total earnings for a driver in platoon is presented.

$J^{er} := JoinEarnings$
$L^{er} := LeaveEarnings$
$Earnings = J^{er} + L^{er}$

The following snippet calculate earnings for a merge or join maneuver for leader or follower.

```
Join Earnings
if number_cars_joining >1:
    calculate merge_earnings
else
    calculate earnings("leader")
    or
    calculate earnings("follower")
```

Similar to above snippet of algorithm, leave maneuver for leader or follower are presented.

```
Leave Earnings
if number_cars_leaving >1:
    calculate split_earnings
else
    calculate earnings("leader")
    or
    calculate earnings("follower")
```

merge_earnings and split_earnings refers to earnings that can be awarded in the merge and split maneuvers respectively for a driver.

Algorithm 1. merge_earnings

1: **if** *platoon_1 leader* **then**
2: *earnings("leader")*
3: **end if**
4: **if** *platoon_2 leader* **then**
5: *earnings("follower")*
6: **end if**
7: **if** *platoon follower* **then**
8: *earnings("follower")*
9: **end if**

Depending on the driver_type whether its leader or follower of the platoon, *earnings* algorithm is used for a platoon maneuver especially for join. Similarly *penalized_earnings* is also defined which is used in the case of the leave maneuver.

5 Driver Safety Reward in Platoons

We implemented a simple six car platoon with Join and Leave Maneuvers in PERMIT. With the data from the PERMIT and the above formulation we calculated the earnings for leader and a follower in both the scenarios. For the

Algorithm 2. split_earnings

if *platoon_1 leader* **then**
    penalized_earnings("leader")
end if
if *platoon_2 leader* **then**
    penalized_earnings("leader")
end if
if *platoon follower* **then**
    penalized_earnings("follower")
end if

Algorithm 3. earnings(driver_type)

Require: $P^{er}_{join} = 0, \delta = 0.01$
for w **do** in *num_of_platoons*:
    if *driver* = *Platoon_Leader* **then**
        $P_{er} = ER_{d-1} + (joined\_cars * \delta)$
        for s **do** in *num_of_states*:
            $S_P = S_P + (L_s * d_s)$
        end for
    else
        $P_{er} = ER_{d-1} + \delta$
        for s **do** in *number_of_states*:
            $S_P = S_P + (d_s)$
        end for
    end if
    $P^{er}_{join} = P^{er}_{join} + S_P + P_{er}$
end for

Algorithm 4. penalized_earnings(driver_type)

Require: $P^{er}_{leave} = 0, \delta = 0.01$
for w **do** in *number_of_platoons*:
    calculate *penalty*
    if *driver* = *Platoon_Leader* **then**
        $P_{er} = ER_{d-1} - (left\_cars * \delta)$
        for s **do** in *num_of_states*:
            $S_P = S_P + (L_s * d_s)$
        end for
    else
        $P_{er} = ER_{d-1} + \delta$
        for s **do** in *num_of_states*:
            $S_P = S_P + (d_s)$
        end for
    end if
    $P^{er}_{leave} = P^{er}_{leave} + S_P + P_{er} + penalty$
end for

leader, the earnings resulted in 26.29 tokens and in the case of follower its 19.34 tokens.

A Rinkeby test network is used as our Ethereum network for implementation [21]. On the rinkeby etherscan network, two smart contracts were deployed, one for tokenization and the other for storing driver records. Tokenization contract will authorize the driver-record contract with driver DSR test tokens and transfer a small number of DSR test tokens to the driver-record contract at first. The driver-record will now be able to credit DSR test tokens to allocated driver's wallets based on their rating. The tokenization contract, for example, created 10000000 (10^7) DSR test tokens and authorized the data-record to spend them. The contract sends 10000 DSR test tokens to data-record, where they will be assigned to the drivers.

The tokenization contract, for example, created 10000000 (10^7) DSR test tokens and authorized the data-record to spend them. The contract sends 10000 DSR test tokens to data-record, where they will be assigned to the drivers. In the Admin wallet, the MetaMask represents 9990000 ($0.999*10^7$) DSR test tokens.

6 Conclusion and Future Work

In this paper, we proposed a model to incentivize the driver behaviour in the case of platooning. For this, we articulated a system to calculate the earnings awarded for a driver in a platoon. This system considers the different maneuvers in platoon namely Join, Leave, Merge, and Split. We used PERMIT, a platoon simulating system to mimic the aforementioned maneuvers. A Rinkeby test network is used to award the test tokens to driver based on their earnings. This paper assumes for the platooning in cars, but can be extended to different homogeneous vehicles and even heterogeneous vehicles.

References

1. Bergenhem, C., Shladover, S., Coelingh, E., Englund, C., Tsugawa, S.: Overview of platooning systems. In: Proceedings of the 19th ITS World Congress, 22–26 October 2012, Vienna, Austria (2012)
2. Piao, J., McDonald, M.: Advanced driver assistance systems from autonomous to cooperative approach. Transp. Rev. **28**(5), 659–684 (2008)
3. Shladover, S.E., et al.: Automated vehicle control developments in the path program. IEEE Trans. Veh. Technol. **40**(1), 114–130 (1991)
4. Alam, A., Besselink, B., Turri, V., Mårtensson, J., Johansson, K.H.: Heavy-duty vehicle platooning for sustainable freight transportation: a cooperative method to enhance safety and efficiency. IEEE Control Syst. Mag. **35**(6), 34–56 (2015)
5. Liang, K.-Y.: Fuel-efficient heavy-duty vehicle platoon formation. PhD thesis, KTH Royal Institute of Technology (2016)
6. Shladover, S.E., Nowakowski, C., Lu, X.-Y., Ferlis, R.: Cooperative adaptive cruise control: definitions and operating concepts. Transp. Res. Rec. **2489**(1), 145–152 (2015)

7. Wolf, M.T., Burdick, J.W.: Artificial potential functions for highway driving with collision avoidance. In: 2008 IEEE International Conference on Robotics and Automation, pp. 3731–3736. IEEE (2008)
8. Singh, P.K., Sharma, S., Nandi, S.K., Singh, R., Nandi, S.: Leader election in cooperative adaptive cruise control based platooning. In: Proceedings of the 1st International Workshop on Communication and Computing in Connected Vehicles and Platooning, pp. 8–14 (2018)
9. Amoozadeh, M., Deng, H., Chuah, C.-N., Zhang, H.M., Ghosal, D.: Platoon management with cooperative adaptive cruise control enabled by VANET. Veh. Commun. **2**(2), 110–123 (2015)
10. Hallé, S., Chaib-draa, B.: A collaborative driving system based on multiagent modelling and simulations. Transp. Res. Part C: Emerg. Technol. **13**(4), 320–345 (2005)
11. Fernandes, P., Nunes, U.: Platooning of autonomous vehicles with intervehicle communications in sumo traffic simulator. In: 13th International IEEE Conference on Intelligent Transportation Systems, pp. 1313–1318. IEEE (2010)
12. Van Arem, B., Cornelie, J.G., Van Driel, R.: Visser: the impact of cooperative adaptive cruise control on traffic-flow characteristics. IEEE Trans. Intell. Transp. Syst. **7**(4), 429–436 (2006)
13. Segata, M., Joerer, S., Bloessl, B., Sommer, C., Dressler, F., Cigno, R.L.: Plexe: a platooning extension for veins. In: 2014 IEEE Vehicular Networking Conference (VNC), pp. 53–60. IEEE (2014)
14. Lopez, P.A., et al.: Microscopic traffic simulation using sumo. In: The 21st IEEE International Conference on Intelligent Transportation Systems. IEEE (2018)
15. Kazerooni, E.S., Ploeg, J.: Interaction protocols for cooperative merging and lane reduction scenarios. In: 2015 IEEE 18th International Conference on Intelligent Transportation Systems, pp. 1964–1970. IEEE (2015)
16. Amoozadeh, M., Ching, B., Chuah, C.-N., Ghosal, D., Zhang, H.M.: Ventos: Vehicular network open simulator with hardware-in-the-loop support. Procedia Comput. Sci. **151**, 61–68 (2019)
17. Mena-Oreja, J., Gozalvez, J.: Permit-a sumo simulator for platooning maneuvers in mixed traffic scenarios. In: 2018 21st International Conference on Intelligent Transportation Systems (ITSC), pp. 3445–3450. IEEE (2018)
18. Rowan, S., Clear, M., Gerla, M., Huggard, M., Goldrick, C.M.: Securing vehicle to vehicle communications using blockchain through visible light and acoustic side-channels. arXiv preprint arXiv:1704.02553 (2017)
19. Calvo, J.A.L., Mathar, R.: Secure blockchain-based communication scheme for connected vehicles. In: 2018 European Conference on Networks and Communications (EuCNC), pp. 347–351. IEEE (2018)
20. Hexmoor, H., Alsamaraee, S., Almaghshi, M.: Blockchain for improved platoon security. Int. J. Inf. **7**(2), 1–6 (2018)
21. Rachamalla, S., Hexmoor, H.: DSRBT-driving safety reward based on blockchain technology. In: Proceedings of 37th International Conference on Computers and Their Applications, vol. 82, pp. 72–81 (2022)

A Mathematica Equational-Logic Proof of a Fundamental Commutativity Theorem in Ortholattice Theory

Jack K. Horner(✉)

Lawrence, KS 66046, USA
jhorner@cybermesa.com

Abstract. The logical structure of the system of "experimental propositions" of quantum mechanics is arguably an ortholattice (more specifically, an orthomodular lattice) in which the incompatible observables (e.g., position and momentum) of quantum mechanics correspond to non-commuting variables in such a lattice. Some of today's automated deduction systems are powerful enough to prove important theorems in ortholattice theory from the ortholattice axioms and definitions alone. Here, I provide an automated equational-logic deduction, solely from the axioms and definitions of ortholattice theory, of a fundamental theorem comprised of seven propositions that jointly characterize commutativity in ortholattice theory. Some of the proof-graphs of those propositions are homomorphic, suggesting that deeper structural relations (e.g., meet/join duality) among those propositions may exist.

Keywords: quantum logic · ortholattice theory · automated deduction

1 Introduction

Ortholattices are of high interest in pure mathematics (see, for example, Beran 1985; Birkhoff 1967; Kalmbach 1983; Grätzer 2011). They are also of significant interest in characterizing the logic(s) of the systems of propositions describing the outcome of measurements in physical theory. In particular, we can think of an *experimental proposition* (EP) about a physical system as an assertion of the form (EPF).

(EPF) Given a specific physical theory T and a measurement theory M consistent with T, at time t physical system S has measurable physical property P with value V(P).

For example, if S is a specific billiard ball and P is the mass of that ball, then.

(BB) In Newtonian physics, at 1345 US Central time on 1 September 2024, S has a mass of 165 grams.

H. R. Arabnia et al. (Eds.): CSCI 2025, CCIS 2970, pp. 267–278, 2027.
https://doi.org/10.1007/978-3-032-28466-2_19

is an experimental proposition (in Newtonian physics). Such propositions can be determined to be true or false by measuring V(P) at t. What is meant by “measurable property”, “measuring V(P) at t”, and how we infer the truth or falsity of an EP evidently depend on what T, M, and the rules of inference are (see for example Tal 2020.)

Within the constraints of a given T and a given M, EPs can be combined by logical connectives, and the result is an EP. Given T and M, we can call the set of finite EPs consistent with T and M the *system of experimental propositions of T*.

In the canonical mathematics of non-quantum physics, there is a homomorphism from observables (such as position and momentum) to the subspaces spanned by linear operators on a vector space (see for example José and Saletan 1998, esp. The Appendix). All the observables of non-quantum physics are *compatible*, i.e., in principle altering one observable does not inherently affect the value of any other observable (see for example von Neumann 1983; Cohen 1989). In non-quantum physics, the linear operators corresponding to all observables commute.

In the canonical mathematics of quantum physics, in contrast, there is a homomorphism from observables to the subspaces spanned by linear operators of an infinite-dimensional Hilbert space over the complex numbers (see for example von Neumann 1983; Akhiezer and Glazman 1961; Jauch 1968, esp. Sections 2–3, 2–4, and 6–4; Cohen 1989). Some pairs of observables in quantum physics (e.g., position and momentum) are not compatible (see for example Bohm 1951, Chaps. 22 and 23). In quantum physics, the linear operators corresponding to incompatible observables do not commute.

In quantum logics, therefore, the commutativity relation plays an important role: the incompatible observables of quantum mechanics (e.g., such as the position and momentum of a particle) are represented by non-commuting variables in quantum logic.

The logic of the system of EPs of non-quantum physics arguably has a Boolean lattice structure (see for example Jauch 1968, esp. Section 5–4; Birkhoff and Von Neumann 1936; Padmanabhan and Rudeanu 2008). In contrast, the logic of the system of EPs of quantum physics arguably has an ortholattice (more specifically, an ortho*modular* lattice) structure (see for example Engesser, Gabbay, and Lehmann 2008; Beran 1985, Chap. III; Kalmbach 1983, pages 4–13; Jauch 1968; Birkhoff 1967). A Boolean lattice is an ortholattice in which the distributive law holds.

If x and y are elements of an ortholattice ‘L such that x = (x ^ y) v (x ^ y’), where ^ denotes lattice meet, v denotes lattice join, and x’ denotes the orthocomplement of x, we write xCy and say that x *commutes* with y (Beran 1985, p. 37). (For a definition of an ortholattice, see Beran 1985, p. 28.) If x = (x v y) ^ (x v y’), we write xDy and say that x *dual commutes* with y (Beran 1985, p. 38).

Beran 1985, Theorem 2.3 (pp. 39–40) summarizes seven elementary relationships among commutativity, dual commutativity, orthocomplementarity, and the partial order (< =, or > =) of ortholattices, as:

i. xCy < − > xCy’
ii. xDy < − > xDy’
iii. xCy < − > x’Dy’
iv. x < = y − > xCy
v. x > = y − > xDy
vi. x < = y’ − > xCy
vii. x > = y’ − > xDy.

where
A and B are expressions in the language in which Theorem 2.3 expressed.
A <−> B means "A if and only if B".
A −> B means "A implies B".

For the purposes of this paper, I define a Fully Automated Derivation Framework (FADF) as a computing system (software + hardware) that accepts a file containing statements of axioms/assumptions of a theory T (in the sense of Chang and Keisler 2012, Sect. 1.4), together with a statement S of what is to be derived from those axioms/assumptions, as input, and emits (in successful cases), without further user input, a derivation of S from those axioms/assumptions. (FADFs were one of the earliest kinds of artificial intelligence (AI; see for example Russell and Norvig 2015, esp. Chap. 9; Wooldridge 2020, pages 72–75; Quaife 1992, page 2.)). I define a fully automated derivation (proof) of S as a derivation produced by an FADF. I define a *proof-assistant* as a computing system that can validate/check a derivation generated outside that system (e.g., by a human), but cannot, without human intervention after derivation start but before derivation completion, generate that derivation. (In this sense, a proof-assistant cannot generate a fully automated derivation.) A non-automated derivation of S is a derivation that uses no automation.

Some of today's automated deduction systems are powerful enough to prove, or to assist in the proof of, important theorems in ortholattice theory (see for example VAMPIRE 2020; Sutcliffe and Suttner 2024a, 2024b, 2025; Megill 2019; Megill and Pavičić 2003). The principal objective of the present paper is to show that.

(E) There is a FADF equational-logic proof of a fundamental characterization of commutativity (Theorem 2.3 of Beran 1985) in ortholattice theory.

2 Method

The platform used in the work described in the present paper was.

Software:

- *Mathematica* (Wolfram Research 2025).
- Windows 10.

Hardware:

- Dell Inspiron 545 desktop with.
- Intel Q8200 quadprocessor clocked at 2.33 GHz.
    - 8 GB RAM
    - 1 TB disk, ~500 MB free

The propositions comprising Theorem 2.3 of Beran 1985 were translated to *Mathematica* expressions, and each of these propositions was passed as an argument to the *Mathematica* function **FindEquationalProof**, executing on the platform described in this Section. In addition, the *Mathematica* "ProofNotebook", "ProofDataset", and "ProofGraph" of each proof produced by **FindEquationalProof** was computed.

2.1 An Overview of Mathematica's Automated Logic Derivation Framework

This section provides an overview of the algorithm implemented in the *Mathematica* function **FindEquationalProof.**

I assume the definitions of *term, value of a term, variable,* and *constant* in Baader and Nipkov 1999, Chapter 3.

A *rewriting system* is a system of R rules that transforms expressions that satisfy some well-defined set of formation rules to another expression that satisfy those formation rules. For the purposes of this paper, I restrict "rewriting system" to be a rewriting system that concerns only *identical terms.* Two terms are said to be *identical* if the values of the terms are equal for all values of variables occurring in them.

A *reduction of a term T to a term T'* is a (typically recursive) rewriting of T to T' using a set of rewriting rules R such that T' is "simpler than" T (given some definition of "simpler than"). A *reduction sequence of a term T to a term T'* is a sequence T0 = T, T1, T2, T3,..., Tn = T', where each Ti is the result of applying R to Ti-1, i = 1, 2,..., n.

If "simpler than" is a partial ordering (Suppes 1972, Df. 21, 72) on a reduction sequence that begins with T and ends with T' in a system with a set of rewriting rules R, "simpler than" induces a reduction order (Baader and Nipkov 1999, 102) on the reduction sequence that begins with T and ends with T'. A term *Tn is in normal form* if no application of R to Tn changes Tn.

A rewriting system is said to be *finitely terminating* if every reduction sequence of any term T produces, in a finite number of iterations, a normal form of T. A rewriting system is said to be *confluent* if the normal forms of all terms in the system are unique.

Some term rewriting systems are both finitely terminating and confluent (Baader and Nipkov 1999, esp. Chapter 9). Such rewriting systems have unique normal forms for all expressions. This permits us to use the output of such a system to determine whether there is an identity between two terms T1 and T2 in the following manner. If T1 and T2 and have the same normal form, then there is an identity between T1 and T2. Otherwise, there is not an identity.

The inference algorithm in *Mathematica*'s equational logic automated deduction framework (ADF), invoked by the *Mathematica* function **FindEquationalProof**, is the Knuth-Bendix completion algorithm (KBC; Knuth and Bendix 1970). KBC attempts to transform a given finite set of identities (an "input" to KBC) to a finitely terminating, confluent term-rewriting system that preserves identity. At initialization, KBC attempts to "orient" the identities supplied in its input according to the Knuth-Bendix reduction order (Baader and Nipkov 1999, Sect. 5.4.4). This results in an initial set of reduction rules. KBC then attempts to complete this initial set of rules with additional rules, obtaining their normal forms, and adding a new rule for every pair of the normal forms in accordance with the reduction order.

KBC may

1. Terminate with success, yielding a finitely terminating, confluent set of rules,
2. Terminate with failure, or
3. Loop without terminating.

Part of *Mathematica*'s equational logic ADF is a descendant of *Waldmeister* (Hillenbrand, Buch, Vogt, and Löchner 1997). For further information on term rewriting, see

Baader and Nipkov 1999. For additional information on KBC, see Knuth and Bendix 1970.

2.2 Implementation of Ortholattice Theory in Mathematica

This section lists the *Mathematica* code developed for the present paper.

Idempotence of meet. Birkhoff 1967, Lemma 1, L1, p. 8

```
axiom1=ForAll[x0,meet[x0,x0]==x0]
∀x0 meet[x0, x0] == x0
```

Idempotence of join. Birkhoff 1967, Lemma 1, L1, p. 8. (Note: idempotence of join can be derived from the absorption axioms.)

```
axiom2=ForAll[x0,join[x0,x0]==x0]
∀x0 join[x0, x0] == x0
```

Commutativity of meet. (Birkhoff 1967, Chap. 1, Lemma 1, L2, p. 8).

```
axiom3=ForAll[{x0,x1},meet[x0,x1]==meet[x1,x0]]
∀{x0,x1} meet[x0, x1] == meet[x1, x0]
```

Absorption axiom. Birkhoff 1967, Chap. 1, Lemma 1, L4, p. 8.

```
axiom4=ForAll[{x0,x1},join[x0,meet[x0,x1]]==x0]
∀{x0,x1} join[x0, meet[x0, x1]] == x0
```

Commutativity of join. (Birkhoff 1967, Chap. 1, Lemma 1, L2, p. 8).

```
axiom6=ForAll[{x0,x1},join[x0,x1]==join[x1,x0]]
∀{x0,x1} join[x0, x1] == join[x1, x0]
```

Associativity of meet. (Birkhoff 1967, Chap. 1, Lemma 1, L3, p. 8).

```
axiom7=ForAll[{x0,x1,x2},meet[meet[x0,x1],x2]==meet[x0,me
et[x1,x2]]]
∀{x0,x1,x2} meet[meet[x0, x1], x2] == meet[x0, meet[x1, x2]]
```

Associativity of join. (Birkhoff 1967, Chap. 1, Lemma 1, L3, p. 8).

```
axiom8=ForAll[{x0,x1,x2},join[join[x0,x1],x2]==join[x0,jo
in[x1,x2]]]
∀{x0,x1,x2} join[join[x0, x1], x2] == join[x0, join[x1, x2]]
```

Lattice axioms and definitions.

```
latticeAxioms={axiom1,axiom2,axiom3,axiom4,axiom6,axiom7,
axiom8}
{∀x0 meet[x0, x0] == x0, ∀x0 join[x0, x0] == x0,
∀{x0,x1} meet[x0, x1] == meet[x1, x0],
∀{x0,x1} join[x0, meet[x0, x1]] == x0,
∀{x0,x1} join[x0, x1] == join[x1, x0],
∀{x0,x1,x2} meet[meet[x0, x1], x2] == meet[x0, meet[x1, x2]]
,
∀{x0,x1,x2} join[join[x0, x1], x2] == join[x0, join[x1, x2]]
}
```

Lattice bounds axioms.

```
axiom9=ForAll[x0,meet[x0,n0]==n0]
∀x0 meet[x0, n0] == n0

axiom10= ForAll[x0,join[x0,n0]==x0]
∀x0 join[x0, n0] == x0

axiom11=ForAll[x0,meet[x0,n1]==x0]
∀x0 meet[x0, n1] == x0
axiom12=ForAll[x0,join[x0,n1]==n1]
∀x0 join[x0, n1] == n1
```

Lattice bound axioms list.

```
latticeBoundsAxioms={axiom9,axiom10,axiom11,axiom12}
{∀x0 meet[x0, n0] == n0, ∀x0 join[x0, n0] == x0,
∀x0 meet[x0, n1] == x0, ∀x0 join[x0, n1] == n1}
```

Orthocomplement axioms.

```
axiom13=ForAll[x0,meet[x0,ocomp[x0]]==n0]
```

$\forall_{x0}$ meet[x0, ocomp[x0]] == n0

```
axiom14=ForAll[x0,join[x0,ocomp[x0]]==n1]
```

$\forall_{x0}$ join[x0, ocomp[x0]] == n1

```
axiom15=ForAll[{x0,x1},ocomp[meet[x0,x1]]==join[ocomp[x0]
,ocomp[x1]]]
```

$\forall_{\{x0,x1\}}$ ocomp[meet[x0, x1]] == join[ocomp[x0], ocomp[x1]]

```
axiom16=ForAll[{x0,x1},ocomp[join[x0,x1]]==meet[ocomp[x0]
,ocomp[x1]]]
```

$\forall_{\{x0,x1\}}$ ocomp[join[x0, x1]] == meet[ocomp[x0], ocomp[x1]]

```
axiom17=ForAll[x0,ocomp[ocomp[x0]]==x0]
```

$\forall_{x0}$ ocomp[ocomp[x0]] == x0

Orthocomplement axiom list.

```
orthocomplementAxioms={axiom13,axiom14,axiom15,axiom16,ax
iom17}
```

{$\forall_{x0}$ meet[x0, ocomp[x0]] == n0,
$\forall_{x0}$ join[x0, ocomp[x0]] == n1,
$\forall_{\{x0,x1\}}$ ocomp[meet[x0, x1]] == join[ocomp[x0], ocomp[x1]]
,
$\forall_{\{x0,x1\}}$ ocomp[join[x0, x1]] == meet[ocomp[x0], ocomp[x1]]
, $\forall_{x0}$ ocomp[ocomp[x0]] == x0}

The ortholattice axioms.

```
ortholatticeAxioms=
Join[latticeAxioms,latticeBoundsAxioms,orthocomplementAxi
oms]
{∀x0 meet[x0, x0] == x0, ∀x0 join[x0, x0] == x0,
∀{x0,x1} meet[x0, x1] == meet[x1, x0],
∀{x0,x1} join[x0, meet[x0, x1]] == x0,
∀{x0,x1} join[x0, x1] == join[x1, x0],
∀{x0,x1,x2} meet[meet[x0, x1], x2] == meet[x0, meet[x1, x2]]
,
∀{x0,x1,x2} join[join[x0, x1], x2] == join[x0, join[x1, x2]]
, ∀x0 meet[x0, n0] == n0, ∀x0 join[x0, n0] == x0,
∀x0 meet[x0, n1] == x0, ∀x0 join[x0, n1] == n1,
∀x0 meet[x0, ocomp[x0]] == n0,
∀x0 join[x0, ocomp[x0]] == n1,
∀{x0,x1} ocomp[meet[x0, x1]] == join[ocomp[x0], ocomp[x1]]
,
∀{x0,x1} ocomp[join[x0, x1]] == meet[ocomp[x0], ocomp[x1]]
, ∀x0 ocomp[ocomp[x0]] == x0}
```

Translation of commutativity expressions from Beran 1985, to *Mathematica*, notation.

```
xCy=(x0==join[meet[x0,x1],meet[x0,ocomp[x1]]])
x0==join[meet[x0,x1],meet[x0,ocomp[x1]]]

xCyprime=(x0==join[meet[x0,ocomp[x1]],meet[x0,ocomp[ocomp
[x1]]]])
x0==join[meet[x0,ocomp[x1]],meet[x0,ocomp[ocomp[x1]]]]

xDy=(x0==meet[join[x0,x1],join[x0,ocomp[x1]]])
x0==meet[join[x0,x1],join[x0,ocomp[x1]]]

xDyprime=(x0==meet[join[x0,ocomp[x1]],join[x0,ocomp[ocomp
[x1]]]])
x0==meet[join[x0,ocomp[x1]],join[x0,ocomp[ocomp[x1]]]]

xprimeDyprime=(ocomp[x0]==meet[join[ocomp[x0],ocomp[x1]],
join[ocomp[x0],ocomp[ocomp[x1]]]])
ocomp[x0]==meet[join[ocomp[x0],ocomp[x1]],join[ocomp[x0],ocomp[ocomp[x1]]]]

xLEy=(meet[x0,x1]==x0)
meet[x0,x1]==x0

xGEy=(meet[x0,x1]==x1)
meet[x0,x1]==x1

xGEyprime=(meet[x0,ocomp[x1]]==ocomp[x1])
meet[x0,ocomp[x1]]==ocomp[x1
```

Axiom 1

We are given that:

`a==join[a,meet[a,b]]`

Axiom 2

We are given that:

`meet[a,b]==meet[b,a]`

Axiom 3

We are given that:

`meet[x0,x1]==x0`

Hypothesis 1

We would like to show that:

`join[meet[x0,x1],meet[x0,ocomp[x1]]]==x0`

Substitution Lemma 1

It can be shown that:

`meet[x1,x0]==x0`

PROOF

We start by taking Axiom 3, and apply the substitution:

`meet[a_,b_]→meet[b,a]`

which follows from Axiom 2.

Substitution Lemma 2

It can be shown that:

`join[meet[x1,x0],meet[x0,ocomp[x1]]]==x0`

PROOF

We start by taking Hypothesis 1, and apply the substitution:

`meet[a_,b_]→meet[b,a]`

which follows from Axiom 2.

Substitution Lemma 3

It can be shown that:

`join[x0,meet[x0,ocomp[x1]]]==x0`

PROOF

We start by taking Substitution Lemma 2, and apply the substitution:

`meet[x1,x0]→x0`

which follows from Substitution Lemma 1.

Conclusion 1

We obtain the conclusion:

`x0==x0`

PROOF

Take Substitution Lemma 3, and apply the substitution:

`join[a_,meet[a_,b_]]→a`

which follows from Axiom 1.

Fig. 1. ProofNotebook for $x < = y \rightarrow xCy$.

3 Proof of Beran 1985, Theorem 2.3

This section describes *Mathematica* equational-logic proofs, their "ProofDatasets", and "ProofGraphs"of each of the seven propositions comprising Theorem 2.3 of Beran 1985. In case Beran's formulation of such a proposition contains a bidirectional implication ($<->$), the proposition is first translated into two one-directional implications.

The following depicts an example of one of those proofs. Figure 1 shows the Mathematica "Proof Notebook" for ($x <= y \rightarrow xCy$). Figure 2 shows the "ProofGraph" for this proposition.

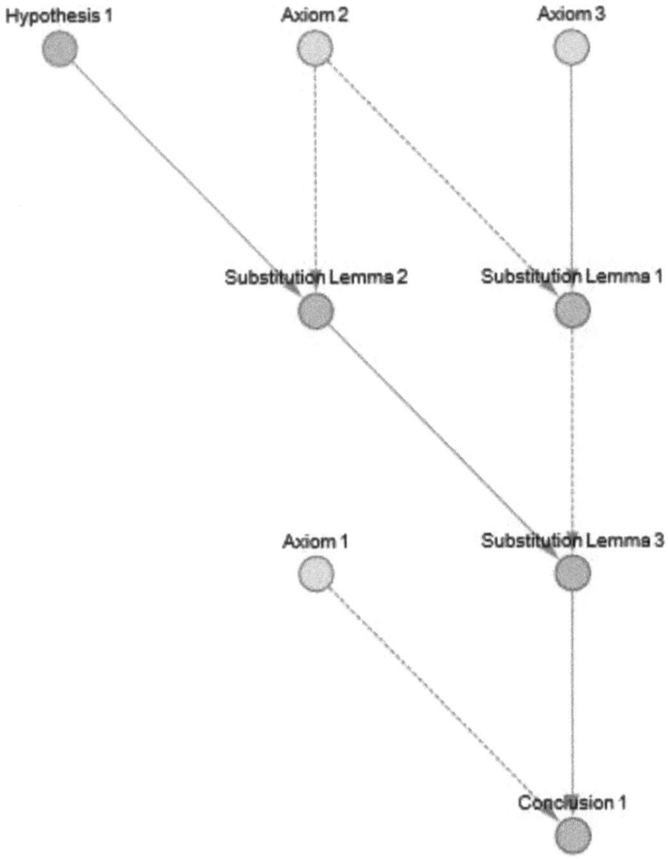

Fig. 2. Proof Graph for $x <= y \rightarrow xCy$.

A PDF file (about 55 pages) of the complete set of proofs, proof datasets, and proof graphs for this paper is available at Horner 2025b.

4 Conclusions and Discussion

Theorem 2.3 of Beran 1985 characterizes, in seven propositions, fundamental relations among commutativity, dual commutativity, and partial order in an ortholattice. The results of Sect. 3 show that Mathematica can produce an automated equational-logic proof of that theorem.

The proof of Theorem 2.3 of Beran 1985 does not use automation and invokes duality theorems (that are proven in Beran 1985). The proofs contained in the present paper use only the axioms and definitions of ortholattice theory.

Inspection of the "proof graphs" contained in the present paper reveals graph-structural homomorphisms among several of the proofs of the propositions comprising Theorem 2.3 (see in particular the proof-graphs in Sects. 4.1, 4.4, and 4.6 in Horner 2025b). Some of these homomorphisms may reflect graph-structural correlates of meet/join duality. In future work I plan to investigate this conjecture.

Disclosure of Interests. The author declares no competing interests.

References

Akhiezer, N.I., Glazman, I.M.: Theory of Linear Operators in Hilbert Space. Vol. I. trans. by M. Nestell. Frederick Ungar (1961)
Baader, F., Nipkow, T.: Term Rewriting and All That. Cambridge University Press, Cambridge (1999)
Beran, L.: Orthomodular Lattices: Algebraic Approach. D. Reidel (1985)
Birkhoff, G.: Lattice Theory, 3rd edn. American Mathematical Society (1967)
Birkhoff, G., von Neumann, J.: The logic of quantum mechanics. Ann. Math. 2nd Ser., **37**, 823–843 (1936)
Bohm, D.: Quantum Theory. Dover reprint, 1989 (1951)
Chang, C.C., Keisler, H.J.: Model Theory, 3rd edn. Dover (2012)
Cohen, D.W.: An Introduction to Hilbert Space and Quantum Logic. Springer, Heidelberg (1989)
Engesser, K., Gabbay, D.M., Lehmann, D. (eds.) Handbook of Quantum Logic and Quantum Structures: Quantum Logic. Elsevier (2008)
Grätzer, G.: Lattice Theory: Foundation. Birkhäuser (2011)
Hillenbrand, T., Buch, A., Vogt, R., Löchner, B.: WALDMEISTER - high-performance equational deduction. J. Autom. Reason. **18**, 265–270 (1997)
Horner, J.K.: A fully automated derivation of the marsden-herman theorem from the foulis-holland theorems in orthomodular lattice theory. In: Proceedings of the 2024 International Conference on Artificial Intelligence. Springer, Heidelberg (2025a)
Horner, J.K.: Supplementary information for this paper (2025b). http://jkhorner.com/QUANTUM_LOGIC/Fundamental_Theorem_Commutativity.pdf
Jauch, J.: Foundations of Quantum Mechanics. Addison-Wesley, Boston (1968)
José, J.V., Saletan, E.J.: Classical Dynamics: A Contemporary Approach. Cambridge (1998)
Kalmbach, G.: Orthomodular Lattices. Academic Press, New York (1983)
Knuth, D.E., Bendix, P.B.: Simple word problems in universal algebras. In: Leech, J. (ed.) Computational Problems in Abstract Algebra, pp. 263–297. Pergamon Press (1970)
Padmanabhan, R., Rudeanu, S.: Axioms for Lattices and Boolean Algebras. World Scientific, Singapore (2008)
Quaife, A.: Automated Development of Fundamental Mathematical Theories. Kluwer, Alphen aan den Rijn (1992)

Russell, S.J., Norvig, P.: Artificial Intelligence: A Modern Approach, 3rd edn. Pearson, Boston (2015)
Sutcliffe, G., Suttner, C.: TPTP BNF Syntax (2024a). https://tptp.org/TPTP/SyntaxBNF.html
Sutcliffe, G., Suttner, C.: The TPTP Problem Library for Automated Theorem Proving (2024b). https://tptp.org
Sutcliffe, G., Suttner, C.: File LAT003–0.ax: axioms for an orthomodular lattice (2025). https://tptp.org/cgi-bin/SeeTPTP?Category=Axioms&File=LAT003-0.ax
Tal, E.: Measurement in science. In: Zalta, E. (ed.) Stanford Encyclopedia of Philosophy (2020). https://plato.stanford.edu/entries/measurement-science/#TheLadMea
VAMPIRE github. VAMPIRE V4.5.1. File vampire_z3_Release_static_master_4764.zip (2020). https://github.com/vprover/vampire/releases/tag/4.5.1
von Neumann, J.: Mathematical Foundations of Quantum Mechanics. (German ed., 1936). Trans. by R. T. Beyer (1949). Reprinted and copyrighted by Princeton University Press (1983)
Wolfram Research. Mathematica Home Edition. v14.3.0 (2025)
Wooldridge, M.:. A Brief History of Artificial Intelligence: What It Is, Where We Are, and Where Are We Going? Flatiron Books (2020)

Toroidal Hertzsprung's Problem

Michael Laszlo(✉) and Sumitra Mukherjee

Department of Computer Science, Nova Southeastern University, Fort Lauderdale, FL, USA
{mjl,sumitra}@nova.edu

Abstract. We consider the problem of placing nK non-attacking kings on an $n \times n$ toroidal board so that each row and each column contains exactly K kings. We show that no solutions exist if $n \leq 4K$. For $n = 4K + 1$, we present an algorithm to generate $2nT_K$ solutions where T_K denotes the number of positive divisors of K. We conjecture that our algorithm generates every solution and show that this conjecture holds if successive rows in a solution are obtained by constant circular shifts.

Keywords: Enumerative combinatorics · Hertzsprung's problem

1 Introduction

There is a time-honored interest in counting the number of ways of placing non-attacking playing pieces on an $n \times n$ board. Pieces considered include kings [9,15], queens [2,8], rooks [10,14], mixed-type pieces [6], and variants of standard chess pieces [7]. In addition to the non-attack constraint, count constraints prescribing the number of pieces that appear in each row and/or column are also sometimes assumed. Where both row and column counts equal one (i.e., the problem's *classic* form), the problem is equivalent to computing the number of permutations of S_n that satisfy the non-attack constraint. Where the playing pieces are kings, this is the classic n-kings problem, also referred to as Hertzsprung's problem [9,15]; where the pieces are queens, this is the classic n-queens problem [2,13]. Some works drop one of the two count constraints. Where (say) the row count must equal some fixed integer k but the column count is unconstrained, [16] presents a method for counting solutions to a class of grid-based puzzles, and [12] gives an explicit formula where the pieces are kings.

Another form of the problem considers topological variants of the standard planar grid: the cylinder (in which the board's left and right edges are glued together) or toroidal (both pairs of opposite edges are glued together) [18]. Topological variants are addressed for the classic n-queens problem in [3,5,17] and classic n-kings problem in [1]; the n-kings problem with row and column counts equal to two is treated algorithmically in [11].

The problem we address in this paper generalizes Hertzsprung's n-kings problem in two ways. First, we count solutions on an $n \times n$ board comprising nK

M. Laszlo and S. Mukherjee—Both authors contributed equally to this work.

H. R. Arabnia et al. (Eds.): CSCI 2025, CCIS 2970, pp. 279–288, 2027.
https://doi.org/10.1007/978-3-032-28466-2_20

non-attacking kings with exactly K kings in each row and in each column. Secondly, we assume the board is toroidal. For the purposes of this paper, we refer to this as the *Toroidal Hertzsprung's Problem* (*THP*). To our knowledge, the toroidal n-kings problem in which row and column counts must equal K has not been addressed except for the classic version of this problem (i.e., where $K = 1$) and, algorithmically, for counting solutions up to $n \leq 15$ where $K = 2$. The present paper does not solve THP generally, but rather focuses on toroidal $n \times n$ boards where $n \leq 4K + 1$.

In Sect. 2 we show that THP has no solutions if $n \leq 4K$. Section 3 presents an algorithm to generate $2nT_k$ solutions for $n = 4K + 1$ where T_K denotes the number of positive divisors of K. Search using a method described in [11] yields no other solutions for $K \leq 6$, which prompts the conjecture that our algorithm generates every solution for all $K \geq 1$. Section 4 shows that our conjecture holds if successive rows in a solution are obtained by constant circular shifts. Section 5 identifies areas of future work.

2 Minimum Board Dimension for Solutions to Exist

In the discussions that follow, row and column indices range from 0 to $n - 1$. We use $i^{\langle x \rangle}$ to denote $(i + x)$ *modulo* n, and $\omega(i)$ to denote the set of column indices occupied by kings in row i. We refer to the requirement that no two kings can occupy cells that share a common edge or vertex – that kings are non-attacking – as the *adjacency constraint.* Theorem 1 establishes the minimum board dimension needed for solutions to exist.

Theorem 1. *THP has no solutions if* $n \leq 4K$.

Proof. Let $\Omega = \omega(i) \cup \omega\left(i^{\langle 1 \rangle}\right)$ be the set of columns occupied by kings in a pair of adjacent rows i and $i^{\langle 1 \rangle}$ of a solution, and $\Omega^c = \{j \mid 0 \leq j \leq n-1,\ j \notin \Omega\}$ the set of columns that do not contain kings. For each $j \in \Omega$, $j^{\langle 1 \rangle} \in \Omega^c$ since kings in adjacent columns attack each other. That is, every column occupied by a king is followed by at least one empty column, and thus $|\Omega^c| \geq |\Omega| = 2K$. Hence $n = |\Omega \cup \Omega^c| = |\Omega| + |\Omega^c| \geq 4K$, and THP has no solutions if $n < 4K$.

When $n = 4K$, $|\Omega| = |\Omega^c| = 2K$. Hence every occupied column in a pair of adjacent rows is followed by exactly one empty column. Thus, either $\Omega = \{2j \mid 0 \leq j \leq 2K - 1\}$ or $\Omega = \{2j + 1 \mid 0 \leq j \leq 2K - 1\}$. The requirement that every column in a solution must contain exactly K kings can be satisfied only for even columns in the former case and only for odd columns in the latter case. Hence, THP has no solutions if $n \leq 4K$.

3 Solution Generator for n = 4K + 1

When $n = 4K + 1$, we have exactly one pair of adjacent columns in a pair of adjacent rows that do not contain kings. We call a solution a *canonical solution* if $\omega(0) \cup \omega(1) = \{2j \mid 0 \leq j \leq 2K - 1\}$ and $0 \in \omega(0)$. That is, a canonical solution has kings in the first $2K$ even columns in the first two rows, with a king in the first column of the first row. We define the *shift* of a canonical solution as $\min(\omega(1))$, the first even column in which a king appears in the second row.

Our algorithm generates a canonical solution with shift $2d$ for each positive divisor d of K. Figure 1 displays the canonical solutions it produces for the four positive divisors d of $K = 6$ where $n = 4K + 1 = 25$. In each canonical solution, kings appear in the first $2K$ even columns of the first two rows with column 0 of row 0 containing a king. The leftmost king in the second row appears in column $2d$, and successive rows are obtained by rotating the previous row right by $2d$.

We present our algorithm using *Python* functions for clarity. Each row of a solution is a binary list of length $4K+1$ with exactly K ones indicating positions of kings in the row. For each positive divisor d of K, we generate a *canonical row* with shift $2d$. A solution is a list of $4K + 1$ rows. Successive rows of a canonical solution with shift $2d$ are generated by rotating the previous row by $2d$. Function `canonical_solution` generates a canonical solution given K and a positive divisor d of K. It uses function `canonical_row` to generate the first row of a canonical solution.

Next, we show that function `canonical_solution` generates a canonical solution with shift $2d$.

```
def canonical_solution(K, d):
    # Returns a canonical solution with shift 2d given K
    # and a divisor d of K.
    assert K        n = 4 * K + 1  # board dimension
    shift = 2 * d  # constant shift for rotating rows
    row = canonical_row(K, d)  # first row of canonical solution
    solution = [row]
    for r in range(1, n):   (# add next n-1 rows iteratively
        row = row[-shift:] + row[:-shift]  # rotate previous row by
shift
        solution.append(row)  # add rotated row to solution
    return solution

def canonical_row(K, d):
    # Returns canonical row given K and a positive divisor d of K.
    A = d * [1, 0]  # list of length 2d containing alternating 1s and 0s
    Z = d * [0, 0]  # list of length 2d containing 0s
    AZ = A + Z  # list of length 4d formed by concatenating lists A and Z
    row = (K // d) * AZ  #list of length 4K formed by repeating list AZ (K/d)
times
    return row + [0]  # canonical row of length 4K+1, with 0 in column 4K
```

Proposition 1. *Every row of a canonical solution is obtained by rotating the previous row by* $2d$.

Proof. By construction, row $i^{\langle 1\rangle}$ is obtained by rotating row i by $2d$ for $0 \le i \le 4K - 1$. Since row $4K$ is obtained by rotating row 0 by $2d(n-1)$, rotating row $4K$ by $2d$ rotates row 0 by $2d(n-1) + 2d = 2dn$ *modulo* $n = 0$. That is, row 0 is obtained rotating row $4K$ by $2d$.

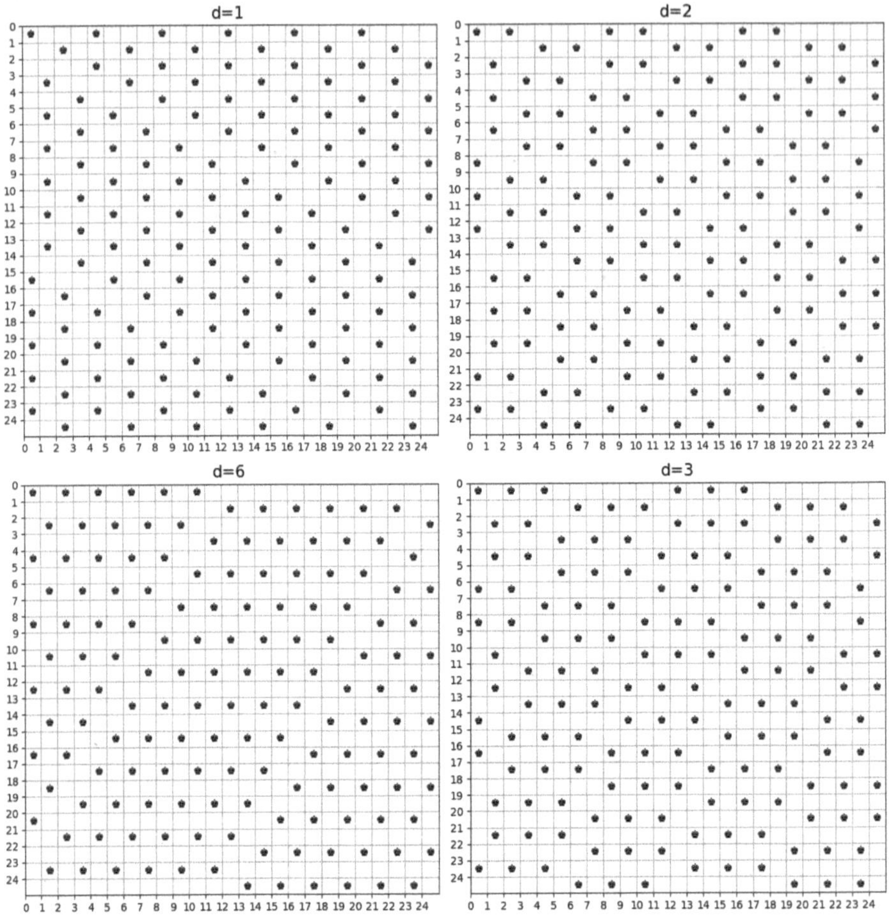

Fig. 1. Canonical solutions corresponding to the four divisors d of $K = 6$ when $n = 4K + 1 = 25$

Proposition 2. *Every row in a canonical solution contains exactly K kings.*

Proof. Row 0 contains K kings since each of the K/d copies of list AZ in it contains exactly d kings. Every other row is obtained by rotating row 0, and the number of kings is invariant under rotation.

Proposition 3. *Every column in a canonical solution contains exactly K kings.*

Proof. Since rows are obtained by constant shifts of $2d$, a king in column j in row 0 appears in column $j^{\langle 2di \rangle}$ in row i. Since d divides K, $j^{\langle 2di \rangle} = j^{\left\langle 2di' \right\rangle}$ only if $\left(i = i'\right)$ *modulo* $4K + 1$. Thus, each of the K columns with kings in row 0 cycles through every other column over the next $n - 1$ rows. Hence, each column in a canonical solution contains exactly K kings.

Proposition 4. *If column j contains a king in row 0, then columns $j^{\langle 2d-1\rangle}$, $j^{\langle 2d\rangle}$, and $j^{\langle 2d+1\rangle}$ do not.*

Proof. Row 0 is a canonical row with kings appearing only in even columns. Hence, the odd columns $j^{\langle 2d-1\rangle}$ and $j^{\langle 2d+1\rangle}$ cannot contain kings. List AZ is a concatenation of lists A and Z, each of length $2d$. Since kings only appear in list A, if j contains a king then column $j^{\langle 2d\rangle}$ belongs to list Z and hence does not contain a king.

Proposition 5. *Kings in adjacent rows of a canonical solution do not attack each other.*

Proof. Kings in the same row do not attack each other since if column j in a canonical row contains a king, then columns $j^{\langle -1\rangle}$ and $j^{\langle 1\rangle}$ do not, and spacing between successive columns with kings is invariant under rotation. To show that kings in adjacent rows do not attack each other, we only need to show that kings in row 0 do not attack kings in row 1 since spacing between successive kings in a pair of adjacent rows is invariant under rotation by a constant shift $2d$. Since row 1 is obtained by rotating row 0 by $2d$, column $j^{\langle 2d\rangle}$ in row 1 contains a king if and only if column j in row 0 contains a king. By Proposition 4, if column j in row 0 contains a king, then columns $j^{\langle 2d-1\rangle}$, $j^{\langle 2d\rangle}$, and $j^{\langle 2d+1\rangle}$ of row 0 do not contain kings. Hence, no king in row 1 is attacked by a king in row 0.

Theorem 2. *Function canonical_solution generates a unique solution for THP for each positive divisor d of the number of kings K.*

Proof. By Proposition 2, every row in a canonical solution contains exactly K kings. By Proposition 3, every column contains exactly K kings. By Proposition 5, kings in adjacent rows do not attack each other. Hence, function `canonical_solution` generates a valid solution for THP. Since the canonical row is unique for each divisor d of K, these solutions are unique.

Function `all_canonical_solutions` generates all T_k canonical solutions for $n = 4K + 1$, where T_K denotes the number of positive divisors of K. We address the question of why this function produces all possible solutions generated by constant circular shift in Sect. 4.

```
from sympy import divisors, divisor_count
def all_canonical_solutions(K):
    # Returns a canonical solution for each positive divisor of K.
    solutions = []
    for d in divisors(K):
        solutions.append(canonical_solution(K, d))
    return solutions
```

We refer to a 2×2 block of cells in some solution as *quad*. We say that a block is *empty* if it contains no kings. The last two columns in rows 0 and 1 of a canonical solution comprise an *empty quad*.

Proposition 6. *Every pair of consecutive rows contains one empty quad.*

Proof. To satisfy the adjacency constraint, kings occupy alternate columns in any pair of consecutive rows. Because the rows span an odd number n of columns, some pair of consecutive columns must be devoid of kings and so defines an empty quad. The empty quad must be unique for otherwise the row pair would lack the capacity for the requisite $2K$ kings.[1]

Proposition 7. *An empty quad is encircled by kings in either a clockwise or counterclockwise orientation as depicted in Fig. 2.*

Proof. No consecutive pair of rows (columns) can contain a 2×3 (3×2) empty block for otherwise it would lack capacity for the requisite $2K$ kings. Once a king is placed in a cell adjacent to an empty quad, the adjacency constraint determines the placement of the remaining three kings.

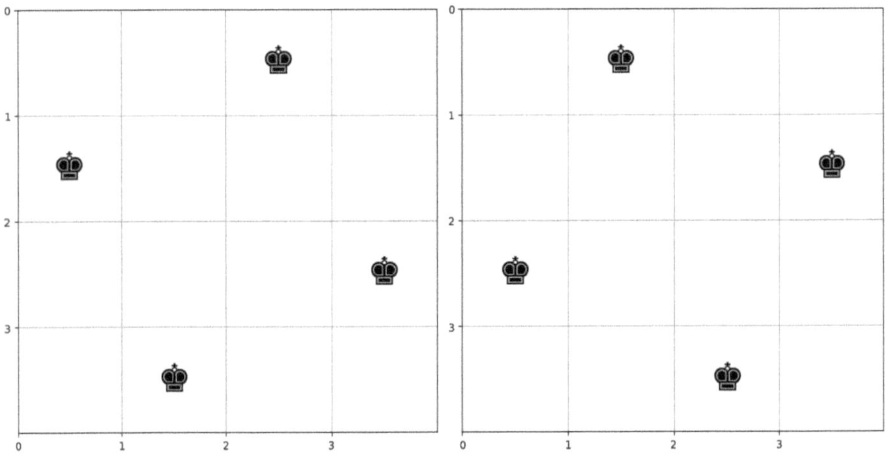

Fig. 2. Empty quads with clockwise and counterclockwise orientations.

[1] The *capacity* of a block is the maximum number of kings it can contain without violating the adjacency constraints. For example, a row of 5 cells has capacity 3, and a quad has capacity 1. Results that reference block capacity depend on the following easily proved theorem: Where $h = 1$ or $h = 2$, an $h \times 2g$ and an $h \times (2g-1)$ block each has capacity g. For example, to see that a pair of rows contains no more than one empty quad, suppose to the contrary, and observe that any two empty quads partition the row pair into two blocks of length $2g$ and $4K-3-2g = 2(2K-g-1)-1$. This implies that the capacity of the row pair is equal to $g+(2K-g-1) = 2K-1$ which is insufficient to contain $2K$ kings.

Theorem 3. *Each canonical solution yields 2n distinct solutions.*

Proof. Rows 0 and 1 of a canonical solution contains an empty quad with counterclockwise orientation. Since the remaining rows obtain from rotational shift, all n empty quads have counterclockwise orientation. Horizontal rotation of the canonical solution by i columns for $0 \leq i < n$ yields n solutions, all of which are distinct since placement of kings in row 0, given by $\omega(0)$, are distinct.

The transpose of a canonical solution is also a solution for THP since it respects row count, column count, and adjacency constraints. Moreover, orientation of empty quads in the transpose is reversed (clockwise) and so is distinct from the n solutions derived by rotation. The transpose in turn gives rise to n more solutions by rotation of rows.

The function `generate_all_solutions` to generate $2nT_k$ solutions for a specified K using standard *numpy* (https://numpy.org/) functions: `A.T` returns the transpose of array `A` and `numpy.roll(X, j, axis=1)` rotates the columns of array `X` by `j`. A function to verify empirically that the configurations produced by `generate_all_solutions` are valid solutions is presented in Appendix A.

```
import numpy as np
def generate_all_solutions(K):
    # Returns all solutions given number of kings K.
    solutions = []
    for solution in all_canonical_solutions(K):
        A = np.array(solution) # canonical solution as numpy array
        for X in [A, A.T]: # solution and its transpose
            for j in range(4*K+1): # rotate columns of solution by j
                solutions.append(np.roll(X, j, axis=1))
    return solutions
```

4 Solution Uniqueness Under Constant Circular Shift

We've assumed so far that the solution generated by `canonical_solution` for given d and K, produced by constant circular shift $2d$, is unique if d divides K and nonexistent otherwise. In this section, we prove that no other solutions produced by constant circular shift are possible. Recall that kings occupy the even cells in the first two rows of a canonical solution: $\omega(0) \cup \omega(1) = \{0,\ 2,\ \ldots,\ 4K-2\}$. To simplify our discussion, we divide the even indexes by 2. So let $C_K = \{0,\ 1,\ \ldots,\ 2K-1\}$. Where $f_d(x) = x + d$, we say that a partition (X, Y) of C_K is *d-valid* for C_K if $f_d(X) = Y$ where $1 \leq d \leq K$. Any assignment of kings to rows 0 and 1 generated by constant circular shift by $2d$ corresponds to a d-valid partition (X, Y) of C_K; specifically, $X = \{\frac{x}{2} \mid x \in \omega(0)\}$ and $Y = \{\frac{y}{2} \mid y \in \omega(1)\}$. The rest of this section develops results concerning partitions of C_K to address the existence of canonical solutions generated by constant circular shifts.

The following theorem shows that the placement of kings in rows 0 and 1 of a canonical solution is unique if d divides K, and that it is produced by `canonical_solution(K, d)`.

Theorem 4. *For each positive divisor d of K, there exists a unique d-valid partition (X, Y) of C_K. Where $K = qd$, this partition is given by $X = \bigcup_{i=0}^{q-1} A_i$ where $A_0 = \{0, \ldots, d-1\}$ and $A_i = \{x + 2d \mid x \in A_{i-1}\}$ for all $i > 0$.*

Proof. Define $B_i = f_d(A_i)$ for all $i \geq 0$. Proof by induction on q. For the basis, let $q = 1$ so that $C_K = \{0, \ldots, 2d-1\}$. We must have $A_0 \cap Y = \varnothing$, for otherwise if $y \in A_0 \cap Y$, then $f_d^{-1}(y) < 0$ implying that $f_d^{-1}(y) \notin C_K$, a contradiction. Hence $A_0 \subset X$. Since $f_d(A_0) = B_0$, we have that (A_0, B_0) is the only possible partition of C. For the inductive step, assume that the theorem holds for all $1 \leq q' < q$, and suppose that $C_K = \{0, \ldots, 2dq - 1\}$. Consider the set $B_{q-1} = \{2dq - d, \ldots, 2dq - 1\}$. We must have $B_{q-1} \cap X = \varnothing$, for otherwise if $x \in B_{q-1} \cap X$, then $f_d(x) > 2dq - 1$ implying that $f_d(x) \notin C_K$, a contradiction. It follows that $A_{q-1} \subset X$ and $B_{q-1} \subset Y$. The remaining elements $X' \subset X$ and $Y' \subset Y$ obtain from the set $C' = \{0, \ldots, 2d(q-1) - 1\}$ which, by the induction hypothesis, is uniquely partitioned by $X' = \bigcup_{i=0}^{q-2} A_i$ and $Y' = \bigcup_{i=0}^{q-2} B_i$. Hence C_K is uniquely partitioned by $X = X' \cup A_{q-1} = \bigcup_{i=0}^{q-1} A_i$ and $Y = Y' \cup B_{q-1} = \bigcup_{i=0}^{q-1} B_i$.

The following lemma is helpful in showing that a d-valid partition of C_K exists only if d divides K.

Lemma 1. *In any d-valid partition (X, Y) of C_K, for all $q \geq 0$, if $y = 2qd + r \in Y$, then $r \in Y$.*

Proof. By induction on q. The basis $q = 0$ is immediate because $r = y \in Y$. For the inductive step, assume it holds for all $0 \leq q' < q$. Since $y \in Y$, $f_d^{-1}(y) = (2q-1)d + r \in X$. This implies that $y' = 2(q-1)d + r \in Y$, for otherwise if $y' \in X$ then $f_d^{-1}(y) = f_d\left(y'\right) \in Y$, a contradiction. Since $y' = 2(q-1)d + r \in Y$, the induction hypothesis implies that $r \in Y$.

Theorem 5. *A d-valid partition for C_K exists only if d divides K.*

Proof. Let (X, Y) be a d-valid partition of $C_K = \{0, \ldots, 2K - 1\}$ where $K = qd + r$ for some positive integer q and integer $0 \leq r < d$. Consider $y = 2dq + 2r - d \in C$. Observe that $y \in Y$, for otherwise if $y \in X$ then $f_d(y) = 2dq + 2r = 2(dq + r) = 2K$ which does not belong to C_K. If $r = 0$, then $y = 2dq - d = 2(q-1)d + d \in Y$, hence due to the previous lemma, we have $y' = d \in Y$, hence $f_d^{-1}\left(y'\right) = 0 \in X$. However, if $1 \leq r < d$, we have $y = 2dq + 2r - d \in Y$, hence $y' = 2r - d \in Y$ due to the previous lemma, and so $f_d^{-1}\left(y'\right) = 2r - 2d \in X$. But then $2r - 2d = 2(r - d) \geq 0$, implying that $r \geq d$, contradicting the assumption that $1 \leq r < d$. It follows that $r = 0$ and d divides K.

The unique d-valid partition for C_K that exists for each positive divisor d of K determines the placement of kings in the first two rows of a canonical solution, and circular shift by $2d$ populates the remaining rows.

5 Conclusion

We've shown that for $n = 4K + 1$, there are at least $2nT_K$ solutions where T_K denotes the number of positive divisors of K, and have presented an algorithm for generating these solutions. We've also shown that under a constant circular shift – whereby each row is obtained from the previous row by a $2d$ circular shift – no other solutions exist. If we drop the circular shift assumption, other solutions may exist, so we treat $2nT_K$ as a lower bound on the number of solutions. Nonetheless, there is evidence that this bound is tight. Motivated by an approach for counting solutions to grid-based puzzles described in [16], an algorithm to enumerate solutions for the popular *Star-Battle* puzzle (see [4]) was developed in [11] for use on planar, cylindrical, and toroidal boards when $K = 2$, where it is used to compute solutions for $n \leq 15$. We adapted this algorithm to count all THP solutions for $1 \leq K \leq 6$ when $n = 4K + 1$ and confirmed that the total number of solutions is given by $2nT_K$. Table 1 presents our results. This leads us to conjecture that our algorithm generates every solution for all $K \geq 1$.

Table 1. When $n = 4K + 1$, the total number of solutions is $2nT_k$.

K	1	2	3	4	5	6
T_K	1	2	2	3	2	4
$n = 4K + 1$	5	9	13	17	21	25
$2nT_k$	10	36	52	102	84	200

To conclude, we identify three areas of future work. First, show that the $2nT_K$ bound is tight. The most promising approach would be to show that every solution is obtained by constant circular shift. Second, express the number of solutions to Hertzsprung's Problem on a cylinder when $n = 4K+1$, for which $2nT_K$ is a provably loose lower bound. Lastly, express the number of solutions for Hertzsprung's Problem on planar, cylindrical, and toroidal boards for $n > 4K$.

A Appendix: Functions to Check Validity

```
def is_valid(a, K):
    #Checks if numpy array a representing solution is valid.
    n = len(a)
    cts = K * np.ones(n) #required row-sum or column-sum array
    if not ((a.sum(axis=0)==cts).all() and (a.sum(axis=1)==cts).all()):
        return False #some row or column does not contain exactly K kings
    #Check for adjacency constraint
    right, down = np.roll(a, 1, axis=1), np.roll(a, 1, axis=0)
    diag1, diag2 = np.roll(right, 1, axis=0), np.roll(right, -1, axis=0)
    return all((a*x).sum()==0 for x in [right, down, diag1, diag2])
```

```
def check_validity(K):
    #Checks if all 2n*T_K solutions generated are valid and distinct.
    n = 4*K + 1
    solution_set = set()
    for a in generate_all_solutions(K):
        if not is_valid(a, K):
            return False
        solution_set.add(tuple(map(tuple, a)))
    T_k = divisor_count(K)
    print(f'K={K}, n={n}: {len(solution_set)} of {2*n*T_k} solutions generated')
```

References

1. Bagno, E., Eisenberg, E., Reches, S., Sigron, M.: Counting king permutations on the cylinder. Enumer. Comb. Appl. (2022)
2. Bell, J., Stevens, B.: A survey of known results and research areas for n-queens. Discret. Math. **309**(1), 1–31 (2009)
3. Bell, J., Stevens, B.: A survey of known results and research areas for n-queens. Discrete Math. **309**(1), 1–31 (2009)
4. Bumgardner, J. (2024). https://krazydad.com/twonottouch/
5. Burger, A.P., Mynhardt, C.M., Cockayne, E.J.: Queens graphs for chessboards on the torus. Australas J. Combin. 213–246 (2001)
6. Chatham, D.: Arrangements of mutually non-attacking chess pieces of mixed type. Recreat. Math. **11**(18), 1–16 (2024)
7. Escamocher, G., O'Sullivan, B.: Leprechauns on the chessboard. Discrete Math. **344**(5) (2021)
8. Gent, I., Jefferson, C., Nightingale, P.: Complexity of n-queens completion. J. Artif. Intell. Res. **59**, 305–331 (2017)
9. Kaplansky, I.: Symbolic solution of certain problems in permutations. Bull. Am. Math. Soc. **50**, 906–914 (1944)
10. Larson, L.C.: The number of essentially different nonattacking rook arrangements. Recreat. Math. **7**(3), 180–181 (1974)
11. Laszlo, M., Mukherjee, S.: Counting star-battle configurations. Math. Comput. Sci. **17**(8) (2023). https://doi.org/10.1007/s11786-023-00558-7
12. Robins, D.P.: The probability that neighbors remain neighbors after random rearrangements. Am. Math. Mon. **87**(2), 122–124 (1980)
13. Sloane, N.J.A.: Sequence a000170: number of ways of placing n nonattacking queens (2024). https://oeis.org/A000170, online Encyclopedia of Integer Sequences, published electronically at https://oeis.org
14. Sloane, N.J.A.: Sequence a000903: number of inequivalent ways of placing n nonattacking rooks (2024). https://oeis.org/A000903, online Encyclopedia of Integer Sequences, published electronically at https://oeis.org
15. Sloane, N.J.A.: Sequence a002464: Hertzsprung's problem (2024). https://oeis.org/A002464, online Encyclopedia of Integer Sequences, published electronically at https://oeis.org
16. Spahn, G.: Enumerating solutions to grid-based puzzles with a fixed number of rows. Math. Comput. Sci. **16**(12) (2022). https://doi.org/10.1007/s11786-022-00530-x
17. Tolpygo, A.: Queens on a cylinder. Quantum, 38–42 (1996)
18. Watkins, J.J.: Across the Board: The Mathematics of Chessboard Problems. Princeton University Press, Princeton, New Jersey (2004)

An LS-DYNA Analysis of the Stress Sensitivity of a Jet-Engine Intake (Fan) Blade to Bird Density in Bird Strikes

Jack K. Horner(✉)

Lawrence, Kansas, USA 66046
jhorner@cybermesa.com

Abstract. Bird strikes on jet engines cause significant economic loss, estimated at over $1 billion worldwide per year. US airworthiness regulations require, among other things, that an aircraft air-intake-fan ("fan") be designed to allow the aircraft to land safely after the fan impacts a "standard-size" bird. Achieving a cost-effective, regulations-compliant engine design typically involves extensive, expensive experimentation. High-fidelity computer modeling of bird-strike damage to jet engines can cost-effectively help to restrict the range of such experimentation and retest. To adequately assess the effects of bird strikes on fans, we need to understand how bird density contributes to those effects. Toward that end, the present study uses the LS-DYNA finite element analysis software system to estimate the stress and strain sensitivity of a fan blade to individual bird density during a bird strike under nominal jet-engine operating conditions. The results are consistent with experiments performed on some actual engines.

Keywords: jet-engine fan · bird-strike · finite element analysis · LS-DYNA

1 Introduction

Though rarely fatal, bird strikes on jet engines cause significant economic loss, estimated at over $1 billion worldwide per year (Allan and Orosz 2001; US Department of Transportation 2025). The US FAA airworthiness regulations (FARs) require, among other things, that an aircraft air-intake-fan (hereafter, "intake fan" or "fan") be designed to allow the aircraft to land safely after the fan impacts a standard-size bird (about 2.1 kg (a large "chicken"); US National Archives 2025; Nizampatnam 2007, Sect. 1.1). No engine that has obviously suffered fan damage is allowed to fly after landing until the damage has been eliminated and the engine re-certified for operations.

Aircraft engine design typically undergoes several build/test cycles before achieving FAR compliance. In addition, if a prototype engine meets the bird-strike FARs, the manufacturer often performs further testing with higher impact velocities or larger bird masses to determine the actual limits of the structure. Achieving a cost-effective, regulations-compliant engine design typically involves extensive, expensive experimentation. Computer modeling of bird-strike damage to jet engines can cost-effectively help

H. R. Arabnia et al. (Eds.): CSCI 2025, CCIS 2970, pp. 289–299, 2027.
https://doi.org/10.1007/978-3-032-28466-2_21

to restrict the range of such experimentation and retest (but cannot replace compliance testing). Such modeling has become an essential adjunct of fan design (Nizampatnam 2007, Sect. 1.1).

A fan/bird-strike event is characterized by high-intensity, short-duration (< 1 millisecond) loading. During impact, both the fan and the bird suffer high, inelastic strain rates that can result in large permanent blade deformation. A complex interaction exists between the impact energy and deformation of the bird and the resulting response of the fan. Any simulation of this regime that is adequate to inform regulation-compliant design and implementation, therefore, must address significant detail of the coupling between the bird and fan (Nizampatnam 2007, Sect. 1.1). A general-purpose multi-physics simulation software package for highly nonlinear transient dynamic finite element analysis (FEA) using explicit time integration is well suited for this kind of modeling.

Finite element (FE) modeling of a fan blade is relatively straightforward because the mechanical properties and geometry of such blades are known (Selezneva et al. 2010). In contrast to the modeling of the fan blade, modeling of the bird in a bird-strike regime is an active research area (see for example Airoldi and Cacchione 2006, Meguid et al. 2008, McCallum and Constantinou 2005; Nizampatnam 2007; McCallum, Shoji, and Akiyama 2013). Most such studies assume the bird is composed of water (see for example Wilbeck 1978). At present, however, there is significant uncertainty about how sensitive fan blades are to assumptions about the density of the bird (see, for example, Selezneva 2010). Variation in that density could directly affect the physics of the impact.

The principal objective of this paper is to assess, using an FEA simulator that is capable of modeling the highly nonlinear interaction between a fan blade and bird, the stress and strain sensitivities of a fan blade to individual bird densities of 900, 1000, and 1100 kg/m^3, respectively, during a FAR-compliant 2.1 kg bird strike under nominal jet engine operating conditions.

Strictly speaking, the present study concerns only the outputs of simulation software under specific conditions. These outputs can provide informed *hypotheses* for empirical tests. Those hypotheses are not a substitute for actual testing.

2 Method

LS-DYNA™ is a general-purpose multi-physics simulation software package for highly nonlinear transient dynamic finite element analysis (FEA) using explicit time integration. LS-DYNA is used by the automobile, aerospace, and construction industries, among others (ANSYS/LS-DYNA 2023).

2.1 The Baseline Setup File

An LS-DYNA fan-blade/bird-strike setup (scenario-configuration) file (*bird.k*) that is available for educational use was obtained from an ANSYS/LS-DYNA website (ANSI/LS-DYNA 2025a) and was used as the baseline/benchmark for this study.

It appears that *bird.k* is not described in any detailed open-source narrative. However, it is relatively straightforward to construct such a narrative by analyzing the content of *bird.k* with the help of the LS-DYNA manuals (see especially ANSYS/LS-DYNA

2025b) and by comparing that information with information about jet engine fans that is available in open sources (see for example Selezneva et al. 2020)

Here are a few features of that narrative. *bird.k* assumes a titanium-alloy fan blade that is similar to fan blades in use as of 2025 (Lesuer 2000). The blade strikes a 2.1 kg featherless bird. The tip of the blade has an initial y-velocity of approximately 0.09 km/s, corresponding to an initial rotation speed of several thousand RPM. The bird is stationary at t_0 and initially impacts the blade at approximately 70% of the blade span (measured from the blade root).

bird.k defines the properties of the blade with LS-DYNA's MAT_PIECEWISE_LINEAR_PLASTICITY specification (see LS-DYNA 2023), assuming a density of 4000 kg/m^3, a Young's modulus of 2 x 10^{11} Pa, and a Poisson ratio of 0.3.

bird.k assumes that the bird can be represented as a 4160-particle cylinder with hemispherical ends obeying Smooth Particle Hydrodynamics (Lavoie et al. 2008). *bird.k* defines the equation of state (EOS) for the bird by an LS-DYNA's EOS_GRUNEISEN specification for water, with the speed of sound set at 1647 m/s.

2.2 Modifications of the Baseline Setup File Made for this Study

File *bird.k* was edited using Microsoft Notepad to change the initial density of the bird to each of 900, 1000, and 1100 kg/m^3, in turn, resulting in three setup files. All other parameters were left as they are in *bird.k*. Physical-units-calibrated 102-frame x-, y-, and z-stress, and x-, y-, and z-strain movies of the strikes, for each of the initial bird densities, were generated by **LS-PrePost 2025 R1, v4.12.6** (for a total of six movies. Each of these movies was examined frame by frame to identify those regions of the blade that were stressed beyond the yield stress of the blade (in the present study, assumed to be that of titanium)).

To perform these analyses, LS-DYNA was run under Windows 10 on a desktop PC under the GUI provided by **LS-RUN 2025 R1 Student Version**. First, LS-DYNA was run from the main LS-RUN window. When that task completed successfully, the LSP icon next to the "d3plot" drop-down on the **LS-RUN 2025 R1** window was clicked. From the FEM drop-down list, the following selections were made

Fringe Component ➔ Post ➔ [Stress | Strain] ➔ [x | y | z]-[stress | strain] ➔ Apply

From the **LS-PrePost** window the following selections were then made

File ➔ Movie ➔ Start

These commands produced stress and strain movies for each of the bird densities noted above.

These movies show color-coded [stress | strain] magnitudes. The red regions in the stress movies display show that the stress has exceeded the yield stress of the blade, i.e., red indicates regions in which the blade deforms but will not restore to its original configuration.

These results were compared to results reported on experiments on actual engines (see for example Selezneva et al. 2010, pages 1–40, 1–42).

Representative frames from each of the movies are shown in Section 3.0. Figure 1 shows the configuration of the simulation at t_0. Figures 2, 3, 4, 5, 6 and 7 illustrate the stresses and strains along the X, Y, and Z axes at various times. Figure 8 is an excerpt of the run log (file lsrun.out.txt) for the case in which the bird density is 1000 kg/m^3.

3 Results

In Figs. 3, 4, 5, 6 and 7, the blade rotates clockwise about the hub (not shown, but is located at the root of the blade on the left side of each of these Figures), parallel to the X-Y plane of each Figure.

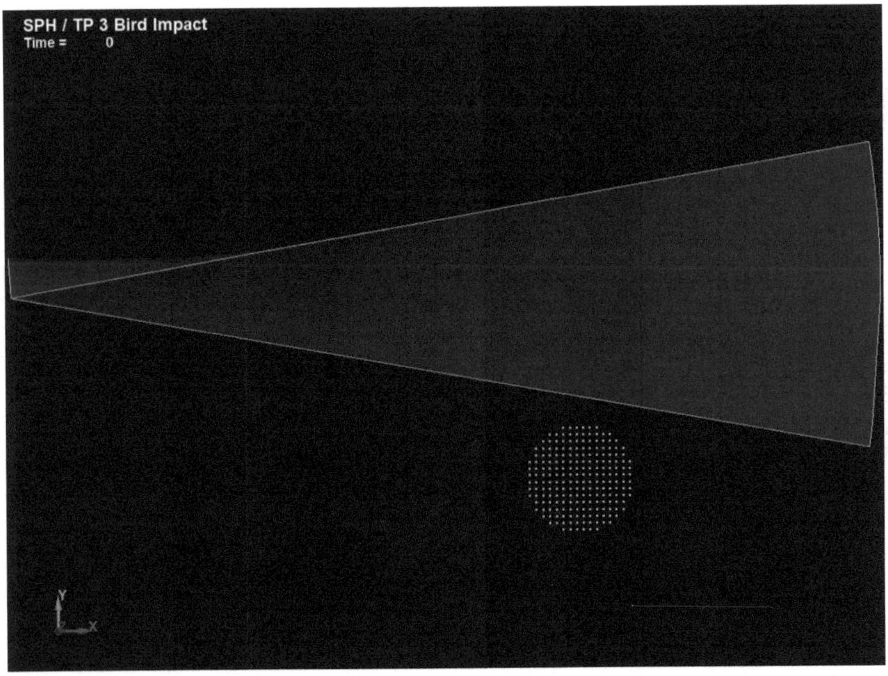

Fig. 1. Bird (yellow dots)-blade(red) collision configuration at t_0. The bird is represented as a cylinder capped with a hemisphere at each end. The main axis of the body of the bird is parallel to the z-axis, i.e., the Figure shows one end of the bird as projected into the X-Y plane.

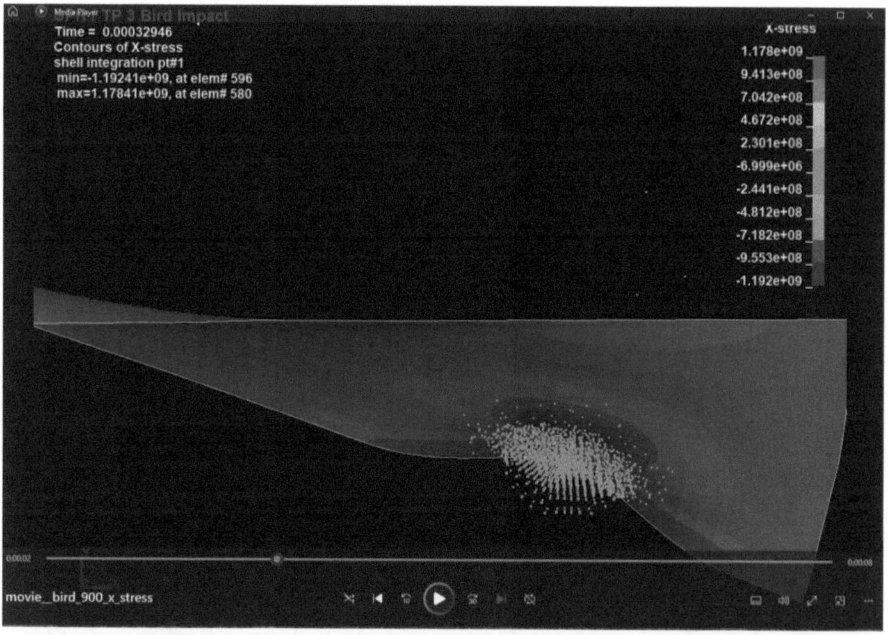

Fig. 2. X-stress distribution in bird-blade collision at approximately 0.3 milliseconds after t_0. Initial bird density = 900 kg/m^3. Stress units are Pa.

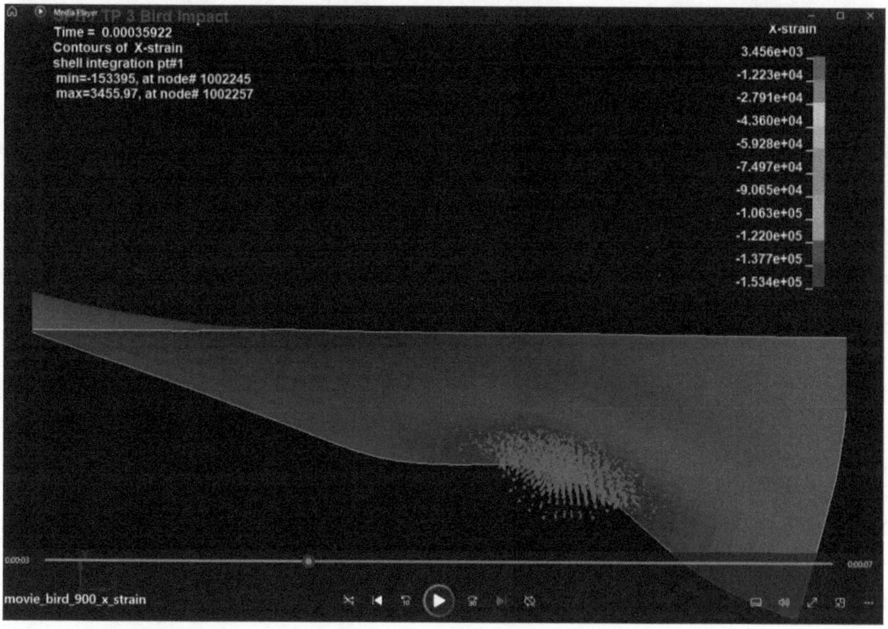

Fig. 3. X-strain distribution in bird-blade collision at approximately 0.35 milliseconds after t_0. Initial bird density = 900 kg/m^3. Strain units are meters displacement from t_0 position.

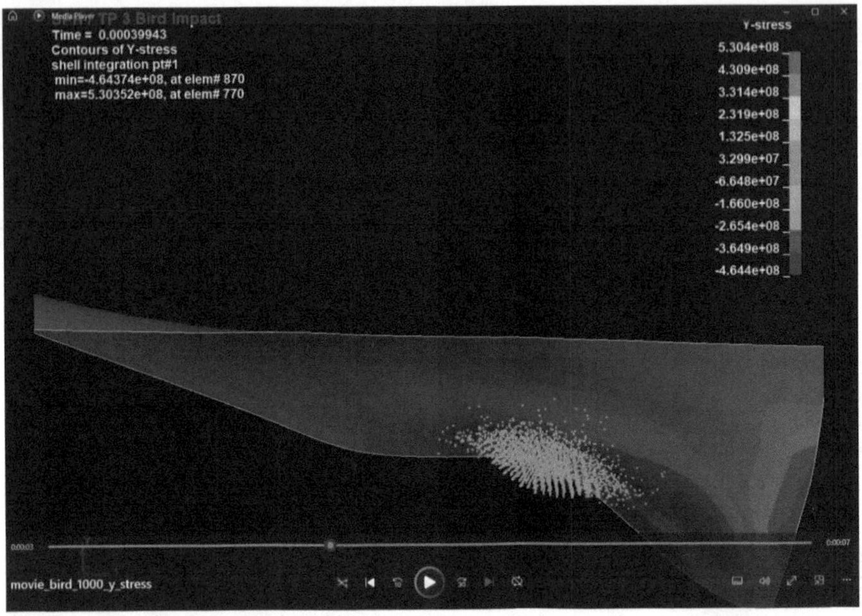

Fig. 4. Y-stress distribution in bird-blade collision at approximately 0.4 milliseconds after t_0. Initial bird density = 1000 kg/m^3. Stress units are Pa.

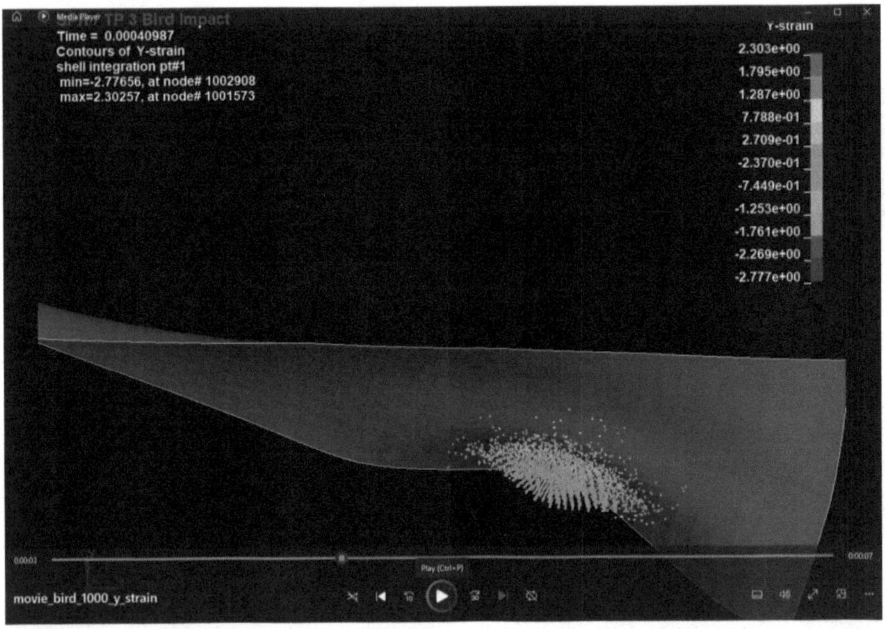

Fig. 5. Y-strain distribution in bird-blade collision at 0.4 milliseconds after t_0. Initial bird density = 1000 kg/m^3. Strain units are meters displacement from t_0 position.

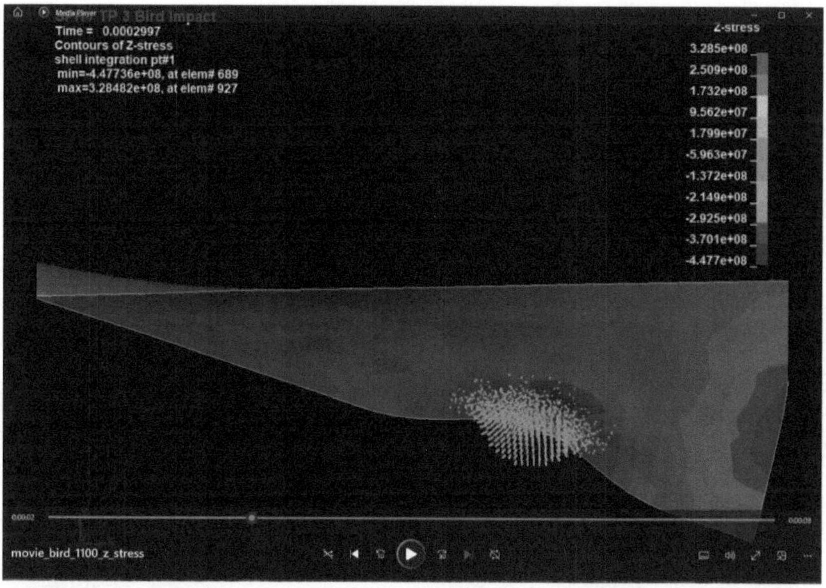

Fig. 6. Z-stress distribution in bird-blade collision at approximately 0.2 milliseconds after t0. Initial bird density = 1100 kg/m^3. Stress units are Pa.

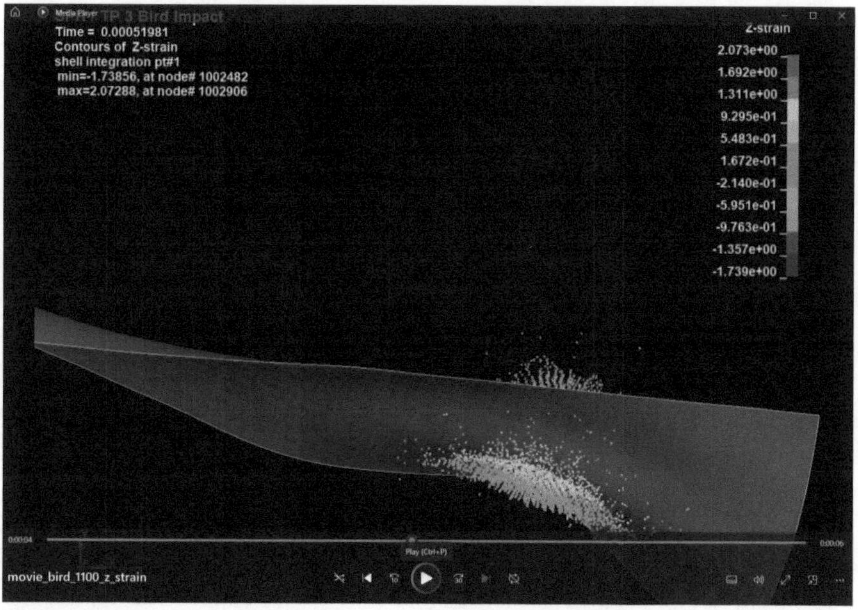

Fig. 7. Z-strain distribution in bird-blade collision at approximately 0.5 milliseconds after t_0. Initial bird density = 1100 kg/m^3. Strain units are meters displacement from t_0 position.

```
. . .

     Date: 10/19/2025      Time: 09:56:39
     ___________________________________________________
     |                                                 |
     |  LS-DYNA, A Program for Nonlinear Dynamic       |
     |  Analysis of Structures in Three Dimensions     |
     |  Date    : 07/09/2025    Time: 09:53:35         |
     |  Version : smp d R16                            |
     |  Revision: R16.1-180-gd50332dbe5                |
     |  AnLicVer: 2025 R2 (20250506+dl-73-g4dde854)    |
     |                                                 |
     |  Features enabled in this version:              |
     |    Shared Memory Parallel                       |
     |    CESE CHEMISTRY EM ICFD STOCHASTIC_PARTICLES  |
     |    FFTW (multi-dimensional FFTW Library)        |
     |    ARPACK (nonsymmetric eigensolver library)    |
     |    ANSYSLIC enabled                             |
     |                                                 |
     |  Platform   : WINDOWS X64                       |
     |  OS Level   : Windows 10  (Build 19045) uum     |
     |  Compiler   : Intel Fortran Compiler 19.0 SSE2  |
     |  Hostname   : DESKTOP-AM4IKPU                   |
     |  Precision  : Double precision (I8R8)           |
     |                                                 |
     |  Unauthorized use infringes Ansys Inc. copyrights |
     |_________________________________________________|

. . .

 Executing with ANSYS license

 Command line options: i=C:\Miscellaneous\BIRD\bird.k\bird.k
                       ncpu=2
                       memory=20m

 Input file: C:\Miscellaneous\BIRD\bird.k\bird.k

 The native file format       : 64-bit small endian
 Memory size from command line:    20000000

 . . .

 initial kinetic energy = 0.10176297E+06

 The LS-DYNA time step size should not exceed     1.307E-06
 to avoid contact instabilities.  If the step size is
 bigger then scale the penalty of the offending surface.

 Memory required to begin solution      :     349K
 Additional dynamically allocated memory:    3127K
                                  Total:    3476K
```

Fig. 8. Excerpts from the LS-DYNA run log *lsrun.out.txt* for bird-edge collision. Initial bird density = 1000 kg/m^3. Text deleted from original *lsrun.out.txt* is represented by ellipsis (...)

4 Conclusions

Several conclusions can be drawn from this study:

```
   . . .
    N o r m a l    t e r m i n a t i o n                         10/19/25
09:57:18

    Memory required to complete solution    :     349K
    Additional dynamically allocated memory:     4575K
                                     Total:     4923K


   . . .


    Problem time        =    1.0005E-03
    Problem cycle       =      1246
    Total CPU time       =          39 seconds (     0 hours    0 minutes 39
seconds)
    CPU time per zone cycle  =       5799.809 nanoseconds
    Clock time per zone cycle=       5704.191 nanoseconds

    Number of CPU's     2
   . . .
    Start time    10/19/2025 09:56:41
    End time      10/19/2025 09:57:18
    Elapsed time       37 seconds for     1246 cycles using  2 SMP threads
                (      0 hour    0 minute   37 seconds)

    N o r m a l    t e r m i n a t i o n                         10/19/25
09:57:18
```

Fig. 8. (*continued*)

1. In every case in this study, the bird/blade collision produced stress that exceeded the yield stress in at least one region of the blade.
2. Within about 0.1 milliseconds after the blade first contacts the bird, the stress at the root of the blade exceeds the yield stress of the blade even though the stress in the initial region of contact between the bird and blade has not exceeded the yield stress. This somewhat unexpected result is likely due to an impulsive torque load on the root that is generated by the strike. (To first approximation, the blade acts as a torque lever whose length is the distance between the strike region and the blade root.)
3. The results are consistent with the results of experiments reported in Selezneva et al. 2010.

All of the movies and corresponding setup files produced in this study are available at Horner 2025.

5 Discussion

As is the case with nearly all non-linear simulation regimes, the results shown in this study should not, without explicit caveats, be extrapolated to other configurations.

The speed of sound in the bird is kept constant (1647 m/s) across all setups used in this study. Were the values of the initial density of the bird intended to reflect physical reality as closely as Selezneva et al. 2010, we would expect the speed of sound to vary by initial density (i.e., by the porosity of water in the case of Selezneva et al. 2010). The concern of the present study, however, is technically only with the sensitivity of blade stress and strain to variation in the purely numerical magnitude of the initial density of the bird. As it turns out, moreover, the range of variation in the magnitude of effective initial density of the bird does remains consistent with simulation and experiments reported in Selezneva et al. 2010.

A physically more realistic model of the collision of a bird with a fan would assume that the fan has multiple blades. Several such simulations have been performed (see for example Selezneva et al. 2010; Schmotin et al. 2009). Those simulations suggest that ~10 fan blades are typically involved in a collision. The present study considers a collision of just one blade with the bird. Although the distribution of stresses and strains in the multi-blade cases is physically more informative, all such configurations studied to date imply that at least one blade is irreversibly deformed, thus requiring the plane to be grounded until the damage is removed.

Acknowledgements. This work benefited from discussions with Bill Spangenberg, Mike Gittings, and Bob Weaver about issues in simulating high stress/rate materials behaviors. The work also benefited from discussions with Eric Winsberg, John Symons, and Art Skidmore about the general relationships between systems being simulated and the results of such simulations.

Disclosure of Interests. The author has no competing interests.

References

Airoldi, A., Cacchione, B.: Modelling of impact forces and pressures in Lagrangian bird strike analyses. Int. J. Impact Eng. **32**, 1651–1677 (2006)

Allan, J.R., Orosz, A.P.: The costs of Birdstrikes to commercial aviation. In: Bird Strike Committee Proceedings. Bird Strike Committee-USA/Canada, Third Joint Annual Meeting, Calgary, Alberta (2001)

ANSYS/LS-DYNA: Overview (2023). https://lsdyna.ansys.com/applications/ .

ANSI/LS-DYNA: LS-DYNA Examples (2025a). https://www.dynaexamples.com/sph/bird .

ANSI/LS-DYNA: Keyword User's Manual. Volume II. Material Models (2025b). https://ftp.lstc.com/anonymous/outgoing/web/ls-dynamanuals/DRAFT/DRAFT_Vol_II.pdf .

Lavoie, M.-A., Gakwaya, A., Ensan, M.N.: Application of the SPH method for simulation of aerospace structures under impact loading. In: Proceedings of the 10th International LS-DYNA Users Conference (2008) https://www.dynalook.com/conferences/international-conf-2008/ImpactAnalysis-4.pdf

Lesuer, D.R.: Experimental investigation of material models for Ti-6Al-4V titanium and 2024-T3 aluminum. Report No. DOT/FAA/AR-00/25. Prepared for U. S. Department of Transportation, Federal Aviation Administration (2000)

McCallum, S.C., Constantinou, C.: The influence of bird-shape in bird-strike analysis. In: Proceedings of the 5th European LS-DYNA Users Conference (2005)

McCallum, S.C., Shoji, H., Akiyama, H.: Development of an advanced multi-material bird-strike model using the smoothed particle hydrodynamics method. Int. J. Crashworthiness. **18**, 579–597 (2013)

Meguid, D.A., Mao, R.H., Ng, T.Y.: FE analysis of geometry effects of an artificial bird striking an aeroengine fan blade. Int. J. Impact Eng. **35**, 487–498 (2008)

Nizampatnam, L.S.: Models and Methods for Bird Strike Load Predictions. Wichita State University, PhD thesis (2007). https://studylib.net/doc/14844390/ .

Schmotin, Y.N., Chupin, P.V., Gabov, et al.: Bird strike analysis of aircraft engine fan. In: Proceedings of the 7th European LS-DYNA Users Conference (2009)

Selezneva, M., Stone, P., Moffat, T., Behdinan, K., Poon, C.: Modeling bird impact on a rotating fan: the influence of bird parameters. In: Proceedings of the 11th International LS-DYNA Users Conference (2010) https://www.dynalook.com/conferences/international-conf-2010/Aerospace-1-4.pdf/view

US Department of Transportation, Federal Aviation Administration: Wildlife Strikes to Civil Aircraft in the United States 1990–2024 (2025). https://www.faa.gov/airports/airport_safety/wildlife/wildlife-strike-report-1990–2024

US National Archives: Code of Federal Regulations, Title 14, Chapter I, Subchapter C, Part 33, Subpart E, Section 33.76 (2025). https://www.ecfr.gov/current/title-14/chapter-I/subchapter-C/part-33/subpart-E/section-33.76 .

Wilbeck, J.S.: Impact behavior of low strength projectiles. University of Dayton Research Institute for the Air Force Flight Dynamics Laboratory, Wright-Patterson Air Force Base. Technical report AFML-TR-77–134 (1978)

On the Emergence of Harmonious Kingdoms

Michael Laszlo(✉) and Sumitra Mukherjee

Department of Computer Science, Nova Southeastern University, Fort Lauderdale, FL, USA
{mjl,sumitra}@nova.edu

Abstract. We consider the problem of placing nK non-attacking kings on an $n \times n$ board, so that each row and each column contains exactly K kings. For any specified K, we show that no solutions exist if $n < 4K$, and that the problem has exactly two solutions if $n = 4K$.

Keywords: Enumerative combinatorics · Generalized Hertzsprung's problem

1 Introduction

Hertzsprung's problem counts the number of ways in which we can place n non-attacking kings on an $n \times n$ board, so that each row and each column contains exactly one king [3]. This problem has been extensively studied and recurrence relations for enumerating solutions are known [5]. Counting the number of solutions to the popular *Star-Battle* puzzle [2] may be considered a generalization of this problem in that each row and each column of the board must contain exactly K non-attacking kings. An algorithm to enumerate solutions for this problem when $K = 2$ is presented in [4] and solutions are computed for $n \leq 15$.

Formally, a *solution* is a placement of nK kings in a $n \times n$ board so that following constraints are satisfied:

1. *Row constraints*: Each row contains exactly K kings.
2. *Column constraints*: Each column contains exactly K kings.
3. *Adjacency constraints*: No two kings are orthogonally or diagonally adjacent.

Given $1 \leq K < n$, we use $C_K(n)$ to denote the number of solutions. Observing that $C_1(n) = 0$ if $n < 4$ and $C_2(n) = 0$ if $n < 8$, we pose the following questions: What is the minimum board-size for a solution to exist? Can we derive a closed form expression for the number of solutions in this minimal board? The fact that we investigate conditions that are conducive to the existence of solutions to this problem of non-attacking kings motivates this paper's title. We show that, for any specified K:

M. Laszlo and S. Mukherjee—Both authors contributed equally to this work.

H. R. Arabnia et al. (Eds.): CSCI 2025, CCIS 2970, pp. 300–305, 2027.
https://doi.org/10.1007/978-3-032-28466-2_22

- **Theorem** 1. $C_K(n) = 0$ if $n < 4K$.
- **Theorem** 2. $C_K(4K) = 2$.

Definitions introduced in the next section are used to prove our theorems in the section that follows. Opportunities for future work are identified in the concluding section.

2 Preliminaries

The rows and columns of an $n \times n$ board are indexed from 0 to $n-1$, where indices increase from bottom to top along rows and from left to right along columns. Given a solution, a *duplex* is a distribution of kings along a pair of adjacent rows j and $j+1$. We refer to the distribution of kings in row j to be the duplex's *lower row,* and in row $j+1$ its *upper row.* A column in a duplex is *occupied* if it contains a king in either row; a column that is not occupied is *empty.* A duplex is said to be *even* if all its occupied columns have even indices, *odd* if all its occupied columns have odd indices, and *mixed* otherwise. Due to symmetry, results about (row) duplexes also apply to column duplexes, which are defined analogously (e.g., a column-duplex spans a pair of adjacent columns and comprises a left and a right column). We will usually use *duplex* to mean row duplex, but will qualify with the adjective *row* or *column* as context requires. One additional definition: a *quad* is a 2×2 block of some solution, and an *empty quad* is a quad that contains no kings.

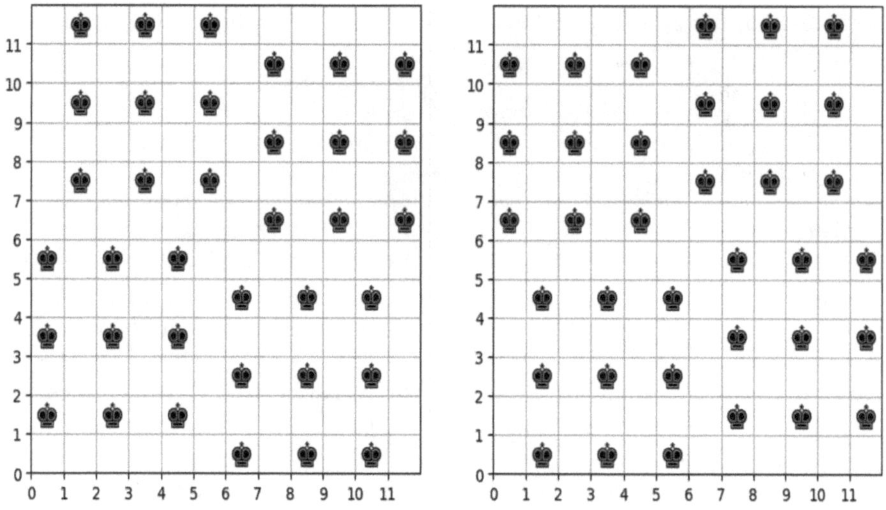

Fig. 1. The pair of solutions for $K = 3$ and $n = 12$.

Figure 1 illustrates the pair of solutions for $K = 3$ and $n = 12$. In the solution on the left, three identical even duplexes span rows 0 to 5, and three identical odd duplexes span rows 6 to 11. In the solution on the right, three identical odd duplexes appear above three identical even duplexes. An empty quad spans columns 5 and 6 and rows 5 and 6 in each solution.

3 Results

In Theorem 1 we show that no solutions exist if $n < 4K$.

Theorem 1. *$C_K(n) = 0$ if $n < 4K$.*

Proof. $1 \leq K < n$, any solution must contain at least one duplex. Adjacency constraints require that the $2K$ occupied columns in a duplex must be separated by at least $2K - 1$ vacant columns. Thus, no solutions exist if the number of columns $n < 2K + (2K - 1) = 4K - 1$.

If $n = 4K - 1$, any solution can be partitioned into $2K - 1$ duplexes and a single row. Each duplex must have all $2K$ even numbered columns occupied and all $2K - 1$ odd numbered columns empty. If $K > 1$, the single row cannot satisfy the column sum constraints in any of the empty odd numbered columns. If $K = 1$ and $n = 3$, the sole odd numbered column, column 1, in the single row cannot contain a king since it is adjacent to both even numbered columns in an adjacent row, at least one of which must contain a king.

The rest of this paper focuses on the case $n = 4K$. For convenience, we use B_K to denote the $4K \times 4K$ board.

Lemma 1. *If $0 < m < 2K$ of the occupied columns in a mixed duplex are even numbered, then:*

1. *the first m even numbered columns are occupied,*
2. *an empty quad spans columns $2m - 1$ and $2m$ of the duplex, and*
3. *the last $2K - m$ odd numbered columns are occupied.*

Proof. Consider the set $I_m = \{2i \mid 0 \leq i \leq m-1\} \cup \{2m+1+2i \mid 0 \leq i \leq 2K-m-1\}$ of indices for occupied columns in the postulated mixed duplex. In any other mixed duplex, some column in I_m must be swapped with a column that is not in it. Adjacency constraints permit only two feasible options, both of which violate the condition that m of the occupied columns in this duplex are even: If we swap the occupied column $2m-2$ and the vacant column $2m-1$, only $m-1$ of the occupied columns are even. If we swap the occupied column $2m+1$ with the vacant column $2m$, then $m+1$ of the occupied columns are even. Thus, I_m is the only possible set of occupied columns as postulated.

Lemma 1 does not specify which rows of a duplex its $2K$ kings belong to. For example, where $m = K = 2$, the lemma does not exclude a duplex whose lower row contains kings in columns 0 and 7 and whose upper row contains kings in columns 2 and 5.

Lemma 2. *An empty quad appears only in a mixed duplex.*

Proof. Both even and odd duplexes contain kings in alternate columns; only a mixed duplex contains a 2-column empty gap.

A row duplex is said to be *aligned* if its lower row has an even index, and *unaligned* otherwise. The $4K$ rows in a solution for board B_K can be partitioned into $2K$ aligned duplexes. In the discussion that follows, we refer to a partition into aligned duplexes as $\mathcal{P}$.

Lemma 3. *$\mathcal{P}$ contains exactly K even duplexes and K odd duplexes.*

Proof. Suppose to the contrary that some duplex in $\mathcal{P}$ is mixed. By Lemma 1, columns 1 and $4K-2$ are both vacant. This requires K odd duplexes and K even duplexes, respectively, to satisfy the column sum constraints for these two vacant columns. Since $\mathcal{P}$ contains exactly $2K$ duplexes, our postulated mixed duplex cannot co-exist with these K even and K odd duplexes.

We refer to the property of the evenness or oddness of an aligned duplex as its *parity*. Because $\mathcal{P}$ contains both even and odd aligned duplexes, there exists some pair of adjacent duplexes of opposite parity. We consider the unaligned duplex D whose lower row D_l belongs to one of the aligned duplexes and whose upper row D_r belongs to the other. We will refer to such a duplex D as a *transition duplex.*

Lemma 4. *Every solution contains exactly one transition row-duplex and one transition column-duplex.*

Proof. Let D be some transition row-duplex D whose rows D_l and D_r belong to aligned duplexes of opposite parity. Because rows D_l and D_r each contain K kings, Lemma 1 applies where $m = K$, which implies that an empty quad Q appears in columns $2K-1$ and $2K$ of duplex D. Lemma 2 says that since Q is empty, the column-duplex E in which Q appears is mixed. Because each of the two columns comprising column-duplex E contains K kings, Q occurs in rows $2K-1$ and $2K$. Combining both inferences, the empty quad Q appears at the center of board B_K. Because the position of quad Q is fixed at this one position by the arbitrary transition duplex D, there exists exactly one transition row-duplex D and one transition column-duplex E.

It follows that the lower half of a solution (rows 0 through $2K-1$) is made of K aligned duplexes of one parity, and the upper half is made of K aligned duplexes of opposite parity. More importantly, because there is exactly one transition across the row-duplexes and one across the column-duplexes, we have the following result:

Lemma 5. *The row-duplexes adjacent to the top and bottom borders of the board – covering rows 0 and 1, and rows $4K-2$ and $4K-1$ – have opposite parity. Likewise, the column-duplexes abutting the left and right borders have opposite parity.*

Lemma 5 is used in an inductive proof that establishes both the main theorem, that $C_K(4K) = 2$ for all $K \geq 1$, and the structure of a solution. To prepare, a couple definitions: Board B_{K+1} can be partitioned into two regions: A *frame* consisting of the pair of aligned row duplexes, and pair of aligned column duplexes, adjacent to the board's border, and an *inner board* B_K. We'll say that an even (odd) row-duplex is *canonical* if K kings appear in successive even (odd) columns in one row, and the remaining K kings appear in successive even (odd) columns in the other row. Canonical column-duplexes are defined analogously.

Theorem 2. *$C_K(4K) = 2$ for all $K \geq 1$. Moreover, the four duplexes comprising each solution's frame are canonical and, ordering the duplexes in clockwise rotation starting from the bottommost, they appear in the order even-odd-odd-even in one solution, and in the order odd-even-even-odd in the other.*

Proof. By induction on K. For the basis, Fig. 2 depicts the two solutions for the board B_1. These are the only two solutions since no configuration that places a king on a corner cell can simultaneously satisfy the row/column count and adjacency constraints, and each solution's frame satisfies the theorem's conditions. For the inductive step, assume that the theorem's conditions hold for B_K, and consider the partition of B_{K+1} into frame F and inner board B_K. Due to Lemma 5, the frame contributes exactly one king to each row and to each column that crosses F. Hence any solution for B_{K+1} contains a solution for its inner board B_K. By the induction hypothesis, there are two such solutions for B_K each of which is bounded by four canonical duplexes. Each canonical duplex can be extended by a duplex that respects the adjacency constraint in only one way, and each such extension is a canonical duplex of same parity. Lastly, this four-duplex extension contributes exactly one king to each row and column of B_K, ensuring that the row and column counts are satisfied in B_{k+1}.

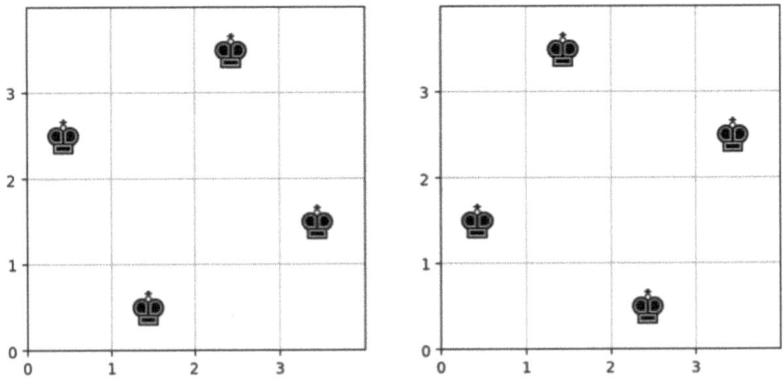

Fig. 2. The pair of solutions for the board B_1.

Every solution has certain characteristics. First, the central quad is empty since every solution emerges from one of the two solutions for B_1. No other empty quad appears because a solution contains only two mixed duplexes which intersect at the central (empty) quad. Diagonal cells are empty because every aligned quad along a diagonal belongs to a row-duplex and column-duplex of opposite parity.

4 Conclusions

We have shown that solutions exist when $n \geq 4K$, and that $C_K(4K) = 2$. An algorithm presented in [4] computes $C_2(n)$ for $n < 16$. Whether closed formed expressions or recurrence relations can be obtained for $C_K(n)$ when $n > 4K$ is an open question.

In the special case when $K = 1$, our problem has been studied on a cylindrical board that is obtained by gluing together the sides of a planar board [1]. The questions that we address in this paper regarding the minimum board-size for a solution to exist, and the number of solutions in this minimal board, are also relevant on other board topologies: on a cylindrical board, and on a toroidal board formed by gluing together the top and bottom edges of a cylindrical board.

References

1. Bagno, E., Eisenberg, E., Reches, S., Sigron, M.: Counting king permutations on the cylinder. Enumer. Combin. Appl. **2**, Article #S4PP5 (2021). https://doi.org/10.54550/ECA2022V2S4PP
2. Bumgardner, J. (2024). https://krazydad.com/twonottouch/
3. Kaplansky, I.: Symbolic solution of certain problems in permutations. Bull. Am. Math. Soc. **50**, 906–914 (1944)
4. Laszlo, M., Mukherjee, S.: Counting star-battle configurations. Math. Comput. Sci. **17**, 8 (2023). https://doi.org/10.1007/s11786-023-00558-7
5. OEIS Foundation Inc.: Hertzsprung's Problem, Entry A002464. In: The On-Line Encyclopedia of Integer Sequences (2015). https://oeis.org/A002464

A Computational Model for Quark Confinement and Thermodynamics in a Casimir-Like Trap

Boya Hou[1], Patrick Bruskiewich[2], Amirreza Sedaghat[3], and Chris Cheng Zhang[3](✉)

[1] Burnaby North Secondary School, Vancouver, Canada
[2] Sager Education Inc., Toronto, Canada
[3] Cansight Technology Corporation, Toronto, Canada
head@roboca.org

Abstract. This paper explores a theoretical and computational model for quark confinement based on the principles of the Casimir effect. We treat the proton as a "Casimir Trap" and derive estimates for quark masses. To further investigate the dynamics of this system, we present a 2D simulation of a piston containing quarks and photons. The simulation models the interactions between particles, including Coulomb forces, Compton scattering, and elastic collisions, within a container with a movable piston. This simulation acts as a crucial bridge, linking the theoretical Casimir framework to the complex dynamics of quark-gluon interactions, a significant challenge in the field. We analyze the thermodynamic properties of the system, such as temperature, pressure, and volume, under various compressive forces on the piston. This work provides a framework for studying the emergent thermodynamic behavior of a quark-gluon plasma-like system within a confined space.

Keywords: Quark Confinement · Casimir Effect · Particle Simulation · Thermodynamics · Compton Scattering · Piston Model

1 Introduction

Quark confinement, the phenomenon wherein quarks and gluons are permanently bound within hadrons and never observed in isolation, stands as a cornerstone of the Standard Model and one of the most profound unsolved problems in non-perturbative Quantum Chromodynamics (QCD) [5]. For over half a century, the QCD Lagrangian has successfully described the strong interaction at high energies, where asymptotic freedom allows for perturbative calculations [6]. However, at the low-energy scales characteristic of hadron interiors, the strong force becomes non-perturbative, and a complete, analytical explanation for the mechanism that confines quarks remains elusive [4]. This long-standing challenge has necessitated the development of effective theoretical models and sophisticated

H. R. Arabnia et al. (Eds.): CSCI 2025, CCIS 2970, pp. 306–318, 2027.
https://doi.org/10.1007/978-3-032-28466-2_23

computational methods to probe the structure of hadrons and the thermodynamics of strongly interacting matter.

Among the various phenomenological frameworks, the MIT Bag Model has been particularly influential, offering an intuitive picture of a hadron as a bubble in the physical vacuum within which quarks propagate freely, confined by an external pressure [7]. While successful in spectroscopic applications, the model's ad-hoc nature presents fundamental limitations, especially when extended to describe the thermodynamics of dense quark matter. Introducing medium effects, such as a density-dependent bag pressure, can lead to thermodynamic inconsistencies within the grand canonical ensemble, including violations of the Euler relation [2]. On the other end of the spectrum, Lattice QCD provides a rigorous, first-principles approach that has confirmed key features of confinement, such as the linear potential between heavy quarks [8]. However, this method is computationally demanding and faces the prohibitive "sign problem" at finite baryon chemical potential, limiting its applicability to the high-density regimes relevant to neutron stars and certain heavy-ion collisions [11]. This creates a compelling need for novel, computationally tractable models that can explore the thermodynamics of confined systems in regimes inaccessible to current first-principles methods.

This paper explores an alternative perspective grounded in the Casimir effect, a physical manifestation of quantum vacuum fluctuations in the presence of boundary conditions [10]. Typically observed as a force between neutral conducting plates, the Casimir effect is a general feature of any quantum field confined to a finite volume and has been previously connected to the dynamics of QCD in small systems and even to the bag model itself [9]. We build upon this foundation by proposing a theoretical model that treats the proton as a "Casimir Trap." In this framework, the energy required for confinement is not an ad-hoc parameter but is identified with the Casimir energy of the quark and gluon fields constrained within the proton's volume. This approach provides a physical basis for the stability of the hadron, effectively re-interpreting the bag model's vacuum pressure as a calculable energy density arising from well-established principles of quantum field theory [3].

To investigate the dynamic and thermodynamic consequences of this model, a direct computational approach is indispensable. The extreme conditions and transient nature of the Quark-Gluon Plasma (QGP) created in relativistic heavy-ion collisions make direct experimental measurement of its bulk properties impossible, necessitating the use of simulations to bridge theory and observation [1]. We therefore introduce a 2D particle simulation designed as a "mesoscale" bridge, connecting the macroscopic Casimir Trap hypothesis to the emergent, collective behavior of its constituents. The simulation models a system of quarks and photons within a container with a movable piston, incorporating key microscopic interactions such as effective Coulomb forces, elastic collisions, and the fundamental quantum process of Compton scattering to facilitate thermalization. This computational framework allows for the direct study of thermodynamic variables—pressure, volume, and temperature—and the system's equation of state, providing a means to explore the thermodynamic implications of a Casimir-based confinement mechanism.

This work thus presents a dual contribution: first, a theoretical framework that provides a physical basis for confinement and yields reasonable estimates for the up and down quark masses; and second, a computational tool to study the emergent thermodynamics of the resulting confined quark-photon system. The remainder of this paper is structured as follows: Sect. 2 details the theoretical formulation of the Casimir Trap model and the setup of our computational simulation. Section 3 presents and discusses the results from the simulation, including system equilibration and its equation of state. Finally, Sect. 4 provides our conclusions and outlines avenues for future work.

2 Methodology

2.1 Theoretical Model of the Proton as a Casimir Trap

There are varied levels of complexity in models for quark confinement. The simplest to accept is the so called "bag model." What the bag may be is that it is in fact a Casimir Trap. If we consider a proton as a Casimir Trap then both the free mass and the constituent mass for the Up and Down quarks draws directly from this hypothesis. We know there are three quarks within a proton.

The mean value of the energy $\langle E \rangle$ on the walls of the Casimir Trap is given by

$$E = \frac{\hbar c \pi^2}{720} \frac{A}{a^3} \tag{1}$$

where $\hbar = 1.05 \times 10^{-34}$ J·s, c is the speed of light, A is the area of the plates and a is the distance between the plates. In our model, we treat a as the effective "hard diameter" of the proton. This simplification allows us to focus on the thermodynamic properties arising from confinement without delving into the full quantum chromodynamics (QCD) of the boundary, providing a valid first-order approximation for the system's bulk properties. In this thesis we use a different sign convention from that used by Dr. Casimir in his 1948 paper. By analogy to the attractive force of a spring (the spring energy is given by $U = \frac{1}{2}kx^2$ and therefore $F = -kx$) for the two Casimir plates $F = -\partial E/\partial r$. The force between the plates is attractive.

All things being equal, if we were to imagine the proton as a sphere with surface area A ($A = \pi a^2$), where a is the diameter of the sphere, then we have an inverse distance expression for the energy, namely

$$E = \frac{\hbar c \pi^3}{720} \frac{1}{a} \tag{2}$$

The measured hard diameter of a proton is $a = 1.5 \times 10^{-15}$ m which means that

$$E = 9 \times 10^{-13} \text{ J} \tag{3}$$

This is an energy on the order of 5.6 MeV and represents the binding energy of the proton as a Casimir Trap for the individual Up and Down quarks.

If we take the $\langle E \rangle$ and divide by the volume of the proton this should provide an estimate for an energy density which we can translate into a temperature, namely

$$\frac{E}{V} = \frac{\hbar c \pi^2}{960} \frac{1}{a^3} \tag{4}$$

If this energy density is "purely electromagnetic", then

$$\frac{E}{V} = \frac{\hbar c \pi^2}{960} \frac{1}{a^3} = \sigma \epsilon \pi a^2 T^4 \tag{5}$$

where is the Stefan-Boltzmann Constant $\sigma = 5.67 \times 10^{-8}$ W/m^{2}K^4, and ϵ is the emissivity ($\epsilon = 1$). We have assumed the proton is a sphere (hence $S = \pi a^2$). This permits an estimate of the internal temperature of the proton as a Casimir Trap.

Let us solve for the internal Temperature T of a proton,

$$T = \left(\frac{\hbar c \pi}{960 \sigma \epsilon a^5} \right)^{\frac{1}{4}} = 2.2 \times 10^{13} \text{ K} \tag{6}$$

We can now estimate the internal energy of the proton as

$$E \sim \frac{1}{2} kT = 948.8 \text{ MeV} \sim 950 \text{ MeV} \tag{7}$$

where k is the Boltzmann Constant $k = 1.38 \times 10^{-23}$ J/K. The factor of 1/2 in this expression is understandable since the only degree of freedom in this model is the radius. We estimate the internal electromagnetic mass-energy of the proton to be 948.8 MeV/c^2, which is comparable to the measured energy of a proton, at 938.3 MeV/c^2.

The difference between the "predicted" and "measured" values is

$$\delta M = 948.8 - 938.3 \text{ MeV/c}^2 \sim 10.5 \text{ MeV/c}^2 \tag{8}$$

which represents the sum of the masses of the three free quarks inside the proton. We therefore hypothesize that for the sum of their free masses is on the order of 10.5 MeV/c^2.

If we accept the hypothesize that the sum of their free masses is on the order of 10.5 MeV/c^2 then we can proceed to use the differences between the measured mass of the neutron and proton to estimate the free masses of the Up (u) and Down (d) quark (refer to Table 1):

This comparison clearly shows us that the mass of the d quark is greater than the mass of the u quark. As a simple model we may consider

$$10.5 = 3m + 1.293 \cong 3m + 1.3 \tag{9}$$

which results in an upper bound for the u quark mass of

$$m \cong 3.1 \text{ MeV/c}^2 \tag{10}$$

Table 1. Mass comparison between Neutron and Proton

Particle	Quarks	Mass (MeV/c^2)
Neutron	ddu	939.565
Proton	duu	938.272
difference	$m_d - m_u$	1.293

and a lower bound to the d quark mass of

$$m_d = 3.1 + 1.3 = 4.4 \text{ MeV/c}^2 \tag{11}$$

These very simple estimates compare favorably with the Standard Values.

Let us imagine a quark bouncing off the walls of the Casimir Trap similar to how a ball might bounce off the inside of a round room. It was shown by both Newton and Huygens that the force experienced by the object is towards the center and hence is centripetal in nature. Let us assume the circular motion of quarks inside the Casimir Trap is governed by a centripetal force

$$F_{\text{cent}} = \frac{mv^2}{R} \sim -\frac{\partial E}{\partial r} \tag{12}$$

where $\langle E \rangle$ is the energy of the Casimir Trap.

$$-\frac{\partial E}{\partial r} = -\frac{\hbar c \pi^3}{720} \frac{1}{a^2} \tag{13}$$

Then (set aside the negative sign for it indicates the direction of the force)

$$\frac{mv^2}{R} \sim \frac{\hbar c \pi^3}{720} \frac{1}{2R^2} \tag{14}$$

When the quark bounds off the wall, consider an "instantaneous" velocity of $v = c$, then solving this expression for the mass we arrive at

$$m \sim \frac{\hbar c \pi^3}{2880} \frac{1}{R} \tag{15}$$

Again the only parameter is the radius R. Let $R = 0.75$ fm, then

$$m \sim 2.8 \text{ MeV/c}^2 \tag{16}$$

Let us assume this is the mass of the u Quark, then we see that the mass of the d Quark becomes

$$m_d = 10.5 - 2 \times 2.8 = 4.9 \text{ MeV/c}^2 \tag{17}$$

Again, these very simple estimates compare favorably with the Standard Values. In this estimate the difference in mass between the free d and free u quarks is 2.1 MeV/c^2.

It is understandable then that the lion's share of the mass of a proton is to be found in the high energy photons within the proton. It is a back of the envelope calculation to determine that the quarks are not traveling at the speed of light but at a rate of speed that is more like Brownian movement in a plasma.

2.2 Simulation Setup

To investigate the thermodynamic properties of a system of quarks and photons confined within a Casimir-like trap, we developed a 2D simulation using Pygame. The simulation consists of a rectangular container with a movable right wall (piston) and a fixed left wall. Inside the container, we have a number of "free" quarks and a number of "wall" quarks attached to the piston. The system also contains a large number of photons.

The simulation proceeds in discrete time steps, Δt. At each step, we update the state of the system by calculating the forces on each particle and updating their positions and velocities. The key components of the simulation are:

Piston Dynamics. The piston is modeled as a massive wall that can move horizontally. Its motion is governed by the net force acting on it, which is the sum of the impulse from particle collisions on its inner surface and an optional external pressure applied to its outer surface. The acceleration of the piston is given by $a_p = F_{net}/m_p$, where F_{net} is the net force and m_p is the mass of the piston.

Quark-Quark Interactions. While the strong force governs quark interactions at short distances, the long-range component of their interaction within the plasma is modeled using an effective Coulomb-like potential. This is a standard simplification in plasma physics appropriate for studying collective behavior:

$$\boldsymbol{F}_{ij} = k_e \frac{q_i q_j}{r_{ij}^2} \hat{r}_{ij} \tag{18}$$

where k_e is the Coulomb constant, q_i and q_j are the charges of the quarks, and r_{ij} is the distance between them. In addition to this force, direct quark-quark collisions are treated as elastic. This Newtonian simplification effectively models the conservation of momentum and energy required for thermalization without introducing the full complexity of QCD scattering, which is beyond the scope of this macroscopic model.

Quark-Photon Interactions (Compton Scattering). The interaction between photons and quarks is modeled using the Compton scattering effect, a fundamentally quantum mechanical process. This is a key feature, not a simplification, of our model. When a photon collides with a quark, its wavelength and direction change, and it imparts an impulse to the quark. The change in the photon's wavelength is given by:

$$\Delta\lambda = \frac{h}{m_q c}(1 - \cos\theta) \tag{19}$$

where h is Planck's constant, m_q is the mass of the quark, c is the speed of light, and θ is the scattering angle. The change in momentum of the photon is transferred to the quark, resulting in a change in the quark's velocity.

Particle-Wall Interactions. Quarks and photons collide with the walls of the container. Collisions with the fixed walls are treated as perfectly elastic. Collisions with the movable piston are modeled to impart an impulse to the piston, contributing to the internal pressure. For a particle with mass m and velocity component v_x perpendicular to the piston, the impulse delivered to the piston is $\Delta p = 2mv_x$.

Thermodynamic Variables. The simulation tracks several macroscopic thermodynamic variables:

- **Volume (V)**: The volume of the container is calculated as the area of the rectangular region, $V = L_x L_y$, where L_x is the position of the piston.
- **Temperature (T)**: The temperature of the system is derived from the total internal energy, which includes both the kinetic energy of the quarks and the energy of the photon gas. The total energy is given by:

$$E_{total} = \sum_{i=1}^{N_q} \frac{1}{2} m_q v_i^2 + \sum_{j=1}^{N_p} \frac{hc}{\lambda_j} \tag{20}$$

  The temperature is then related to the average energy per particle, reflecting the thermal state of the entire system.
- **Pressure (P)**: The pressure on the piston is calculated from the time-averaged impulse per unit area delivered by the particles colliding with it:

$$P = \frac{1}{A\Delta t} \sum \Delta p \tag{21}$$

  where A is the area of the piston and the sum is over all collisions in the time interval Δt.

3 Results and Discussion

The simulation allows us to study the emergent thermodynamic behavior of the confined quark-photon system. We have conducted several experiments to analyze the system's properties, beginning with its evolution towards a stable equilibrium.

3.1 System Equilibration and Thermalization

A fundamental prerequisite for any thermodynamic analysis is that the system must be able to reach a state of thermal equilibrium. To verify this, the simulation was initialized with a static piston wall (zero external pressure) and allowed to evolve freely. Figure 1 plots the evolution of the system's kinetic temperature and internal pressure as a function of simulation time.

As shown in the figure, the system undergoes an initial, highly dynamic phase lasting approximately 1.0 s of simulation time, during which the kinetic

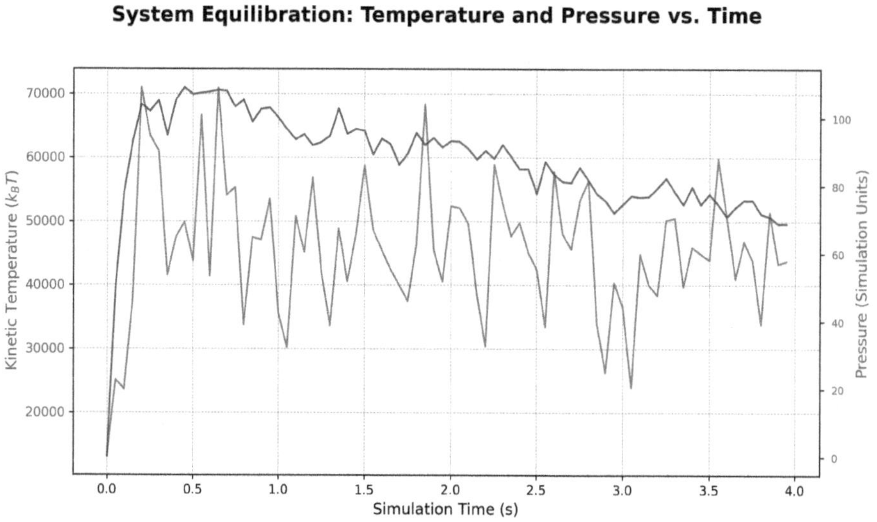

Fig. 1. Evolution of kinetic temperature (left axis, red) and internal pressure (right axis, blue) over time in an isolated system. After an initial transient phase of approximately 1.0 s, both quantities stabilize and fluctuate around a steady-state average, indicating the system has reached thermodynamic equilibrium. (Color figure online)

temperature and pressure rapidly increase as energy is distributed among the particles. Following this period, the macroscopic variables cease to show any systematic drift and instead exhibit stationary fluctuations around a well-defined average value. This behavior is characteristic of a system that has successfully reached a state of thermodynamic equilibrium.

Further evidence of thermalization is provided by the velocity distribution of the constituent particles. At equilibrium, the speeds of the quarks are expected to follow a Maxwell-Boltzmann distribution. Figure 2 shows a histogram of the quark speeds recorded at the end of the simulation run, once the system had stabilized.

The simulation data closely matches the characteristic shape of the Maxwell-Boltzmann distribution, confirming that the frequent interactions—Compton scattering between quarks and photons, and elastic collisions between quarks—are effective at thermalizing the system. Having established that our simulation can produce a stable, equilibrated quark-photon plasma, we can proceed with confidence to analyze its thermodynamic properties under different conditions.

3.2 Equation of State

To investigate the thermodynamic properties of the confined plasma, a series of simulations were run where a constant external pressure was applied to the piston, allowing the system to reach a new equilibrium state at a different volume. By varying the magnitude of the applied external pressure, we can map the

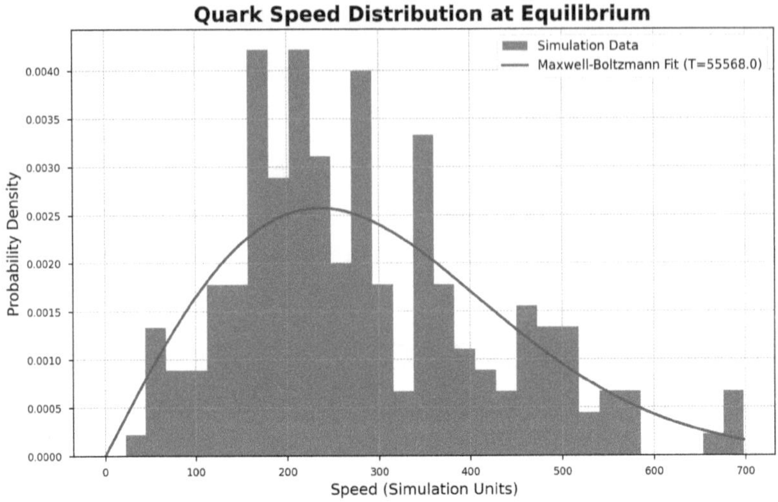

Fig. 2. Probability distribution of quark speeds at equilibrium. The histogram of the simulation data (blue bars) is well-described by a Maxwell-Boltzmann distribution fit (red curve), confirming that the system has achieved a state of thermal equilibrium. (Color figure online)

relationships between the system's pressure (P), volume (V), and temperature (T), effectively determining its equation of state. The results of this analysis are summarized in Fig. 3.

The P-V diagram in Fig. 3(a) reveals a distinct inverse relationship, consistent with the behavior of a compressible gas: as the volume is decreased by external pressure, the internal pressure of the quark-photon plasma increases. This result is a direct computational verification that the collective interactions within the Casimir Trap give rise to a state of matter that exerts pressure on its boundaries.

The thermodynamic response of the system extends to its temperature, as shown in Fig. 3(b). As the system is compressed to smaller volumes, the work done on it by the piston increases the total internal energy, which manifests as a higher kinetic temperature. This behavior is analogous to the adiabatic heating of a conventional gas.

Finally, the relationship between pressure and temperature, shown in Fig. 3(c), further characterizes the system's equation of state. The data suggests an approximately linear trend, akin to Gay-Lussac's law for an ideal gas. The deviations from a perfect linear fit can be attributed to the complex inter-particle interactions and the relativistic nature of the photon gas, which are not accounted for in the simple ideal gas model. These results collectively demonstrate that the confined quark-photon system in our simulation exhibits a coherent, predictable thermodynamic behavior that can be analyzed and characterized by a well-defined equation of state.

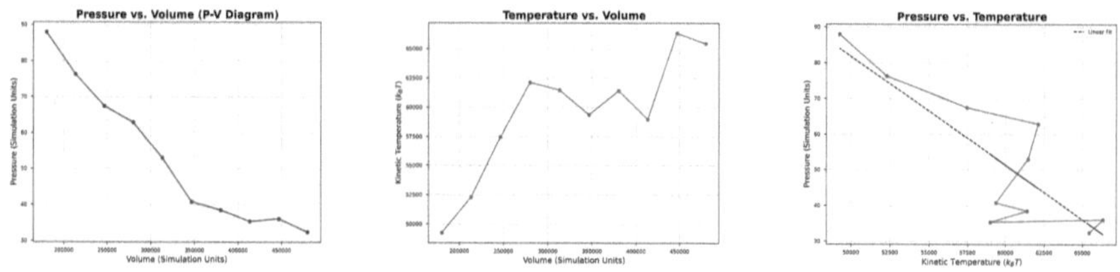

(a) Pressure vs. Volume (P-V)
(b) Temperature vs. Volume (T-V)
(c) Pressure vs. Temperature (P-T)

Fig. 3. Thermodynamic relationships of the simulated quark-photon plasma. Each data point represents an equilibrium state under a specific constant external pressure. (a) The P-V diagram shows a clear inverse relationship. (b) The T-V diagram demonstrates adiabatic heating upon compression. (c) The P-T diagram shows an approximately linear, positive correlation.

3.3 Dynamic Response to External Pressure

Beyond the static equilibrium properties, our simulation framework allows for the study of the system's dynamic response to external stimuli. To investigate this, we conducted a simulation in two distinct phases: an initial equilibration phase in an isolated container, followed by the sudden application of a large, constant external pressure to the piston. The resulting time-series data for the system's volume, temperature, and internal pressure are shown in Fig. 4.

The simulation begins with an equilibration period from $t = 0\,s$ to $t = 5\,s$. During this time, the volume remains constant at its maximum extent, and the temperature stabilizes as the system thermalizes. The internal pressure is low, punctuated by sporadic spikes corresponding to random, collective particle collisions against the piston face.

At $t = 5.0\,s$, indicated by the dashed vertical line, the external pressure is applied. The system's response is immediate and dramatic. The piston is driven inward, causing a sharp decrease in the container volume. The internal pressure exhibits a large, instantaneous spike as the piston motion begins, followed by continued high-amplitude fluctuations during the chaotic compression phase. The system eventually settles into a new equilibrium state at a significantly smaller volume, where the time-averaged internal pressure now balances the applied external force.

Interestingly, the kinetic temperature of the system shows a gradual decrease throughout the compression and relaxation phase. While adiabatic compression typically leads to heating, the observed cooling in our model can be attributed to the specific energy-exchange dynamics between the relativistic photon gas and the non-relativistic quarks. As the volume shrinks, the increased frequency of photon-wall collisions and Compton scattering events may provide an efficient mechanism for energy dissipation, leading to a lower final equilibrium temperature. This non-trivial result highlights the complexity of the thermodynamics within the simulated plasma and warrants further investigation. This entire

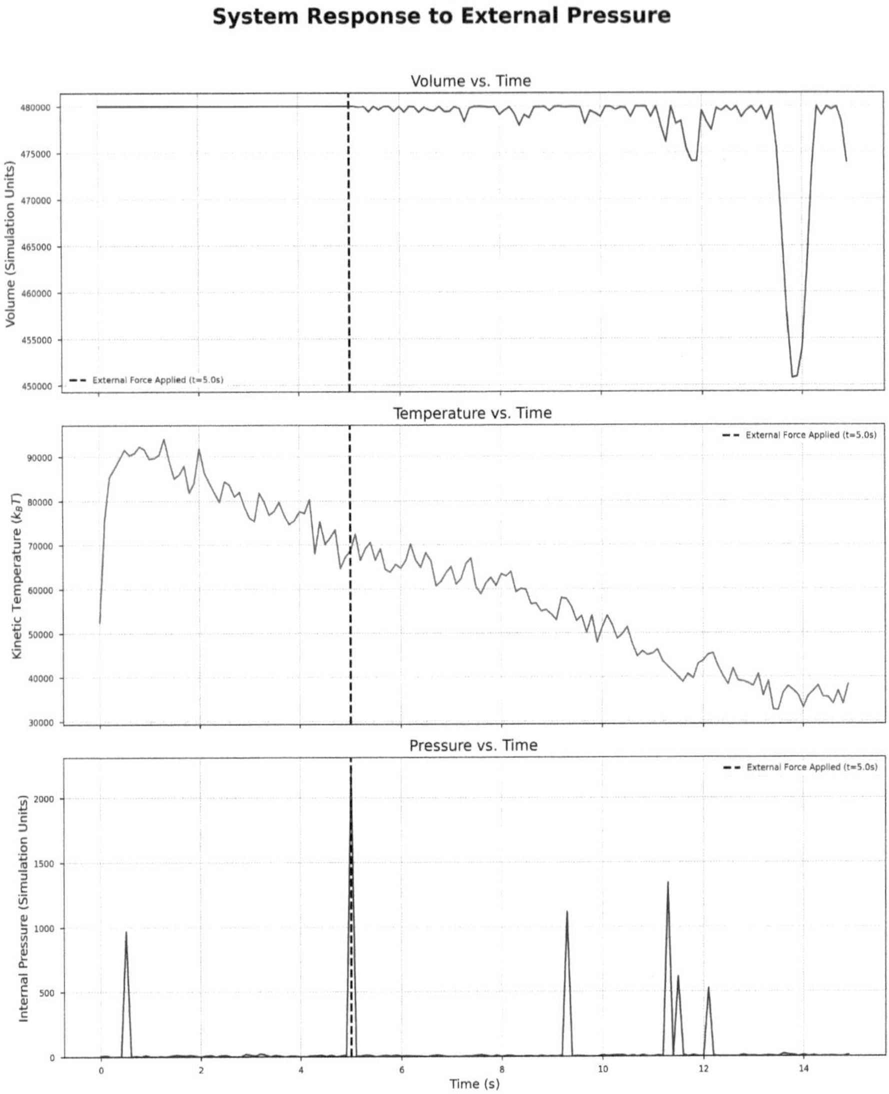

Fig. 4. Time-series analysis of the system's response to a suddenly applied external pressure at t = 5.0 s. The plots show Volume (top), Temperature (middle), and internal Pressure (bottom) over the full duration of the simulation. The system transitions from an initial equilibrium to a new, compressed equilibrium state.

process demonstrates the stability and responsiveness of the model, providing a direct visual and quantitative account of how the confined quark-photon gas behaves under external work.

4 Future Work

The current work establishes a foundational model for exploring the thermodynamics of a confined quark-photon system; however, it also opens several compelling avenues for future investigation. A more rigorous statistical analysis of the simulation data is a clear next step, which would enable the calculation of transport coefficients, correlation functions, and the specific heat to more fully characterize the plasma's properties. Furthermore, a systematic study of the system's behavior while varying key parameters—such as the quark-to-photon number ratio, the strength of the effective Coulomb interaction, and the piston's mass—would clarify the model's robustness and its sensitivity to initial conditions.

An important extension of this research involves exploring the system's dependence on particle composition. Future simulations will investigate configurations containing fewer than three quarks, such as two-quark (meson-like) systems. This will allow us to study the stability of such states within the Casimir Trap framework and examine whether the model provides a thermodynamic basis for the observed confinement of color-charged particles into color-neutral hadrons. Ultimately, the long-term goal is to refine the model, potentially by extending it to three dimensions, to allow for more direct quantitative comparisons with experimental data from heavy-ion collisions and results from first-principles lattice QCD calculations, further bridging the gap between phenomenological models and the fundamental theory of the strong interaction.

5 Conclusion

In this paper, we have presented a theoretical model of the proton as a Casimir Trap, which provides reasonable estimates for the masses of the up and down quarks. To further explore the dynamics of such a confined system, we have developed a 2D simulation of a piston containing quarks and photons. The simulation incorporates key physical interactions, including effective Coulomb forces, quantum mechanical Compton scattering, and elastic collisions.

Our simulation allows us to study the emergent thermodynamic properties of the system, such as temperature, pressure, and volume. We have demonstrated that the system reaches thermodynamic equilibrium and have begun to investigate its equation of state and response to external pressure. This work represents a promising step towards understanding the complex behavior of matter under extreme conditions, such as those found inside protons and in the early universe. Future work will focus on a more detailed statistical analysis of the simulation data and a systematic exploration of the parameter space.

References

1. Bhaduri, P.P., et al.: Smallest drop of QGP: thermodynamic properties of p-pb collisions. Phys. Rev. C **109**, 014904 (2024). https://doi.org/10.1103/PhysRevC.109.014904
2. Bhattacharyya, A., Chowdhury, S.K., Sanyal, S.: Medium effects in the MIT bag model for quark matter: self-consistent thermodynamical treatment. arXiv preprint arXiv:2311.07720 (2023)
3. Bimonte, G., et al.: Advances and prospects in casimir physics. Axioms **13**, 156 (2024). https://doi.org/10.3390/axioms13030156
4. Brambilla, N., et al.: QCD and strongly coupled gauge theories: challenges and perspectives. Eur. Phys. J. C **74**, 2981 (2014). https://doi.org/10.1140/epjc/s10052-014-2981-5
5. Bruskiewich, P.: Quark confinement. ResearchGate (2024). https://doi.org/10.13140/RG.2.2.14873.43365
6. Greensite, J.: Quantum Chromodynamics. World Scientific, Singapore (2003). https://doi.org/10.1142/4736
7. Hasenfratz, P., Kuti, J.: The quark bag model. Phys. Rep. **40**, 75–179 (1978)
8. Lüscher, M.: Lattice QCD — from quark confinement to asymptotic freedom. Lecture Notes (2002)
9. Qiu, Z., Gao, F., Liu, Y.X.: Dual chiral density wave induced oscillating casimir effect. Phys. Rev. D **110**, 014039 (2024). https://doi.org/10.1103/PhysRevD.110.014039
10. Rodriguez-Lopez, P., et al.: Science and technology of the casimir effect. Phys. Today **74**, 42 (2021). https://doi.org/10.1063/PT.3.4655
11. Tawfik, A.N., et al.: Pressure distribution inside nucleons in a Tsallis-MIT bag model. Universe **10**, 183 (2024). https://doi.org/10.3390/universe10040183

Using Llama 3.3 to Discover Novel Hypergraph Coarsening Heuristics

Yin Yee Williams(✉)

Imperial College London, United Kingdom SW7 2AZ, UK
yk18@ic.ac.uk

Abstract. We employ a large language model to generate novel hypergraph coarsening heuristics. The heuristics are evaluated by replacing the default coarsening heuristic within KaHyPar, a state-of-the-art multi-level hypergraph partitioner. We find that the LLM-generated heuristics discover improved solutions to around half of the hypergraphs in the benchmark hypergraph partitioning dataset.

Keywords: Multi-level Hypergraph Partitioning · Hypergraph Coarsening Heuristic · Large Language Models

1 Introduction

The utility of large language models (LLMs) as a research tool is an open question. We investigate the ability of Llama 3.3 70B Versatile to suggest novel hypergraph coarsening heuristics. In our framework, the tendency of LLMs to hallucinate is beneficial, as the aim is to produce novelty, as opposed to recalling facts. We tightly limit the scope of the LLM, and rely on an existing state-of-the-art partitioner to provide the scaffolding for evaluating our results against a known benchmark.

2 Definitions and Notation

Hyperedge. *Given a set of vertices V, a hyperedge e is defined as a subset of V. The null set $\emptyset$ as well as the set of all vertices V are valid hyperedges. The set of all hyperedges is denoted E.*
Vertex Weights and Hyperedge Weights. *Vertices and hyperedges can be weighted or unweighted. The scalar γ_v denotes the weight of vertex v; likewise ω_e denotes the weight of hyperedge e. The set of vertex weights is denoted Γ, and the set of hyperedge weights is denoted Ω.*

H. R. Arabnia et al. (Eds.): CSCI 2025, CCIS 2970, pp. 319–331, 2027.
https://doi.org/10.1007/978-3-032-28466-2_24

Hypergraph. *An unweighted hypergraph is a dyad* (V, E) *where* V *is a set of* n *unweighted vertices and* E *is a set of* m *unweighted hyperedges. The definition for weighted hypergraphs expands to a triad or a quad, accordingly, to include either* Γ, Ω, *or both.*

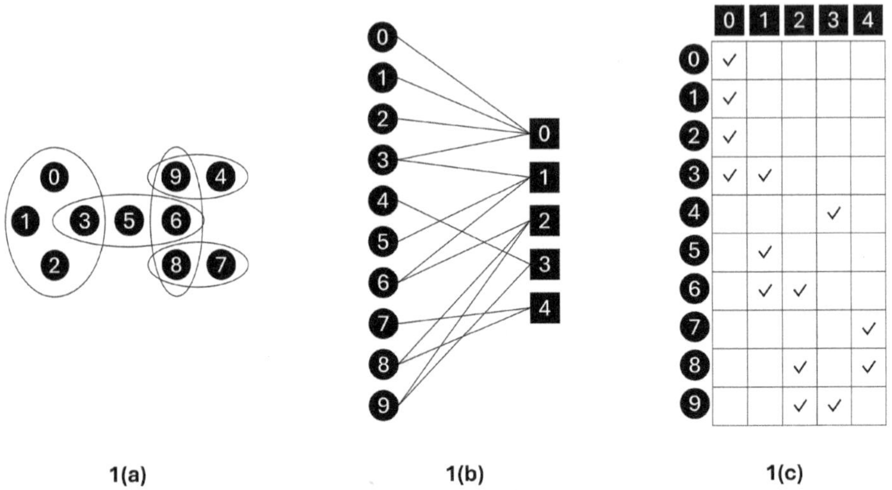

Fig. 1. Three ways of representing a hypergraph. Figure 1(a) shows a hypergraph as a set of sets. Vertices are depicted as solid circles, whilst hyperedges are depicted as ellipses. Figure 1(b) shows the same hypergraph in the form of a bipartite graph, where the vertices are depicted on the left as solid circles and the hyperedges are depicted on the right as solid squares. The lines connecting circles and squares indicate hyperedge membership. In Fig. 1(c), the same hypergraph is represented as an incidence matrix, where the solid circles and the solid squares depict vertices and hyperedges, respectively, as in Fig. 1(b). In this representation, the tick marks indicate hyperedge membership. Alternatively, the tick marks can be replaced by 1's and the empty elements be filled with 0's.

Hyperedge Size. *The size of a hyperedge is the number of vertices it contains, denoted by* $|e|$.
Number of Pins. *The number of pins in a hypergraph is the sum of all the hyperedge sizes, i.e.* $\Sigma|e|$ *for all* $e \in E$.
Partition. *A* k*-way partition* P *of a hypergraph* $H(V, E)$ *is an allocation of its vertices to* k *blocks,* $b_1, b_2, ..., b_k$, *such that* $b_1 \cup b_2 \cup ... \cup b_k = V$ *and* $b_i \cap b_j = \emptyset$.

Block weight *The weight of block* b_i *of a partition* P *is the sum of the weights of the vertices allocated to that block, i.e.* $\Sigma\gamma_j$ *for all* $v_j \in b_i$.

Balanced Partition. *A k-way partition P of an unweighted hypergraph is perfectly balanced if every block contains $\frac{n}{k}$ vertices. For a hypergraph with vertex weights $H(V, E, \Gamma)$, a partition is perfectly balanced if the weight of each block is $\frac{\Sigma\gamma}{k}$ for all $\gamma \in \Gamma$. We can relax the definition of balanced by introducing a predefined tolerance value ϵ such that if the weight of $b_i \leq (1+\epsilon)\frac{\Sigma\gamma}{k}$ for all $b_i \in P$, the partition is accepted as balanced.*
Cut-Set. *A hyperedge e belongs to the cut-set Q if there exist vertices $u \in e$ and $v \in e$ such that $u \in b_i$ and $v \in b_j$ where $i \neq j$.*
Balanced Hypergraph Partitioning Problem. *Given an unweighted hypergraph $H(V, E)$ and an integer k, return the ϵ-balanced k-way partition P^* that minimises the size of the cut-set, $|Q|$. In the same way, given a hypergraph with hyperedge weights $H(V, E, \Omega)$ and an integer k, return the ϵ-balanced k-way partition P^* that minimises the weight of the cut-set. The balanced hypergraph partitioning problem is known to be NP-hard* [6].

The multi-level hypergraph partitioning framework is the dominant paradigm for solving balanced hypergraph partitioning problems. The multi-level framework proceeds in three stages: coarsening, initial partitioning, and refinement. During the coarsening stage, the original hypergraph is simplified by merging together vertices. The term multi-level applies to the coarsening stage, with each level producing a smaller hypergraph. An initial partition is computed on the coarsest hypergraph. As the coarsening process is reversed, the partition is iteratively refined using the Fiduccia–Mattheyses (FM) [3] algorithm. KaHyPar [10] is a state-of-the-art open-source multi-level hypergraph partitioner written in C++.

3 Methodology

The FunSearch framework proposed by Romera-Paredes et al. [8] provides the basis of the methodology. We use the FunSearch framework to generate and evolve novel hypergraph coarsening heuristics. To do this, we implemented a single-threaded version of the framework[1] that sits within a copy of KaHyPar [10] (Fig. 2).

During the experiment, the LLM is provided with example heuristics and requested to produce a better heuristic, written in C++. KaHyPar is re-compiled every time a new heuristic is generated, and the generated heuristic is evaluated during the experiment by using the modified version of KaHyPar to produce solutions for a small set of medium-sized hypergraphs (Table 1).

The heuristic provided in the very first prompt is the Heavy-Edge heuristic[2] that is used by default in KaHyPar. Thereafter, the example heuristics are selected by sampling from the database. The sampling is biased towards better-performing and more concise heuristics.

[1] Available at https://github.com/yinyee/kahypar.
[2] The Heavy-Edge heuristic was proposed by Karypis and Kumar [5] in 1998 and has remained popular ever since.

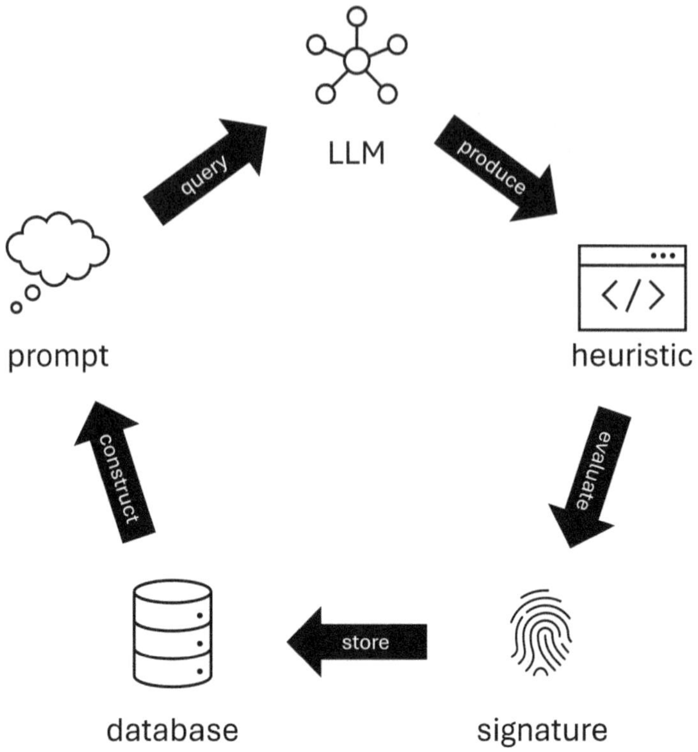

Fig. 2. A prompt containing examples is sent to the LLM. The LLM-generated heuristic is inserted into KaHyPar and indicatively evaluated on a small number of hypergraphs, producing a signature. The heuristic is stored in the database using its signature as a key. Stored heuristics are retrieved from the database to be used as examples for the prompt.

Heuristics are grouped according to their signature, a set of values that represent a particular heuristic's advantage over the Heavy-Edge coarsening heuristic on the aforementioned small set of hypergraphs used for online evaluation. The signature of a cluster is an ordered collection of values that capture the improvement in the number of cut hyperedges or the improvement in the balance of the resulting partition for each of the hypergraphs in the online evaluation set. If there is no improvement in either the number of cut hyperedges or the balance of the resulting partition, the value of zero is recorded for that particular hypergraph. The sum of the values in its signature is used to weight the sampling process, which is described in detail in the caption accompanying Fig. 4.

The large language model we study is the Llama 3.3 70B Versatile [7] provided by Groq [4] (Fig. 5).

Table 1. Hypergraphs in the online evaluation set.

Hypergraph	Number of vertices	Number of hyperedges	Number of pins
bundle1.mtx.hgr	10,581	10,581	770,901
Maragal_6.mtx.hgr	10,152	21,251	537,694
mult_dcop_01.mtx.hgr	25,187	25,187	193,276
sat14_6s11-opt.cnf.dual.hgr	97,312	33,276	227,060
Median of benchmark dataset	103,266	139,455	1,429,872

rating_score_policy.h

```
RatingType HeavyEdgeScore::score(const Hypergraph& hypergraph,
                                 const HyperedgeID he,
                                 const Context&) {

    return static_cast<RatingType>(hypergraph.edgeWeight(he)) /
        (hypergraph.edgeSize(he) - 1);
}
```

Fig. 3. The Heavy-Edge heuristic, provided as the example heuristic in the very first prompt to the LLM.

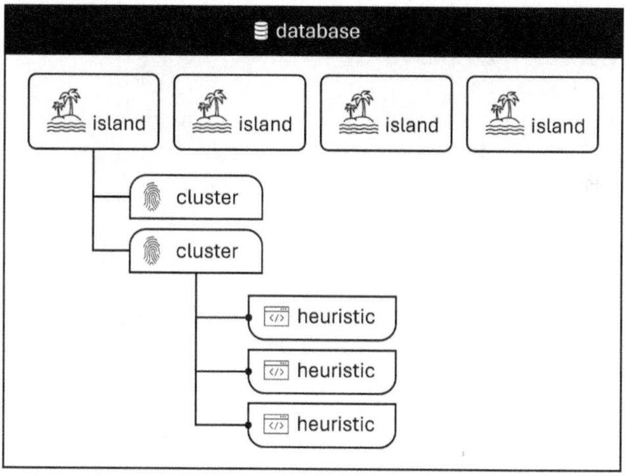

Fig. 4. The database is a collection of **islands**, each of which contains **clusters** of **heuristics**. When constructing a prompt with k example heuristics, one island is selected through uniform random sampling. Next, k clusters are selected with replacement; the sampling is biased in favour of clusters with higher scores. The score of a cluster is the sum of the values of the cluster's signature. Finally, a single heuristic is selected per cluster. Sampling at this level is biased in favour of shorter heuristics, where the length of a heuristic is the length of the C++ function without whitespaces. The value of k used is 2, except in the early stages of the experiment where there are fewer than two clusters in an island.

SYSTEM PROMPT

You are a brilliant and helpful mathematician and computer scientist. You are interested in multi-level hypergraph partitioning. Please completely enclose code between triple backticks (```) in your response.

USER PROMPT

Here are *k* algorithms for producing edge rating scores used for coarsening hypergraphs in the multi-level hypergraph partitioning paradigm:

```
RatingType EdgeScore_v0::score(…){…}

RatingType EdgeScore_v1::score(…){…}
```

After a hypergraph is coarsened, an initial partition is found. The initial partition is greedily refined as the coarsened hypergraph is uncoarsened so that we obtain a partition on the original hypergraph.

The partition on the original hypergraph is scored using the modified inverse ratio cut score, where the numerator is the product of the sum of the weight of the vertices in each part, and the denominator is the number of cut hyperedges plus one.

Please can you provide only one implementation of the EdgeScore function in C++ that performs better with respect to the modified inverse ratio cut score than the *k* provided. Please use the following function definition:

```
RatingType EdgeScore::score(const Hypergraph& hypergraph, const HyperedgeID he, const
Context& context)
```

Fig. 5. The system and user prompts provided to Llama3.3.

4 Experimental Results

The results presented in this section are from a single run, which produced a total of 30 heuristics[3]. The majority (24) of the heuristics generated can be grouped into four families with the following forms:

$$score_e = \begin{cases} \omega_e & \text{if } |e| \leq 2 \\ \frac{\omega_e}{(|e|+c)^r} & \text{if } |e| > 2 \end{cases} \qquad \text{(Type A)}$$

$$score_e = \begin{cases} \omega_e & \text{if } |e| \leq 2 \\ \omega_e(|e|+c)^r & \text{if } |e| > 2 \end{cases} \qquad \text{(Type B)}$$

$$score_e = \begin{cases} \omega_e & \text{if } |e| \leq 2 \\ \frac{\omega_e}{\log_2(r|e|+c)} & \text{if } |e| > 2 \end{cases} \qquad \text{(Type C)}$$

$$score_e = \begin{cases} \omega_e & \text{if } |e| \leq 2 \\ \frac{\omega_e}{\log_{10}(r|e|+c)} & \text{if } |e| > 2 \end{cases} \qquad \text{(Type D)}$$

Note from Fig. 3 that the Heavy-Edge heuristic is a close relative of the Type A family, with $r = 1$ and $c = -1$.

[3] The heuristics are available at https://github.com/yinyee/kahypar/tree/main/heuristics.

From each family, we select one heuristic on which to perform a full evaluation. The selection is based on its performance on the online evaluation set. The parameter values provided to KaHyPar are the same for both online and full evaluation runs (see Fig. 3), except that the online evaluation runs use the default value of -1 for the random seed (Table 2).

Table 2. Heuristics selected from each family and their parameter values.

Family	Id	r	c
Type A	llama-1	0.7	-1
Type B	llama-26	0.2	0
Type C	llama-14	1	1
Type D	llama-6	10	0

We perform a full evaluation of the selected heuristics using the hypergraph partitioning benchmark dataset compiled by Schlag [9], by producing ten runs per hypergraph for each heuristic, resulting in 4,880 runs for each heuristic[4].

Table 3. The parameter values provided to KaHyPar for each run.

Parameter	Option	Value
Number of parts	-k	2
Imbalance limit	-e	0.03
Objective	-o	cut
Mode	-m	direct
Random seed	–seed	[0, 9]
Time limit	–time-limit	900
Config file	-p	`cut_kKaHyPar_llm.ini`

There is more than one way of defining which heuristic is the best. We first consider the definition of *best* as the heuristic which produces a partition with the smallest cut-set out of ten attempts.

The pie chart in Fig. 6 allocates each of the hypergraphs in the benchmark dataset to a category, based on how many and which heuristics were able to produce a partition with the smallest cut-set. For just under half (45%) of the dataset (219 hypergraphs), the best solution is found at least once by all four Llama-generated heuristics as well as the Heavy-Edge (HE) heuristic.

For the remaining hypergraphs, the HE heuristic finds the best solution only a third of the time (89 out of 269 hypergraphs). Together, the Llama-generated

[4] The results of those runs are available at https://zenodo.org/records/18164672.

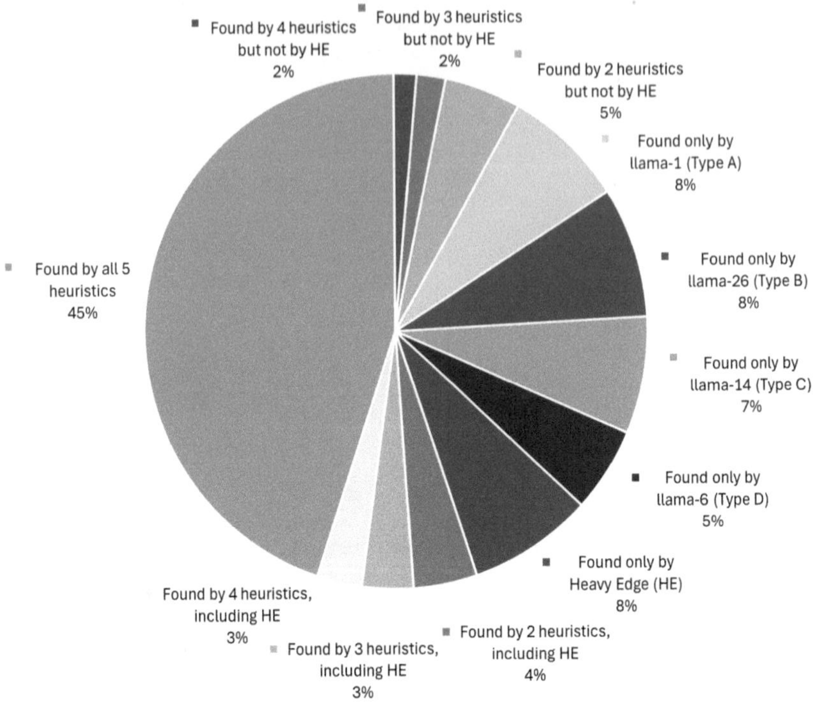

Fig. 6. Diversity produces better results. None of the heuristics significantly outperform the others; notably this applies also to the default coarsening heuristic used by KaHyPar (HE). Yet, taken together, the Llama-generated heuristics account for improvements to about a third of the benchmark dataset, suggesting that a portfolio of coarsening heuristics may be a superior approach to employing a single heuristic.

heuristics account for improvements to about a third (37%) of the entire benchmark dataset, suggesting that a portfolio of coarsening heuristics may be a superior approach to employing a single heuristic.

The pie chart in Fig. 6 presents a viewpoint comparing the HE heuristic against the Llama-generated heuristics. Setting aside the 219 hypergraphs for which the best solution was found by every heuristic, we subject each heuristic to the same comparison. The bar chart colours in Fig. 7 carry the same meaning as the pie chart in Fig. 6. The pink and purple pie slices from Fig. 6 are represented collectively by the new palest orange stacked column.

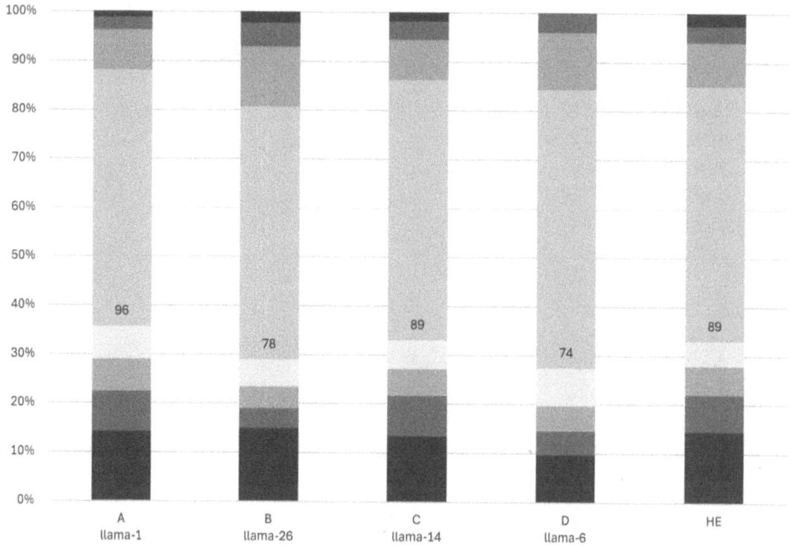

Fig. 7. Each heuristic compared in turn against the other four heuristics using the definition of *best* as the solution that results in the smallest cut-set. The colours in this chart carry the same meaning as in Fig. 6. The pink and purple pie slices from Fig. 6 are represented collectively by the palest orange stacked column. The numbers on the chart are the sum of the green stacked columns, meaning that particular heuristic produced the best solution, whether singularly or together with one or more other heuristics.

The results in Fig. 7 can be interpreted as a ranking of the heuristics, with Llama-1 (Type A) coming in first place, HE and Llama-14 (Type C) tie for second place, Llama-26 (Type B) third, and Llama-6 (Type D) in last position.

Setting New Per-instance Results for Smallest Cut-Set Size

Bustany et al. [2] published their results on the ISPD98 circuit benchmark suite [1], which is a subset of the benchmark dataset used to produce the results above. For ten out of eighteen hypergraphs, the novel heuristics produce a new best solution, with respect to size of cut-set.

Additionally, we observe that Llama-6 (Type D) produces the highest number of best solutions. We conjecture that there exists classes of hypergraphs for which certain families of heuristics are capable of producing superior solutions (Fig. 8).

Hypergraph	Spec Part	New	llama-1 Type A	llama-26 Type B	llama-14 Type C	llama-6 Type D	HE
ISPD98_ibm01.hgr	203	202	✓	✓	✓	✓	✓
ISPD98_ibm02.hgr	341	337	✓				
ISPD98_ibm03.hgr	954	954			✓	✓	
ISPD98_ibm04.hgr	580	580				✓	✓
ISPD98_ibm05.hgr	1719	1707					✓
ISPD98_ibm06.hgr	976	977					
ISPD98_ibm07.hgr	910	900	✓	✓		✓	✓
ISPD98_ibm08.hgr	1140	1140	✓	✓	✓	✓	✓
ISPD98_ibm09.hgr	625	620	✓	✓	✓	✓	✓
ISPD98_ibm10.hgr	1258	1313					
ISPD98_ibm11.hgr	1065	1062	✓	✓	✓	✓	✓
ISPD98_ibm12.hgr	1934	1932	✓		✓		
ISPD98_ibm13.hgr	837	848					
ISPD98_ibm14.hgr	1842	1842		✓		✓	
ISPD98_ibm15.hgr	2743	2741	✓	✓	✓	✓	✓
ISPD98_ibm16.hgr	1975	1898				✓	
ISPD98_ibm17.hgr	2314	2310	✓	✓	✓	✓	✓
ISPD98_ibm18.hgr	1521	1704					

Fig. 8. Comparison of size of cut-set produced by SpecPart [2] (2022) versus HE and the heuristics produced by Llama3.3. The highlighted cells indicate a smaller cut-set compared to SpecPart, and the tick marks indicate which of the heuristics achieved that result.

Another definition of *best* takes into account how well-balanced the partitions are, in addition to the size of the cut-set. For this, we propose the *modified inverse ratio cut* metric, which is slightly altered from Wei and Cheng [11].

$$\text{Modified inverse ratio cut} = \frac{\Pi_0^k |b_i|}{|Q| + 1}$$

This is a maximising form of the metric, where the denominator is largest when both parts are perfectly balanced and where the numerator increases as the size of the cut-set decreases.

Looking at the results through this lens, we find that the number of *best* solutions found by only one heuristic increases from 179 to 235, or from 37% to 48% of the benchmark dataset. All heuristics see an increase, but the increase is most pronounced for Llama-1 (Fig. 9).

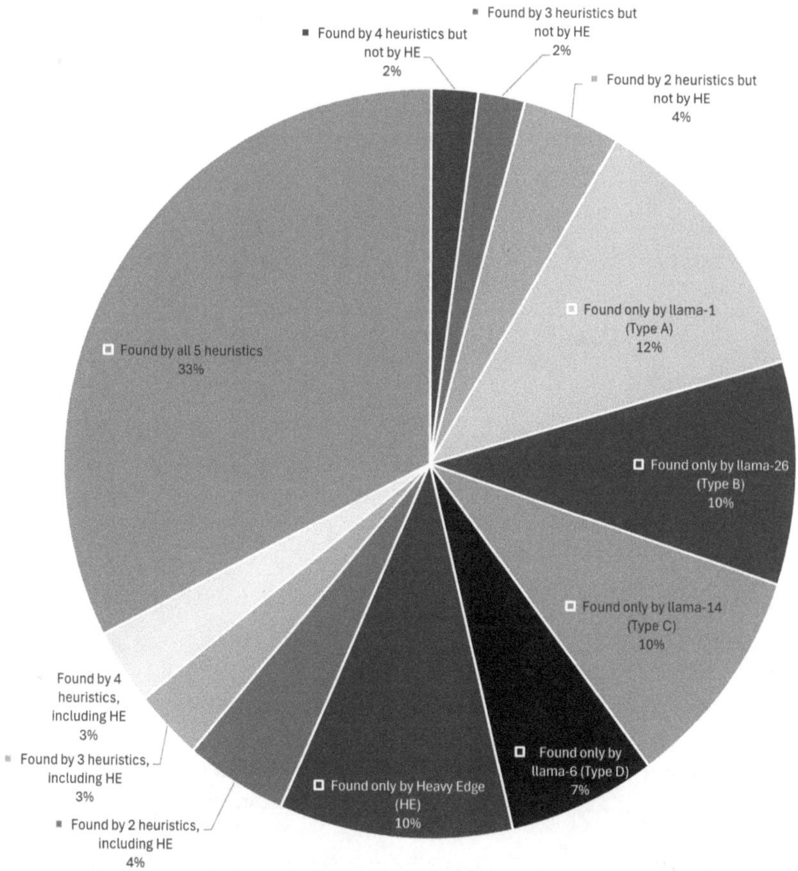

Fig. 9. Using the modified inverse ratio cut as a metric of quality increases the heterogeneity of the best solutions found. Improvements in solution quality from more balanced partitions are obscured when the quality metric used is the size of the cutset.

Using the same method for comparing each heuristic against the others as in Fig. 7, but with *best* defined as the partition with the largest modified inverse ratio cut value, we find that the ranking of the heuristics changes slightly. Llama-1 (Type A) extends its lead in the first place, but Llama-14 (Type C) pips HE to second place. Llama-6 (Type D) and Llama-26 (Type B) have swapped places to fourth and fifth, respectively (Fig. 10).

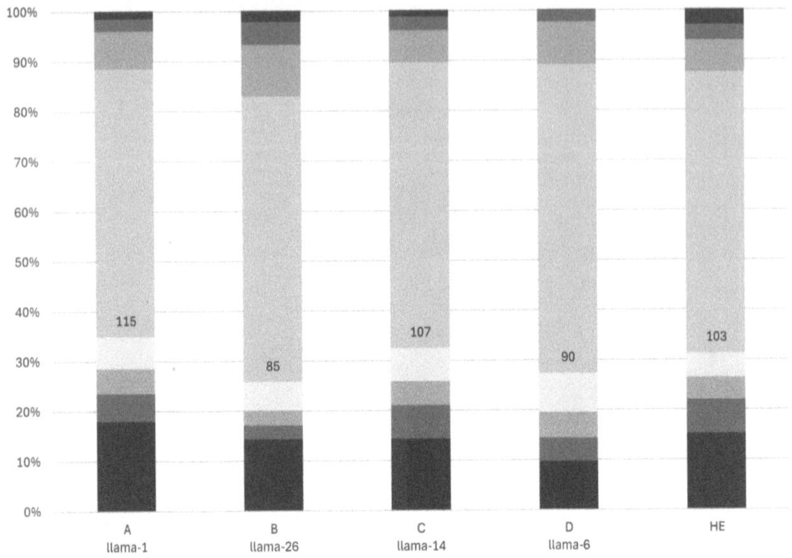

Fig. 10. The ranking of the heuristics is affected by taking balance into account. Llama-1 extends its lead, whilst Llama-14 pips HE to second place.

5 Conclusions and Future Work

We are able to produce higher-quality results for just under half of the benchmark dataset, by using different coarsening heuristics than the default Heavy-Edge heuristic.

Developing new families of heuristics appears to be a fruitful line of enquiry, and the use of LLMs has been helpful in that endeavour. We also note that employing a portfolio of coarsening heuristics is likely to yield higher-quality partitions than relying on a single heuristic.

Given that Llama-1 and HE are both variants of the same family (Type A), the results indicate that there could be merit in trying to fine-tune the parameter values for a set of hypergraphs, particularly if the hypergraphs originate from the same source and are therefore more likely to have similarities in structure.

Acknowledgments. This research was produced during a studentship funded by Amadeus IT Services UK Ltd.

References

1. Alpert, C.J.: The ISPD98 circuit benchmark suite. In: Proceedings of the 1998 international symposium on Physical design, pp. 80–85 (1998)
2. Bustany, I., Kahng, A.B., Koutis, I., Pramanik, B., Wang, Z.: SpecPart: a supervised spectral framework for hypergraph partitioning solution improvement. In: Proceedings of the 41st IEEE/ACM International Conference on Computer-Aided Design, pp. 1–9 (2022)

3. Fiduccia, C.M., Mattheyses, R.M.: A linear-time heuristic for improving network partitions. In: Proceedings of the 19th Design Automation Conference (1982)
4. GROQ: Llama 3.3 70B model card (2024). https://console.groq.com/docs/model/llama-3.3-70b-versatile
5. Karypis, G., Kumar, V.: A fast and high quality multilevel scheme for partitioning irregular graphs. SIAM J. Sci. Comput. **20**(1), 359–392 (1998)
6. Karypis, G., Kumar, V.: Multilevel K-way hypergraph partitioning. In: Proceedings of the 36th Annual ACM/IEEE Design Automation Conference, pp. 343–348 (1999)
7. Meta: Llama 3.3 model card (2024). https://github.com/meta-llama/llama-models/blob/main/models/llama3_3/MODEL_CARD.md
8. Romera-Paredes, B., et al.: Mathematical discoveries from program search with large language models. Nature **625**(7995), 468–475 (2024)
9. Schlag, S.: Benchmark hypergraphs and detailed experimental results of "k-way hypergraph partitioning via n-level recursive bisection". https://zenodo.org/records/30176. https://doi.org/10.5281/zenodo.30176
10. Schlag, S., Heuer, T., Gottesbüren, L., Akhremtsev, Y., Schulz, C., Sanders, P.: High-quality hypergraph partitioning. ACM J. Exp. Algorithmics **27**, 1–39 (2023)
11. Wei, Y.C., Cheng, C.K.: Ratio cut partitioning for hierarchical designs. IEEE Trans. Comput. Aided Des. Integr. Circuits Syst. **10**(7), 911–921 (2002)

Internet of Things and Applications

An Efficient Feature Selection Framework Using Genetic Algorithms for AI-Driven IDS in IoT Environments

José Carlos García-Merino[1,2(✉)], Llanos Tobarra[2], Antonio Robles-Gómez[2], Rafael Pastor-Vargas[2], and Pedro Vidal-Balboa[3]

[1] Programa de Doctorado en Sistemas Inteligentes Escuela Internacional de Doctorado UNED (EIDUNED), Universidad Nacional de Educacion a Distancia (UNED) Madrid, Madrid, Spain
jc.garcia@scc.uned.es

[2] Departamento de Sistemas de Comunicación y Control, ETSI Informática, Universidad Nacional de Educación a Distancia (UNED) Madrid, Madrid, Spain
{llanos,arobles,rpastor}@scc.uned.es

[3] Consorcio CA UNED en Ponferrada / INTECCA Ponferrada, Ponferrada, Spain
pvidal@intecca.uned.es

Abstract. The increasing concern over Internet of Things (IoT) cybersecurity, driven by the growing number of connected devices and their vulnerability to attacks, has intensified the demand for effective Intrusion Detection Systems (IDSs). In recent years Artificial Intelligence (AI) solutions has emerged as a common approach in modern cybersecurity systems, offering powerful tools for detecting complex patterns and threats. However, the limited computing capabilities of IoT environments require models that are not only accurate but also resource-efficient. Feature selection (FS) plays a key role in this context, but traditional techniques such as the χ^2 test and mutual information focus only on predictive performance, overlooking important practical concerns like inference time and model size. In this work, we propose a novel FS approach based on a custom scoring function that combines accuracy with penalties for inference time and model size. This metric is integrated into a Genetic Algorithm framework to guide the search for optimal feature subsets. The proposed method is evaluated on the NF-ToN-IoT-v2 dataset against standard techniques. Results demonstrate that our approach achieves competitive predictive performance while significantly reducing inference time and model size, highlighting the relevance of considering deployment-oriented metrics in the FS process.

Keywords: Artificial Intelligence (AI) · Intrusion Detection Systems (IDSs) · Feature Selection · Internet of Things (IoT) · Model Efficiency

H. R. Arabnia et al. (Eds.): CSCI 2025, CCIS 2970, pp. 335–344, 2027.
https://doi.org/10.1007/978-3-032-28466-2_25

1 Introduction

Advances in digital technology and the development of increasingly smaller electronic components have led to the widespread adoption of Internet of Things (IoT) technologies across a broad spectrum of areas. However, this growth has not been accompanied by improvements in system protection and security mechanisms. The information provided by these devices is often sensitive and therefore requires careful protection. The wide variety of IoT standards and computational constraints, as well as the complexity of IoT environments, highlight the need for new techniques and methodologies to analyze IoT networks from a security perspective.

AI-based IDSs solutions have become increasingly common in modern cybersecurity systems. However, as highlighted by Rafique et al. [14], most existing research focuses on improving predictive performance, often ignoring the importance of model efficiency for practical deployment in IoT environments. For example, Deep Learning (DL) approaches have gained popularity in recent years due to their ability to leverage large volumes of data and eliminate the need for manual feature engineering [17]. However, such models are often computationally expensive, which hinders their applicability in IoT scenarios [8,12]. Consequently, it is essential to adopt AI techniques that are efficient in terms of computational burden. In this respect, Wu et al. [18] compared various AI models for anomaly detection in IoT networks -including DL and traditional Machine Learning (ML) approaches— and concluded that Random Forests (RF) offered the best trade-off between accuracy, memory consumption, and inference speed. Similarly, Eid et al. [10] reached the same conclusion, highlighting RFs as the best-performing compromise among competing approaches. However, although RFs provide solid results, there remains potential to further refine processes and optimize room to reduce complexity and optimize resource usage for real-world deployment.

In this respect, despite the well-known sensitivity of ML algorithms to the quality and relevance of the features used for training, relatively little attention has been paid to the role of FS during the model preprocessing stage [1]. Furthermore, although some authors have explored the use of FS techniques for model improvement (see for example [2,3,9]), they do not take into account model efficiency. However, as discussed above, the effectiveness of FS should not be measured solely by the improvement of model accuracy, but also by its impact on model efficiency. To address this gap, this work introduces a custom scoring function that considers not only classification accuracy but also model size and inference time. The incorporation of this metric into a Genetic Algorithm (GA) framework allows to guide the FS process towards feature subsets that result in efficient models without compromising predictive power. The effectiveness of the suggested approach is assessed by fitting RF models on the NF-ToN-IoT-v2 dataset, a benchmark dataset with a standardized set of features specifically designed for intrusion detection in IoT environments. The proposed methodology is also compared with many of the state-of-the-art FS techniques existing in the literature.

The rest of the paper is organized as follows. First, the established FS methods in the literature are briefly reviewed in Sect. 2. Then, the custom scoring function introduced in this work is described in Sect. 3. Details of the NF-ToN-IoT-v2 dataset are discussed in Sect. 4. Numerical results are presented in Sect. 5. Finally, Sect. 6 concludes the paper with future directions.

2 Feature Selection Approaches

The purpose of this section is to briefly describe some of the main FS methods commonly employed in the ML literature.

1. **χ^2 test.** The χ^2 test is a simple and widely used statistical method for determining whether a pair of random variables are statistically independent. The χ^2 statistic quantifies the extent to which the observed data deviate from the expected distribution under the assumption of independence. In the context of FS, the test is applied to each feature individually, producing a score that reflects the strength of its statistical dependence on the target variable. Predictors with higher χ^2 values are considered more informative, as they reflect stronger statistical dependence on the target [11].
2. **Mutual Information.** Mutual Information (MI) is a non-parametric statistical measure used to quantify the amount of shared information between two random variables. Formally, MI measures the divergence between the joint distribution $P(X, Y)$ and the product of the marginal distributions $P(X)P(Y)$. When applied to FS, MI is computed between each predictor and the target variable, producing a score that quantifies how much knowing a particular feature reduces uncertainty about the target [6].
3. **Permutation Feature Importance.** Permutation Feature Importance (FI) is a metric used to assess the impact of randomly shuffling the values of a feature on the accuracy of a model. If the permutation process causes a noticeable increase in prediction error, it indicates that the model relies on that feature to make accurate predictions. Conversely, if the shuffle has little or no effect, the feature is likely unimportant [6].
4. **Principal Component Analysis.** Principal Component Analysis (PCA) stands for a dimensionality reduction method that transforms the input features into a new set of orthogonal variables called principal components. These components are constructed as linear combinations of the original features and are ranked according to the proportion of variance they explain in the dataset [4].
5. **Metaheuristic Algorithms.** Metaheuristic algorithms are optimization techniques designed to efficiently explore large and complex search spaces, often inspired by natural or physical processes. In this work, a GA is employed as a representative example of the application of metaheuristic approaches for FS. Within the GA framework, feature subsets are encoded as binary vectors of length equal to the number of features, where a '1' denotes that the feature is selected and a '0' that it is not. An initial population of random vectors then

evolves over generations based on a metric that evaluates model performance. Random mutation enhances exploration, and the process stops upon meeting a convergence criterion, usually a predefined fitness level or number of generations [13]. To avoid bias in the FS process, the algorithm is performed solely on the training set, with cross-validation guiding the evaluation of candidate solutions.

The main advantage of metaheuristic algorithms is their flexibility. The fitness function used to evaluate each subset of features is not restricted to traditional metrics such as accuracy or F1-score, but can also incorporate deployment-oriented criteria, such as inference time or model size. However, this possibility remains largely unexplored in the literature [16]. To address this gap, this work proposes a methodology that naturally integrates the search for efficient feature subsets into the optimization process. A simple custom metric designed to account for both predictive performance and computational efficiency is incorporated into the GA to guide feature selection. The scoring function used in the numerical experiments is detailed below in Sect. 3.

3 Genetic Algorithm with Cost-Oriented Metric for Feature Selection

To jointly evaluate predictive performance, inference speed, and model size, we define the following metric:

$$\text{score} = \frac{\text{accuracy}}{\text{accuracy}_{\text{base}}} - \lambda_{\text{inf}} \cdot \frac{t_{\text{inf}}}{t_{\text{base}}} - \lambda_{\text{size}} \cdot \frac{s}{s_{\text{base}}} \tag{1}$$

where *accuracy* refers to the classification performance of the model trained on the selected feature subset (see Eq. 2); t_{inf} denotes the time required for the model to make a prediction and s indicates the model size in megabytes (MB), computed by saving the trained model to disk. The subscript "base" refers to the values obtained from a predefined baseline model used for normalization, which in this work is the model trained on the full set of features. Finally, λ_{inf} and λ_{size} are weighting hyperparameters that control the penalization of inference time and model size in the final score.

In this work, only inference time and model size have been included in the scoring function. Nonetheless, it is important to note that the proposed framework is flexible and could easily be extended to include other resource-related metrics, such as energy consumption or RAM memory usage, which are particularly relevant in IoT environments.

4 NF-ToN-IoTv2 Dataset

Selecting adequate datasets is essential when designing AI-based IDSs. However, most IoT cybersecurity works utilize generic datasets that fail to reflect the specific characteristics of IoT traffic. According to the review by Aversano et al. [7], the most commonly employed datasets in the literature are NSL-KDD, UNSW-NB15, and KDD Cup 99, none of which were specifically designed for real IoT scenarios. Furthermore, it is key to ensure diversity in both data volume and attack types to conduct a robust analysis. For these reasons, the NF-ToN-IoT-v2 dataset was in this study selected as the benchmark. NF-ToN-IoT-v2 is an enhanced NetFlow-based dataset derived from the original ToN-IoT dataset [15]. ToN-IoT is a realistic dataset collected from a large-scale heterogeneous network that includes a wide range of IoT devices and services [5]. The dataset covers a comprehensive set of cyberthreats, including DDoS/DoS, injection, scanning, password attacks, ransomware, XSS, MITM, and backdoor attacks, in addition to benign traffic. The preference for the NetFlow version is motivated by two factors: first, the adoption of a standardized NetFlow feature set, which supports more generalizable analyses; and second, the demonstrated improvement in multi-class classification performance of ML models when using NetFlow-based features in comparison to the original ones.

5 Numerical Results

This section presents a case study to demonstrate the effectiveness of the proposed cost-oriented feature selection methodology. To this end, a GA was used to select features for training a RF model employing the custom metric defined in Eq. 1. The hyperparameters λ_{inf} and λ_{size} were set to 0.2 and 0.1, respectively. The GA was configured with a population size of 60 and evolved over 5 generations. The results have been compared against those obtained using the standard techniques introduced in Sect. 2. To ensure a fair comparison across methods, three feature rankings were generated using the χ^2 test, MI, and FI criteria, as described in Sect. 2. Similarly, PCA was applied to compute the principal components of the original feature space. Each method was then evaluated incrementally using a sequential forward selection strategy, where features (or components) were added one by one to the RF model based on their ranking. Finally, a GA was executed to identify feature subsets that maximize accuracy, following the classic approach. All resulting models were assessed in terms of accuracy, inference time, and model size on disk using a hold-out validation scheme (80% training, 20% testing). Since inference time cannot be measured with absolute precision, the prediction step was repeated ten times on the test set, and the median value of all runs was recorded as the model inference time.

After this process, the best-performing models produced by each method were identified, prioritizing either the ones with highest accuracy and those which present the best trade-off between accuracy and resource usage. The selected models were then evaluated in more detail using precision, recall, and F1-score metrics, defined in Eq. 2:

Table 1. Performance comparison on the NF-ToN-IoT-v2 dataset: proposed vs. standard FS methods. For each classic method, results are reported for the most accurate model (Acc.) and the best accuracy-efficiency balanced configuration (Bal.).

Method	#Feat.	Acc.	Prec.	Rec.	F1	Size (MB)	Inf (s)	Train (s)
Baseline	40	97.87	97.88	97.87	97.85	979	2.67	118
χ^2 test (Acc.)	38	97.87	97.88	97.87	97.86	973	2.61	120
χ^2 test (Bal.)	14	97.73	97.74	97.73	97.72	1054	2.37	111
MI (Acc.)	36	97.87	97.88	97.87	97.86	968	2.57	116
MI (Bal.)	26	97.81	97.81	97.81	97.80	778	2.36	118
FI (Acc.)	26	97.87	97.89	97.87	97.86	955	2.38	114
FI (Bal.)	11	97.75	97.78	97.75	97.75	440	1.57	63
PCA	17	97.14	97.15	97.14	97.12	988	2.56	362
GA (Classic)	36	97.87	97.88	97.87	97.86	970	2.55	113
GA (Proposed)	**10**	**97.67**	**97.70**	**97.67**	**97.67**	**403**	**1.44**	**58**

$$\textbf{Accuracy} = \frac{TP + TN}{TP + TN + FP + FN} \qquad \textbf{Precision} = \frac{TP}{TP + FP}$$
$$\textbf{F1-score} = 2 \cdot \frac{\text{Precision} \cdot \text{Recall}}{\text{Precision} + \text{Recall}} \qquad \textbf{Recall} = \frac{TP}{TP + FN} \tag{2}$$

where TP stands for true positives, TN for true negatives, FP for false positives, and FN for false negatives. All the numerical tests have been conducted in a computer Intel(R) Core(TM) i9-10900X CPU@3.70 GHz with 64 GB of RAM memory.

Table 1 summarizes the performance of the evaluated feature selection methods in terms of classification accuracy, model size on disk, inference time, and training time. The proposed approach achieves an excellent trade-off between predictive performance and resource efficiency, outperforming all baseline techniques in both inference speed and model compactness. In particular, the accuracy loss compared to the baseline model is only 0.2%, while inference time and model size are reduced by 46.5% and 58.8%, respectively. Among the models obtained by classical techniques, the FI–based model with 11 selected features provided by far the most favorable balance. However, the model derived from the proposed approach still outperforms it in both inference time and model size by over 8%, with a minimal accuracy loss of less than 0.1%.

It is also worth noting that the model built using the proposed feature subset achieves the lowest training time among all evaluated configurations. In general, training time is less critical, since models are typically trained on high-performance machines before being deployed in IoT environments. However, this improvement, together with the smaller data dimensionality, could ease the use of training strategies closer to the edge.

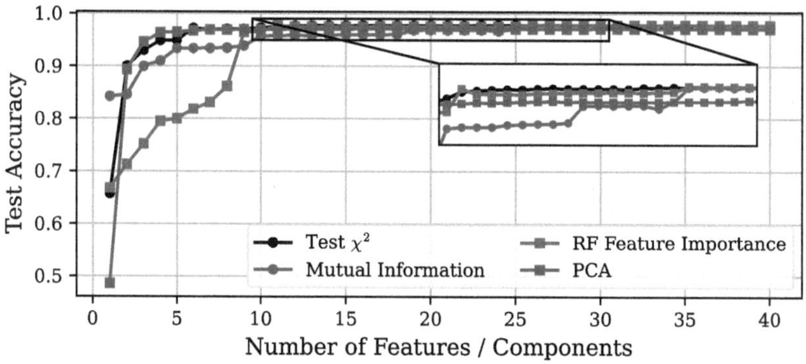

Fig. 1. Test accuracy vs number of features or principal components using different DR techniques on the NF-ToN-IoTv2 dataset.

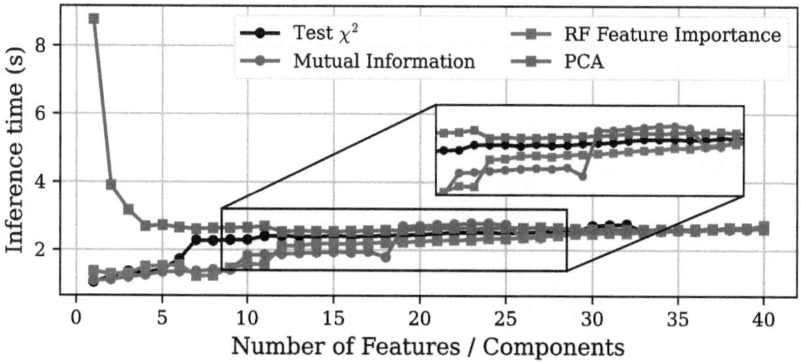

Fig. 2. Inference time vs number of features or principal components using different DR techniques on the NF-ToN-IoTv2 dataset.

A more comprehensive view of the results can be obtained by examining the evolution of the relevant metrics during the sequential forward selection process applied to the classical methods described above. In this respect, Figs. 1, 2, and 3 depict the evolution of model performance for the χ^2 test, MI, FI, and PCA approaches in terms of accuracy, inference time, and model size, respectively. Note that models with similar predictive performance can differ substantially in terms of resource consumption. This fact further supports the assertion that model efficiency should be explicitly considered during the FS process.

5.1 Deployment on Raspberry Pi

Beyond the previous analyses in a high-performance setting, additional tests were conducted on devices with limited resources to evaluate deployment in a realistic IoT environment. A Raspberry Pi equipped with an ARM Cortex-A72 processor (quad-core, 64-bit, 1.8 GHz) and 8 GB of RAM was used to assess the

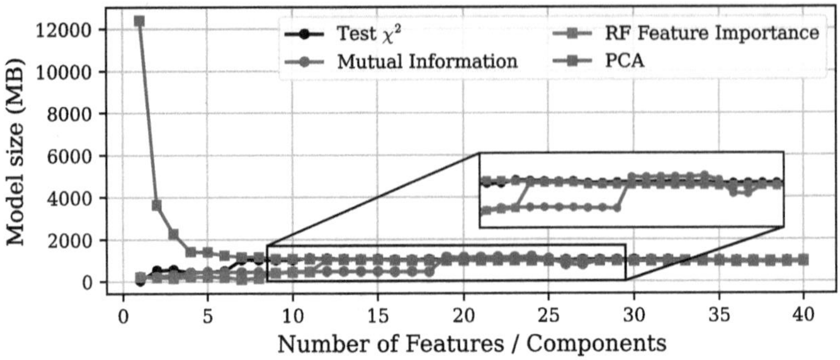

Fig. 3. Test accuracy vs number of features or principal components using different DR techniques on the NF-ToN-IoTv2 dataset.

selected models. The evaluation focused on two aspects: inference time and RAM consumption, the latter measured considering both the actual memory in use and the memory allocated during execution. To ensure reliable measurements, each model was run ten times and the median values for inference time and memory usage were retained. Results are summarized in Table 2.

Table 2. Inference time and RAM usage comparison on Raspberry Pi device.

Method	Inf. Time [s]	Used RAM [MB]	Allocated RAM [MB]
Base RF	32.87	537	2349
FI (Bal.)	19.11	364	1339
Proposed	17.49	315	1292

The deployment tests on the Raspberry Pi demonstrate that the efficiency gains observed in the high-performance environment are preserved after deployment on IoT environments. The proposed approach nearly halves both inference time and memory usage with respect to the baseline RF. Against FI (Bal.), the best model obtained by classical methods, inference time is reduced inference time by about 8% and memory usage by 13%. These results emphasize the importance of including efficiency measures in the FS process.

6 Conclusions and Future Work

This work presents a novel feature selection approach for developing lightweight IDSs for protecting IoT environments. The proposed approach is based on a custom scoring function that balances predictive power with inference time and model size on disk. By integrating this metric into a GA framework, we have been able to identify feature subsets that improve the deployment efficiency of machine learning models while maintaining their predictive performance on the NF-ToN-IoTv2 dataset. The appropriateness of the proposed approach has been assessed against various of the most popular feature selection approaches reported in the literature. However, despite the robustness of the results, some limitations of the methodology must be acknowledged. First, despite efforts to reduce its variability, inference time cannot be measured with absolute precision, and small variations may occur across runs. Secondly, this preliminary study does not include a systematic method for selecting the optimal values of the weights λ_{inf} and λ_{size}.

It should be stressed that this work represents a first step towards customizing FS according to the specific resource constraints of a system, such as memory usage or energy consumption. Future research will involve developing systematic strategies for optimizing the hyperparameters λ_{inf} and λ_{size}. Furthermore, the generalizability of the proposed approach will be assessed by testing it with alternative classification algorithms, such as XGBoost or Support Vector Machines.

Acknowledgments. Authors would like to acknowledge the support of the Spanish Government (through the Cybersecurity Institute of Spain, INCIBE), with the Strategic Research Project "Analysis of mobile applications from the perspective of data protection: Cyber-protection and Cyber-risks of citizen information" and the International Chair "Smart Rural IoT and Secured Environments", within the context of the Recovery, Transformation and Resilience Plan financed by the European Union (NextGenerationEU/PRTR). This is part of one of the research activities of the international chair "Smart Rural IoT and Secured Environments". The authors also thank the support of the UNED Associated Center Consortium in Ponferrada, INTECCA, as well as the UNED CiberCSI research group, and the UNED CiberGID innovation group.

Disclosure of Interests. The authors have no competing interests to declare that are relevant to the content of this article.

References

1. Abawajy, J., Darem, A., Alhashmi, A.A.: Feature subset selection for malware detection in smart IoT platforms. Sensors **21**(4), 1374 (2021)
2. Abellana, D.P.M., Roxas, R.R., Lao, D.M., Mayol, P.E., Lee, S.: Ensemble feature selection in binary machine learning classification: a novel application of the evaluation based on distance from average solution (EDAS) method. Math. Probl. Eng. **2022**(1), 4126536 (2022)
3. Albulayhi, K., Abu Al-Haija, Q., Alsuhibany, S.A., Jillepalli, A.A., Ashrafuzzaman, M., Sheldon, F.T.: IoT intrusion detection using machine learning with a novel high performing feature selection method. Appl. Sci. **12**(10), 5015 (2022)
4. Alhowaide, A., Alsmadi, I., Tang, J.: PCA, Random-forest and Pearson correlation for dimensionality reduction in IoT IDS. In: 2020 IEEE International IOT, Electronics and Mechatronics Conference (IEMTRONICS), pp. 1–6. IEEE (2020)
5. Alsaedi, A., Moustafa, N., Tari, Z., Mahmood, A., Anwar, A.: TON_IoT telemetry dataset: a new generation dataset of IoT and IIoT for data-driven intrusion detection systems. IEEE Access **8**, 165130–165150 (2020)

6. Altmann, A., Toloşi, L., Sander, O., Lengauer, T.: Permutation importance: a corrected feature importance measure. Bioinformatics **26**(10), 1340–1347 (2010)
7. Aversano, L., Bernardi, M.L., Cimitile, M., Pecori, R.: A systematic review on deep learning approaches for IoT security. Comput. Sci. Rev. **40**, 100389 (2021)
8. Daghero, F., et al.: Adaptive random forests for energy-efficient inference on microcontrollers. In: 2021 IFIP/IEEE 29th International Conference on Very Large Scale Integration (VLSI-SoC), pp. 1–6. IEEE (2021)
9. Doost, P.A., Moghadam, S.S., Khezri, E., Basem, A., Trik, M.: A new intrusion detection method using ensemble classification and feature selection. Sci. Rep. **15**(1), 13642 (2025)
10. Eid, A.M., Soudan, B., Nassif, A.B., Injadat, M.: Comparative study of ml models for IIoT intrusion detection: impact of data preprocessing and balancing. Neural Comput. Appl. **36**(13), 6955–6972 (2024)
11. Fatima, M., Rehman, O., Rehman, I.M.: LI-IDS: an approach towards a lightweight ids for resource-constrained IoT. In: 2023 International Conference on Smart Applications, Communications and Networking (SmartNets), pp. 1–6. IEEE (2023)
12. Ghaffari, A., Jelodari, N., pouralish, S., derakhshanfard, N., Arasteh, B.: Securing internet of things using machine and deep learning methods: a survey. Cluster Comput.**27**(7), 9065–9089 (2024)
13. Kunhare, N., Tiwari, R., Dhar, J.: Intrusion detection system using hybrid classifiers with meta-heuristic algorithms for the optimization and feature selection by genetic algorithm. Comput. Electr. Eng. **103**, 108383 (2022)
14. Rafique, S.H., Abdallah, A., Musa, N.S., Murugan, T.: Machine learning and deep learning techniques for internet of things network anomaly detection–current research trends. Sensors **24**(6), 1968 (2024)
15. Sarhan, M., Layeghy, S., Portmann, M.: Towards a standard feature set for network intrusion detection system datasets. Mobile Netw. Appl. **27**(1), 1–14 (2022)
16. Singh, R., Ujjwal, R.: Feature selection methods for IoT intrusion detection system: comparative study. In: Computational Intelligence: Select Proceedings of InCITe 2022, pp. 227–236. Springer (2023)
17. Tsimenidis, S., Lagkas, T., Rantos, K.: Deep learning in IoT intrusion detection. J. Netw. Syst. Manage. **30**(1), 8 (2022)
18. Wu, X.W., Cao, Y., Dankwa, R.: Accuracy vs efficiency: machine learning enabled anomaly detection on the Internet of Things. In: 2022 IEEE International Conference on Internet of Things and Intelligence Systems (IoTaIS), pp. 245–251. IEEE (2022)

A Novel Platform Model for IoT Devices M2M Communication in 6G Network

Tariq Daradkeh[1(✉)], Salah Harb[2], and Yanal Alahmad[3]

[1] Computer Science, Yarmouk University, Irbid, Jordan
tariq.daradkeh@yu.edu.jo

[2] Lawrence Berkeley National Laboratory, 1 Cyclotron Rd, Berkeley 94720, CA, USA
sharb@lbl.gov

[3] Software Engineering Department, Jordan University of Science and Technology, Irbid, Jordan
ynalahmad@just.edu.jo

Abstract. Machine-to-Machine (M2M) communication is a cornerstone of the Internet of Things (IoT), enabling autonomous device-to-device interaction. However, existing protocols such as MQTT, CoAP, and OneM2M face limitations in interoperability, scalability, and adaptability, especially as the number of connected devices grows to tera-scale levels. Current models are often centralized, static, and not optimized to exploit the capabilities of sixth-generation (6G) networks, which provide ultra-high speed, ultra-low latency, and distributed computational resources. In this paper, we propose a novel unified platform model for massive M2M communication within 6G networks. The architecture integrates edge computing with distributed intelligence to dynamically allocate computational tasks across end devices, edge nodes, and data centers. This approach minimizes latency and communication cost while enhancing scalability, traffic aggregation, and energy efficiency. The proposed model addresses the shortcomings of existing solutions by combining interoperability, flexibility, and efficient resource utilization, providing a foundation for future large-scale IoT deployments in 6G environments.

Keywords: 6G · M2M · IoT · Automation · Edge Computing

1 Introduction

By increasing the number of connected devices to the internet with the advent of sixth-generation (6G) networks, the famous term Internet of Things (IoT) has been introduced, and communication automation is becoming a mandatory issue in machine-to-machine (M2M) communication. Today, there are more than 50 million connected devices that interact automatically [1]. This M2M has been adopted by smart technology, which allows devices to communicate and act independently without human help. By using Artificial Intelligence (AI) and Machine Learning (ML), these devices can also make decisions independently.

H. R. Arabnia et al. (Eds.): CSCI 2025, CCIS 2970, pp. 345–356, 2027.
https://doi.org/10.1007/978-3-032-28466-2_26

There are many examples of using M2M communication in the IoT environment, such as industrial manufacturing, where M2M is used in remote equipment management through systems like Supervisory Control and Data Acquisition (SCADA). Moreover, it has spread to other fields such as healthcare (smart wearables monitoring vital signs), business (supply chain tracking), security (surveillance), and facility management (for example, smart buildings or university campus automation using vending machines or cleaning robots, etc.).

Mainly, the common communication protocols used between interacting machines are Message Queuing Telemetry Transport (MQTT) and Constrained Application Protocol (CoAP), which are well-suited for low-mobility and low-power devices with limited network bandwidth. There are two techniques of data acquisition models from IoT sensors. The first is polling, where a data collector continuously requests information from the sensor (data siloing). However, this method generates excessive and unnecessary traffic. On the other hand, there is the publish–subscribe model, where the sensor pushes data using a method called report by exception, in which it sends updates only when a change occurs [2].

The Industrial IoT (IIoT) requires secure, efficient, and agile messaging techniques that enable interoperability of M2M intercommunication. In addition, it demands fast scalability and massive communication convergence over new 6G infrastructure by expanding computational resources closer to end devices through edge computing. Existing M2M communication protocols, such as MQTT and MQTT/Sparkplug B [2], do not provide the needed functionalities. MQTT has issues with interoperability, where all IoT devices and data silos must use the same messaging, encoding, and format. MQTT/Sparkplug addresses this by using a common underlying MQTT framework called a broker, which unifies the messages and acts as a centralized standard communication interface, solving the inconsistency in MQTT implementations. However, this model still has many issues and limitations in IoT deployment, including a single point of failure, reliance on a single standardized message encoding and format, and challenges with fast scalability to support the expected integration of tera-scale devices into 6G networks.

In this work, we address secure massive M2M communication over 6G networks by integrating edge computing, where scalability and message convergence can be guaranteed through our proposed communication model.

This paper is organized as follows: related work is presented in Sect. 2; the proposed model architecture and components are described in Sect. 3; the experimental setup and operational algorithm are explained in Sect. 4; the experimental results are presented in Sect. 5; and finally, the conclusion and future work are discussed in Sect. 6.

2 Related Work

In the following chronologically ordered survey papers [3–7], M2M communication is covered from different aspects. The challenges of M2M communication include the need for low energy consumption, standardization to ensure interoperability, messaging volume, speed and aggregation, supported technologies and vendor unification, and finally network efficiency and utilization.

In [3], the authors discussed the M2M communication system model, applications, traffic types, technology standardization, and challenges and future research directions. One of the main relevant points highlighted is that within the interoperability challenge, the Internet Engineering Task Force (IETF) enhanced CoAP through its Constrained RESTful Environments (CoRE) working group, which introduced key points for standardization. However, many limitations remain with this progress, as the variety of device types, geographical locations, and user interface types can still pose significant challenges. Matching application interactions will require a unified messaging and interface gateway.

In [4], the authors focused on the technological ecosystem for handling M2M communication between devices. This research investigated all aspects of M2M communication, starting from the data link layer up to the application layer. The emphasis was on protocol architectures and models that support M2M communication across different platforms, middleware, and frameworks. M2M communication depends on the device network connection technology, protocol standards, application tasks, service middleware, common frameworks, or the serving cloud platform. The main interoperability solution proposed in this work is to integrate messaging protocols with the web using the Representational State Transfer (REST) model, where CoAP and MQTT are enhanced by integrating REST APIs with the publish–subscribe model.

In [5–7], the works focused on M2M communication taxonomy from different aspects. In [5], the authors made a significant effort by introducing multiple classes of taxonomy such as device types, medium access, connection technologies, applications, standards, architectures, services, and features. Meanwhile, in [6, 7], the authors concentrated on one aspect of the M2M communication layer, namely the data-link layer and the Media Access Control (MAC). Wireless access connection protocols such as contention-based approaches and resource reservation approaches were addressed in [7]. On the other hand, ultra-low latency in LTE cellular network architecture and models was discussed in [6], emphasizing the conditions for M2M communication in connected devices, including direct device-to-device (D2D) communication and linking devices with cellular technology.

As a recap of the literature, we can conclude that most prior works have treated M2M communication as a standalone context, linking it to various network service layers, but none have tackled it in conjunction with 6G technology management and operation. This includes expanding computational resources while considering ultra-high communication speed, link capacity, and the massive aggregated traffic that must be collected and processed. Google has developed its own MQTT broker [8] to enable connections with client-end devices. As a cloud application moderator, it links at the application level, but this effort again does not provide a holistic view of M2M communication in the context of 6G networks. Finally, a significant effort has been made by a union of seven

Standard Development Organizations (SDOs) from various countries to develop a unified M2M communication standard, represented by the oneM2M project [9]. The project considers multiple aspects of M2M communication, such as interoperability, scalability, reliability, energy efficiency, and device performance. Its main standardization contribution is unifying the protocol standards of M2M communication while addressing these conditions.

In this work, we consider massive M2M communication in the context of 6G networks by proposing a unified model architecture that accounts for application weight, computational capacity, and device locations.

3 M2M Model Architectures and Components

First of all, we have to emphasize that the network links in 6G do support ultra-high network links that make the connection capacity and throughput super fast, and the network delay or latency negligible. With this network setup, we do not focus on low-speed link communication protocols such as MQTT or CoAP. We need to focus on protocol portability, interoperability, agility, and integration with 6G network components in a way that can utilize the existing infrastructure and its arrangement.

As Fig. 1 depicts the 6G network architecture, we can see that computation resources are decoupled and distributed over all layers. We can use end devices to perform high and complex computational tasks since the capabilities are available, but if we need more resources, the task will be moved to higher computational resources and hosted by the edge layer close to the end devices in a way that minimizes communication cost and latency. Or we can move the task to be served by data center servers that have super computation, storage, network, and services resources that can handle resource expansion and provide automatic resource management and scaling.

The presented diagram in Fig. 1 explains the 6G network components, applications, and M2M communication protocols. The M2M protocols can be used in any 6G layer, where MQTT, CoAP, CoRE, and OneM2M are used to facilitate machine-to-machine or device-to-device communications. The applications that can utilize the collected data include SCADA, time-series data historian databases, data analytics solutions, ERP systems, industrial solutions such as Manufacturing Execution Systems, and the proposed standard OneM2M communication. In addition, there is a unified cloud link, which is Google MQTT broker, built using the REST protocol. We refer to any unified interface as a unified broker in the diagram, which is used in all the following protocols: OneM2M, CoAP, and CoRE architectures.

The MQTT protocol [10] has a lightweight publish/subscribe architecture that is considered a direct IoT messaging protocol to support data streaming over wire-limited resource devices. OneM2M, CoAP, and CoRE protocols are also similar to MQTT but use a REST (Request/Response)–based model to support IoT intercommunication. Overall, all of the aforementioned protocols are built on similar architectures (publish/subscribe or request/response), with or without using the HTTP-based REST model, as Fig. 2 depicts.

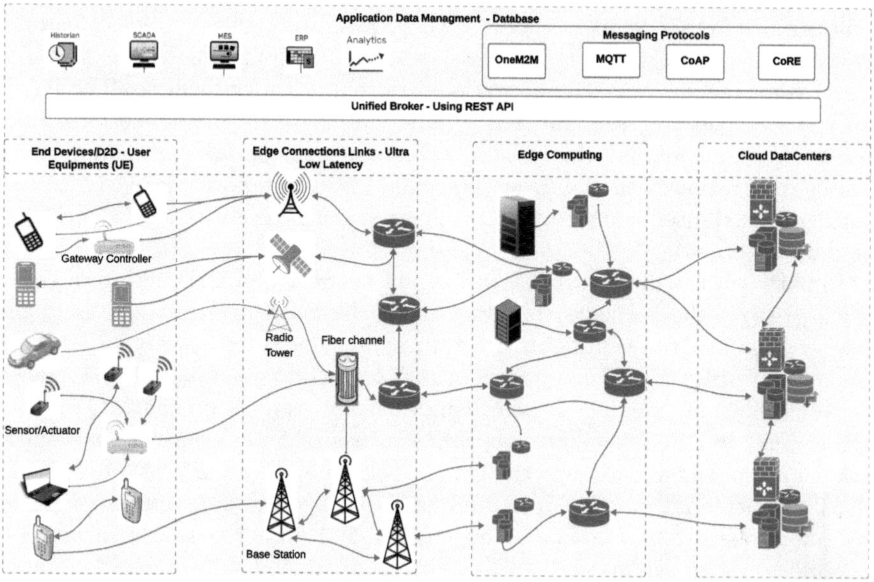

Fig. 1. General 6G Network Architecture with M2M Application Integration.

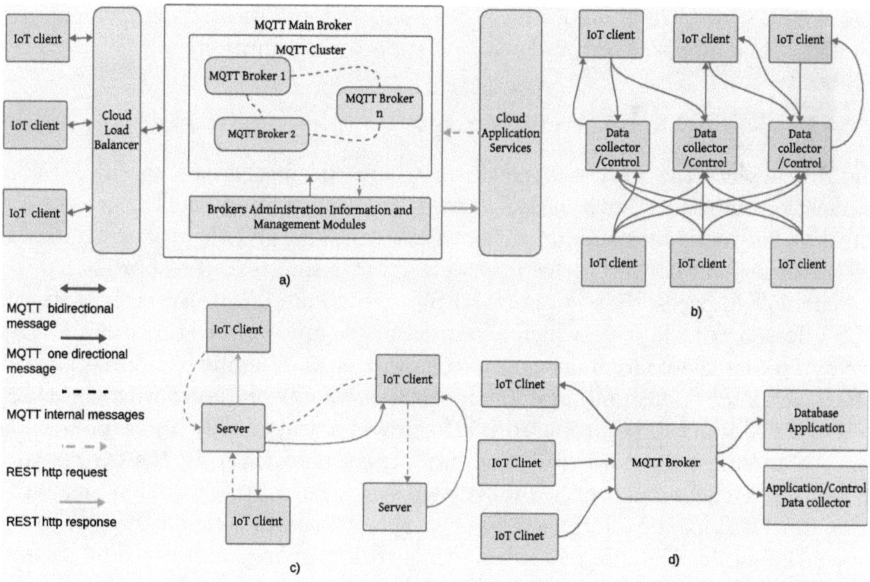

Fig. 2. Comparison between four protocols architecture. a) Google MQTT Broker Design using REST, b) Direct device to device MQTT, c) CoAP with REST, d) MQTT with broker.

In Fig. 2 part (a), the Google MQTT broker cloud integration architecture is depicted, where IoT devices connect with MQTT messaging through a load balancer that evenly distributes the messages into an MQTT broker cluster. This cluster is managed by an administration service unit that provides a REST API, and the messages can be redirected to different cloud application services using REST. We can see that this model serves one part of M2M communication, namely integration with cloud services, but it does not consider network capabilities or distributed computational resources, since the entire model is built on the data center side.

In part (b), the direct MQTT messaging between the D2D and data collectors is shown, where a mesh topology is the best arrangement in this model. However, the model has a highly dynamic arrangement between the M2M components but still faces many issues. In the diagram, we assume all IoT devices are from the same vendor and can communicate using a unified MQTT message format, but in reality, the network consists of heterogeneous components that cannot communicate directly. On the other hand, in parts (c) and (d), the REST protocol and unified MQTT broker solve the interoperability issues, but both models still rely on static arrangements that do not reflect localization or network resource distribution.

Our goal in this work is to overcome these limitations and propose a novel architecture that, like the OneM2M model, provides unification while also being aware of 6G network components and capabilities. In this way, it can enhance the model structure (simple design), support fast intercommunication, efficiently utilize aggregated traffic and data processing, and enable easy deployment with low latency, minimal delay, and reduced energy consumption.

3.1 Proposed Model Architecture

Figure 3 depicts the proposed platform where the model can handle all communication messages, such as IoT legacy devices using an MQTT broker with single or bidirectional communication messages. The MQTT broker will handle the messages and convert them into web REST requests and responses.

The IoT devices that support REST can communicate directly with the REST Resources Mapper, which acts as a messaging router and resource mapper by linking the communicating parties to the same unified resource location URL. All cloud operation application services can also interact with the REST Resources Mapper. The proposed platform will act as a distributed model that contains manifest files listing all the distributed platforms over the 6G network, and each IoT client can use the publish–subscribe model in a way that ensures all communication parties find the closest replica to the IoT client. This approach makes the model scalable, adaptive to network changes, and aware of resource capabilities and network capacity. The message data model is a simple JSON format that contains a data array vector with a maximum of 100 elements, resulting in a maximum size of 3,863 bytes and a minimum size of 90 bytes with one message element. We use a single HTTP method, POST. POST is used to send data to the server in addition to the request message body.

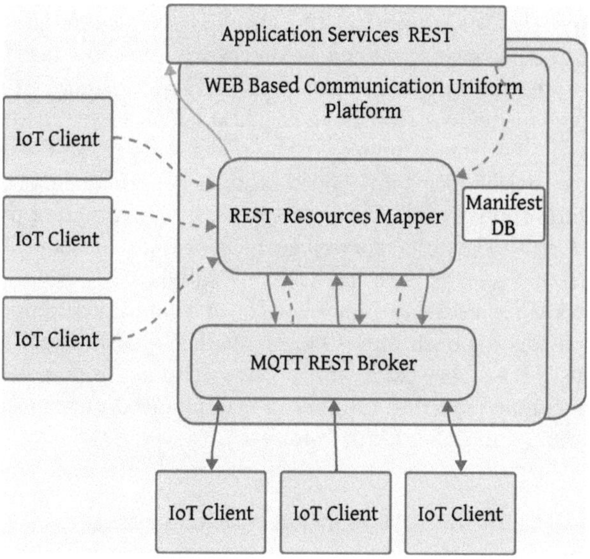

Fig. 3. The proposed platform model.

3.2 Model Metrics Analysis and Evaluation

In this section, we present a simple mathematical representation to analyze the platform serving operation functionalities and performance. Using simple notation, the number of communication parties (IoT devices) is denoted as D, the number of messages per device as dm, and the total number of aggregate messages as $M = dm \times D$. The performance P is measured by dividing the utilized network bandwidth ubw by the full capacity L, $P = ubw \div L$. Once the utilization value exceeds 96%, the latency increases exponentially. By comparing the total number of messages M and the performance P in the proposed model with the four other protocols (Google MQTT, direct MQTT, CoAP, and MQTT broker), we find that the proposed model has R replicas distributed over the network. The next section will explain the simulation setup and results. In addition, we measure the communication end-to-end delay $eted$ and the message latency ml, which is the request time.

4 Experiments Setup and Simulation

To evaluate the proposed M2M communication model platform, we implemented and tested it within the OMNeT++ version 6.2.0 and INET-4.5.4 simulation framework [11]. The simulations were executed on a workstation equipped with an Intel Core i9-14900K CPU and 128 GB DDR5 RAM, with additional GPU acceleration provided by an NVIDIA RTX 4070 Ti (12 GB GDDR6X) for parallelized processing tasks.

We simulated the 6G network using OMNeT++, where the assumption is that in 6G the media access latency is very low (ultra-low latency). We modeled the wireless channels as ideal links with 10 Gbps speed. The IoT connected devices are arrays of devices that have regional radio access mediums. We used different numbers of wireless devices wD, as Fig. 4 depicts. The backbone network architecture has 40 Gbps link speed, and there is one connected edge server farm (ECS1) with 100 servers. The edge network has two computational server farms (ES1 and ES2) with 1000 servers each. Four data centers (DS1, DS2, DS3, DS4) contain 10,000 servers each, as listed in Table 1.

The simulation experiments modeled the five aforementioned models, and we use abbreviations for each one (Google MQTT – GM, direct MQTT – DM, CoAP – CP, MQTT broker – MB, and proposed unified model – UM). We evaluated four performance metrics: number of accumulated messages, performance, latency, and end-to-end delay $(M, P, ml, eted)$.

Table 1. The simulation characteristics.

IoT Devices	ECS	ES	DS	Wireless Speed	Network Speed
Varies (1000, 10 000, 100 000, 1 000 000)	100	1000	10 000	10 Gbps	40 Gbps

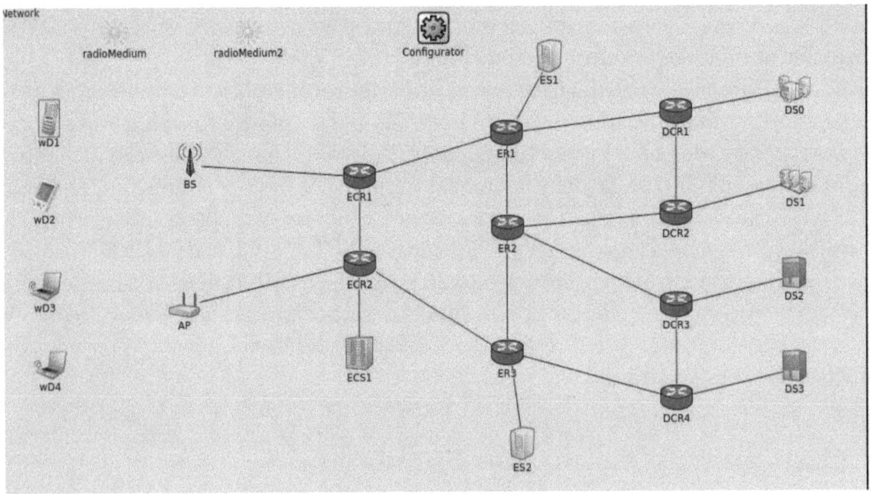

Fig. 4. The simulation environment in OMNeT++.

The messaging in the simulation environment occurs as a uniform random distribution, where half of the IoT devices have destinations that can be either another IoT device (D2D) or a data collector for application services in the

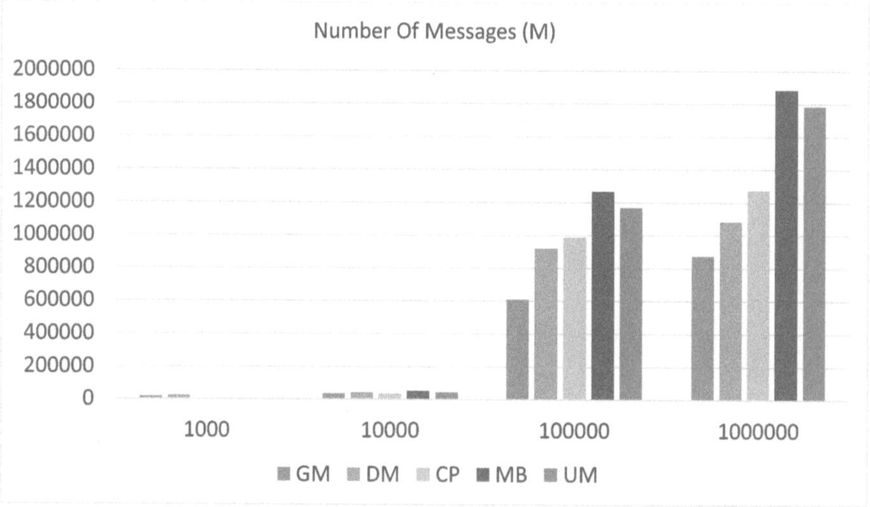

Fig. 5. Total number of messages.

data center servers. The operation of the publish–subscribe model must be done through the broker agent, which increases the number of interchanged messages. However, the request–response model allows direct interaction between the communicating devices. For the Google Cloud model, we implemented a data sink collector as a link to the cloud service with dynamic scaling for resources, as cloud services offer elastic provisioning. On the other hand, we set a maximum value for scaling as the upper boundary for provisioned resources. In the direct-to-direct messaging model, each device communicates directly with the target device. In the broker architecture, the broker agent works as a proxy for communicators, where devices must go through the broker.

In our proposed model, the platform has two engines: the legacy MQTT broker that supports MQTT messaging and the REST Resources Mapper that handles web interaction messaging. The platform has the capability to unify interoperability and support both polling and interrupt-driven data triggers. The HTTP POST function can push data from the end device to the data controller and at the same time receive operation commands (in the case of a sensor with an actuator). In this case, there is no need to deploy a web server on the end device side, which is limited in computational resources and battery life. The model is very simple to deploy, as only the HTTP POST function is required on the device side.

5 Results

In this section, we present the simulation results for device-to-device communication operations in the 6G environment. The setup was run for 150 s of simulation time, and the performance metrics were collected and compared with each other.

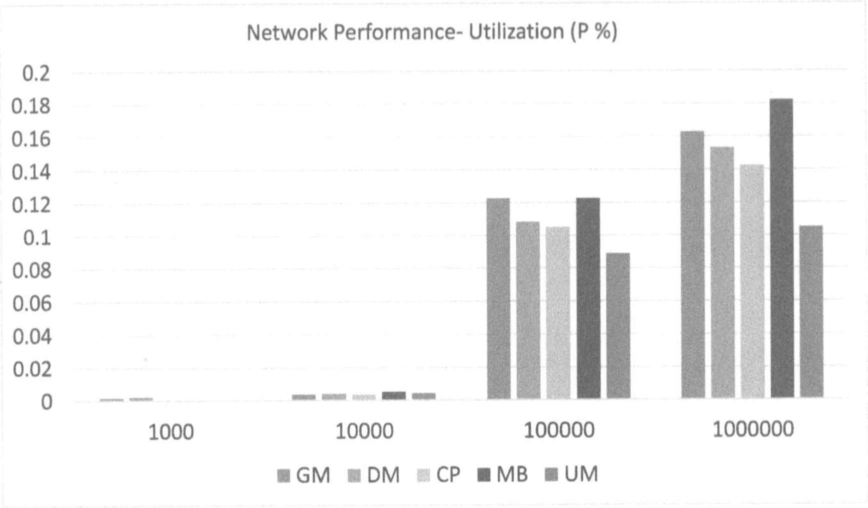

Fig. 6. Model performance.

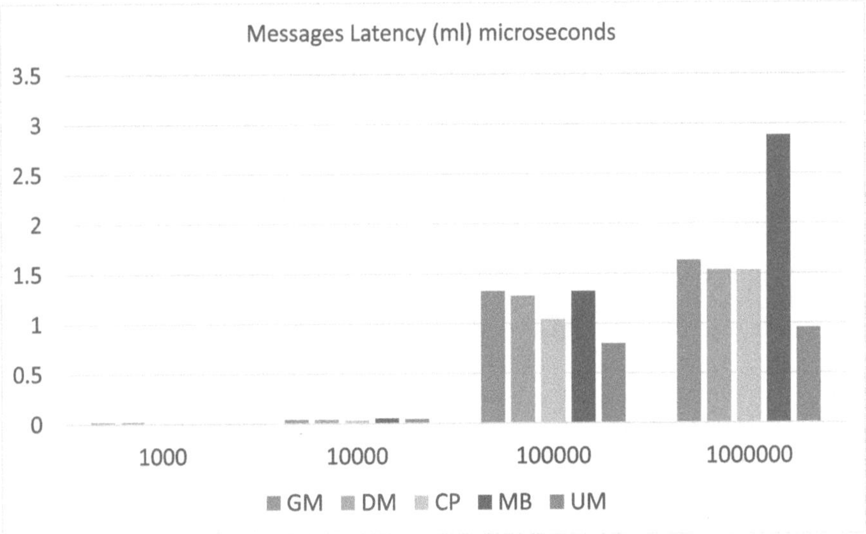

Fig. 7. Message latency (request delay).

In Fig. 5, the total number of accumulated messages required to complete a full communication transaction is depicted. We can clearly see that with the publish–subscribe broker setup, more messages are exchanged. The MQTT broker has the highest number of messages because the operation requires both subscription and publishing to the target device. In our model, we have the MQTT broker and the REST Resources Mapper, which both act as intermedi-

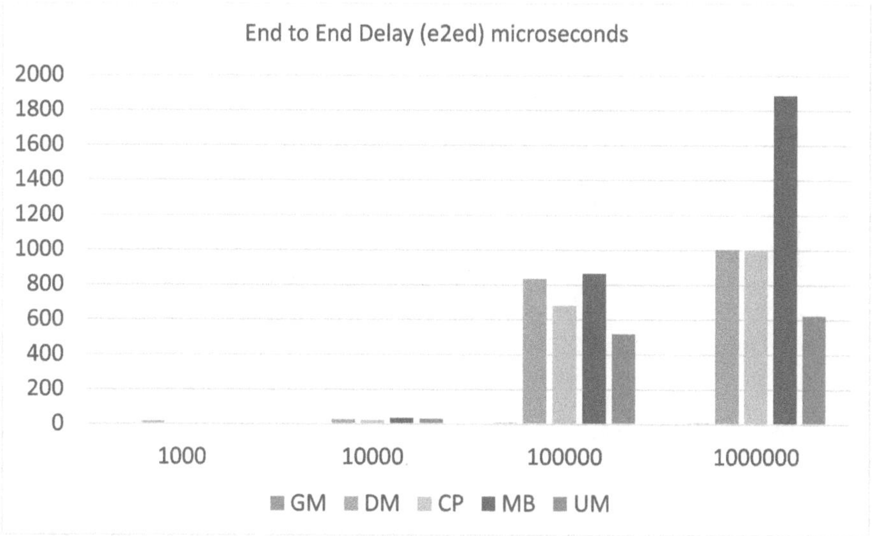

Fig. 8. End-to-end delay.

ate agents, making the number of messages close to the MQTT broker model. In the Google model, we observe fewer interactions because it is a one-way data collector. With CoAP and direct MQTT, there are fewer exchanged messages because the interaction occurs directly.

The overall performance metrics, which measure utilization, are presented in Fig. 6. As the whole 6G network has ultra-high-speed links (10 Gbps and 40 Gbps), the overall utilization never reached the critical ratio of 96%. This means no congestion or slowness occurred, with the maximum ratio being 19%.

The message latency and end-to-end delay are presented in Figs. 7 and 8. The maximum message latency is 3 microseconds, and the end-to-end delay is 1900 microseconds. We can see that the delay is negligible because the overall 6G network and all models are operating on top of it. However, the proposed model achieves the lowest message latency and end-to-end delay due to its location-aware feature and scaling capabilities.

6 Conclusion and Future Work

In summary, we developed a novel platform that enables interoperable machine-to-machine communication by utilizing the new 6G network capabilities and resource power. Our proposed model has the agility to integrate any device type by supporting MQTT and REST-based interaction. The performance metrics we used to evaluate the system in the 6G network are the accumulated number of messages M, performance P, message latency ml, and end-to-end delay $e2ed$. Our proposed model shows the best overall performance results. Future work will focus on testing the model at a larger scale, deploying it in real-world environments.

References

1. "White paper on machine-to-machine communication (M2M) " (2020). white Paper
2. "Industrial-strength mqtt/sparkplug b, building industrial MQTT networks at scale with edge computing" (2022). white Paper
3. Verma, P. K., et al.: "Machine-to-machine (M2M) communications: a survey." J. Netw. Comput. Appl. **66**, 83–105 (2016). Available: https://www.sciencedirect.com/science/article/pii/S1084804516000990
4. Ali, A., Shah, G. A., Farooq, M. O., Ghani, U.: "Technologies and challenges in developing machine-to-machine applications: a survey. J. Netw. Comput. Appl. **83**, 124–139 (2017). [Online]. Available: https://www.sciencedirect.com/science/article/pii/S1084804517300620
5. Amodu, O. A., Othman, M.: Machine-to-machine communication: an overview of opportunities. Comput. Netw. **145**, 255–276 (2018). Available: https://www.sciencedirect.com/science/article/pii/S138912861830851X
6. Kar, U. N., Sanyal, D. K.: An overview of device-to-device communication in cellular networks. ICT Express **4** (4), 203–208 (2018). Available: https://www.sciencedirect.com/science/article/pii/S2405959517301467
7. Gaurav, A., Pandey, A. K.: A survey: hybrid medium access control protocol for M2M communication. In: 2018 International Conference on Wireless Communications, Signal Processing and Networking (WiSPNET), pp. 1–6 (2018)
8. Google Cloud, "MQTT broker architecture for connected devices." https://cloud.google.com/architecture/connected-devices/mqtt-broker-architecture (2025). Accessed 08 Sept 2025
9. oneM2M "ONEM2M developers API." https://www.onem2m.org/using-onem2m/developers/api#n2a (2025). Accessed 08 Sept 2025
10. OASIS, "Oasis standard." https://docs.oasis-open.org/mqtt/mqtt/v5.0/mqtt-v5.0.html (2019). Accessed 08 Sept 2025
11. OMNeT++, "Omnet++," https://omnetpp.org/ (2025). Accessed 16 Sept 2025

LSO-IFC: A Novel Lightweight and Secure Offloading Scheme for IoV in Fog–Cloud Confederation

Yashar Salami(✉)

Faculty of Computer and Information Technologies, Cappadocia University, Ürgüp, Nevşehir, Turkey
yashar.salami@kapadokya.edu.tr

Abstract. The rapid expansion of the Internet of Vehicles (IoV) has underscored the growing need for efficient and secure computational offloading schemes within fog–cloud Confederation environments. Although existing approaches have achieved secure offloading (SO), they often incur high computational and communication costs, limiting their applicability in resource-constrained vehicular networks. To address this challenge, this paper introduces LSO-IFC, a novel lightweight and secure offloading scheme tailored for IoV. The proposed design integrates elliptic curve cryptography (ECC), hash functions, and mutual authentication mechanisms to establish secure channels for offloading. Security validation is conducted through formal analysis using the AVISPA tool under the Dolev–Yao adversarial model, complemented by informal evaluation, with results demonstrating strong resilience against a broad spectrum of threats. Furthermore, performance analysis shows that LSO-IFC reduces the number of operational phases to three, significantly minimizing bit consumption, cryptographic overhead, and communication costs compared to the NSO-VFC protocol. LSO-IFC emerges as a practical, lightweight, and deployable solution for SO in large-scale IoV scenarios within fog–cloud Confederation.

Keywords: IOV · Secure · Fog · Authentication

1 Introduction

The evolution of distributed computing has reshaped data processing and management across heterogeneous platforms [1, 2]. Cloud–fog integration addresses the scalability and performance limits of traditional cloud systems by pushing computation closer to end-users, thereby reducing latency and improving availability for real-time services [3–6]. Driven by the growth of IoT and its sensor-rich ecosystems, IoV extends this paradigm by treating vehicles as intelligent, mobile nodes that offload large data volumes to nearby fog infrastructures to meet strict requirements on latency, energy, cost, and reliability [7–14]. However, despite extensive research on secure offloading (SO) in edge and fog–cloud environments, many existing works either prioritize security at the expense of a lightweight design or focus on performance while neglecting the stringent resource

H. R. Arabnia et al. (Eds.): CSCI 2025, CCIS 2970, pp. 357–371, 2027.
https://doi.org/10.1007/978-3-032-28466-2_27

constraints of IoV networks [15]. In 2025, Yi et al. proposed SOCA, a ChaCha20-based SO strategy for multi-access edge computing [16];. However, it improved security and performance; ChaCha20 and multi-objective optimization introduced substantial overhead, so it is not lightweight. Likewise, Li et al. (2025) presented a deep reinforcement learning–based multi-slot SO framework for vehicular edge computing [17], and Lu et al. (2025) developed a blockchain-based SO scheme with multi-agent reinforcement learning [18]; however, the combined computational cost of reinforcement learning and blockchain transactions remains too high for latency-sensitive, resource-constrained IoV environments. In the same year, Salami presented SO-ITS, a SO scheme for intelligent transportation systems in fog–cloud federation [19]. The scheme demonstrated robustness against common attacks and achieved low communication overhead. Nonetheless, although offloading was securely performed, the framework offered only moderate computational efficiency, leaving the lightweight requirement partially unaddressed in large-scale vehicular environments. Yao et al. (2025) extended SO to space–air–ground integrated networks [20]. The proposed framework strengthens confidentiality against aerial eavesdroppers using artificial noise and trajectory control; however, its reliance on UAVs, satellites, and iterative optimization introduces excessive complexity, making it unsuitable for lightweight IoV scenarios. Khajehvand et al. (2024) developed SOS-FCI, an ECC-based authentication framework for fog–cloud IoT [21]. Earlier schemes achieved SO and resisted attacks, but intensive ECC operations caused high computational and communication overhead, so they were not lightweight. Likewise, the 2024 NSO-VFC scheme for IoV [22] provides secure offloading but still incurs substantial costs, limiting its suitability for resource-constrained vehicular networks. Other studies, such as those by Gao et al. (2025) and Zheng et al. (2025), employed UAV-mounted intelligent reflecting surfaces and reconfigurable intelligent surfaces, respectively, to achieve SO [23, 24]. While existing schemes achieve SO, their reliance on heavy infrastructure and computations prevents a lightweight design, highlighting the need for a lightweight and SO approach for IoV in fog Confederations.

1.1 Paper Organization

The rest of this paper is organized as follows. Section 2 introduces the network model, architecture, threat model, and assumptions. Section 3 details the LSO-IFC scheme and its operational phases. Section 4 presents security analysis using both formal and informal methods. Section 5 evaluates performance in terms of computation, communication, and security comparisons. Finally, Sect. 6 concludes the paper and outlines future research directions.

1.2 Paper Contribution

The main contributions of this paper are as follows: we propose LSO-IFC, a lightweight secure offloading scheme for IoV in fog–cloud federations that provides strong security with low computational overhead, making it suitable for resource-constrained environments. Its security is validated through AVISPA-based formal verification under the Dolev–Yao model and informal analysis against common attacks, and its three-phase design reduces communication rounds, cryptographic operations, and bit overhead; moreover, comparison with NSO-VFC shows that LSO-IFC offers comparable or stronger security while significantly lowering computational and communication costs, yielding a practical and scalable solution for large-scale IoV deployments.

2 Network Model

This section presents the network architecture, the threat model, and the underlying assumptions of the proposed scheme.

2.1 Network Architecture

The LSO-IFC framework adopts a three-tier architecture to improve processing and communication efficiency. At the top layer, powerful cloud servers act as centralized controllers, handling complex tasks and securely interacting with the trusted authority and lower layers. The middle layer consists of fog nodes, organized under a fog coordinator, which provide localized computation, manage task distribution, and ensure smooth integration between the cloud and vehicular levels. At the bottom layer, resource-constrained IoVs offload computationally intensive tasks to nearby fog nodes, which process them with low latency. Figure 1 illustrates the architecture of the LSO-IFC system.

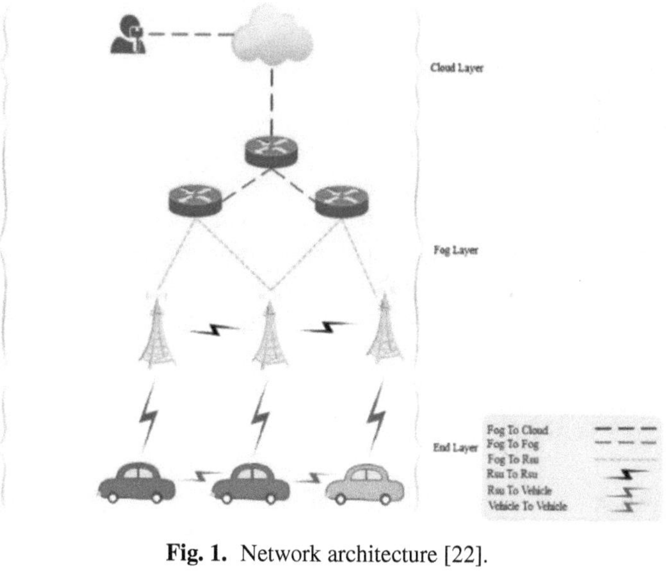

Fig. 1. Network architecture [22].

2.2 Network Threat Model

The LSO-IFC scheme is analyzed under the Dolev–Yao (DY) adversarial model [25], in which all channels are insecure, and the attacker fully controls the network. The adversary can intercept, modify, inject, delete, and replay messages between IoV and RSU entities, and may attempt to compromise both long-term and session keys. This strong model captures a wide range of standard cryptographic and protocol attacks that threaten integrity, confidentiality, mutual authentication, and credential security. Evaluating LSO-IFC in this hostile setting shows its robustness against a powerful adversary with unrestricted access to the communication medium.

2.3 Network Assumption

The LSO-IFC scheme is built on explicit system and security assumptions under a strong adversarial model. All communication channels are insecure, and a computationally powerful attacker can mount active and passive attacks (e.g., eavesdropping, impersonation, brute force). Still, no legitimate entity is assumed to collude with the adversary. IoVs have no prior knowledge of RSU, fog, or cloud identities, while mutual authentication is pre-established between the fog layer, RSUs, and cloud servers; the fog center centrally manages task offloading and cryptographic keys for pre-authenticated devices. Cloud servers are assumed secure, all entities are time-synchronized, session keys are ephemeral, and standard cryptographic primitives with secure elliptic-curve parameters are used, ensuring that elliptic-curve point multiplication remains computationally infeasible to invert.

2.4 Problem Statement

IoVs, as mobile and resource-constrained devices, often offload heavy computations to external infrastructures, which exposes them to significant security risks in fog–cloud environments. Prior works commonly assume fully trusted fog nodes, overlooking practical vulnerabilities. This paper explicitly considers well-known security threats and emphasizes the need for mutual trust and reliable identity verification before offloading. To this end, we propose a lightweight yet robust authentication scheme for SO that ensures secure identity validation against active and passive adversaries while keeping computational overhead low, making it suitable for resource-constrained IoV settings.

3 LSO-IFC Scheme

In this section, we provide a detailed description of the proposed LSO-IFC scheme, including the adopted notations, initialization process, and operational phases such as registration, login, authentication, and offloading.

3.1 Notation

Table 1 shows the Notation used in the LSO-IFC.

Table 1 Notation used in the proposed scheme.

Notation	Description
F, R, Vi	*Fog Rsu, IoV*
IdFi. IdRi, Idvi	*Identity of Fi, Ri, Vi.*
IdVisf	*Identity Suggested IOV to offload*
$r1$, $r2$	*Random number for Ecc*
$R1$	*Random number IdRi*
Pow	*Idvi processing power*
Pr	*Processing requirements*
Pwvi	*Password of Idvi*
NFR	*Number of offloading requests Idvi*
SFIdVisf	*Successful offloading IdVisf*
BuIdVisf	*Busy IdVisf. It can be true or false.*
SKVR	*The session key in the Login and authentication phase*
Skof	*The session key in the offloading phase*
TFi. TRi, Tvi, TVisf	*Timestamp IdFi. IdRi, Idvi, IdVisf*
ΔT	*Message transmission delay*
$h(\cdot)$	*Hash function*
$\oplus$	*XOR*
$\|$	*Concatenation operation*

3.2 Initialization Phase

During the initialization phase, the entities *IdRi* and *IdVisf* agree on an elliptic curve $E : y^2 = x^3 + ax + b(\text{mod}p)$ over the finite field $\mathbb{Z}_p.$, where $p > 2^{160}$ and $4a^3 + 27b^2 \equiv /0(\text{mod}p)$, ensuring that the curve is non-singular. *IdRi* selects a random private key X_R and computes its public key $P_R = X_R \cdot G$, while *IdVisf* chooses a private key X_{visf} and computes $P_{visf} = X_{visf} \cdot G$, where G. is a generator point of large prime order $n > 2^{160}$. (i.e., $n \cdot G = O$). This initialization ensures that both parties share consistent elliptic-curve domain parameters and hold valid private–public key pairs before the subsequent protocol phases, as illustrated in Fig. 2.

Initialization
$IdRi, IdVisf$ *Chooses:* $(y^2 = x^3 + ax + b) \pmod p$
where $4a^3 + 27b^2 \pmod p \neq 0$, *voer* $Zp\ (P > 2^{160})$
XR: Id_{Ri} *secret key;* $P = (XR.G)$*: IdRi public key*
Xvisf: Id_{Visf} *secret key;* $P = (Xvisf.G)$*: IdVisf public key*
G: generator point; $r.G = 0, r > 2^{160}$

Fig. 2. nitialization phase.

3.3 Register Phase

Step 1: At the beginning of the registration procedure, the Idvi generates several values. It first computes $P = h\ (Pwvi \oplus Idvi)$, $Pi = h\ (Idvi \parallel Tvi \parallel \oplus P \oplus Pi \| Pow \oplus Pi)$, $Psw = h\ (Pi \oplus TVi)$, and Send $\{Idvi, TVi, Psw, P, Pi, Pow, IdRi\}$ to *IdRi*.

Step 2: Once the request is received, IdRi first validates the timestamp TVi to ensure freshness Check if TVi $< \Delta$T. Then it performs the following calculations $Pi' = h\ (Idvi \parallel Tvi \parallel \oplus Pwvi \oplus Pi \| Pow \oplus Pi)$, *if* $Pi = Pi'$ it computes P $sw' = Psw = h\ (Pi \oplus TVi)$, and if $Psw = Psw'$ *Computing* $HRr = h\ (((IdRi \parallel IdFi \parallel TRi) \oplus Pi) \oplus Psw)$, $Rr = (((IdRi \parallel IdFi \parallel TRi) \oplus Pi) \oplus Psw)$ xt, a random number R1 is chosen and used in session key generation $PK = h\ (((IdRi \oplus XR) \parallel (IdRi \oplus IdFi) \oplus R1) \oplus (Ps \oplus Psw)))$, $Idvi' = PK.\ G$, $Pkr = h\ (IdRi \| HRr \| Rr \| Idvi')$, $Pkr^* = Pkr.\ G$, $HRPr^* = h\ (Rr, Idvi', HRr, TRi)$, *Storage* Pkr^*, $Idvi'$, *pi*, *Pow* in memory, and Send $\{Rr, Idvi', HRr, HRPr^*, TRi, Idvi\}$ to the Idvi.

Step 3: Upon receiving the response, Idvi verifies the timestamp if TRi $< \Delta$T). If valid, it computes $HRr' = h\ (((IdRi \parallel IdFi \parallel TRi) \oplus Pi) \oplus Psw))$, Check if HRr = HRr′, HRPr*′ = h (Rr, Idvi′, HRr, TRi), Check if HRPr* = HRPr*′, Store {Idvi′, HRr} in memory. Figure 2 shows the registration phase (Fig. 3).

<table>
<tr><th>Idvi</th><th>IdRi</th></tr>
<tr><td>Registration phase
Computing
P=h (Pwvi⊕ Idvi)
Pi=h (Idvi || Tvi ||⊕P⊕ Pi||Pow⊕ Pi)
Psw=h (Pi ⊕ TVi)
Send {Idvi, TVi, Psw, P, Pi, Pow, IdRi} to IdRi.
→
Via open channel

Check if TRi < ΔT
Computing
HRr' =h (((IdRi || IdFi || TRi) ⊕Pi) ⊕Psw))
Check if HRr = HRr'
HRPr*' =h (Rr, Idvi', HRr, TRi)
Check if HRPr*= HRPr*'
Storge {Idvi', HRr} in memory.</td><td>Check if TVi < ΔT
Computing
Pi'=h (Idvi || Tvi ||⊕Pwvi⊕ Pi||Pow⊕ Pi)
Check if Pi=Pi'
Computing
Psw'=Psw=h (Pi ⊕ TVi)
Check if Psw = Psw'
Computing
HRr =h (((IdRi || IdFi || TRi) ⊕Pi) ⊕Psw)
Rr = (((IdRi || IdFi || TRi) ⊕Pi) ⊕Psw)
Select a random number R1
PK=h (((IdRi ⊕ XR) || (IdRi⊕ IdFi) ⊕R1) ⊕ (Ps ⊕Psw)))
Idvi' = PK. G
Pkr=h (IdRi|| HRr||Rr||Idvi')
Pkr*= Pkr. G
HRPr* =h (Rr, Idvi', HRr, TRi)
Storage Pkr*, Idvi', pi, Pow in memory.
Send {Rr, Idvi', HRr, HRPr*, TRi, Idvi} to the Idvi.
←
Via open channel</td></tr>
</table>

Fig. 3. Registration phase.

3.4 Login and Authentication Phase

Step 1: In the login procedure, the Idvi selects a random number r1 and computes Pvr1 = (r1.G), Pvr2 = h (r1. Idvi′), Pvr3 = h ((Pvr1 ⊕ Pvr2 ⊕ TVi) || (IdRi⊕ Idvi) || HRr), and sends {Pvr1, Pvr2, Pvr3, TVi, Idvi, Idvi′, IdRi, HRr} to IdRi.

Step 2: After receiving the login request, IdRi first checks the timestamp TVi < ΔT If valid, Computing $HRr' = h(((IdRi \,||\, IdFi \,||\, TRi) \oplus Ps) \oplus Psw))$ Check if HRr = HRr′, Pvr3′ = h ((Pvr1 ⊕ Pvr2 ⊕ TVi) || (IdRi⊕ Idvi) || HRr) if Pvr3 = Pvr3′ and verify if Idvi′ = Idvi*′. Computing Pvr2* = h (Pvr1. Idvi′) if Verify Pvr2* = Pvr2. If the checks are correct, IdRi chooses a random number r2 and computes $Pvr4 = r2.\ G$; $Pvr5\ r2.\ Pkr^*$, $Pvr6 = h\ ((Pvr4 \oplus Pvr5) \,||\, ((IdRi \oplus TRi) \,||\, Idvi'))$ and Finally, IdRi sends the message {Pvr4, Pvr5, Pvr6, IdRi, TRi, Idvi} to Idvi.

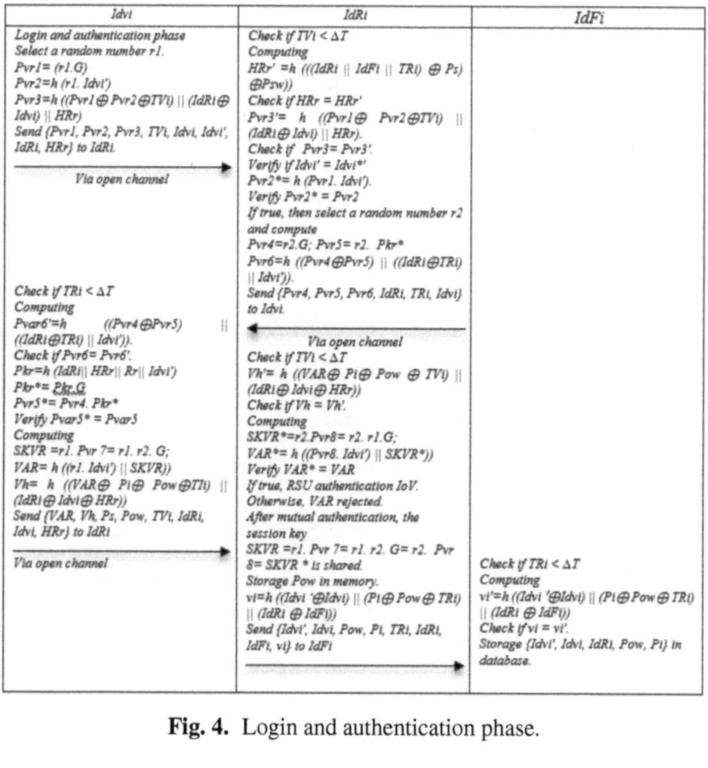

Fig. 4. Login and authentication phase.

Step 3: Upon receiving the response, Idvi validates the timestamp. If fresh, it computes $Pvar6' = h\ ((Pvr4 \oplus Pvr5) \,||\, ((IdRi \oplus TRi) \,||\, Idvi'))$, Check if $Pvr6 = Pvr6'$, $Pkr = h\ (IdRi||\ HRr||\ Rr||\ Idvi')$, $Pkr^* = Pkr.\ G$, $Pvr5^* = Pvr4.\ Pkr^*$, Verify Pvar5* = Pvar5, the session key is computed as, $SKVR = r1.\ Pvr\ 7 = r1.\ r2.\ G$; Additionally, two validation parameters are generated: $VAR = h\ ((r1.\ Idvi') \,||\, SKVR))$, $Vh\ h\ ((VAR \oplus Pi \oplus Pow \oplus TIi) \,||\, (IdRi \oplus Idvi \oplus HRr))$, Finally, Idvi transmits {*VAR, Vh, Ps, Pow, TVi, IdRi, Idvi, HRr*} to IdRi.

Step 4: After receiving the response, IdRi checks the timestamp validity. If correct, it computes $Vh^{'} = h\ ((VAR \oplus Pi \oplus Pow \oplus TVi) \parallel (IdRi \oplus Idvi \oplus HRr))$, and verifies that $Vh = Vh^{'}$. Then it computes $SKVR^{*} = r2.\ Pvr8 = r2.\ r1.\ G;\ VAR^{*} = h\ ((Pvr8.\ Idvi^{'}) \parallel SKVR^{*}))$, if Verify $VAR^{*} = VAR$ If true, RSU authentication IoV. Otherwise, VAR rejected. After mutual authentication, the session key $SKVR = r1.\ Pvr\ 7 = r1.\ r2.\ G = r2.\ Pvr\ 8 = SKVR^{*}$ is shared. Storage Pow in memory. $vi = h\ ((Idvi^{'} \oplus Idvi) \parallel (Pi \oplus Pow \oplus TRi) \parallel (IdRi \oplus IdFi))$.

Finally, IdRi sends $\{Idvi^{'}, Idvi, Pow, Pi, TRi, IdRi, IdFi, vi\}$ to IdFi.

Step 5: Upon receiving the data, IdFi checks the timestamp and Tri. It then computes $vi^{'} = h\ ((Idvi^{'} \oplus Idvi) \parallel (Pi \oplus Pow \oplus TRi) \parallel (IdRi \oplus IdFi))$, and if verifies vi = vi′, the values $\{Idvi^{'}, Idvi, IdRi, Pow, Pi\}$ are securely stored in the database. Figure 4 shows the Login and authentication phase.

3.5 Offloading Phase

Step 1: The IdviI initiates the offloading process by computing $R\ f = h\ (Idvi^{'} \oplus Pr \oplus TVi)$, Send $\{Rf, Idvi^{'}, Pr, TVi, IdRi\}$ to $IdRi$.

Step 2: Upon receiving the request, IdRi checks the timestamp if valid TVi < ΔT it recomputes $Rf^{'} = h\ (Idvi^{'} \oplus Pr \oplus TVi)$. If $Rf = Rf^{'}$, it computes $PK = h\ (((IdRi \oplus XR) \parallel (IdRi \oplus IdFi) \oplus R1) \oplus (Ps \oplus Psw)))$, $Idvi^{'} * = PK.\ G$, Verify $Idvi^{'} * = Idvi^{'}$ and Send $\{Idvi^{'} *, Idvi^{'}, Pr, TRi, IdFi\}$ to $IdFi$

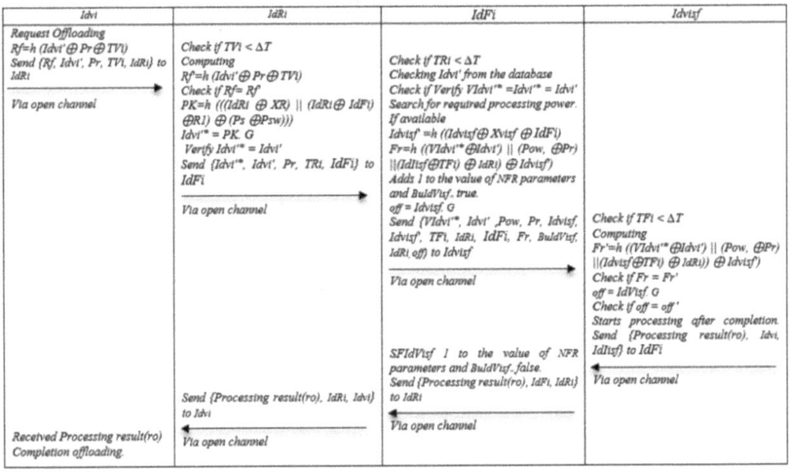

Fig. 5. Offloading phase.

Step 3: After receiving the message, IdFi checks the validity of the timestamp, $TRi < \Delta T$. It then verifies that VIdvi'* = Idvi'* = Idvi', the identity, from its database. If processing resources are available, Compute Idvisf' = h ((Idvisf⊕ Xvisf ⊕ IdFi) FR = h ((VIdvi'* ⊕ Idvi') || (Pow, ⊕Pr) ||(IdIisf⊕TFi) ⊕ IdRi) ⊕ Idvisf'), adds 1 to the value of *NFR* parameters, and $BuldVisf = true.\ Off = Idvisf.\ G$ and Send $\{VIdvi^{'} *, Idvi^{'}, Pow, Pr, Idvisf, Idvisf^{'}, TFi, IdRi, IdFi, FRH, BuldVisf, IdRi, off\}$ to $Idvisf$.

Step 4: When Idvis receives the offloading request, it first verifies the timestamp TFi < ΔT It recomputes $Fr' = h\ ((VIdvi' * \oplus Idvi') \parallel (Pow, \oplus Pr) \parallel (Idvisf \oplus TFi) \oplus IdRi)) \oplus Idvisf')$ and checks whether Fr = Fr′. It also validates the offloading parameter off = IdVisf. G, If the checks are correct, the offloading computation begins. After completion, the result is returned as Send {.} to IdFi.

Step 5: Finally, IdFi consolidates the processing result and forwards it back to IdRi, which delivers it to Idvi. This finalizes the offloading phase with verified processing and secure communication between all parties. Figure 5 shows the Offloading phase.

4 Security Analysis

In this section, we analyze the security of the proposed LSO-IFC scheme through formal verification with the AVISPA tool under the Dolev–Yao model and complementary informal evaluation against common cryptographic attacks, demonstrating its robustness and lightweight security guarantees.

4.1 Formal Analysis

This section presents the formal verification of the LSO-IFC scheme, conducted using the AVISPA framework.

4.2 AVISPA Tool

The AVISPA framework is a widely used tool for the formal verification of security protocols, enabling automated analysis of whether a scheme is secure or vulnerable under the DY intruder model, which assumes complete control over the communication medium [26, 27]. Protocols are specified in the role-based HLPSL language and automatically translated by *hlpsl2if* into an Intermediate Format (IF) for analysis [28]; AVISPA then verifies the protocol using four backends: OFMC, CL-AtSe, SATMC, and TA4SP.

4.2.1 Simulation Results

The proposed LSO-IFC scheme is formally verified with the AVISPA tool under three scenarios: resistance to replay attacks, active intruder behavior, and man-in-the-middle (MITM) attacks. Simulation results from both the OFMC and CL-AtSe backends Figs. 6 and 7 classify the protocol as SAFE, confirming its robustness against replay, MITM, and active intruder attacks and supporting its practical applicability in adversarial network environments.

```
% OFMC
% Version of 2006/02/13
SUMMARY
 SAFE
DETAILS
 OUNDED_NUMBER_OF_SESSIONS
PROTOCOL
/home/span/span/testsuite/results/ LSO-IFC. If
GOAL
 as_specified
BACKEND
 OFMC
COMMENTS
STATISTICS
 parseTime: 0.00s
 searchTime: 0.29s
 visited Nodes: 17 nodes
depth: 12 plies
```

Fig. 6. OFMC results.

```
SUMMARY
 SAFE
DETAILS
BOUNDED_NUMBER_OF_SESSIONS
 TYPED_MODEL
PROTOCOL
/home/span/span/testsuite/results/ LSO-IFC. If
GOAL
 as_specified
BACKEND
 CL-AtSe
STATISTICS
 Analysed: 51 states
 Reachable: 21states
 Translation: 0.19 seconds
 Computation: 0.8 seconds
```

Fig. 7. CL-AtSe results.

4.3 Informal Analysis

An informal security evaluation confirms that LSO-IFC is resilient against a broad range of cryptographic attacks. Timestamped messages, validated against a maximum delay bound ΔT, effectively prevent replay attacks. Mutual authentication based on ECC operations, fresh random nonces, and identity-bound parameters mitigates MITM and impersonation attempts. Robust elliptic-curve–derived session keys, together with non-transmitted secrets (XR, Xvisf, and other private values), protect against brute-force attacks and session key leakage. Passwords are never transmitted directly; instead, they are embedded in salted hash chains combined with identities and timestamps, thereby thwarting both online and offline guessing as well as rainbow-table attacks. Even if verifier data is compromised at the fog or cloud layer, the absence of critical secrets and fresh nonces prevents the adversary from constructing valid authentication requests. Sniffer attacks expose only hashed or ECC-derived values, which are unusable without the underlying secrets. Collectively, these mechanisms ensure confidentiality, integrity, and secure mutual authentication against the considered threats.

5 Performance Analysis

In this section, we evaluate the performance of the LSO-IFC. The analysis compares the communication cost, computation cost, and security features of this protocol with those of existing protocols to highlight its lightweight nature. The LSO-IFC scheme operates in three Sessions: (1) the registration phase, (2) the offloading phase, and (3) the login and authentication phase. For comparison, the NSO-VFC scheme [22] It is employed as a baseline, with the primary objective of showing how LSO-IFC achieves a lightweight reduction in computational and communication overhead.

5.1 Communication Cost and Total Bits

To estimate the communication cost in LSO-IFC, we use 160-bit integers, a 160-bit SHA-1 hash function, and ECC with a 160-bit key size. Each elliptic curve point P is represented as $P = (P_x, P_y)$, where both coordinates are 160 bits (320 bits total). In addition, 32 bits are assigned to the timestamp, 32 bits to the identity, and 32 bits to the control flag. In Session 1, the messages exchanged between Idvi and IdRi include: {*Idvi, TVi, Psw, P, Pi, Pow, IdRi*}, {*Rr, Idvi', HRr, HRPr*, TRi, Idvi*}. Session 1 requires (32 + 32 + 160 + 160 + 160 + 32 + 32) + (32 + 320 + 160 + 160 + 32 + 32) = 1344 bits In Session 2, the messages exchanged between Idvi, IdRi, and IdFi include: {*Pvr*1, *Pvr*2, *Pvr*3, *TVi, Idvi, Idvi', IdRi, HRr*}, {*Pvr*4, *Pvr*5, *Pvr*6, *IdRi, TRi, Idvi*}, {*VAR, Vh, Ps, Pow, TVi, IdRi, Idvi, HRr*} {Idvi′, Idvi, Pow, Pi, TRi, IdRi, IdFi, vi}, Session 2 requires (320 + 320 + 160 + 32 + 32 + 320 + 32 + 160) + (320 + 320 + 160 + 32 + 32 + 32) + (160 + 160 + 160 + 32 + 32 + 32 + 32 + 160) + (320 + 32 + 32 + 160++32 + 32 + 160) = 3808bits. In Session 3, the messages exchanged between Idvi, IdRi, IdFi and IdIisf include {*Rf, Idvi', Pr , TVi, IdRi*} {*Idvi' * , Idvi', Pr , TRi, IdFi*},{VIdvi′*, Idvi′, Pow, Pr, Idvisf, Idvisf′, TFi, IdRi, IdFi, Fr, BuIdVisf, IdRi, off},{Processing result(ro), Idvi, IdIisf}, {Processing result(ro), IdFi, IdRi} {Processing result(ro), IdRi, Idvi} Session 3 requires:(160 + 320 + 160 + 32 + 32) + (320 + 320 + 160 + 32 + 32) + (320 + 320 + 32 + 32 + 32 + 160 + 32 + 32 + 32 + 160 + 32 + 32 + 32) + (32 + 32 + 32) + (32 + 32 + 32) + (32 + 32 + 32) = 3104 bits. As summarized in Table 2, the LSO-IFC scheme is compared with the NSO-VFC protocol in terms of exchanged messages, communication cost, and the number of operational phases. LSO-IFC operates with only three Sessions, requiring 1344, 3808, and 3104 bits, respectively. By contrast, NSO-VFC involves six Sessions with substantially higher communication costs of 2016, 4160, 768, 2656, 2976, and 544 bits. This comparison clearly highlights the superiority of LSO-IFC. With half the number of Sessions compared to NSO-VFC, LSO-IFC significantly reduces communication rounds, minimizes overall bit consumption, thereby offering a more lightweight and efficient solution for IOV environments.

Table 2. Comparison of communication cost in terms of the number of bits

Comparison		*Message count*	*Total cost (in bits)*
LSO-IFC	(*Session* 1)	2	1344
	(*Session* 2)	4	3808
	(*Session* 3)	6	3104
NSO-VFC	(*Session* 1)	2	2016
	(*Session* 2)	4	4160
	(*Session* 3)	3	768
	(*Session* 4)	5	2656
	(*Session* 5)	6	2976
	(*Session* 6)	5	544

5.2 Computation Cast

To evaluate the computation cost of LSO-IFC, we use three metrics: T_{fh} for a hash operation (average 0.0023), T_{fr} for random number generation (average 0.539), and T_{fecm} for elliptic curve point multiplication (average 2.226), as reported in [29]. To avoid human calculation errors, all results are additionally verified using the E3C tool [30].The NSO-VFC protocol is composed of six Sessions. It is important to note that in the sixth Session, hash operations and random number generation are not involved. At the same time, elliptic curve point multiplication cannot be included in the evaluation at this stage. Therefore, the computation cost analysis focuses only on the first five Sessions where these cryptographic operations are actively utilized. The computation cost of the proposed LSO-IFC scheme is evaluated against the NSO-VFC protocol by considering the number of cryptographic operations performed in each Session. As shown in Table 3, LSO-IFC operates with only three Sessions. In Session 1, it performs Tfh11 + Tfr1 + Tfecm2, resulting in a total cost of 5.0163 ms. In Session 2, it requires Tfh14 + Tfr2 + Tfecm5, with a cost of 12.2402 ms. In Session 3, it executes Tfh6 + Tfr2 + Tfecm3, resulting in a cost of 7.7698 ms. By contrast, NSO-VFC is structured into six Sessions, thus involving more cryptographic operations and consequently higher computation costs. In Session 1, it executes Tfh13 + Tfr1 + Tfecm2, resulting in a duration of 5.0209 ms. Session 2 requires Tfh20 + Tfr2 + Tfecm4 with a cost of 10.028 ms, while Session 3 only involves three hash functions, resulting in a negligible cost of 0.0069 ms. In Session 4, the protocol uses Tfh9 + Tfecm4, with a cost of 8.9247 ms, and Session 5 executes Tfh16 + Tfr2 + Tfecm4, resulting in a cost of 10.0188 ms. Session 6 does not involve measurable cryptographic operations. This comparison clearly demonstrates the superiority of the LSO-IFC scheme, which achieves a significantly lower computation cost with fewer Sessions compared to NSO-VFC. By reducing both the number of operational phases and the total cryptographic overhead, LSO-IFC ensures. This reduction in Computation cost makes LSO-IFC highly suitable for resource-constrained IoV environments, where minimizing computational burden is critical. The detailed results are summarized in Table 3.

Table 3. Comparison Computation cost.

Comparison		*Total cost*	*Total cost (ms)*
LSO-IFC	*)Session 1(*	*Tfh*11, *Tfr*1, *Tfecm*2	5.0163 *ms*
	)Session 2(	*Tfh*14, *Tfr*2, *Tfecm*5	12.2402 *ms*
	)Session 3(	*Tfh*6, *Tfr*2, *Tfecm*3	7.7698
NSO-VFC	*(Session 1)*	*Tfh*13 + *Tfr*1 + *Tfecm*2	5.0209*ms*
	(Session 2)	*Tfh*20 + *Tfr*2 + *Tfecm*4	10.028 *ms*
	(Session 3)	*Tfh*3	0.0069 *ms*
	(Session 4)	*Tfh*9 + *Tfecm*4	8.9247 *ms*
	(Session 5)	*Tfh*16 + *Tfr*2 + *Tfecm*4	10.0188 *ms*
	(Session 6)	*NA*	*NA*

5.3 Comparison of Security Features

Table 4 provides a comparative analysis of security requirements between NSOS-FCV and LSO-IFC. The findings indicate that both schemes deliver strong resilience against critical attack vectors, including impersonation (LSO2), man-in-the-middle (LSO1), replay (LSO3), brute-force (LSO4), stolen-verifier (LSO5), session key leakage (LSO6), guessing (LSO7), and rainbow-table attacks (LSO8). Furthermore, both schemes guarantee mutual authentication (LSO9), secure session key agreement (LSO10), confidentiality (LSO12), and robustness against sniffer attacks (LSO11). Their security assurances are supported through formal validation with AVISPA (LSO14) and informal analysis (LSO15). Importantly, both schemes also support SO (LSO13), which plays a crucial role in cloud–fog–IoV environments by ensuring that computational tasks delegated to external nodes are executed without exposing sensitive information. The main differentiator, however, is the lightweight property (LSO16), exclusively achieved by the proposed LSO-IFC scheme. This feature makes LSO-IFC particularly suitable for resource-constrained IoV deployments by reducing computational and communication overhead while preserving comprehensive security guarantees. Consequently, LSO-IFC matches NSOS-FCV in its coverage of diverse threats but surpasses it in efficiency and deployability, making it a more practical solution for real-world applications.

Table 4 Comparison of security features.

Scheme	LSO1	LSO2	LSO3	LSO4	LSO5	LSO6	LSO7	LSO8	LSO9	LSO10	LSO11	LSO12	LSO13	LSO14	LSO15	LSO16
NSOS-FCV	Yes	Yes	Yes	Yes	Yes	Yes	Yes	Yes	Yes	Yes	Yes	Yes	Yes	Yes	Yes	No
LSO-IFC	Yes	Yes	Yes	Yes	Yes	Yes	Yes	Yes	Yes	Yes	Yes	Yes	Yes	Yes	Yes	Yes

6 Conclusion

This paper introduced LSO-IFC, a novel lightweight and SO scheme designed for IoV within fog–cloud Confederation. Unlike existing approaches that achieve SO at the expense of high computational and communication overhead, LSO-IFC successfully balances strong security guarantees with a lightweight design. By integrating ECC, hash functions, and mutual authentication mechanisms, the scheme ensures confidentiality, integrity, and mutual trust among participating entities. Formal analysis using the AVISPA tool under the Dolev–Yao adversarial model, complemented by informal evaluation, confirms LSO-IFC's robustness against a broad range of security threats. Moreover, comparative performance analysis demonstrates that LSO-IFC reduces operational phases, communication cost, and computational complexity, thereby outperforming the NSO-VFC protocol. Future research will focus on extending the scheme to dynamic and heterogeneous IoV environments, incorporating adaptive trust management, and exploring its integration with emerging technologies.

References

1. Sharma, N., Dhiman, P.: Design of secure and unique addressing with mutual authentication scheme in IoT networks. Peer-to-Peer Netw. Appl. **18**(1), 50 (2025)

2. Salami, Y., Khajehvand, V.: LSKE: lightweight secure key exchange scheme in fog federation. Complexity. **2021**, 4667586 (2021). https://doi.org/10.1155/2021/4667586
3. Ahmed, I., Syed, M.A., Maaruf, M., Khalid, M.: Distributed computing in multi-agent systems: a survey of decentralized machine learning approaches. Computing. **107**(1), 2 (2025)
4. Salami, Y., Khajehvand, V., Zeinali, E.: Efficiency simultaneous key exchange-cryptography extraction from public key in fog-cloud federation-based secure offloading for automatic weather stations observing systems. Nivar. **47**(120–121), 153–165 (2023). https://doi.org/10.30467/nivar.2023.416270.1261
5. Chuan, W.C., Manickam, S., Ashraf, E., Karuppayah, S.: Challenges and opportunities in fog computing scheduling: a literature review. IEEE Access (2025)
6. Salami, Y., Khajehvand, V., Zeinali, E.: SAIFC: a secure authentication scheme for IOV based on fog-cloud federation. Secur. Commun. Netw. **1**, 1–19 (2023). https://doi.org/10.1155/2023/9143563
7. Ghafari, R., Mansouri, N.: Reinforcement learning-based solution for resource management in fog computing: a comprehensive survey. Expert Syst. Appl., 127214 (2025)
8. Salami, Y., Ebazadeh, Y., Khajehvand, V.: CE-SKE: cost-effective secure key exchange scheme in fog federation. Iran J. Comput. Sci. **4**(3), 1–13 (2021)
9. Salami, Y., Hosseini, S.: BSAMS: Blockchain-based secure authentication scheme in meteorological systems. Nivar. **47**(120–121), 181–197 (2023). https://doi.org/10.30467/nivar.2023.415722.1260
10. Chaganti, K.C.: A scalable, lightweight AI-driven security framework for IoT ecosystems: optimization and game theory approaches. IEEE Access (2025)
11. Salami, Y., Khajehvand, V., Zeinali, E.: LSMAK-IOV: lightweight secure mutual AKE scheme in fog-based IoV. In: 2024 10th International Conference on Artificial Intelligence and Robotics (QICAR), pp. 1–5. IEEE (2024). https://doi.org/10.1109/QICAR61538.2024.10496659
12. Li, C., Dong, M., Fu, Y., Yu, F.R., Cheng, N.: Integrated sensing, communication, and computation for IoV: challenges and opportunities. IEEE Commun. Surv. Tutorials (2025)
13. Salami, Y.: SOBT-UF: secure offloading in Blockchain infrastructure for intelligent transportation systems using 5G-enabled UAVs within a fog-edge computing federation. In: 2024 19th Iranian Conference on Intelligent Systems (ICIS), pp. 217–222. IEEE (2024). https://doi.org/10.1109/ICIS64839.2024.10887460
14. Wang, L., Deng, X., Zhang, H., Wan, S., Min, G.: Multi-task oriented efficient computational offloading orchestrator for IoT applications in Mobile edge computing. IEEE Internet Things J. (2025)
15. Salami, Y.: SOBV-FEF: secure lightweight data offloading base in blockchain technology for internet of vehicles enabled handover UAVs within a fog-edge federation. Sci. Rep. (2025). https://doi.org/10.1038/s41598-025-31114-x
16. Yi, M., Yang, P., Xie, J., Fang, C., Li, B.: Soca: secure offloading considering computational acceleration for multi-access edge computing. Wirel. Netw. **31**(2), 1021–1035 (2025). https://doi.org/10.1007/s11276-024-03813-2
17. Li, Z., Gong, J., Xiong, X., Wang, D.: Multi-slot secure offloading and resource management in VEC networks: a deep reinforcement learning-based method. IEEE Access. **13**, 4533–4546 (2025). https://doi.org/10.1109/ACCESS.2024.3524636
18. Lu, X., et al.: Blockchain-enabled secure offloading for VEC: a multi-agent reinforcement learning approach. IEEE Trans. Dependable Secur. Comput. **22**(3), 2978–2995 (2025). https://doi.org/10.1109/TDSC.2024.3523561
19. Salami, Y.: SO-ITS: a secure offloading scheme for intelligent transportation systems in federated fog-cloud. Iran J. Comput. Sci. (2025). https://doi.org/10.1007/s42044-025-00318-9

20. Yao, Y., Zhou, Q., Song, L., Huang, S., Yue, X.: Optimization of secure offloading data for space-air-ground integrated networks oriented to Mobile edge computing. IEEE Internet Things J. (2025). https://doi.org/10.1109/JIOT.2025.3597917
21. Salami, Y., Khajehvand, V., Zeinali, E.: SOS-FCI: a secure offloading scheme in fog–cloud-based IoT. J. Supercomput. **80**(1), 570–600 (2024). https://doi.org/10.1007/s11227-023-05499-3
22. Salami, Y., Khajehvand, V., Zeinali, E.: A new secure offloading approach for internet of vehicles in fog-cloud federation. Sci. Rep. **14**(1), 5576 (2024). https://doi.org/10.1038/s41598-024-56141-y
23. Gao, A., Zhang, Q., Wang, Y., Zhang, J., Zhang, Z.: Mission completion time minimization for secure offloading in aerial-IRSs assisted MEC systems. IEEE Trans. Veh. Technol. (2025). https://doi.org/10.1109/TVT.2025.3590004
24. Zheng, T., et al.: Reconfigurable intelligent surface-enabled green and secure offloading for Mobile edge computing networks. IEEE Internet Things J. (2025). https://doi.org/10.1109/JIOT.2025.3591020
25. Dolev, D., Yao, A.C.: On the security of public key protocols. IEEE Trans. Inf. Theory. **29**(2), 198–208 (1983). https://doi.org/10.1109/TIT.1983.1056650
26. Viganò, L.: Automated security protocol analysis with the AVISPA tool. Electron. Notes Theor. Comput. Sci. **155**, 61–86 (2006)
27. Von Oheimb, D.: The high-level protocol specification language HLPSL developed in the EU project AVISPA. In: Proceedings of the APPSEM 2005 Work, pp. 1–17 (2005)
28. Gotsman, A., Massacci, F., Pistore, M.: Towards an Independent Semantics and Verification Technology for the HLPSL. **135**, 59–77 (2005). https://doi.org/10.1016/j.entcs.2005.06.004
29. Kilinc, H.H., Yanik, T.: A survey of SIP authentication and key agreement schemes. IEEE Commun. Surv. Tutorials. **16**(2), 1005–1023 (2014). https://doi.org/10.1109/SURV.2013.091513.00050
30. Salami, Y., Khajehvand, V., Zeinali, E.: E3C: a tool for evaluating communication and computation costs in authentication and key exchange protocol. Iran J. Comput. Sci., 1–11 (2024). https://doi.org/10.1007/s42044-024-00176-x

A Method for Estimating Residential Situations Based on Sequential Analysis of Acoustic Information Using a Pretrained Audio Event Classifier

Ryoya Yumoto[1(✉)] and Katsumi Wasaki[2]

[1] Graduate School of Science and Technology, Shinshu University, Nagano, Japan
25w6104g@shinshu-u.ac.jp

[2] Department of Electrical and Computer Engineering, Faculty of Engineering, Shinshu University, Nagano, Japan
wasaki@cs.shinshu-u.ac.jp

Abstract. This study presents a system that proposes a non-invasive sensing approach based on audio classification to estimate daily living situations in residential spaces. Using YAMNet for audio event classification, acoustic information was segmented and sequenced over time, enabling the estimation of ambient moods and the detection of hazardous sounds through probabilistic state transition analysis. For mood estimation, we designed a hierarchical framework that characterizes moods using the number of states, category occurrence ratios, and dominant states. The proposed method was evaluated using YouTube videos as audio input, demonstrating both classification accuracy and the plausibility of mood estimation. In particular, the consistency between dominant acoustic states, their transition diagrams, and the actual video content demonstrated the practical feasibility of the proposed method. Compared with conventional vision- or wearable-based monitoring, the proposed method reduces privacy concerns and user burden while providing meaningful situational awareness. These findings suggest that our approach can serve as a foundation for future smart home and assistive applications.

Keywords: Audio classification · Anomaly detection · YAMNet · State transition analysis · Context-aware computing

1 Introduction

With the rapid development of AI and IoT technologies, various methods have been proposed to estimate residents' situations using cameras and sensors. These approaches are also being applied in the field of smart homes, where they are expected to improve the efficiency and safety of daily life. However, camera- or sensor-based behavior estimation methods raise significant privacy concerns, particularly when cameras are deployed in private spaces, and they often impose

H. R. Arabnia et al. (Eds.): CSCI 2025, CCIS 2970, pp. 372–377, 2027.
https://doi.org/10.1007/978-3-032-28466-2_28

burdens when residents are required to wear devices, making them invasive sensing solutions [7]. In this study, we focus on everyday sounds and environmental noises occurring in residential spaces, and propose a non-invasive situation estimation method that leverages audio classification techniques. Sounds generated in living spaces contain diverse information about residents' daily situations and their combinations. By recording and analyzing such sounds, it becomes possible to infer occupants' behaviors and contexts.

2 Pretrained Model for Environmental Sound Classification: YAMNet

YAMNet is an audio event classifier that takes waveform inputs and independently predicts scores for 521 audio event classes defined in the AudioSet ontology [3]. It adopts the MobileNet V1 architecture [4] and is trained on the large-scale AudioSet corpus. Because of its lightweight design and low computational cost, YAMNet is well suited for real-time in-home environmental sound classification.

The model accepts normalized mono audio at 16 kHz within the range $[-1.0, +1.0]$. Internally, the waveform is divided into sliding windows of 0.9 s with a 0.48 s hop, and the core network is executed on each frame batch; one of the outputs, scores, is a float32 tensor of shape $(N, 521)$ representing per-frame class scores [6].

Although YAMNet provides fine-grained predictions of concrete acoustic events, it does not directly model abstract concepts such as situations or moods. To estimate such higher-level contexts, a sequential model over classified audio slices is required, as discussed later. Beyond YAMNet, large-scale pretrained audio backbones such as PANNs demonstrate strong transfer to tagging and detection tasks when trained on AudioSet-scale data [5].

3 Proposed Method for Estimating Residential Situations

In this study, audio events unlikely to occur in daily life were excluded from the analysis. Furthermore, audio events classified by YAMNet were grouped into three categories—*human*, *object*, and *environment*—based on their sources, and hazardous sounds were also labeled in advance. The overall flow of the proposed situation estimation method is as follows:

1. **Audio Preprocessing and Segmentation.** To capture short-term environmental changes and situations, the preprocessed audio data are divided into fixed-length segments, e.g., 10 s per slice. This duration is sufficient to capture short-term variations that reflect residential situations.
2. **Classification of Each Segment and Labeling.** Each segment is input into YAMNet to calculate the likelihoods of audio event classes. The top two labels are extracted and combined as an unordered pair to define the *state* of

the segment. This enables the audio data to be interpreted as specific, human-understandable situations. The pre-assigned categories and hazardous labels are also attached at this step.

3. **State Transition Probability Analysis and Visualization.** Based on the classification results, state-to-state transition probabilities are computed to construct a probabilistic state transition model. This analysis reveals how states change over time and allows us to confirm characteristic patterns and sequences. Visualizing the state transitions is crucial for understanding residents' behaviors and environmental changes, as well as for predicting anomalies or specific situations in residential spaces. By representing the temporal relationships of classification results quantitatively, deeper insights into residential situations can be obtained.
4. **Ambient Mood Estimation from State Information.** Finally, the results are analyzed to estimate the *mood* of the environment. As illustrated in Fig. 1, the mood estimation model is defined hierarchically by the number of states, the ratio of categories, and the dominant audio event classes, moving from abstract to concrete levels of description. Our hierarchical mood descriptors align with established perceptual dimensions in soundscape research (e.g., pleasantness–eventfulness) derived from controlled listening experiments [1].

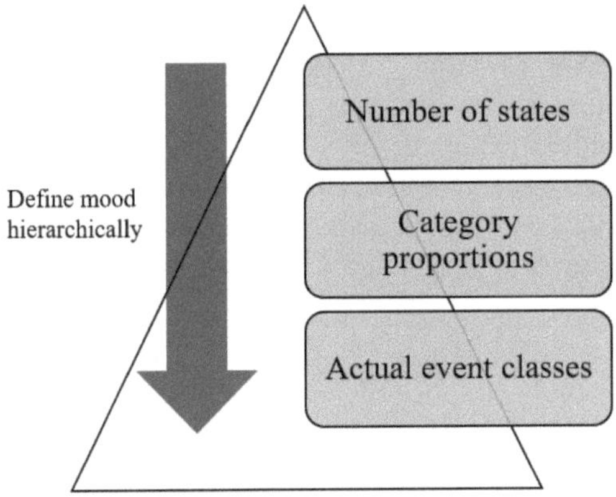

Fig. 1. Hierarchical model of mood estimation (defined by number of states, ratio of categories, and order of event classes, in descending order of abstraction).

For evaluation, a trial experiment was conducted using YouTube videos converted into audio input. Figure 2 shows an example timeline: a 12-minute household dining scene with artificially inserted anomalous scenes in the final portion.

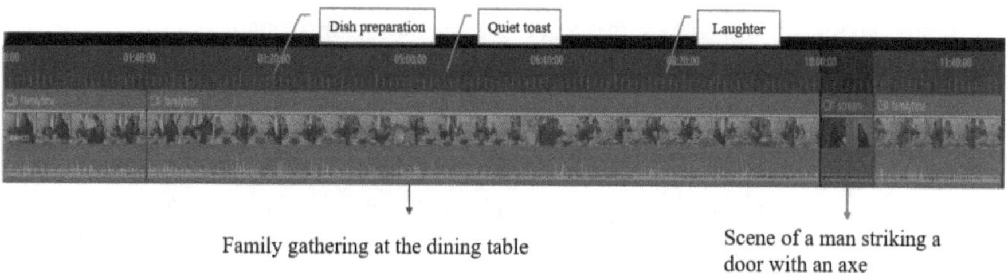

Fig. 2. Timeline of YouTube video used in the trial experiment (12 min in total. Most parts consist of dining scenes, with anomalous scenes inserted at the end).

4 System Overview and Preliminary Experiment

4.1 System Overview

To verify the classification performance of YAMNet and evaluate the validity of the proposed method, we implemented an analysis program that integrates the entire workflow from audio input through preprocessing, segmentation, classification, state transition modeling, mood estimation, and visualization. For audio event classification, we used the pretrained YAMNet model obtained from TensorFlow Hub [6].

The constructed probabilistic state transition model not only represents states and their transition probabilities but also visually distinguishes nodes by coloring them according to the audio event categories (human, object, environment). In addition, nodes corresponding to hazardous sounds are highlighted so that dangerous events can be easily identified.

For the preliminary experiment, we used audio extracted and edited from YouTube videos. As shown in Fig. 2, simple ground-truth labels were manually annotated along the timeline, and the predicted classification results were compared with these annotations to evaluate accuracy.

Mood estimation was implemented by rule-based conditional branching. Following the hierarchical definition, descriptive terms were selected in order from higher-level to more concrete levels, and combined into natural phrases. In this process, selection conditions were designed based on the number of states, the category distribution, and the dominant state.

4.2 Preliminary Experiment

Figure 3 shows the probabilistic state transition model obtained from the preliminary experiment. The following summarizes the estimation results from the perspective of classification scores.

Focusing on the `Speech_Music` node, the following observations were made. Approximately 90% of the transitions looped back to itself, while transitions to other states were rare. The main transition targets included `Speech_Laughter` and Music_Dishes, pots, and pans. Hazardous sounds such as Door_Slam and Arrow_Thump, thud were also detected.

When compared with the actual video, the Speech_Music state corresponded to the presence of conversation and background music, and the high self-transition probability was consistent with the video content. The states Speech_Laughter and Music_Dishes, pots, and pans matched the laughter at 07:40 and dish preparation at 03:25, respectively. However, no state corresponding to the clinking sound of toasts at 05:40 was observed. Regarding Speech_Domestic animals, pets, no pets appeared in the video, and there was no matching segment. The classification score showed "Speech: 0.85, Domestic animals, pets: 0.01," indicating that the speech label was dominant. For hazardous sounds, the detected Door_Slam and Arrow_Thump, thud corresponded to the axe scene in the video.

Based on the number of states, the proportions of categories, and the dominant state, the estimated mood was characterized as "a calm yet moderately lively Speech_Music space." This estimation was consistent with the actual video content and was therefore considered reasonable.

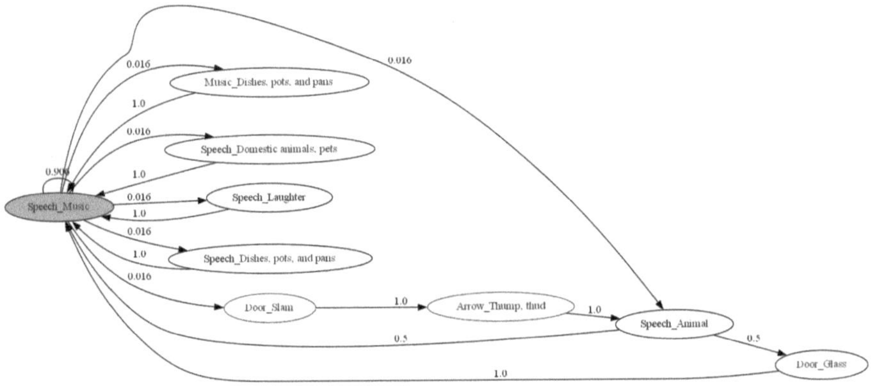

Fig. 3. Example of probabilistic state transition model.

5 Conclusion and Future Work

The results of the preliminary experiment were generally favorable; however, in some cases, extreme bias in classification scores led to misclassification. This phenomenon is considered to be an effect of the softmax function, which forces scores to compete for probability mass. To address this issue, it will be necessary to introduce thresholding when obtaining classification results. In the trial video, in addition to background music and speech, there were very short sounds such as bottles being placed, contact between chopsticks and dishes, and YouTube sound effects. Due to their brevity, these sounds were excluded from manual labeling in the 12-minute video. Consequently, they were not reflected in the classification results, and the estimated mood remained reasonable, suggesting that such short sounds have only limited influence on overall mood estimation.

On the other hand, for the purpose of anomaly detection, the difficulty of classifying short sounds could reduce estimation accuracy. Thus, the unit duration of audio segmentation requires further investigation. Regarding mood estimation, the current definitions remain somewhat abstract. To achieve more concrete estimations, future work should investigate subdividing the available labels or expanding the hierarchical structure of mood definitions. Additionally, incorporating approaches from other research domains may contribute to a more refined framework for mood representation. Our observation that short transients are harder to capture is consistent with findings from rare sound event detection benchmarks (e.g., baby cry, glass break) [2].

References

1. Axelsson, Ö., Nilsson, M.E., Berglund, B.: A principal components model of soundscape perception. J. Acoust. Soc. America **128**(5), 2836–2846 (2010). https://doi.org/10.1121/1.3493436
2. DCASE Challenge: Tut rare sound events 2017 development dataset. https://zenodo.org/records/401395 (2017). baby cry, glass break, gunshot mixtures for rare event detection
3. Gemmeke, J.F., et al.: Audio Set: an ontology and human-labeled dataset for audio events. In: Proceedings of IEEE ICASSP, pp. 776–780 (2017). https://doi.org/10.1109/ICASSP.2017.7952261
4. Howard, A.G., et al.: MobileNets: efficient convolutional neural networks for mobile vision applications (2017). arXiv:1704.04861 arXiv preprint
5. Kong, Q., Cao, Y., Iqbal, T., Wang, Y., Wang, W., Plumbley, M.D.: PANNs: large-scale pretrained audio neural networks for audio pattern recognition. IEEE/ACM Trans. Audio Speech Lang. Process. **28**, 2880–2894 (2020). https://doi.org/10.1109/TASLP.2020.3030497
6. TensorFlow Hub: Yamnet: A pretrained deep net that predicts 521 audio event classes. https://tfhub.dev/google/yamnet/1 (2023)
7. Zeng, E., Mare, S., Roesner, F.: End user security and privacy concerns with smart homes. In: Proceedings of the 13th Symposium on Usable Privacy and Security (SOUPS). USENIX Association (2017). https://www.usenix.org/system/files/conference/soups2017/soups2017-zeng.pdf

Intelligent Safety Warning System for Bridge Crane Operation Scenarios

Zijian Lei[1], Yaokun Chen[2], Fujia Wei[2], Xueyu Zhao[2], and Xinggan Peng[2](✉)

[1] China Machinery Innovation Technology Development (Chongqing) Co., Ltd., Chongqing 400041, China
[2] CMCU Engineering Co., Ltd., Chongqing 400039, China
{zhaoxueyu,pxg}@cmcu.cn

Abstract. Bridge cranes, vital in industrial workshops, pose safety risks when workers inadvertently enter hazardous zones during heavy-load operations—hindering smart manufacturing progress. To mitigate this, we propose a YOLOv11-based real-time intrusion prevention system integrating visual detection, edge computing, and alarms. A custom dataset addresses industrial data scarcity. Our system achieves 0.795 $mAP_{50:95}$and 0.940 F_1 score, outperforming traditional detectors. This work delivers a deployable, scalable safety framework tailored for intelligent factories, advancing proactive protection for high-risk industrial equipment.

Keywords: YOLOv11 · Intrusion Detection · Deep Learning · Smart Factory Safety · Bridge Crane Monitoring

1 Introduction

The bridge crane safety system provides real-time alerts for hazardous operator behaviors—primarily unauthorized entry under suspended loads or into crane movement zones. This is critical in workshops where multi-ton loads move amid dense equipment, overlapping operations, and complex lighting.

Key challenges include dynamic lighting, occlusion, and cluttered backgrounds. While AI vision systems offer contactless, real-time monitoring with strong adaptability, most existing research targets generic scenes, leaving crane-specific safety monitoring underdeveloped—leading to high false/missed detection rates.

This study proposes a YOLOv11-based safety framework for intrusion prevention in crane hazard zones. Integrating detection with physical alarms and real-time decision-making, it enables closed-loop risk-to-protection control. Contributions:

- A deployable intrusion-prevention framework validated in real industrial environments.

Y. Chen—This author contributed equally to this work.

H. R. Arabnia et al. (Eds.): CSCI 2025, CCIS 2970, pp. 378–388, 2027.
https://doi.org/10.1007/978-3-032-28466-2_29

- A high-quality dataset capturing real-world challenges: lighting changes, occlusion, and dynamic backgrounds.
- High adaptability and reusability across industrial sites, reducing deployment barriers.

2 Literature Review

Object detection is central to safety systems in bridge crane environments, aiming to locate and identify intruders from visual data. Current methods fall into two-stage [1] and one-stage detectors [2,3]. While two-stage models offer high accuracy, their computational cost limits real-time use. Single-stage models, with end-to-end regression and feature pyramids, better balance speed and precision for industrial deployment. Applications include construction [4,5], railways [6,7], aviation [8], coal mines [9,10], and parking [11].

However, these methods require adaptation to bridge crane conditions. Dynamic lighting—including reflections, shadows, and sudden brightness shifts—severely degrades detection [12]. Low-contrast targets, such as workers in uniforms blending with metal backgrounds, further challenge feature extraction and classification [13], especially for small, distant objects [14–16]. Thus, robustness to lighting and low contrast is essential for reliable detection in crane environments.

3 Proposed System

This study proposes a bridge crane safety management framework based on object detection, comprising three modules: a visual detection unit, an edge computing unit, and an alarm device, with the specific installation method shown in Fig. 1. Installed on both sides of the trolley frame, the visual module captures real-time video using Hikvision cameras and transmits it wirelessly to the edge computing unit, which analyzes the data to detect personnel intrusions and trigger alarms. The alarm system, consisting of a warning bell and projection light, executes alerts to prevent accidental entry into hazardous zones, thereby reducing safety risks. In practice, the framework operates through three main steps:

1. the visual module captures real-time video near the crane hook and transmits it wirelessly.
2. the edge computing unit performs object detection to identify potential threats.
3. if an intrusion is detected, the alarm system triggers a bell and projects warning lights to alert personnel.

Figure 2 shows the complete process of the system from monitoring, processing to early warning.

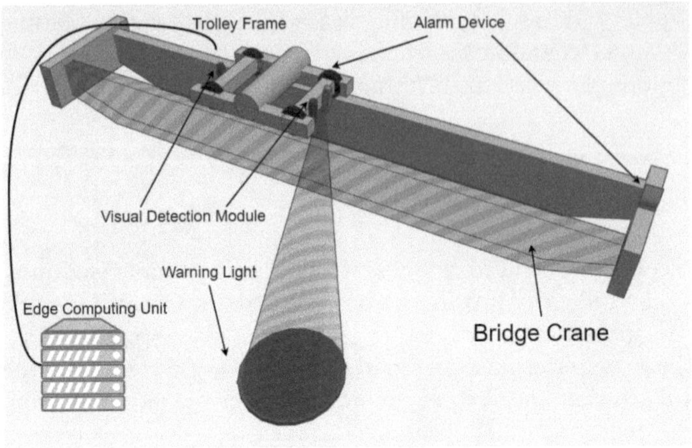

Fig. 1. System Equipment Architecture

3.1 Dataset

Reliable object detection in industrial settings critically hinges upon the development of datasets tailored to specific operational environments. Current studies encounter two primary challenges. Firstly, factory settings present intricate visual noise including fluctuating illumination and equipment-induced clutter; secondly, publicly available datasets are predominantly devoid of industrial context-specific samples. To overcome this, we built a dedicated bridge crane safety monitoring dataset containing five categories.

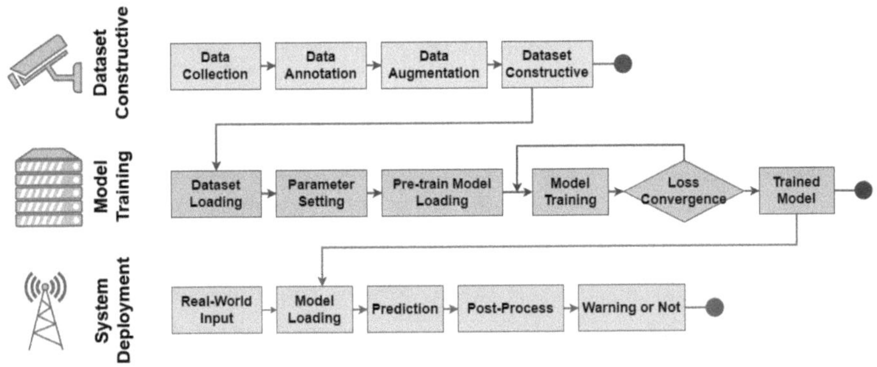

Fig. 2. Flowchart for bridge crane safety management

Dataset Environment This study examines bridge crane operation scenarios where complex visual environments pose significant challenges for object detection. Safety helmets and uniforms blend with steel backgrounds, reducing contrast and edge clarity. Dynamic lighting causes reflections and shadows, while similar background structures and multi-scale targets complicate detection, demanding robust feature discrimination and anti-interference capabilities.

Dataset Annotation To enhance the detection of individuals with weak edges, we specifically designed and constructed a set of fine-grained categories in the dataset, as illustrated in Fig. 3. These include **person**, **mold**, **rope**, **blue label**, and **finger difference indicator**. This design aims to improve the model's anti-interference capability through detailed visual monitoring and negative sample reinforcement. Additionally, it establishes a visual-semantic mapping that guides the model to focus on distinguishing features across these manually designed classes, better meeting the demands of security monitoring. Example annotation results are shown in Fig. 4.

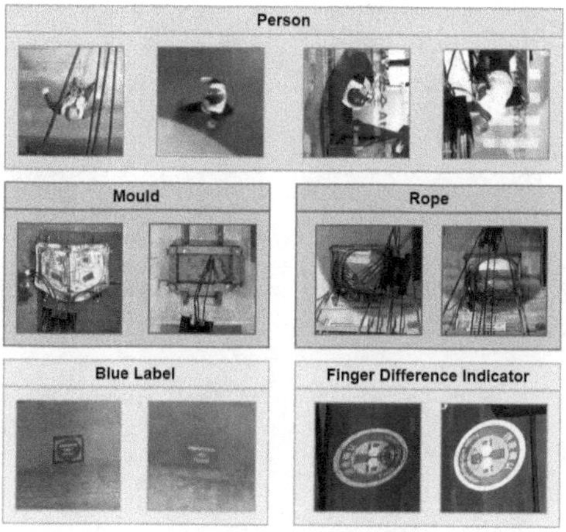

Fig. 3. Five categories of specific images in the proposed dataset

Dataset Augmentation Data augmentation is essential in deep learning for enhancing model generalization and robustness by diversifying training samples while preserving semantic integrity. This study employs an Ultralytics-based augmentation pipeline that strategically combines geometric transformations, photometric adjustments, and occlusion modeling to simulate real-world industrial variations. Geometric operations such as rotation, translation, scaling, mosaic composition, and perspective distortion strengthen spatial invariance. Photometric enhancements including hue saturation value modulation, contrast variation, and Gaussian blurring improve resilience to lighting fluctuations. Controlled occlusion is introduced through randomly placed rectangular masks covering 10–30% of each image, ensuring partial but never complete target concealment. To further broaden data diversity, one thousand annotated aerial imagery samples of human subjects are standardized in format and seamlessly fused into the training set for cross-domain enrichment. This approach harmonizes universal augmentation principles with domain-specific refinements, yielding significantly improved detection accuracy under the challenging visual conditions typical of industrial factory environments.

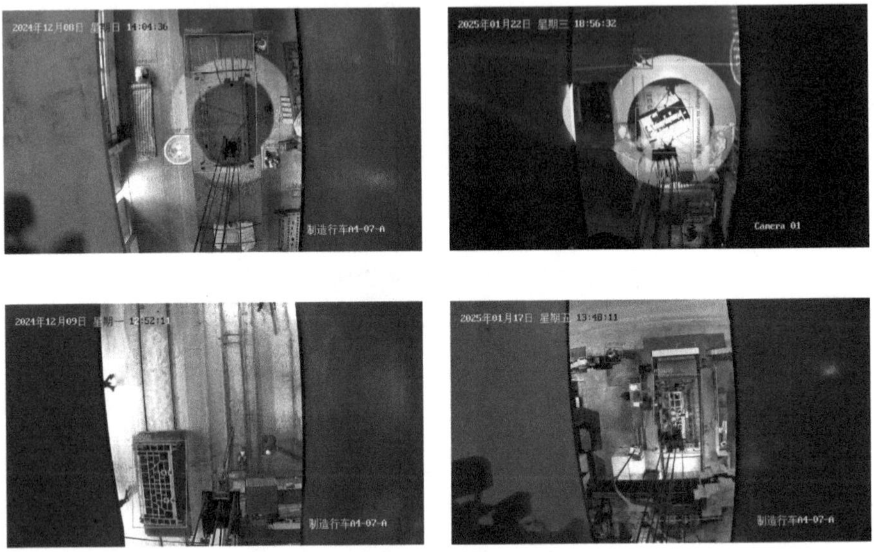

Fig. 4. Annotated dataset image

Dataset Configuration As shown in Table 1, the original dataset comprises 3000 images, which increased to 4,000 through integration of person-category samples from the UAV dataset. The dataset was partitioned into 2800 training samples, 800 validation samples, and 400 test samples. The variety of data was improved via the application of data enhancement techniques: geometric spatial transformations, photometric aberrations, and occlusion simulation approaches. This process ultimately generated 5,200 enhanced data instances. The 4000 enhanced training samples underwent geometric transformations to establish orientation/scale-invariant feature learning, photometric variations strengthened illumination-robust reasoning capabilities, and occlusion patterns via stochastic masking guided the network to prioritize locally significant target characteristics. Validation set construction utilized stratified sampling with 15% cross-domain UAV samples per batch, systematically testing model generalization across distribution shifts. Performance assessment objectivity was guaranteed by maintaining pristine, non-augmented data distributions in the final test set (Fig. 5).

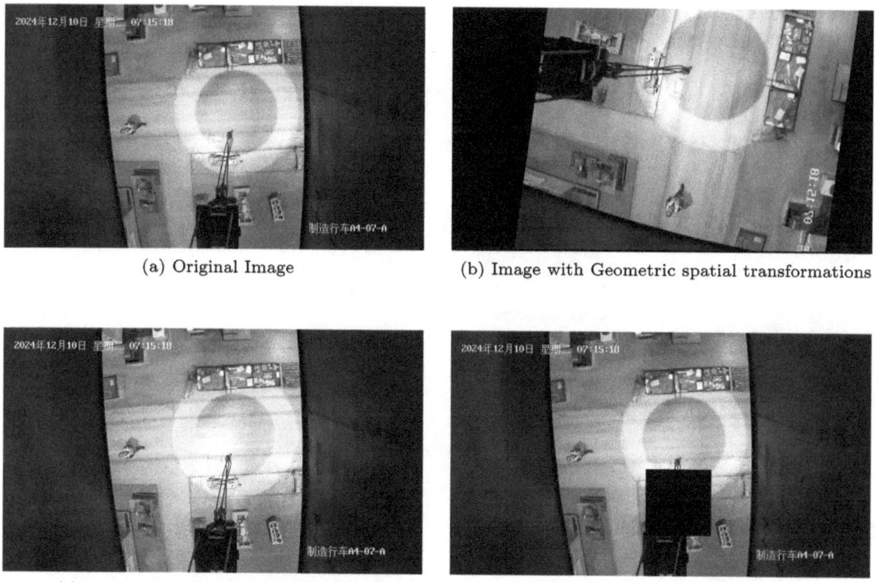

(a) Original Image

(b) Image with Geometric spatial transformations

(c) Image with Photometric distortions

(d) Image with Occlusion simulation

Fig. 5. Augmented dataset images

Table 1. Dataset Split Summary

	Training	Valid	Test	Total
Original Data	2000	600	400	3000
UAV	800	200	0	1000
Geometric Spatial Transformations	400	0	0	400
Photometric Distortions	400	0	0	400
Occlusion Simulation	400	0	0	400
Enhanced Dataset	4000	800	400	5200

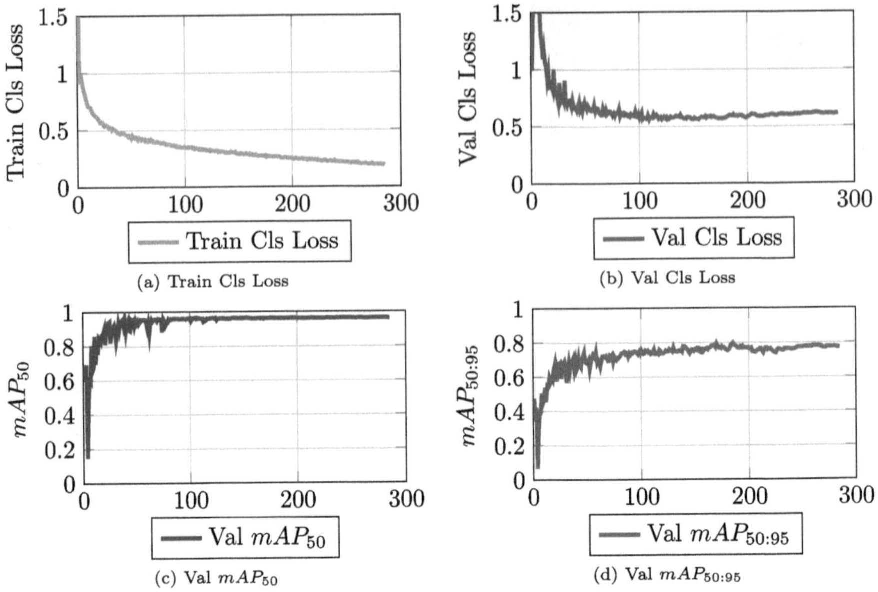

Fig. 6. YOLOv11 Training result curve

4 Results and Discussion

This study aims to design a safe and reliable solution for bridge cranes to detect whether operators accidentally enter hazardous areas during operation. The YOLOv11 object detection model was selected as the core tool for achieving this goal due to its excellent detection performance and efficient inference speed. In order to validate the effectiveness of the dataset developed in this research, the performance evaluation results of the YOLOv11 model on this dataset will be demonstrated. Moreover, considering that the system will be deployed in resource-constrained industrial production environments, this paper selects two widely used models in the industrial field, YOLOv5 and YOLOv8, as benchmarks for comparison.

4.1 Training Detail

Training results are visualized in Fig. 6. Figure 6(a) shows the training classification loss curve indicating improved class discrimination. Figure 6(b) displays the test classification loss which slightly increased after reaching its minimum, suggesting early overfitting. Figure 6(c) and 6(d) present mAP metrics that steadily improved and stabilized, confirming effective model convergence.

Table 2. Performance Comparison of YOLOv5, YOLOv8 and YOLOv11

Model	Recall [11]	Precision [11]	F1 Score [17]	mAP_{50} [16]	$mAP_{50:95}$ [16]
YOLOv5x	0.861	0.936	0.897	0.945	0.753
YOLOv8x	**0.947**	0.908	0.927	**0.962**	0.773
YOLOv11x	0.940	**0.940**	**0.940**	0.961	**0.795**

4.2 Result Analyses

The YOLOv5, YOLOv8, and YOLOv11 models were trained in comparable environments, with their performance subsequently assessed on the test dataset. As presented in Table.2, the evaluation outcomes for the three models include comprehensive breakdowns of five critical metrics: Recall, Precision, F1 Score, mAP50, and mAP50-95. In summary, while all three models achieved strong target accuracy, their metric-specific performances highlighted unique strengths, underscoring the distinctive design philosophies and optimization emphases inherent to each architecture.

Specifically, the YOLOv8 model achieved the highest scores in Recall and mAP_{50}, reaching 0.947 and 0.962, respectively. This demonstrates that YOLOv8 exhibits a notable superiority over YOLOv5 in target coverage and localization accuracy under low IoU thresholds. Especially in challenging industrial environments like bridge crane work zones, its dynamic labeling allocation strategy and enhanced feature pyramid architecture significantly boost the detection rates of small objects, such as far-field pedestrians. Nevertheless, when compared with YOLOv11, the performance edge of YOLOv8 is not substantially pronounced.

In comparison, YOLOv11 achieved the highest scores in the remaining three scores, F1 Score, and $mAP_{50:95}$ particularly excelling with a remarkable 0.795 in the most comprehensive metric, $mAP_{50:95}$. This result clearly demonstrates that YOLOv11 not only maintains high accuracy under lenient IoU thresholds but also sustains excellent detection performance within stricter IoU ranges. The results suggest that YOLOv11 exhibits superior precision in identifying personnel, especially demonstrating remarkable resilience when detecting edge-ambiguous targets like partially occluded or motion-blurred individuals.

Figure 7 intuitively compares the performance of three models. While YOLOv8 slightly outperforms YOLOv11 in Recall and mAP_{50} by 2%, YOLOv11 significantly improves Precision by 3.6% and F1 Score by 2.8%. It also shows a clear advantage in $mAP_{50:95}$, surpassing YOLOv8 by 4.3% and YOLOv5 by 7.8%. These results confirm YOLOv11's superior overall stability and accuracy, making it more suitable for industrial safety applications such as bridge crane danger zone detection, where high precision and reliability are critical.

4.3 Visualization of Experimental Results

This subsection demonstrates the performance of the bridge crane anti-intrusion system in real scenarios through image examples, focusing on detection accuracy,

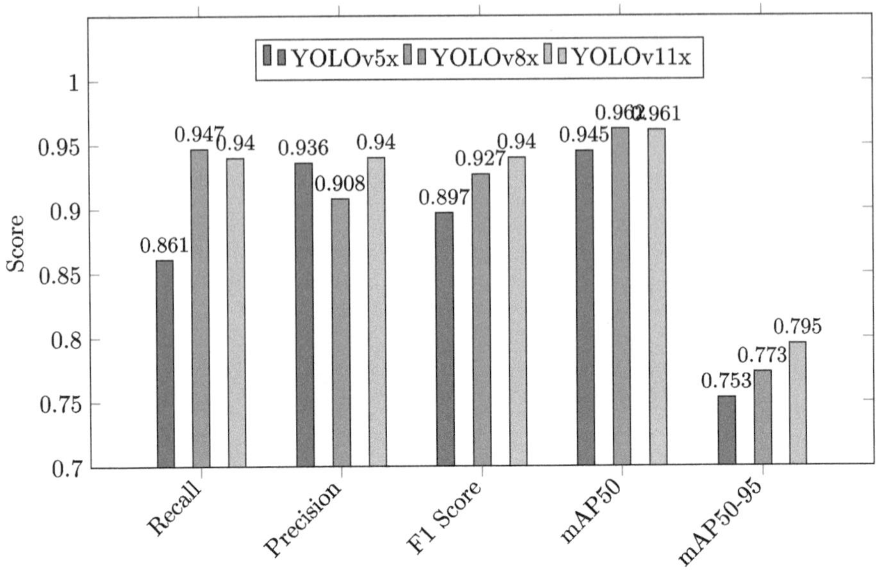

Fig. 7. Performance Comparison of YOLOv5, YOLOv8 and YOLOv11

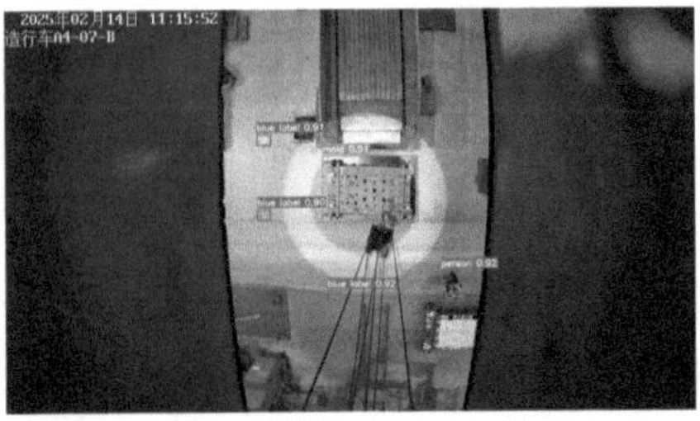

(a) Normal Scenes

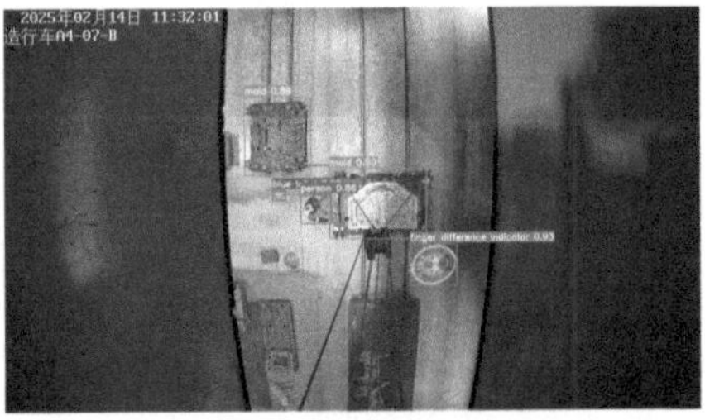

(b) Normal Scenes with Alarm

(c) Complex Sences

(d) Complex Sences with Alarm

Fig. 8. Real world scenes

bounding box precision, and false alarm rate. Specifically, Fig. 8(a) shows operators safely working outside the warning zone with high confidence scores, indicating strong object recognition. Figure 8(b) captures a moment when an operator unknowingly approaches the danger zone, and the system promptly issues an alert, demonstrating its responsiveness. Figure 8(c) shows robust detection under environmental interference, with confidence above 0.8. Finally, Fig. 8(d) illustrates accurate intrusion detection and alarm triggering in complex settings. These results confirm the system's reliability and accuracy in industrial environments.

5 Conclusion

This study developed an intelligent video intrusion alert system for bridge crane operations by integrating advanced object detection with physical alarms, forming a closed-loop safety mechanism. Using YOLOv11 as the core model, significant improvements in accuracy and efficiency were achieved over YOLOv5 and YOLOv8, especially in detecting occluded or blurred personnel, as shown by higher F1 Score and $mAP_{50:95}$ results. The system enhances industrial safety by not only protecting workers but also correcting unsafe behaviors through real-time alerts. Future work should focus on improving robustness under dynamic lighting, enhancing low-contrast detection, exploring multi-modal fusion, and expanding datasets to boost adaptability for safer, more efficient industrial operations.

References

1. Cao, C., et al.: An improved faster R-CNN for small object detection. IEEE Access **7**, 106838–106846 (2019)
2. Redmon, J., Divvala, S., Girshick, R., Farhadi, A.: You only look once: unified, real-time object detection. In: 2016 IEEE Conference on Computer Vision and Pattern Recognition (CVPR), pp. 779–788. IEEE, Las Vegas, NV, USA (2016)
3. Chen, Z., Wu, K., Li, Y., Wang, M., Li, W.: SSD-MSN: an improved multi-scale object detection network based on SSD. IEEE Access **7**, 80622–80632 (2019)
4. Tran, S.V.-T., et al.: A human detection approach for intrusion in hazardous areas using 4D-BIM-based spatial-temporal analysis and computer vision. Buildings **13**(9), 2313 (2023)
5. Huang, H., Hu, H., Xu, F., Zhang, Z., Tao, Y.: Skeleton-based automatic assessment and prediction of intrusion risk in construction hazardous areas. Saf. Sci. **164**, 106150 (2023)
6. Li, C., Xie, Z., Qin, Y., Jia, L., Chen, Q.: A multi-scale image and dynamic candidate region-based automatic detection of foreign targets intruding the railway perimeter. Measurement **185**, 109853 (2021)
7. Chang, C.-C., Huang, K.-H., Lau, T.-K., Huang, C.-F., Wang, C.-H.: Using deep learning model integration to build a smart railway traffic safety monitoring system. Sci. Rep. **15**(1), 4224 (2025)
8. Zhang, T., Zhang, Z., Zhu, X., Chen, B., Li, J., Zhong, Y.: Aircraft engine danger areas incursion detection using keypoint detection and IoT. Alex. Eng. J. **93**, 7–21 (2024)

9. Xu, Z., Li, J., Meng, Y., Zhang, X.: CAP-YOLO: channel attention based pruning YOLO for coal mine real-time intelligent monitoring. Sensors **22**(12), 4331 (2022)
10. Xu, Z., Li, J., Zhang, M.: A surveillance video real-time analysis system based on edge-cloud and FL-YOLO cooperation in coal mine. IEEE Access **9**, 68482–68497 (2021)
11. Zhao, N., Wang, K., Yang, J., Luan, F., Yuan, L., Zhang, H.: CMCA-YOLO: a study on a real-time object detection model for parking lot surveillance imagery. Electronics **13**(8), 1557 (2024)
12. Zhao, K., Miao, X., Xia, J., Li, J.: CP-RDM: a new object detection algorithm for casting and pouring robots. Meas. Sci. Technol. **35**(12), 126009 (2024)
13. Wang, H., Zhao, J., Wang, H., Hu, C., Peng, J., Yue, S.: Attention and prediction-guided motion detection for low-contrast small moving targets. IEEE Trans. Cybern. **53**(10), 6340–6352 (2023)
14. Liu, S., Cao, L., Li, Y.: Lightweight pedestrian detection network for UAV remote sensing images based on strideless pooling. Remote Sens. **16**(13), 2331 (2024)
15. Ku, B., Kim, K., Jeong, J.: Real-Time ISR-YOLOv4 based small object detection for safe shop floor in smart factories. Electronics **11**(15), 2348 (2022)
16. Hua, W., Chen, Q.: A survey of small object detection based on deep learning in aerial images. Artif. Intell. Rev. **58**(6), 162 (2025)
17. Khan, M., Raza, M.A., Abbas, G., Othmen, S., Yousef, A., Jumani, T.A.: Pothole detection for autonomous vehicles using deep learning: a robust and efficient solution. Front. Built Environ. **9**, 1323792 (2024)

Next-Gen IoT Security: Lightweight Cryptographic Protocols Powering the Internet of Everything

Majed. S. Alsayfi(✉)

Department of Cybersecurity, College of Computer Science and Engineering, Taibah University, Medina 42353, Saudi Arabia
msayfi@taibahu.edu.sa

Abstract. The Internet of Everything (IoE) represents a rapidly evolving paradigm that interconnects people, processes, and data through billions of intelligent devices. Securing communication across these heterogeneous and resource-constrained Internet of Things (IoT) environments remains a critical challenge, as traditional cryptographic protocols often fail to meet the stringent energy, memory, and processing limitations of embedded systems. This study provides a comprehensive review of lightweight cryptographic protocols specifically designed for IoE ecosystems. The analysis encompasses symmetric, asymmetric, and hash-based cryptographic primitives, emphasizing their performance, security strength, and implementation feasibility. Recent literature is systematically evaluated based on throughput, energy consumption, CPU utilization, and overall efficiency to benchmark protocol suitability under constrained conditions. The review further explores unresolved challenges, including standardization gaps, quantum computing threats, and integration complexity within IoE architectures. Advances such as adaptive security models, post-quantum cryptography, and hardware–software co-design approaches are highlighted as emerging solutions. The findings advocate hybrid, intelligent, and scalable lightweight cryptographic frameworks that can ensure robust, efficient, and sustainable security across future IoE applications.

Keywords: Internet of Everything (IoE · lightweight cryptography · IoT security · resource-constrained devices · cryptographic protocols · secure communication

1 Introduction

The Internet of Everything (IoE) is a broader evolution of the Internet of Things (IoT), extending beyond device interconnectivity to encompass people, processes, data, and things as a unified intelligent system [1]. While IoT focuses primarily on connecting physical devices to the internet, IoE emphasizes contextual integration and intelligent interaction among all entities in the network, enabling decision-making at the edge and fostering responsive environments [2]. In essence, IoE integrates computing into every

H. R. Arabnia et al. (Eds.): CSCI 2025, CCIS 2970, pp. 389–405, 2027.
https://doi.org/10.1007/978-3-032-28466-2_30

aspect of the environment, enabling seamless communication and intelligent behavior across diverse nodes [3]. These nodes—often limited in size, power, memory, and processing capability—support a wide range of applications including healthcare, transportation, industrial automation, agriculture, and smart homes [4]. Projections indicate that by 2030, over 50 billion devices will be interconnected, forming the backbone of the IoE infrastructure [5].

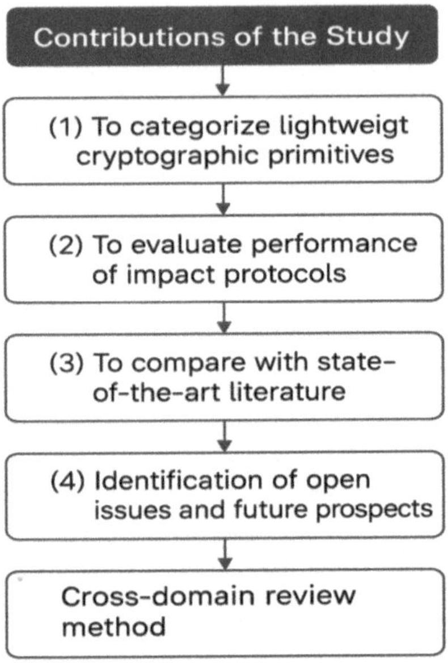

Fig. 1. Core contributions of the study on lightweight cryptographic protocols for IoE security.

However, the widespread deployment of such resource-constrained IoT devices introduces significant security challenges. Many devices operate in untrusted or unattended environments and are incapable of supporting traditional, resource-intensive cryptographic mechanisms [6]. Unlike those in traditional cryptography, the lightweight ones are Common threats in this domain include eavesdropping, data tampering, identity spoofing, and denial-of-service (DoS) attacks [7] Ensuring secure data exchange without compromising device performance is critical to the reliability and scalability of IoE systems. To address these challenges, lightweight cryptographic protocols have emerged as a cornerstone of secure IoE architectures [8]. These protocols are optimized for constrained environments by trading off some cryptographic strength in favor of computational and energy efficiency. They are typically designed to be hardware-friendly, employing streamlined algorithmic structures suitable for limited-resource devices.

This review focuses on lightweight cryptographic methods specifically tailored for IoT devices and their applicability in the broader IoE ecosystem. The key contributions of this paper are as follows:

- -A systematic taxonomy of lightweight cryptographic primitives, detailing their architectural features and design strategies.
- -A comprehensive performance evaluation of representative protocols across multiple criteria, including computational cost, memory usage, and energy efficiency.
- -A critical synthesis of these protocols in the context of recent literature, highlighting their practical strengths and limitations.
- -An analysis of the current challenges—such as standardization gaps and emerging threats like quantum computing—and an outlook on future research directions.

Our review is grounded in theoretical insight and cross-domain analysis, synthesizing technical evaluations with context-aware interpretations. Figure 1 provides a visual summary of the paper's four key contributions. Through this integrative approach, we aim to provide a domain-centric understanding of lightweight cryptography in the era of IoE. While numerous lightweight cryptographic primitives exist, a unified, hybrid architecture optimized for the specific constraints of IoE remains underexplored. This study contributes a conceptual framework and discusses future directions for empirical evaluation.

2 Security Requirements in IoE Environments

In the extensive, interconnected world of IoE, security is more than just about protecting data; it is multidimensional, covering the principles of data confidentiality, system integrity, service availability and user privacy [9]. Considering that IoE devices are commonly installed in mission critical operational contexts, they are predominantly autonomous and that the consequences of a security breach can be dramatic, from devastating financial losses, to even more significantly, human life being put at risk [10].

Confidentiality is a principal requirement that has been developed basically to avoid unauthorized publication of personal or proprietary information. Think about the health industry, for example: IoT-connected monitors are relaying intimate medical information 24/7 [11]. Not only is it the right thing to do, it's also the law under strict guidelines, like those found in the Health Insurance Portability and Accountability Act (HIPAA) in the United States [12]. To address this demand while not burdening the limited computing and power of IoE devices, lightweight encryption schemes are crucial for delivering strong confidentiality assurance at minimal performance cost [13]. Security is also important: the information transmitted must be resistant to any unauthorized changes in the form of maintaining data integrity from source to destination [14]. This is especially crucial for mission-critical scenarios such as industrial control systems where corrupted signals can cause significant breakdowns for the manufacturing flow, or bring dangerous operation exceptions, or even be the cause of actual hazardous results in the physical world [15]. In order to prevent these attacks, lightweight secure hash functions and MAC are often used, providing easy alternatives to check that the data received has not been tampered [16]. Authentication protocols are critical to IoE, and serve as the front line of defense for dealing with the user and allowing only authorized devices and true users to communicate and access some sensitive service of the network. This fundamental security capability is essential for verifying the identities of all involved parties, and to establish a secure and trustworthy operating mode [17]. In order to respond to ubiquitous threats such as spoofing or impersonation, in which threat actors try to impersonate the appearance of legitimate entity/device, mutual authentication schemes are often used.

In contrast to a classic one-way authentication, in which only one of the parties (communicating entities) authenticates the other, mutual authentication necessitates, that both parties verify their identity to communicate. This mutual authentication increases the security status of the network correspondingly. These mutual directional authentication schemes are normally based on complex cryptographic algorithms. This might include cryptographic handshakes (also known as digital handshakes), which are a sequence of messages exchanged between two parties to establish a secure connection and/or identity of the other party, and the use of digital signatures [18]. Digital signatures are a strong means of authenticating a message and verifying its integrity and the identity of its signer.

They make use of asymmetric cryptography; a signature is generated with a private key and a complementary public key is then used to verify the signature. By introducing these facets, a baseline of trust is laid between communicating parties before any exchange of data or execution of commands, greatly reducing the chance of unauthorized use or malicious activity. This preemptive verification is critical to preserving the integrity and security of IoE ecosystems, which encompass everything from smart homes to critical industrial control systems [19].

Availability is also a very important security requirement, though it is often underemphasized. Attackers may attempt to conduct attacks, including denial of service (DoS) attacks, on critical IoE devices or services, making them unavailable to legitimate users [9]. As a result, ultra lightweight security protocols need to be designed with a built-in capability to withstand such kind of disruptive threats in order to operate and offer services consistently. This robustness must be achieved at a very low level of communication overhead and fast response times which are fundamental to real-time IoE application [13].

In the end, resource constraints greatly influence the construction of an efficient security solution for IoE environments. The brute force of these mechanisms must be carefully baked into the device design to materially reduce power consumption, save precious memory, and speed up processing time. These structural prerequisites are not simply merely "nice to have" but fundamental enablers which determine the actual construction and development of domain-specific optimized lightweight cryptographic solutions addressing the specific needs of the Internet of Everything. Table 1. Summarizing security requirements in the IoE environment.

Table 1. Summary of Security Requirements in IoE Environments with Supporting Studies.

Requirement	Relevance to IOE	Lightweight Cryptographic Relevance	Supporting Studies
Confidentiality	Critical for sensitive data (e.g., healthcare, finance).	Lightweight encryption (e.g., block ciphers, stream ciphers).	[8, 9]
Integrity	Ensures trustworthy communication in safety-critical systems.	Lightweight MACs, secure hash functions.	[10, 11]
Authentication	Prevents spoofing, impersonation, and unauthorized device access.	Efficient authentication protocols, digital signatures.	[8, 13]
Availability	Required for real-time applications (e.g., smart grids, surveillance).	DoS-resilient protocols, minimal overhead mechanisms.	[9, 10]
Resource Efficiency	Essential due to constrained power and memory in IoE devices.	Optimization of cryptographic operations for low-resource use.	[6, 13]

3 Lightweight Cryptographic Primitives

In the scope of IoE, lightweight type of complex algorithm sets up a cornerstone for secure and efficiency of the communication. These primitives have been designed to operate in the constrained environment of power, processing and memory requirements of typical IoT devices. They can be classified as symmetric algorithms, asymmetric algorithms, and hash functions, playing different roles for security in lightweight protocols. The relative characteristics of symmetric algorithms, asymmetric algorithms, and lightweight hash functions across five key metrics: key size, energy consumption, speed, memory usage, and security level. Symmetric primitives offer high speed and low resource consumption, making them ideal for constrained devices. Asymmetric algorithms, while more secure and scalable, require greater computational resources. Lightweight hash functions are optimized for minimal memory and energy usage while maintaining acceptable security, particularly for data integrity and authentication tasks.

3.1 Symmetric Algorithms

Symmetric key encryption algorithms, in which encryption and decryption use the same key, are frequently used in IoT as they have relatively low computational and power consumption costs. PRESENT [20], LED [21] and the Simon/Speck families [22] demonstrate high-efficiency lightweight encryption standards. These algorithms are characterized by small key length, good performance and resistance to a range of cryptanalysis techniques. For example, PRESENT [20] is an ultra-lightweight block cipher and is the ISO/IEC 29192 standard for RFID, whose efficient implementation with few gates is a good match to hardware ciphers. Simon and Speck, which were both designed by the NSA, offer parameterizable settings that balance speed and footprint, and are therefore of interest for use in embedded environments [23]. Works like Mohammed et al. [10] demonstrated that PRESENT and Simon are efficient on 8-bit microcontrollers, such that they are suitable for massive IoT setups.

3.2 Asymmetric Algorithms

Asymmetric encryption involves pairs of public and private keys and has especially important applications to secure key distribution and to digital signatures. Yet, classical protocols such as RSA are extremely resource-consuming for IoT devices. Elliptic Curve Cryptography (ECC) can provide an efficient answer: there, the same degree of security can be reached with much shorter keys. ECC-protocols such as ECDSA (Elliptic Curve Digital Signature Algorithm) and ECIES (Elliptic Curve Integrated Encryption Scheme) have been becoming popular in IoT systems [16].

While ECC operations require more resources compared to symmetric operations, recent hardware acceleration techniques and algorithmic enhancement have enhanced its viability. M. Jammula et al. [24] found that ECC is quite heavyweight but provides robust security and scalability in resource constrained environments, especially when mutual authentication is required.

In symmetric ciphertext mode, PRESENT and SPECK proved efficient and in asymmetric encryption scheme, ECDSA and ECIES showed strong security. Lightweight hash functions such as SPONGENT and PHOTON were further compared with performance of low-resource-available platforms.

As well, the CAESAR competition (Competition for Authenticated Encryption: Security, Applicability, and Robustness) was also discussed, which was organized by the cryptographic research community in response to recent advancements. The goal of the competition was to find authenticated encryption methods that were better than AES-GCM and appropriate for devices with limited resources. For lightweight applications, Ascon was chosen as the top recommendation in 2019 [25], with ACORN coming in second. Efficiency, security, and ease of implementation are given top priority in Ascon's architecture, which has since been chosen by NIST as the lightweight cryptography standard in 2023 [26].

CAESAR is not the only ongoing initiative of the National Institute of Standards and Technology (NIST) in the United States. First, to choose safe and effective algorithms for contexts with limitations, the NIST Lightweight Cryptography (LWC) standardization procedure was started in 2015. Ascon's position as the top contender for practical IoT and IoE deployments was further cemented in 2023 when it was formally declared as the NIST standard for Authenticated Encryption with Associated Data (AEAD) with optional hashing for lightweight applications [27]

In order to find and standardize cryptographic algorithms that are resistant to quantum computer assaults, the NIST Post-Quantum Cryptography (PQC) project is a multi-phase international competition. A number of finalist algorithms (such as CRYSTALS-Kyber and Dilithium) have been chosen for standardization, albeit the process is still ongoing.

As the Internet of Everything transitions to long-term data protection in the post-quantum era, these approaches become even more relevant. One interesting avenue for the future is the possible combination of quantum-resistant and lightweight encryption, especially in hybrid security models [28].

3.3 Lightweight Hash Functions

Hash functions are fundamental cryptographic objects that are crucial for preserving the integrity and authenticity of data in digital communications. In the constrained environment of the IoT, these algorithms must find an uneasy balance: they should be extremely small and fast yet still achieve strong collision resistance and preimage security — properties that make it difficult for a malicious actor to counterfeit data or reconstruct their original inputs from their hashed outputs [14]. Conventional cryptographic hash functions, originally developed for more powerful computing systems, are generally too expensive, in terms of computational power, memory, and energy consumption, for the constrained resources of a significant number of IoT devices.

In response to this, the requirement for special purpose lightweight hash functions is established, which are designed extremely carefully to be hardware friendly, consume very low energy and have a very small memory footprint. Examples, such as SPONGENT [29] and PHOTON [11], represent this design philosophy. For instance, SPONGENT uses a special sponge instance, and processes with bit-level permutations. It is designed in a way that shapes the performance into a device with less than 2 KB RAM, making it a good candidate for deeply embedded scenarios. On the other hand, PHOTON is known for its high security against differential attacks with a small circuit size [11]. This property makes it a potential candidate for inclusion in authentication and integrity-check modules in IoT (where every gate and every microjoule (μJ) is valuable).

It has been proven at many seller's sites that those are the light-weight solutions. For instance, Liu et al. [16] that it is actually possible to retain resulting bounded sum-privacy, i.e., even these purpose-built hash functions are cryptographically strong enough. Notably, they do so consume less than half the power that traditional, more resource-heavy protocols like SHA-2 or MD5 would require, emphasizing their importance towards a more secure and energy-efficient IoT [30].

In summary, each kind of lightweight cryptographic primitive has its own advantages according to the application needs. Symmetric techniques are best suited to high-speed, low-power applications. The asymmetric techniques are suitable for secure key exchange, and identity verification, which is at the cost of more resource consumption. Lightweight hash functions also give the necessary data integrity and authentication primitives. The joint use of these primitives, allows to build efficient and sturdy cryptographic protocols for IoE security.

3.3.1 Multi-dimensional Comparative Framework

Representative lightweight cryptographic algorithms are evaluated across four key dimensions:

Security Strength: Resistance to cryptanalytic attacks [20, 31]. **Energy Efficiency**: Power consumption per operation [25, 32]. **Hardware Cost**: Measured in gate equivalents (GE) [20, 31]. **IoE Suitability**: Alignment with device constraints and application requirements [33, 34]. The comparative analysis showed at Table 2, highlights how each protocol addresses trade-offs in energy, security, and hardware efficiency, guiding protocol selection for diverse IoE environments.

Table 2. Comparative Analysis of Lightweight Cryptographic Algorithms Based on Security, Energy Efficiency, Hardware Cost, and IoE Domain Suitability

Algorithm	Security Strength	Energy Efficiency	Hardware Cost	Ideal IoE Domain	Ref
PRESENT	Moderate	High	Very Low	RFID, Sensors	[20, 31]
SIMON	High	Moderate	Low	Industrial IoT	[25, 31]
SPECK	High	High	Low	Mobile Devices	[25, 31]
PHOTON	Moderate	Very High	Very Low	Smart Meters	[25, 35]
LED	Moderate	Moderate	Moderate	Embedded Systems	[20, 21]
Spongent	Low	High	Very Low	Tiny Wearables	[25, 26]

3.3.2 Domain-Based Protocol Taxonomy

Figure 2 presents a taxonomy that aligns lightweight cryptographic protocols with specific IoE application domains:

Smart Healthcare: Demands low-latency and high-confidentiality protocols. Recommended: SPECK, PHOTON [37]. Industrial IoT (IIoT): Requires robustness and extended operational uptime. Recommended: SIMON, LED [36, 37]. Consumer Devices: Prioritizes low power and cost-efficiency. Recommended: PRESENT, Spongent [31]. Autonomous Systems: Depend on secure, real-time decision-making. Recommended: SPECK, LED [20, 31]. This domain-oriented classification bridges the gap between algorithm design and domain-specific deployment needs.

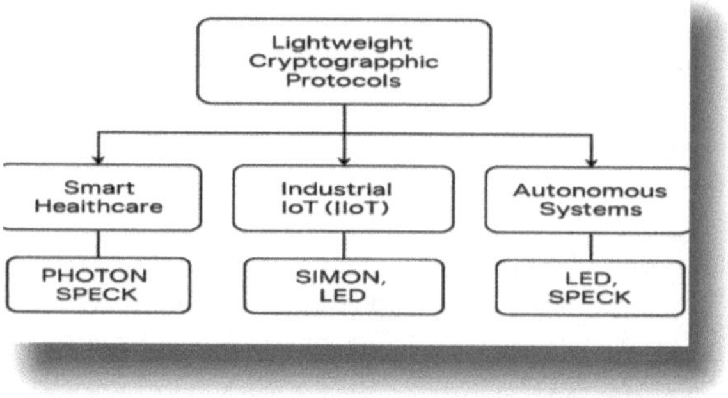

Fig. 2. Domain-oriented taxonomy of lightweight cryptography in IOE.

3.3.3 Research Gaps and Open Challenges

Based on the comparative insights, several key research challenges remain:

Limited development of post-quantum lightweight cryptography [37]. Insufficient empirical benchmarking in real-world edge-IoT settings [38]. Inconsistent standardization of authentication mechanisms across heterogeneous IoE ecosystems [33]. These issues underscore the need for continued innovation and evaluation frameworks tailored to the IoE paradigm.

4 Comparative Studies and Related Work

In order to understand where lightweight cryptographic solutions work well and where they do not in IoT and IoE, there is a need for a review of existing protocols across certain technical categories and to relate them back to the previous work. Because of the extremely restricted capabilities of many IoT devices – such as small processing power, low memory, and limited battery life – cryptographic protocols should be not only secure but also computationally practical. With the advent of the IoE, which is expected to grow to tens of billions of connected endpoints, the security framework needs to be scalable yet at the same time efficient while minimally impacting the users.

As a result, the evaluation of cryptographic protocols becomes a problem that must consider several factors, including algorithmic complexity, computational cost in hardware and software, latency, throughput and resistance to known attacks. These characteristics are usually interconnected; hence comparative analysis is an essential step for determining the protocol that gives the best trade-off between security and efficiency in realistic scenarios. This section combines a comparative analysis of popular lightweight protocols with a synthesis of recent literature to reflect on advances, limitations and open research questions. Table 3. summarizes the comparative studies

A previous review papers studied lightweight cryptography from different points of views. Kumar et al. [39] analyzed low-energy cryptographic primitives based on performance/energy trade-offs without going into too much depth with respect to the protocol-level integration. Mohammed et al. [8] addressed symmetric key algorithms; evaluated several light block ciphers by showing their adequacies on embedded and wireless sensor node. Amrita, et al. [11] proposed a taxonomy of lightweight cryptographic primitives and protocols, but their study provided no insight into the performance of specific deployments.

M. H. Miraz et al. [7] discussed HW/SW co-design for lightweight cryptography but only provided some results how these designs perform in heterogeneous IoT environments. LAKEE Protocol: In [7, 10], 161 Nabavi Razavi and Iyengar presented a lightweight authenticated 162 key exchange protocol for power-constrained devices in IoT, called LAKEE protocol. Based on elliptic curve cryptography, LAKEE can achieve the low overhead of both communication and computation with strong security, and so it is applicable to resource limited scenarios [40]. Post-Quantum Cryptography for Internet of Things: Fernandez-Carames presented a high-level overview of post-quantum cryptosystems relevant to IoT, highlighting demand for quantum-resistant algorithms in upcoming IoT rollouts. The work in [41] notes the challenges and possible solutions for the integration of PQC with limited-resource IoT devices.

Efficiency and Security Analysis: Lightweight cryptographic algorithms such as AES-128, SPECK, and ASCON were assessed in a recent study on IoT devices with limited resources. By evaluating execution time, memory usage, latency, throughput, and security resilience, the study offered important new information on the trade-offs between security and performance [42]. Lightweight Cryptography algorithm survey: Suryateja et al. provided an extensive survey of lightweight cryptographic algorithms for IoT, they examined multiple well-known algorithms and gave taxonomies to facilitate for researchers the understanding of this field [43]. Existing studies must be also compared and related among each other in order to have an overview of the performance and limitations on lightweight cryptographic approaches in an IoT/IoE scenario across different technical issues. Because most IoT devices have very limited processing, memory and therefore battery life, cryptographic protocols should not only be secure but also should require an affordable amount of computation power. The IoE also adds challenges for the security infrastructure, which must now scale to accommodate billions of Internet-connected endpoints without impacting system performance or being intrusive to the user.

Table 3. Comparative Analysis of Lightweight Protocols

Protocol	Type	CPU Usage	Memory Footprint	Energy Cost	Security Level	IoT Suitability
PRESENT	Symmetric	Very Low	Very Low	Very Low	Medium	High
SPECK	Symmetric	Low	Low	Low	Medium	High
ECC-ECDSA	Asymmetric	High	Medium	High	High	Moderate
HIGHT	Symmetric	Low	Low	Low	Medium	High
SPONGENT	Hash	Very Low	Very Low	Very Low	Medium	High

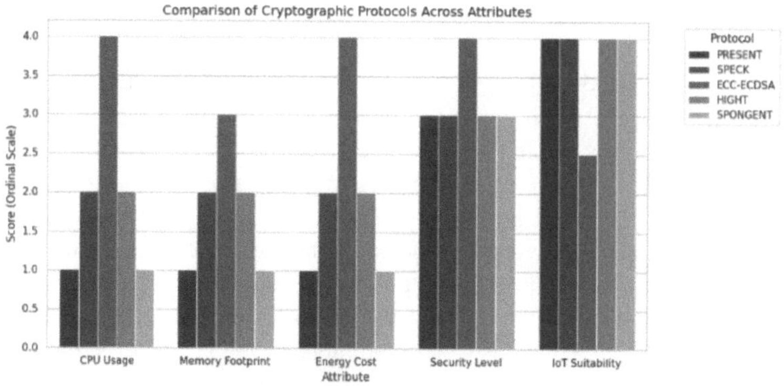

Fig. 3. Comparative analysis of lightweight protocols.

Figure 3. Visual comparison of lightweight cryptographic protocols (PRESENT, SPECK, ECC-ECDSA, HIGHT, SPONGENT) based on CPU usage, memory footprint, energy cost, security level, and IoT suitability. This chart highlights trade-offs for protocol selection in resource-constrained IoT/IoE environments. Thus, the analysis and comparison of cryptographic protocols must consider several dimensions, including algorithmic complexity, hardware and software size and latency, performance and resilience to known attacks. Moreover, these dimensions are not independent, justifying the comparative analysis as an essential part to identify protocols providing the best trade-off between security and performance in practice. This part also combines comparative analysis of the state-of-the-art lightweight protocols with recent studies in the

literature to present advancements, trade-offs and future research issues. In contrast to these, our work provides a more holistic view by addressing not only algorithmic classifications and protocol benchmarking but also practical relevance. This wider view allows researchers and developers to choose or customize lightweight cryptographic protocols to suit their IoE application domains. Standard benchmarking frameworks are important for assessing lightweight cryptographic methods' practical effectiveness in addition to our comparative analysis and literature synthesis. Performance indicators that are essential for deployment in limited environments are provided by a number of well-known initiatives. The NIST Lightweight Cryptography (LWC) Microcontroller Benchmarking Project measures memory utilization, performance, and energy consumption to evaluate cryptographic candidates on popular microcontroller systems including the ARM Cortex-M and AVR. In order to enable equitable performance comparisons under controlled circumstances, the eBACS (ECRYPT Benchmarking of Cryptographic Systems) initiative benchmarks cryptographic software implementations according to throughput and efficiency [44, 45].

5 Challenges

The development of lightweight cryptography has made progress, there still are some open issues. One significant issue is the lack of universally accepted standards for lightweight cryptography, making interoperability across platforms and vendors a critical concern [46]. In the absence of standard security policies, various security solutions are developed, which may not be interoperable, and they might threaten the integrity of the IoE systems [32]. A newfound issue being raised is that of quantum computing. Even old and well-studied light weight algorithms building on integer factorization or discrete logarithms can turn out to be weak against quantum attacks [47]. Post-quantum cryptography is a nascent field and expanding it in lightweight contexts remain an open research issue [9]. The nature of IoE environments is also highly dynamic, which leads to extra challenges. IoE networked devices are mostly mobile and context-aware and adaptive cryptographic schemes which are capable of reacting on dynamic threats and operational condition, are needed. The current static schemes are unable to satisfy these dynamic needs [14]. Finally, the practical dimension of usability and integration represent additional obstacles. Several lightweight cryptography primitives need extensive modifications on firmware, communication protocols, or hardware level [48]. This adds to the development complexities and the associated costs that may restrict penetration.

6 Mitigating Challenges

The increasing popularity of IoT and advance towards the Internet of Everything bring forth many technical and operational difficulties in the deployment of secure and efficient cryptographic protocols. Luckily, scientists and engineers have suggested an array of methods to alleviate these challenges. In this section we cover essential corresponding mitigations that discuss the major challenges in lightweight cryptographic design. One of the major problems is that there is no standard so far, thus it is quite difficult to interoperate among different platforms and devices. To overcome this challenge, various international standards such as ISO/IEC 29192 prescribe cryptosystems which are both secure and lightweight, among them are dedicated lightweight cryptographic algorithms specifically designed for restricted environments [39].

Likewise, encouraging IoT developers and vendors to adhere to them can help prevent many of the compatibility and reliability problems that plague systems in the IoE environment [58]. Despite the resource consumption, the current research trend on PQC optimization work indicates that integrated lightweighted implementation can be achieved in the future [59].

Another direction of mitigation is to design adaptative cryptographic schemes. The traditional static security models cannot adapt to the dynamic and mobile nature of context-aware IoE devices. Solutions based on artificial intelligence (AI), behavior-based anomaly detection and real-time adaptive systems are promising in dynamically changing cryptographic algorithm parameters to adapt to new situations and threats.

As a way of managing the complexity of integration, the trend nowadays is more towards hardware-software codesign. These techniques integrate cryptographic hardware accelerators with flexible software APIs, which permit protocols to be embedded more efficiently in actual systems.

Finally, energy and resource constraints can be addressed using ultra-lightweight ciphers and tunable cryptographic parameters. For example, LAKEE and SPECK are designed to provide developers with the freedom to choose appropriate configurations considering security requirements and hardware constraints. This technique favors configurable and homogeneous installments for a variety of IoT applications.

7 Conclusion

The purpose of this review was to present an in-depth and structured survey of lightweight cryptographic schemes targeted at the IoE. As the IoE grows making room for billions of interconnected devices, preserving secure connectivity and promoting computation and energy efficiency garner tremendous importance. We started by discussing the basic security needs in IoE settings, including privacy, integrity, authentication, availability and resource-efficiency. These requirements are the basic building blocks that any lightweight cryptographic design should fulfill.

The paper then presented important cryptographic primitives: symmetric-key algorithms, asymmetric algorithms, hash functions, and in the context of an IoT constrained environment. Comparison with the state of the art was provided in terms of key size, memory size, energy consumption and security level.

In symmetric ciphertext mode, PRESENT and SPECK proved efficient and in asymmetric encryption scheme, ECDSA and ECIES showed strong security. Lightweight hash functions such as SPONGENT and PHOTON were further compared with performance of low-resource-available platforms. To put this in context, we compared and contrasted with existing work and recent literature, discussing works on performance evaluation, post-quantum readiness, and cryptographic taxonomies. These works established and made apparent that improvements and challenges still exist, especially in terms of performance vs. security trade-offs.

Main challenges identified are standardization issues, new quantum threats, fitting within dynamic IoE surroundings, and how to implement light-weight security into various systems. To mitigate these concerns, we suggested certain solutions, such as advocating ISO/IEC standards, investigating the post-quantum cryptography, employing AI-based adaptive mechanisms, treating the hardware/software co-design aspect, and using tunable lightweight protocols like LAKEE.

Mentioned mitigation strategies could be structured in an energy conversion mitigation framework, and were supported by recent literature, highlighting that overcoming these issues was possible with further interdisciplinary research and joint equipment development efforts. Although the current work is conceptual, the proposed hybrid framework lends itself to implementation using open-source IoT testbeds and lightweight encryption libraries. Future validation will include performance profiling, energy consumption benchmarking, and cryptographic robustness assessment. To support future research and development in this area, we recommend the following directions:

Design and evaluation of hybrid cryptographic schemes that combine AI adaptability with post-quantum resilience. Practical implementation and testing of lightweight cryptographic protocols in real-world IoT and IoE deployments. Development of open benchmarking frameworks and datasets for consistent evaluation of lightweight cryptographic solutions. Investigation into cross-layer integration strategies that combine hardware acceleration with energy-aware software controls. Continued effort toward global standardization and interoperability across diverse IoE ecosystems. To sum up, lightweight cryptographic schemes are not just a technical requirement, but a fundamental block to guarantee the stability of the IoE in the years to come. Dedicated innovations backed with sound standards and empirical validation will be necessary to ensure secure, scalable and sustainable global connectivity. Given the trajectory, embedding lightweight security with machine learning for dynamic adaptation could constitute the next wave of innovation in IoE security.

Statements and Declarations: The authors declare that they have no known competing financial interests or personal relationships that could have appeared to influence the work reported in this paper.

Availability of Data and Materials Not applicable.

Competing Interests The authors declare that they have no competing interests.

A Clinical Trial Number in the Manuscript. Not applicable

Ethics, Consent to Participate, and Consent to Publish Declarations: not applicable.

References

1. da Costa, V.C.F., Oliveira, L., de Souza, J.: Internet of everything (IoE) taxonomies: a survey and a novel knowledge-based taxonomy. Sensors. **21**(2), 568 (2021). https://doi.org/10.3390/S21020568
2. Hashem, E.M., Shalaby, M.F.Y., Mabrouk, M.S., Marzouk, S.Y., Marzouk, Y.: Integrative software design for hospital information systems. Mansoura Eng. J. **48**(4), 7 (2023). https://doi.org/10.58491/2735-4202.3057
3. Kumar, S., Gade, R.S.R.: Experimental evaluation of Cisco ASA-5510 intrusion prevention system against denial of service attacks. J. Inf. Secur. **03**(02), 122–137 (2012). https://doi.org/10.4236/JIS.2012.32015
4. Cisco Annual Internet Report-Cisco Annual Internet Report (2018–2023) White Paper-Cisco. [Online]. Available: https://www.cisco.com/c/en/us/solutions/collateral/executive-perspectives/annual-internet-report/white-paper-c11-741490.html. Accessed 21 May 2025
5. Safitra, M.F., Lubis, M., Fakhrurroja, H.: Counterattacking cyber threats: a framework for the future of cybersecurity. Sustainability. **15**(18), 13369 (2023). https://doi.org/10.3390/SU151813369
6. (PDF) Lightweight Cryptographic Protocols for Secure IoMT Communication in Edge Networks. [Online]. Available: https://www.researchgate.net/publication/389100333_Lightweight_Cryptographic_Protocols_for_Secure_IoMT_Communication_in_Edge_Networks. Accessed 22 May 2025
7. Miraz, M.H., Ali, M., Excell, P.S., Picking, R.: A Review on internet of things (IoT), internet of everything (IoE) and internet of Nano things (IoNT). In: 2015 internet technologies and applications, ITA 2015 - Proceedings of the 6th International Conference, pp. 219–224 (2017). https://doi.org/10.1109/ITechA.2015.7317398
8. El Khaddar, M.A., Boulmalf, M., El Khaddar, M.A., Boulmalf, M.: Smartphone: the ultimate IoT and IoE device. Smartphones Appl. Res. Perspect. (2017). https://doi.org/10.5772/INTECHOPEN.69734

9. Khan, W.Z., Rafique, W., Haider, N., Hakak, S., Imran, M.: Internet of Everything: Enabling Technologies, Applications, Security and Challenges (2022). https://doi.org/10.36227/TECHRXIV.21341796
10. Devarakonda, S., Halgamuge, M.N., Mohammad, A.: Critical issues in the invasion of the internet of things (IoT): security, privacy, and other vulnerabilities. Res. Anthology Privatizing Securing Data, 1672–1694 (2021). https://doi.org/10.4018/978-1-7998-8954-0.CH081
11. Amrita, C.P., Ekwueme, I.H., Adam, Dwivedi, A.: Lightweight cryptography for internet of things: a review. EAI Endorsed Trans. Internet Things. **10** (2024). https://doi.org/10.4108/EETIOT.5565
12. Edemekong, P.F., Annamaraju, P., Afzal, M., Haydel, M.J.: Health insurance portability and accountability act (HIPAA) compliance. In: Encyclopedia of Public Health: Principles, People, and Programs: Volume 1–2, vol. 1, pp. 295–297 (2024). https://doi.org/10.1097/00006527-200323020-00009
13. IEEE Xplore Full-Text PDF: [Online]. Available: https://ieeexplore.ieee.org/stamp/stamp.jsp?arnumber=9328432. Accessed 23 May 2025
14. Guo, J., Peyrin, T., Poschmann, A.: The PHOTON Family of Lightweight Hash Functions
15. Recommended Practice: Developing an Industrial Control Systems Cybersecurity Incident Response Capability (2009)
16. Khan, M.N., Rao, A., Camtepe, S.: Lightweight cryptographic protocols for IoT-constrained devices: a survey. IEEE Internet Things J. **8**(6), 4132–4156 (2021). https://doi.org/10.1109/JIOT.2020.3026493
17. Szymoniak, S., Piątkowski, J., Kurkowski, M.: Defense and security mechanisms in the internet of things: a review. Appl. Sci. (Switz.). **15**(2) (2025). https://doi.org/10.3390/APP15020499
18. (PDF) Masking Host Identity on Internet: Encrypted TLS/SSL Handshake. [Online]. Available: https://www.researchgate.net/publication/348426936_Masking_Host_Identity_on_Internet_Encrypted_TLSSSL_Handshake. Accessed 25 May 2025
19. Tran, S.N., Hoang, V.T., Bui, D.H.: A hardware architecture of NIST lightweight cryptography applied in IPSec to secure high-throughput low-latency IoT networks. IEEE Access. **11**, 89240–89248 (2023). https://doi.org/10.1109/ACCESS.2023.3306420
20. Bogdanov, A., et al.: PRESENT: an ultra-lightweight block cipher. In: Lecture Notes in Computer Science (Including Subseries Lecture Notes in Artificial Intelligence and Lecture Notes in Bioinformatics), vol. 4727, pp. 450–466. LNCS (2007). https://doi.org/10.1007/978-3-540-74735-2_31
21. Mendel, F., Rijmen, V., Toz, D., Varıcı, K.: Differential Analysis of the LED Block Cipher, pp. 190–207 (2012). https://doi.org/10.1007/978-3-642-34961-4_13
22. Beaulieu, R., Shors, D., Smith, J., Treatman-Clark, S., Weeks, B., Wingers, L.: The SIMON and SPECK Families of Lightweight Block Ciphers. Cryptology ePrint Archive (2013). [Online]. Available: https://eprint.iacr.org/2013/404. Accessed 23 May 2025
23. Mohammad Shah, I.N., Ismail, E.S., Samat, F., Rahman, N.N.A.: Modified generalized Feistel network block cipher for the internet of things. Symmetry. **15**(4), 900 (2023). https://doi.org/10.3390/SYM15040900
24. Jammula, M., Vakamulla, V.M., Kondoju, S.K.: Performance evaluation of lightweight cryptographic algorithms for heterogeneous IoT environment. J. Interconnection Netw. **22** (2022). https://doi.org/10.1142/S0219265921410310;PAGEGROUP:STRING:PUBLICATION
25. Koteshwara, S., Das, A., Parhi, K.K.: Architecture optimization and performance comparison of nonce-misuse-resistant authenticated encryption algorithms. IEEE Trans. Very Large Scale Integr. VLSI Syst. **27**(5), 1053–1066 (2019). https://doi.org/10.1109/TVLSI.2019.2894656

26. Status Report on the Final Round of the NIST Lightweight Cryptography Standardization Process: NIST Releases IR 8454 | NIST. [Online]. Available: https://www.nist.gov/news-events/news/2023/06/status-report-final-round-nist-lightweight-cryptography-standardization. Accessed 12 July 2025
27. Lightweight Cryptography: CSRC. [Online]. Available: https://csrc.nist.gov/projects/lightweight-cryptography. Accessed 12 July 2025
28. Post-Quantum Cryptography: CSRC. [Online]. Available: https://csrc.nist.gov/Projects/post-quantum-cryptography. Accessed 12 July 2025
29. Bogdanov, A., Knežević, M., Leander, G., Toz, D., Varici, K., Verbauwhede, I.: Spongent: a lightweight hash function. In: Lecture Notes in Computer Science (Including Subseries Lecture Notes in Artificial Intelligence and Lecture Notes in Bioinformatics), vol. 6917, pp. 312–325. LNCS (2011). https://doi.org/10.1007/978-3-642-23951-9_21
30. Bogdanov, A., Leander, G., Paar, C., Poschmann, A., Robshaw, M.J.B., Seurin, Y.: Hash functions and RFID tags: mind the gap. In: Lecture Notes in Computer Science (Including Subseries Lecture Notes in Artificial Intelligence and Lecture Notes in Bioinformatics), vol. 5154, pp. 283–299. LNCS (2008). https://doi.org/10.1007/978-3-540-85053-3_18
31. Beaulieu, R., Shors, D., Smith, J., Treatman-Clark, S., Weeks, B., Wingers, L.: The SIMON and SPECK Families of Lightweight Block Ciphers. Cryptology ePrint Archive (2013). [Online]. Available: https://eprint.iacr.org/2013/404. . Accessed 13 July 2025
32. Rana, M., Mamun, Q., Islam, R.: Balancing security and efficiency: a power consumption analysis of a lightweight block cipher. Electronics. **13**(21), 4325 (2024). https://doi.org/10.3390/ELECTRONICS13214325
33. Lakshminarayana, S., Praseed, A., Thilagam, P.S.: Securing the IoT application layer from an MQTT protocol perspective: challenges and research prospects. IEEE Commun. Surv. Tutor. **26**(4), 2510–2546 (2024). https://doi.org/10.1109/COMST.2024.3372630;SERIALTOPIC:TOPIC:ACM-PUBTYPE>OTHER-PERIODICAL;WGROUP:STRING:ACM
34. Bhardwaj, I., Kumar, A., Bansal, M.: A review on lightweight cryptography algorithms for data security and authentication in IoTs. In: 4th IEEE International Conference on Signal Processing, Computing and Control, vol. 2017, pp. 504–509. ISPCC 2017 (2017). https://doi.org/10.1109/ISPCC.2017.8269731
35. Berger, T.P., D'Hayer, J., Marquet, K., Minier, M., Thomas, G.: The GLUON family: a lightweight hash function family based on FCSRs. In: Lecture Notes in Computer Science (Including Subseries Lecture Notes in Artificial Intelligence and Lecture Notes in Bioinformatics), vol. 7374, pp. 306–323. LNCS (2012). https://doi.org/10.1007/978-3-642-31410-0_19
36. Kumar, A., Jain, A.K., Dua, M.: A comprehensive taxonomy of security and privacy issues in RFID. Complex Intell. Syst. **7**(3), 1327–1347 (2021). https://doi.org/10.1007/S40747-021-00280-6
37. Nguyen, H.P., Chen, Y.: Lightweight, post-quantum secure cryptography based on Ascon: hardware implementation in automotive applications. Electronics. **13**(22), 4550 (2024). https://doi.org/10.3390/ELECTRONICS13224550
38. Nabavirazavi, S., Iyengar, S.S.: LAKEE: A Lightweight Authenticated Key Exchange Protocol for Power Constrained Devices (2022). [Online]. Available: https://arxiv.org/pdf/2210.16367. Accessed 13 July 2025
39. Fernandez-Carames, T.M.: From pre-quantum to post-quantum IoT security: a survey on quantum-resistant cryptosystems for the internet of things. IEEE Internet Things J. **7**(7), 6457–6480 (2024). https://doi.org/10.1109/JIOT.2019.2958788
40. Radhakrishnan, I., Jadon, S., Honnavalli, P.B.: Efficiency and security evaluation of lightweight cryptographic algorithms for resource-constrained IoT devices. Sensors. **24**(12), 4008 (2024). https://doi.org/10.3390/S24124008

41. Suryateja, P.S., Rao, K.V.: A survey on lightweight cryptographic algorithms in IoT. Cybern. Inf. Technol. **24**(1), 21–34 (2024). https://doi.org/10.2478/CAIT-2024-0002
42. Introduction. [Online]. Available: https://bench.cr.yp.to/. Accessed 12 July 2025
43. Gaj, K., Kaps, J.P., Amirineni, V., Rogawski, M., Homsirikamol, E., Brewster, B.Y.: ATHENa - automated tool for hardware evaluation: toward fair and comprehensive benchmarking of cryptographic hardware using FPGAs. In: Proceedings - 2010 International Conference on Field Programmable Logic and Applications, pp. 414–421. FPL 2010 (2010). https://doi.org/10.1109/FPL.2010.86
44. (PDF) Recent Advances and Trends in Lightweight Cryptography for IoT Security. [Online]. Available: https://www.researchgate.net/publication/345144109_Recent_Advances_and_Trends_in_Lightweight_Cryptography_for_IoT_Security. Accessed 25 May 2025
45. (PDF) Data Integration and Interoperability in IOT: Challenges, Strategies and Future Direction. [Online]. Available: https://www.researchgate.net/publication/377805078_Data_Integration_and_Interoperability_in_IOT_Challenges_Strategies_and_Future_Direction. Accessed 25 May 2025
46. Sharma, T., Shivangi, Sharma, R.: Post-quantum cryptography for navigating challenges and exploring opportunities. Int. J. Res. Rev. Appl. Sci. Hum. Technol., 14–21 (2025). https://doi.org/10.71143/Q1NHVW93
47. Aslan, B.: Energy consumption analysis of ISO/IEC 29192–2 standard lightweight ciphers. Appl. Sci. **15**(7), 3928 (2025). https://doi.org/10.3390/APP15073928
48. Yalli, J.S., Hasan, M.H., Badawi, A.: Internet of things (IOT): origin, embedded technologies, smart applications and its growth in the last decade. IEEE Access. (2024). https://doi.org/10.1109/ACCESS.2024.3418995

Trends, Challenges, and Future Directions of Wireless Sensor Networks in Underground Mining

Khomotso Modika, Topside E. Mathonsi(✉), and Tshimangadzo M. Tshilongamulenzhe

Department of Information Technology, Tshwane University of Technology, Pretoria, South Africa
{MathonsiTE,TshilongamulenzheTM}@tut.ac.za

Abstract. Wireless Sensor Networks (WSNs) enabled by Internet of Things (IoT) technologies play a critical role in underground mining by improving occupational safety, environmental monitoring, and operational efficiency. These networks support continuous sensing of hazardous gases, temperature, humidity, structural conditions, and enable real-time localization of miners and equipment. Despite their potential, the deployment of WSNs in underground environments remains challenging due to severe signal attenuation, high installation and maintenance costs, limited energy resources, and growing cybersecurity risks. To address these challenges, this paper presents a PRISMA 2020–compliant systematic literature review of recent research on WSN technologies applied to underground mining environments. The review synthesizes studies published between 2019 and 2025 and provides an application-driven taxonomy of WSN solutions, communication technologies, security mechanisms, and energy-efficient strategies. Unlike prior narrative surveys, this work offers an evidence-based comparison of low-power wireless technologies, including LoRa/LoRaWAN, ZigBee, Bluetooth Low Energy (BLE), Wi-Fi, and Ultra-Wideband, highlighting trade-offs between energy efficiency, reliability, coverage, and security under harsh mining conditions. The results reveal dominant research trends, identify critical gaps in secure localization and integrated energy–security optimization, and propose a focused research roadmap aligned with Industry 4.0 and sustainable mining objectives. The findings provide actionable insights for researchers, mining engineers, and safety planners, supporting the design of scalable, resilient, and secure WSN-driven monitoring systems for next-generation underground mining operations.

Keywords: Wireless Sensor Networks · Underground Mines · Internet of Things · Sensor Nodes · Security · Energy-Efficient

1 Introduction

Underground mining environments are inherently hazardous, with geotechnical failures, gas explosions, ventilation breakdowns, and equipment malfunctions posing persistent risks to worker safety and operational continuity. Such incidents can result in severe

H. R. Arabnia et al. (Eds.): CSCI 2025, CCIS 2970, pp. 406–423, 2027.
https://doi.org/10.1007/978-3-032-28466-2_31

injuries, long-term health complications, and fatalities, underscoring the critical need for reliable monitoring and communication infrastructures. Robust communication systems are therefore essential not only for efficient mine operations but also for real-time hazard detection, emergency response coordination, and rescue operations. In this context, Wireless Sensor Networks (WSNs) have emerged as a key enabling technology for enhancing safety and situational awareness in underground mines by supporting continuous monitoring of environmental and physiological parameters in areas that are difficult or unsafe for direct human access [1]. Through real-time data acquisition and transmission, WSNs enable proactive risk identification and timely intervention, thereby significantly improving occupational health and safety outcomes.

Conventional positioning and communication technologies such as the Global Navigation Satellite System (GNSS), including the Global Positioning System (GPS), are unsuitable for underground mining environments due to severe signal blockage, weak reception, and limited coverage, as they rely on satellite-based line-of-sight communication designed for surface applications [2]–[3]. To overcome these limitations, underground mining operations increasingly adopt low-power wireless communication protocols such as LoRa/LoRaWAN, ZigBee, and Bluetooth Low Energy (BLE). These technologies support long-range communication, energy-efficient operation, and self-healing mesh networking, enabling reliable data transmission in obstructed tunnels and harsh subterranean conditions.

Beyond safety monitoring, WSNs play a vital role in improving energy efficiency and operational sustainability in underground mines. By enabling context-aware control of infrastructure systems such as lighting, ventilation, and equipment, WSNs allow resources to be activated or deactivated based on real-time conditions, including the presence of personnel or machinery [4]. This adaptive operation reduces unnecessary energy consumption, lowers operational costs, and supports predictive maintenance by identifying equipment degradation before critical failures occur.

Despite their demonstrated benefits, the deployment of WSNs in underground mining environments remains challenging. High installation and maintenance costs, severe signal attenuation caused by geological formations, and the constrained availability of power resources significantly affect system performance and scalability [5, 6]. Moreover, the decentralized architecture of WSNs, coupled with limited computational, memory, and energy resources, exposes sensor nodes to a wide range of security threats, including unauthorized access, data tampering, and denial-of-service attacks. These vulnerabilities make the design of robust, lightweight, and energy-efficient security mechanisms particularly critical for safety- and mission-critical mining applications [7, 8].

Addressing these technical, economic, and security challenges is essential for the development of secure, scalable, and resilient WSN-based monitoring systems capable of supporting the long-term safety, sustainability, and productivity of underground mining operations. Ensuring high system availability and reliability not only enhances real-time hazard detection but also enables predictive maintenance of critical infrastructure, improves operational efficiency, and facilitates compliance with increasingly stringent safety regulations and industry standards.

Motivated by these challenges, this paper presents a PRISMA 2020–compliant systematic literature review of WSN technologies applied to underground mining environments. The study systematically synthesizes recent research to identify dominant trends, assess communication and security solutions, and highlight unresolved research gaps. Unlike prior narrative reviews, this work provides an evidence-based, application-driven analysis that integrates performance, energy efficiency, and security considerations to support the design of next-generation underground mining systems.

The remainder of this paper is organized as follows. Section 2 presents the contributions of this study. Section 3 outlines the PRISMA-based research methodology. Section 4 describes WSN components and data flow in underground mining environments. Section 5 compares traditional monitoring approaches with WSN-based solutions. Section 6 discusses deployment challenges, while Sect. 7 reviews suitable wireless transmission media. Section 8 examines security and energy-efficient mechanisms, and Sect. 9 highlights the Lessons Learned from the review. Section 10 presents the discussions found in the paper. Finally, the paper concludes by identifying key limitations in existing research and proposing future directions toward resilient, secure, and cost-effective WSN deployments that meet the safety, regulatory, and operational demands of modern underground mining.

2 Contributions of this Paper

This paper makes the following unique contributions to the body of knowledge on WSNs for underground mining:

1. PRISMA-compliant systematic review**:** Unlike prior narrative surveys, this study rigorously applies the PRISMA (Preferred Reporting Items for Systematic Reviews and Meta-Analyses) framework to systematically identify, screen, and analyze recent literature (2019–2025) on WSNs in underground mining.
2. Application-driven taxonomy**:** The paper proposes a structured taxonomy that classifies WSN research according to *application domains* (safety monitoring, localization, environmental sensing, energy management, and security), *communication technologies*, and *performance objectives*.
3. Evidence-based comparison of technologies**:** By synthesizing results across selected studies, the paper quantitatively and qualitatively compares ZigBee, BLE, LoRa, Wi-Fi, and UWB in terms of energy efficiency, coverage, reliability, and suitability for harsh underground environments.
4. Integrated security-energy perspective**:** The review uniquely consolidates security mechanisms and energy-efficient strategies, highlighting trade-offs and complementarities that are often treated independently in existing literature.
5. Research gaps and future roadmap**:** Based on systematic evidence, the study identifies underexplored areas such as PRISMA-validated performance benchmarking, secure localization, and AI-assisted adaptive routing for underground mines, and proposes a focused research agenda aligned with Industry 4.0 and sustainable mining.

3 Research Methodology

2.1 Review Design

This study follows the PRISMA 2020 guidelines to ensure transparency, reproducibility, and methodological rigor in conducting a systematic literature review of WSN technologies applied to underground mining environments.

2.2 Data Sources and Search Strategy

A comprehensive search was conducted across the following digital libraries:

IEEE Xplore, SpringerLink, Elsevier ScienceDirect, MDPI, Wiley, and Google Scholar.

Search strings combined keywords using Boolean operators: "Wireless Sensor Networks" OR "WSN" OR "IoT sensors" AND "underground mining" OR "subsurface" OR "coal mine" AND ("safety" OR "monitoring" OR "energy efficiency" OR "security".

2.3 Inclusion and Exclusion Criteria

Inclusion criteria: Peer-reviewed journal and conference papers; Publications between 2019 and 2025; Studies addressing WSN deployment, performance, security, or applications in underground mining; Articles written in English

Exclusion criteria: Studies focused solely on surface mining or non-mining environments; Non-technical reports, theses, patents, and tutorials; Papers lacking experimental, simulation, or application-level validation

2.4 Study Selection Process

The PRISMA selection process consisted of four stages:

1. Identification – Records identified from IEEE Xplore, SpringerLink, Elsevier, MDPI, Wiley, and Google Scholar.
2. Screening – Title and abstract screening; removal of duplicates.
3. Eligibility – Full-text assessment based on the 2019–2025 timeframe and mining-specific technical validation.
4. Inclusion – Final studies selected for Thematic Synthesis (Performance, Security, Energy, etc.).

2.5 Data Extraction and Synthesis

From each selected study, the following data were extracted:

1. Application domain
2. Communication technology
3. Performance metrics (latency, energy consumption, packet delivery ratio)
4. Security mechanisms
5. Deployment challenges

A thematic synthesis approach was adopted to aggregate findings and identify trends and gaps.

3.1 Applications of WSNs in Underground Mines

Some of the key applications of WSNs in underground mines, together with their features, support critical applications that enhance safety, efficiency, and automation. They enable miner tracking by using a combination of wearable sensors, positioning technologies, and wireless communication, and digital access control by integrating with Radio Frequency Identification (RFID), biometrics, and smart cards to restrict entry to hazardous areas, as well as health and compliance monitoring. Lighting systems use motion sensors for automatic control, reducing energy use and improving visibility. WSNs are also used for equipment monitoring, allowing remote tracking and predictive maintenance of machinery. Lastly, vehicle detection and collision avoidance systems that help prevent underground accidents by sensing nearby objects and personnel [9].

(a) **Health Monitoring, Tracking, and Localization of Miners:**

Wearable devices in Fig. 1 continuously monitor physiological parameters, including heart rate, body temperature, blood pressure, and motion. Collected data is transmitted wirelessly to surface-level monitoring systems or designated medical units. In critical scenarios, automated alerts are generated to initiate timely rescue or medical intervention. For spatial localization, each miner is equipped with an RFID tag or BLE sensor that interfaces with distributed wireless nodes within the mine. This architecture enables real-time personnel tracking, enhancing emergency response capabilities, and supporting operational efficiency.

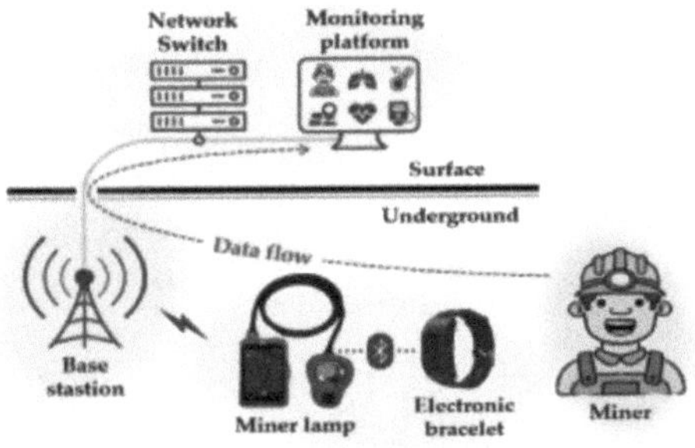

Fig. 1. A monitoring system for miners' status [10].

(b) **Advantages of WSNs in underground mines.**

As highlighted in [11–13], WSNs provide numerous advantages in underground mining operations, including the following:

(i) **Digital Access Control Systems:** Wireless sensors are integrated with access points (e.g., gates or tunnels) and equipped with RFID, biometric scanners, and smart cards to control and monitor the entry and exit of personnel. These systems improve safety by ensuring only authorized personnel have access to hazardous zones.

(ii) **Lighting Automation and Control**: Intelligent lighting systems in underground mines utilize motion detection and scheduled operations to automatically control illumination. This approach reduces energy consumption while improving operational safety in frequently accessed areas.

(iii) **Equipment Monitoring and Control:** Heavy machinery and critical equipment such as drills, conveyor systems, and ventilation fans are equipped with sensors that monitor parameters including vibration, temperature, and operational status. These sensors enable remote monitoring and control, support predictive maintenance strategies, and provide real-time fault alerts to minimize downtime and enhance safety.

(iv) **Ventilation and Gas Monitoring**: Sensors continuously monitor temperature, humidity, airflow, and hazardous gas levels such as methane (CH_4), carbon monoxide (CO), and oxygen (O_2) and transmit this data to central systems for automated ventilation control and safety alerts.

(v) **Environmental Monitoring:** Wireless nodes continuously monitor dust, noise, and seismic activity to maintain workplace safety and ensure regulatory compliance.

(vi) **Vehicle Detection and Collision Avoidance Systems:** Underground vehicles utilize sensors and wireless communication modules to identify nearby personnel, equipment, or obstacles, thereby improving safety and reducing the risk of collisions in areas with limited visibility.

4 Components of WSNs in Underground Mines

4.1 Figure 2 Illustrates the Defined Architecture Required for Efficient Communication in WSNs, Comprising Specific Components and their Respective Functions which Are Listed and Explained below:

(i) **Sensing Unit and ADC:** In underground mines, compact battery-powered nodes collect environmental data and are strategically deployed to ensure comprehensive monitoring. Data is transmitted via a multi-hop mesh network, enabling communication across multiple nodes and maintaining connectivity even if a node fails. It is then forwarded through the transmission layer to base stations or centralized systems for processing, analysis, and timely response. An Analog-to-Digital Converter (ADC) converts raw sensor signals into digital data for processing by the node's processor [14].

(ii) **Communication Unit: Transceiver** The transceiver enables bidirectional wireless communication using technologies such as IEEE 802.15.4 and LoRa. ZigBee supports low-data-rate mesh networking, while LoRa provides long-range, low-power connectivity, even in dense rock and tunnel environments [15]. These technologies are selected for their reliability, energy efficiency, and ability to maintain robust data delivery under challenging underground conditions.

(iii) **Miner & Equipment Localization**: This block provides real-time tracking and positioning of miners and equipment using methods like RSSI, time-based ranging, and beacon-assisted positioning. It improves safety through evacuation planning, collision avoidance, and rapid emergency response, while also enhancing operational efficiency [16].

(iv) **Power Supply**: In underground mining WSNs, power supply is a critical design consideration because nodes are often deployed in remote, hard-to-access areas where maintenance or battery replacement is impractical. As noted in [17], these networks prioritize energy efficiency, typically using long-life lithium batteries or energy-harvesting solutions to ensure sustained low-power operation, supporting system longevity, reliability, and overall efficiency.

(v) **Data Processing, Storage, and Analysis**: According to [18], the cloud or server layer serves as the above-ground platform where all collected mining data is securely stored, processed, and analyzed. It enables real-time decision-making through dashboards, alerts, and predictive analytics such as gas trend monitoring, temperature spike detection, and Machine Learning (ML) based anomaly or maintenance prediction.

(vi) **Network Managed system (NMS):** This component ensures the stability and adaptability of underground mining WSNs by managing efficient routing, fault tolerance, and load balancing. It detects and recovers from failures, distributes traffic to prevent congestion, and supports dynamic reconfiguration to adapt to changing mining conditions. These capabilities maintain secure, consistent, and energy-efficient communication [19].

(vii) **Node Mobility:** The Node Mobility component manages sensor node movement, ensuring routing, localization, and data delivery adapt dynamically as nodes attached to miners or machinery move through tunnels. Mobility awareness preserves connectivity and data reliability in changing underground environments [20].

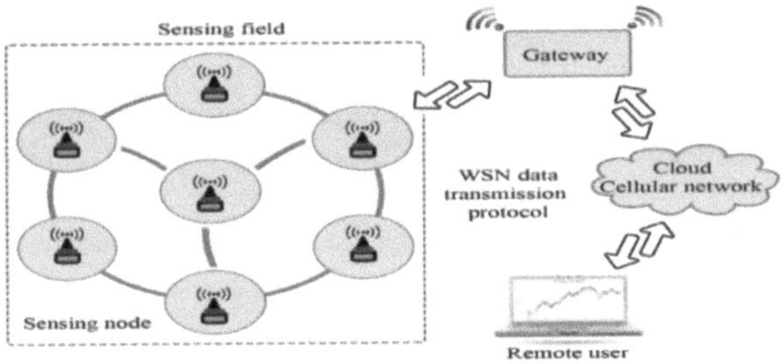

Fig. 2. Basic WSN architecture [21].

4.2 Flow of Data Collected from the Sensor Nodes

Figure 3 examines the flow of data in WSNs is designed to overcome challenges like limited power, harsh conditions, and poor signal propagation. Below is a step-by-step explanation based on the main components:

In a WSN deployed in underground mines, the data flow follows a coordinated process to ensure efficient monitoring and timely response. Initially, data is generated at the SNs, where environmental parameters are sensed and locally processed before transmission. These SNs then wirelessly transmit data to relay nodes using technologies such as ZigBee, LoRa, or proprietary RF protocols.

The collected data is forwarded to a gateway or sink node, which serves as a central hub for aggregating information from multiple sensors or relay nodes. This aggregated data is then prepared and transmitted to the surface control center. At the control center, live data is visualized, stored, and analyzed in real time. Based on this analysis, control systems can take immediate action if hazardous conditions are detected, such as triggering alarms, shutting down equipment, or automatically increasing ventilation to ensure safety within the mine.

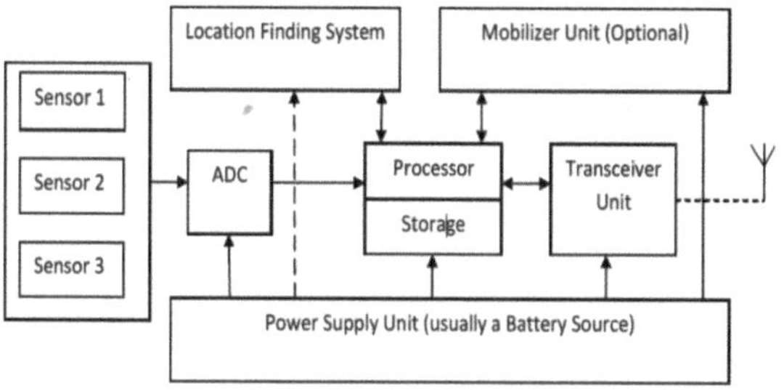

Fig. 3. Typical Internal Architecture of a WSN Node [22].

5 Comparative Analysis: Traditional Monitoring vs. WSN-Based Systems

Compared to traditional manual and wired systems, WSN-based monitoring provides superior performance in underground mining through real-time data transmission, flexibility in deployment, and reduced maintenance efforts. These systems enhance both safety and operational efficiency by enabling continuous environmental and equipment monitoring [23]. Table 1 provides a comparative analysis between traditional monitoring methods and WSN-based systems in the context of underground mining operations. It highlights four key criteria: response time, system reliability, energy consumption, and operational safety.

Traditional methods exhibit slower response times due to manual processes, whereas WSN-based systems offer faster responses through real-time data collection. In terms of reliability, traditional methods are prone to human error and system delays, while WSNs provide higher reliability via automation. Energy consumption is typically higher in manual systems because of constant human involvement, but WSNs are more energy-efficient by design. Lastly, operational safety is limited under traditional monitoring due to reliance on periodic checks, whereas WSNs enhance safety through continuous and remote surveillance, significantly improving real-time hazard detection and response.

Table 1. Comparative Analysis: Traditional monitoring vs. WSN-based systems [24, 25].

No	Criteria	Traditional Monitoring Methods	WSN-Based Systems
1.	Response Time	Slower response due to manual processes	Faster response enabled by real-time data
2.	System Reliability	Prone to human error and system lags	Higher reliability with automated monitoring
3.	Energy Consumption	Typically higher due to continuous manual input	Lower, as WSNs optimize energy usage
4.	Operational Safety	Limited, relies on periodic checks	Enhanced through continuous, remote monitoring

6 Challenges of WSNs in Underground Mines

According to [26, 27], WSNs face several critical challenges when deployed in underground mining environments, including:

(a) **High Installation and Maintenance Costs**: Deploying underground WSNs is considerably more costly than above-ground implementations due to the need for advanced hardware and the complexities of installation, maintenance, and signal propagation in underground environments.
(b) **Limited Energy Supply**: SNs in underground mines typically operate on battery power, making recharging or replacement challenging due to the harsh and inaccessible environment. This constraint necessitates energy-efficient hardware and communication protocols, directly impacting network longevity, reliability, and maintenance costs.
(c) **Signal Attenuation and Interference**: Wireless communication in underground environments faces major obstacles due to severe signal attenuation, electromagnetic interference from heavy mining machinery, and physical barriers such as rock formations and metallic structures. These conditions significantly weaken signal strength, increase transmission delays, and compromise overall network reliability and throughput.
(d) **Security Vulnerabilities**: Due to the limited power, memory, and processing capacity of sensor nodes, they are highly susceptible to security vulnerabilities. These constraints can result in unauthorized access, data breaches, and operational disruptions, creating significant risks to both safety and productivity in underground mining environments.
(e) **Complex Network Topology:** The dynamic and irregular structure of underground mines, characterized by complex tunnel layouts and numerous physical obstacles, complicates the deployment and maintenance of sensor networks. These conditions create significant challenges in achieving consistent coverage, reliable connectivity, and network scalability.

(f) **Scalability to Support Future Integration:** Due to frequent changes in workforce size and work areas driven by production demands and mine expansion, the wireless monitoring system must support seamless integration of new users and devices without compromising reliability. Additionally, the network should dynamically adapt by incorporating new access points and wearable devices to maintain consistent coverage and communication across the evolving mining topology.

7 Overview of Wireless Transmission Media for WSNs Applicable to Underground Mining

Technologies for WSNs in underground mines are explained further below. These transmission media enable reliable data transfer between sensor nodes and the control center, ensuring effective signal propagation in harsh environments.

(a) **ZigBee Subnet:** In the context of underground mining, [28] identifies ZigBee as a low-power, low-data-rate wireless communication protocol built on the IEEE 802.15.4 standard. Its suitability for applications demanding long battery life and secure networking makes it valuable in such environments. ZigBee is commonly applied underground for environmental monitoring, including the measurement of vibration and dust levels, owing to its energy efficiency and capacity to support large numbers of devices.
(b) **Wi-Fi Subnet:** As noted in [29], Wi-Fi offers high-speed wireless connectivity and is widely utilized in surface mining facilities for data-intensive applications. However, its use in underground environments is constrained by high power consumption and severe signal attenuation caused by geological formations. Despite these limitations, Wi-Fi remains viable for real-time data transmission in areas where supporting infrastructure is available.
(c) **Bluetooth and BLE Subnet:** Bluetooth and Bluetooth Low Energy (BLE) operate in the 2.4 GHz ISM band and are primarily used for short-range communication. Traditional Bluetooth offers moderate power consumption and data rates suitable for Personal Area Networks (PANs), whereas BLE is optimized for ultra-low power operation and supports multiple simultaneous connections. In underground mining environments, BLE is increasingly utilized for indoor localization and personnel tracking due to its energy efficiency, scalability, and compatibility with wearable devices, as highlighted in [30].
(d) **Ultra-Wide Band (UWB) Subnet:** Ultra-Wideband (UWB) technology leverages a broad frequency spectrum to deliver high data rates over short distances. In underground mining environments, UWB is particularly advantageous for bandwidth-intensive applications such as video surveillance and real-time monitoring. Its low power consumption and strong obstacle penetration capabilities make it well-suited for complex subterranean conditions. Furthermore, UWB's high throughput supports multimedia transmission, enhancing situational awareness and safety measures, as noted in [31].
(e) **Smart Grid Subnet (SGS):** The Smart Grid subnet in underground mines incorporates Advanced Metering Infrastructure (AMI) to enable real-time monitoring and control of energy consumption. This system facilitates remote equipment management, data-driven optimization of energy usage, and integration of renewable

energy sources alongside energy storage solutions. By leveraging these capabilities, the Smart Grid subsystem supports sustainable mining operations and contributes to reducing operational costs, as noted in [32].

(f) **Body Area Network (BAN) Subnet:** Body Area Networks (BANs) comprise wearable devices that continuously monitor miners' physiological parameters and transmit data wirelessly to centralized systems for health assessment and emergency response. Integrating BANs into underground mining operations enhances worker safety by providing real-time health data and enabling rapid medical intervention when required [33].

(g) **Long-Range (LoRa):** LoRa is a physical layer technology designed for long-range, low-power communication with strong interference resistance. In underground mines, LoRa nodes can monitor parameters such as methane concentration, humidity, and worker location, transmitting this data to a LoRa gateway, which typically forwards it via fiber or Ethernet to a control center. Its energy efficiency and ability to penetrate geological structures make LoRa highly suitable for maintaining safety and operational awareness in hazardous underground environments [34].

8 Overview of Security and Energy-Efficient Mechanisms for WSNs Applicable to Underground Mines

In underground mining environments, standards such as IEEE 802.15.4 and LoRaWAN offer flexible communication frameworks that combine dynamic scheduling with robust encryption to enhance both data transmission and security. When adapted to the demanding conditions of underground operations, these standards enable reliable, prioritized, and energy-efficient WSNs through efficient scheduling and low-power communication protocols, significantly increasing the operational longevity of WSNs deployed in underground settings [35].–[36].

(a) **Security mechanisms of WSNs in underground mines**

To ensure robust, energy-efficient security in underground WSNs, several mechanisms are critical for safeguarding data confidentiality, integrity, and availability, as highlighted in [37]–[41]:

(i) **Data Encryption (lightweight cyphers):** To ensure the confidentiality of sensor data transmitted between SNs, gateways, and surface control systems, lightweight encryption algorithms such as XTEA, Speck, Simon, PRESENT, LEA, HIGHT, and ChaCha20 are used. These algorithms are specifically designed for low-power, resource-limited devices and offer reduced computational overhead while maintaining robust security. Their efficiency makes them well-suited for underground WSN deployments, where energy preservation and reliable data protection are critical operational requirements.

(ii) **Authentication Protocols:** Strong node and user authentication mechanisms prevent unauthorized access to the network. This is crucial in underground scenarios where compromised nodes could endanger personnel or disrupt critical monitoring functions.

(iii) **Data Integrity Assurance:** The use of Message Authentication Codes (MACs) and cryptographic hash functions ensures the integrity of transmitted data by detecting any tampering or corruption, which could result from harsh environmental interference or cyberattacks.

(iv) **Secure Routing Protocols (INSENS):** Routing protocols tailored for hostile environments like INSENS (Intrusion-tolerant routing) help protect against network-layer attacks such as black hole, sinkhole, or wormhole attacks that could disrupt data delivery paths in the mine.

(v) **Intrusion Detection Systems (IDS):** IDS solutions adapted for WSNs continuously monitor traffic and node behaviour, enabling real-time detection of anomalies or malicious activity within the mine networks.

(vi) **Efficient Key Management Schemes:** Given the dynamic nature of mining environments, where nodes may be frequently added, removed, or relocated, secure and scalable key management is essential for maintaining encrypted communication.

(vii) **Redundancy and Secure Data Aggregation:** Redundant node deployment and secure data aggregation improve the reliability and accuracy of monitoring data in harsh underground environments. Even when some nodes fail or become unreachable, secure aggregation protects combined sensor readings from tampering, reduces communication overhead, and enhances overall fault tolerance, therefore ensuring consistent, high-quality information for safety and operational decision-making.

(viii) **Physical Security Measures**: Physical threats are also significant in underground mines, where exposed or remote nodes are vulnerable to tampering or destruction. To mitigate these risks, tamper-resistant hardware and node-concealment techniques are applied to protect devices, helping maintain the integrity and safety of critical data such as gas levels and equipment status.

(b) **Energy-efficient mechanisms required for WSNs deployment in underground mines**

In underground mines, where sensor access and battery replacement are difficult, energy-efficient WSN mechanisms are vital.

(i) Duty cycling, using protocols like Sensor-Medium Access Control (S-MAC) and Timeout-Medium Access Control (T-MAC), allows nodes to sleep when idle, conserving power.

(ii) Energy-efficient routing like LEACH and PEGASIS minimizes transmission costs. Data aggregation reduces redundancy and furthermore, saves bandwidth. Topology control adjusts node activity to maintain connectivity. MAC protocol to lower energy use by avoiding idle listening.

(iii) Technologies like LoRa and Zigbee offer long-range, low-power communication. Energy harvesting, though complex underground, can extend node life.

(iv) Cluster head rotation distributes energy load fairly. Cross-layer optimization and context-aware sensing improve efficiency further [44]–[43].

9 Lessons Learned

WSNs have gained significant attention in recent years, particularly in sectors such as environmental monitoring, industrial automation, and underground mining [46]–[44]:

(i) **Safety Monitoring and Early Warning System**: Provides a reliable early warning and decision-making tool by delivering real-time data on environmental conditions and worker safety. This helps with timely disaster alerts, operational control, informed management decisions, and effective responses to accidents.
(ii) **Scalable Sensor Network Deployment**: The system supports flexible setup and scalability, allowing sensor nodes to be installed in remote or hard-to-access areas without needing complex infrastructure.
(iii) **Minimizing Human Exposure to Hazardous Areas**: It lowers the risk to human workers by reducing the need for them to enter hazardous or unreachable areas to collect data.
(iv) **Improved Efficiency via Sensor Network Automation:** They improve operational efficiency by automating data collection, processing, and communication between sensor nodes and central control systems.
(v) **Optimizing Power Usage in Sensor Networks:** The system uses energy-efficient communication protocols, which help extend the life of the network, particularly important in environments with limited power supply like underground mines.
(vi) **Remote Access and Network Management:** They enable remote access, diagnostics, and control, allowing system administrators to monitor and manage the network from surface-level control rooms or other remote locations.
(vii) **Historical Data Logging and Predictive Analysis**: They provide historical data logging and trend analysis, enabling predictive maintenance and long-term safety improvements by identifying recurring risks or equipment failures before they escalate.

10 Discussion

The results show that safety and environmental monitoring dominate current research, accounting for the largest share of studies [1, 8, 10]. In contrast, secure localization and integrated energy–security optimization remain comparatively underexplored, indicating a significant research gap. As evidenced in [12, 15, 34], ZigBee and LoRa are the most frequently adopted technologies due to their low power consumption and robustness in obstructed environments. Wi-Fi is primarily limited to localized, infrastructure-supported zones [31], while UWB is preferred for high-precision localization [32].

Studies such as [15, 17, 42] demonstrate that LoRa achieves superior energy efficiency at the cost of higher latency, whereas ZigBee offers balanced performance suitable for dense sensor deployments. These findings highlight the importance of application-aware protocol selection. Lightweight cryptographic schemes are the most commonly implemented due to node constraints [37, 40]. However, fewer studies integrate intrusion detection or secure routing simultaneously, revealing a lack of holistic security architectures for underground mining environments. The findings corroborate prior surveys such as Muduli et al. [8] but extend them by systematically quantifying trends and explicitly linking protocol choices to performance outcomes. Unlike earlier narrative reviews, this PRISMA-based analysis reveals that most studies evaluate technologies in isolation, underscoring the need for cross-layer and system-level validation in future work.

11 Future Research Directions

Future research in WSNs for underground mining should focus on enhancing the reliability and adaptability of monitoring systems by incorporating diverse sensor inputs, such as gas levels, temperature, humidity, vibration, equipment status, and geotechnical parameters. Simulation tools such as NS-3, QualNet, and OMNeT++ can emulate underground environments to evaluate network performance, scalability, and resilience prior to deployment. Prototyping of advanced WSN in the form of wearable health monitoring systems specifically tailored for miners, capable of tracking physiological indicators such as heart rate, fatigue levels, and exposure to harmful gases in real time, should be evaluated.

These devices should be reinforced to withstand harsh underground conditions and seamlessly integrate with WSN infrastructures. However, careful selection, training, and integration of individual models is crucial, as complex ensemble systems can increase computational requirements, which must be managed for real-time deployment in resource-constrained underground settings. Another essential direction involves refining and strengthening cybersecurity frameworks to safeguard sensitive health, operational, and location data from potential breaches, ensuring compliance with regulatory requirements and maintaining worker trust. Noting that WSNs are susceptible to attacks such as eavesdropping, node capture, data tampering, spoofing, and denial-of-service. Limited node processing power makes implementing strong encryption and authentication challenging. Any security breach could compromise safety-critical data such as gas levels or equipment status.

Addressing these technical and infrastructural challenges will not only accelerate the broader adoption of WSN-based surveillance and safety systems in mining but will also open pathways for their application in other high-risk, remote, or hard-to-reach industries, such as oil and gas, construction, and offshore operations. These future research directions could significantly enhance underground mining operations by improving environmental monitoring, worker safety, and equipment management, promoting more efficient, data-driven decision-making in challenging and dynamic mining environments.

12 Conclusion

This paper presented a comprehensive and up-to-date PRISMA 2020–compliant systematic review of WSN technologies applied to occupational health and safety in underground mining environments. While underground mining remains a key contributor to national and global economic development, the persistent exposure of miners to hazardous conditions continues to raise serious safety and health concerns. The findings of this review demonstrate that WSN-based monitoring systems offer a reliable and scalable approach for enhancing miner safety, environmental awareness, and operational efficiency in these high-risk environments.

The review systematically analyzed existing WSN applications, system architectures, and core components designed for underground mining, with particular emphasis on health monitoring, environmental sensing, and real-time localization of personnel and equipment. By synthesizing recent studies, this work highlighted the advantages of low-power wireless technologies in improving awareness, supporting emergency response, and enabling predictive maintenance, while also identifying limitations related to signal propagation, energy constraints, system scalability, and deployment cost. In addition, the review provided insights into wireless communication behavior in underground environments, offering guidance for the selection and design of suitable communication protocols and data transmission strategies.

Despite notable progress, several research gaps remain. Future studies should focus on the integration of heterogeneous environmental, geotechnical, and physiological sensors to improve the accuracy, adaptability, and contextual awareness of underground monitoring systems. Advanced simulation and modeling tools should be leveraged to evaluate system performance under diverse geological and operational conditions prior to large-scale deployment. Furthermore, the development of lightweight, ergonomic, and robust wearable health-monitoring devices capable of capturing real-time physiological and exposure metrics remains a critical research priority.

Security and privacy also require sustained attention. Given the decentralized and resource-constrained nature of WSNs, strengthening cybersecurity frameworks to protect sensitive health, operational, and location data is essential. Future research should explore energy-efficient cryptographic mechanisms, intrusion detection strategies, and secure communication protocols tailored to underground mining environments.

This systematic review provides a structured knowledge base and research roadmap for researchers, mining engineers, and safety practitioners. By addressing the identified technical, security, and operational challenges, future WSN-driven solutions can significantly contribute to safer, more resilient, and sustainable underground mining operations that align with evolving industry standards and regulatory requirements.

Acknowledgments. The authors would like to gratefully thank the Tshwane University of Technology for providing financial assistance. I declare that there are no conflicts of interest related to this paper's publication.

References

1. Sadeghi, S., Soltanmohammadlou, N., Nasirzadeh, F.: Applications of wireless sensor networks to improve occupational safety and health in underground mines. J. Saf. Res. **83**, 825 (2022). https://doi.org/10.1016/j.jsr.2022.07.016
2. Cui, Y., Liu, S., Liu, Q.: Navigation and positioning technology in underground coal mines and tunnels: a review. J. South. Afr. Inst. Min. Metall. **121**(6), 279–288 (2021). https://doi.org/10.17159/2411-9717/1065/2021
3. B. Kartik, P. Manimaran: IoT based Smart Helmet for Hazard Detection in mining industry, arXiv preprint, Apr. (2023). Available: https://arxiv.org/abs/2304.10156
4. P. Obioma, O. Agbodike, J. Chen, L. Wang: ISLS: An IoT-Based Smart Lighting System for Improving Energy Conservation in Office Buildings," arXiv, (2025). Available: https://arxiv.org/abs/2503.13474

5. Rojas, R.L., Peña, Á., Garcia, J.: AI-driven predictive maintenance in mining: a systematic literature review on fault detection, digital twins, and intelligent asset management. Appl. Sci. **15**(6), 3337 (2025). https://doi.org/10.3390/app15063337
6. Abbasi, A., Younis, M.: A survey on clustering algorithms for wireless sensor networks. Comput. Commun. **30**(14–15), 2826–2841 (2007). https://doi.org/10.1016/j.comcom.2007.05.024
7. Olakanmi, O.O., Dada, A.: Wireless sensor networks (WSNs): security and privacy issues and solutions. In: Khatib, M., Alsadi, S. (eds.) Wireless Mesh Networks– Security, Architectures and Protocols. IntechOpen (2020). https://doi.org/10.5772/intechopen.84989
8. Muduli, K., Mishra, D.P., Jana, P.K.: Application of wireless sensor network for environmental monitoring in underground coal mines: a systematic review. J. Netw. Comput. Appl. **106**, 4867 (2024). https://doi.org/10.1016/j.jnca.2017.12.022
9. Wang, S., Chen, Z., Yan, Q., Yang, B., Peng, L., Jia, Z.: A mobile malware detection method using behavior features in network traffic. J. Netw. Comput. Appl. **133**, 1525 (2019). https://doi.org/10.1016/j.jnca.2018.12.014
10. Jiang, Y., Chen, W., Zhang, X., Zhang, X., Yang, G.: Real-time monitoring of underground miners status based on mine IoT system. Sensors. **24**(3), 739 (2024). https://doi.org/10.3390/s24030739
11. Ahmad, S.J., Ahmed, M.M., Unissa, I.: Optimization of environmental data to improve safety in coal mine using wireless sensor networks. Indian J. Sci. Technol. **15**(19), 956–964 (2022). https://doi.org/10.17485/IJST/v15i19.686
12. Moridi, M., et al.: Development of underground mine monitoring and communication system integrated ZigBee and GIS. Int. J. Mining Sci. Technol. **25**(5), 811818 (2015). https://doi.org/10.1016/j.ijmst.2015.07.017
13. Bolam, S.M., et al.: Remote patient monitoring with wearable sensors following knee arthroplasty. Sensors. **21**(15), 5143 (2021). https://doi.org/10.3390/s21155143
14. Inoth, V., Balaji, V.: A wireless sensor network for coal mine safety powered by modified localization algorithm. Process Saf. Environ. Prot. **185**, 132–144 (2025). https://doi.org/10.1016/j.psep.2024.12.071
15. Musonda, S.K., Ndiaye, M., Libati, H.M., Abu-Mahfouz, A.M.: Reliability of LoRaWAN communications in mining environments: a survey on challenges and design requirements. J. Sens. Actuator Netw. **13**(1), 16 (2024). https://doi.org/10.3390/jsan13010016. Available: https://www.mdpi.com/2224–2708/13/1/16
16. Zhang, X., Chang, B.: Efficient cooperative target node localization with optimization strategy based on RSS for wireless sensor networks. Comput., Mater. Continua. **82**(3), 5079–5095 (2025)
17. Evangelakos, E.A., Kandris, D., Rountos, D., Tselikis, G., Anastasiadis, E.: Energy sustainability in wireless sensor networks: an analytical survey. J. Low Power Electron. Appl. **12**(4), 65 (2022). https://doi.org/10.3390/jlpea12040065
18. Khriji, S., Trabelsi, A., Fourati, M., Alimi, A.M., Abid, M.: Design and implementation of a cloud-based event-driven architecture for real-time data processing in wireless sensor networks. J. Supercomput. **78**, 33743401 (2022). https://doi.org/10.1007/s11227-021-03955-6
19. Guo, H., & Ben, B.: Reinforcement Learning-Enabled Reliable Wireless Sensor Networks in Dynamic Underground Environments. arXiv. https://arxiv.org/abs/1908.05804 (2019)
20. Temene, N., Sergiou, C., Georgiou, C., Vassiliou, V.: A survey on mobility in wireless sensor networks. Ad Hoc Netw. **125**, 102726 (2022)
21. Wang, C., Tavares, A., Fonseca, J., Soares, F., Li, Z.: Real-time condition assessment of a painted megalithic cave using wireless sensor network. Tunn. Undergr. Space Technol. **120**, 104270 (2022). https://doi.org/10.1016/j.tust.2021.104270

22. Sharma, H.: Modulations and Error Correction Codes for Wireless Sensor Networks. ResearchGate (2019) Available at: https://www.researchgate.net/publication/337486796_Modulations_and_Error_Correction_Codes_for_Wireless_Sensor_Networks
23. Wang, H., Wang, J.: Data aggregation algorithms in wireless sensor networks for underground mines: a review. J. Netw. Comput. Appl. **153**, 102530 (2020). https://doi.org/10.1016/j.jnca.2020.102530
24. Wang, C., Xiao, Z., Wang, Y., Wang, F., Li, Z.: Big data acquisition for underground infrastructure condition assessment. In: Data-Centric Engineering, vol. 5. Cambridge University Press (2024). https://doi.org/10.1017/dce.2024.16
25. Zhang, M., Li, X., Liu, Y.: Comparative performance analysis of WSN-based and traditional monitoring methods in industrial environments. IEEE Access. **11**, 35210–35223 (2023). https://doi.org/10.1109/ACCESS.2023.3267023
26. Zhang, H., Li, X., Wang, Y.: Comparative analysis of traditional and wireless sensor network-based monitoring systems in underground mining safety. Sensors. **21**(9), 3124 (2021). https://doi.org/10.3390/s21093124
27. Saleh, H.M., Marouane, H., Fakhfakh, A.: A comprehensive analysis of security challenges and countermeasures in wireless sensor networks enhanced by machine learning and deep learning technologies. Int. J. Saf. Secur. Eng. **14**(2), 373–386 (2024). https://doi.org/10.18280/ijsse.140206
28. Surenther, I., Sridhar, K.P., Kingston Roberts, M.: Maximizing energy efficiency in wireless sensor networks for data transmission: a deep learning-based grouping model approach. Alex. Eng. J. **83**, 53–65 (2023). https://doi.org/10.1016/j.aej.2023.10.016
29. Begum, B.A., Nandury, S.V.: Data aggregation protocols for WSN and IoT applications – a comprehensive survey. J. King Saud Univ. – Comput. Inf. Sci. **35**(2), 651–681 (2023). https://doi.org/10.1016/j.jksuci.2023.01.008
30. Majumdar, P., Mitra, S., Bhattacharya, D., et al.: Enhancing sustainable 5G powered agriculture 4.0: summary of low power connectivity, internet of UAV things, AI solutions and research trends. Multimed. Tools Appl. **84**, 17389–17433 (2025). https://doi.org/10.1007/s11042-024-19728-1
31. Chetty, B.L., Walingo, T.M., Kruger, C.P., Isaac, S.J.: Experimental investigation into deploying a Wi-Fi6 mesh system for underground gold and platinum mine stopes. Mining. **4**(3), 567–587 (2024). https://doi.org/10.3390/mining4030032
32. Zare, M., Battulwar, R., Seamons, J., Sattarvand, J.: Applications of wireless indoor positioning systems and Technologies in Underground Mining: a review. Min., Metall. Explor. **38**(1–2), 230–246 (2021). https://doi.org/10.1007/s42461-021-00476-x
33. Jiang, W., Han, B., Habibi, M.A., Schotten, H.D.: The road towards 6G: a comprehensive survey. IEEE Open J. Commun. Soc. **2**, 334–366 (2021). https://doi.org/10.1109/OJCOMS.2021.30576
34. Almasabi, S., Shaf, A., Ali, T., Zafar, M., Irfan, M., Alsuwian, T.: Securing smart grid data with Blockchain and wireless sensor networks: a collaborative approach. IEEE Access. **12**, 19181–19198 (2024) https://ieeexplore.ieee.org/document/10419089
35. Elappila, M., Chinara, S., Parhi, D.R.: Survivability Aware Channel allocation in WSN for IoT applications. Pervasive Mobile Comput. **61**, 101107 (2020). https://doi.org/10.1016/j.pmcj.2019.101107
36. Branch, P., Li, B., Zhao, K.: A LoRa-based linear sensor network for location data in underground mining. Telecom. **1**(2), 68–79 (2020). https://doi.org/10.3390/telecom1020006
37. L. Scalambrin, A. Zanella, X. Vilajosana: LoRa Multi-Hop Networks for Monitoring Underground Mining Environments," arXiv preprint https://arxiv.org/abs/2310.20515 (2023). Available: https://arxiv.org/abs/2310.20515

38. Theissen, M., Kern, L., Hartmann, T., Clausen, E.: Use-case-oriented evaluation of wireless communication Technologies for Advanced Underground Mining Operations. Sensors. **23**(7), 3537 (2023) Available: https://www.mdpi.com/1424–8220/23/7/3537
39. Ray, S., Mishra, K.N., Dutta, S.: Sensitive data identification and security Assurance in Cloud and IoT based networks. Int. J. Comput. Netw. Inf. Secur. **5**, 11–27 (2022). https://doi.org/10.5815/ijcnis.2022.05.02. [Online]. Available: http://www.mecs-press.org/ijcnis/ijcnis-v5-n5/IJCNIS-V5-N5-2.pdf
40. Urooj, S., Lata, S., Ahmad, S., Mehfuz, S., Kalathil, S.: Cryptographic data security for reliable wireless sensor network. Alex. Eng. J. **72**, 37–50 (2023). https://doi.org/10.1016/j.aej.2023.03.061
41. Patel, R.N., Bhat, N.V.: A review on energy efficient MAC protocols in wireless sensor networks. Int. J. Eng. Adv. Technol. (IJEAT). **9**(2), 2090–2094 (2019). https://doi.org/10.35940/ijeat.B3415.129219
42. Suresh, B., Prasad, G.S.C.: An energy efficient secure routing scheme using LEACH protocol in WSN for IoT networks. Meas.: Sens. **30**, 100883 (2023). https://doi.org/10.1016/j.measen.2023.100883
43. Ansari, A.A., Gera, P., Mishra, B., et al.: A secure authentication framework for WSN-based safety monitoring in coal mines. Sādhanā. **45**, 98 (2020). https://doi.org/10.1007/s12046-020-1319-2
44. Chen, W., Wang, X.: Coal mine safety intelligent monitoring based on wireless sensor network. IEEE Sensors J. **21**(22), 25465–25471 (2021). https://doi.org/10.1109/JSEN.2020.3040280

Enhancing Security of Wireless IoT Networks Through Intelligent Detection and Prevention of MiTM and DDoS Attacks

Sara Alsloly(✉)

College of Computing and Information Technology, University of Bisha, Bisha, Saudi Arabia

Abstract. This research introduces a dynamic framework that enhances the protection of Internet of Things (IoT) environments operating within Software-Defined Networking (SDN). The proposed design employs adaptive intelligence to detect and suppress both Man-in-the-Middle (MiTM) and Distributed Denial-of-Service (DDoS) attacks in real time. A machine-learning workflow built around the Random Forest (RF) algorithm was deployed to evaluate live network traffic and identify malicious activities with high precision. In addition, a self-learning ARP-verification module continuously validates MAC-to-port bindings to counter spoofing attempts. The system was implemented in a Mininet-WiFi testbed controlled through Ryu on Ubuntu 18.04, achieving approximately 95% detection accuracy while maintaining low controller overhead. Collectively, the framework strengthens IoT scalability and reliability by coupling lightweight traffic management with autonomous, data-driven defensive logic.

Keywords: Internet of Things (IoT) · Software-Defined Networking (SDN) · Cybersecurity · Random Forest Algorithm · DDoS and MiTM Attacks · Wireless Networks · Machine Learning

1 Introduction

The widespread growth of the Internet of Things (IoT) has introduced major security concerns, particularly within wireless ecosystems. Because IoT devices typically possess limited computational and storage capabilities, they are highly exposed to cyber threats such as Man-in-the-Middle (MiTM) and Distributed Denial-of-Service (DDoS) intrusions. To address these persistent challenges, the present study integrates Software-Defined Networking (SDN) and Machine Learning (ML) to build an intelligent, responsive defense mechanism. SDN provides centralized traffic governance and network programmability, while ML contributes pattern-recognition capabilities that enable real-time anomaly detection. The combined architecture was realized using the Ryu controller and Mininet-WiFi on Ubuntu 18.04 to assess detection accuracy, scalability, and performance efficiency.

H. R. Arabnia et al. (Eds.): CSCI 2025, CCIS 2970, pp. 424–431, 2027.
https://doi.org/10.1007/978-3-032-28466-2_32

1.1 Problem Statement

As IoT adoption grows, the continued use of weak and outdated protocols such as Telnet and HTTP leaves networks increasingly exposed to threats like MiTM and DDoS attacks. Conventional security solutions often fail to adapt to these sophisticated intrusions. Consequently, this study applies SDN and ML integration to improve IoT security through continuous monitoring, intelligent attack identification, and automated mitigation strategies.

1.2 Research Aim

The primary goal of this research is to design and implement a secure IoT network that utilizes Software-Defined Networking (SDN) and Machine Learning (ML) techniques for detecting, preventing, and mitigating cyber threats, particularly MiTM and DDoS attacks.

1.3 Research Objectives

- To design an SDN-driven framework that enhances the protection of IoT devices in wireless environments.
- To employ ML algorithms for identifying and countering DDoS and MiTM attacks.
- To examine network flow patterns and classify anomalies through intelligent data models.
- To formulate automated mitigation mechanisms for real-time prevention of cyber intrusions.
- To assess overall system scalability and detection accuracy using Mininet-WiFi integrated with the Ryu controller.

1.4 Scope and Limitation of the Project

The scope of this work centers on improving IoT security in wireless networks through the integration of SDN and ML technologies. The main focus is on detecting and mitigating DDoS and MiTM attacks within a controlled test environment. Experiments were carried out in a simulated setup using Ubuntu 18.04, Mininet-WiFi, and the Ryu SDN controller. This research is limited to virtual testing; therefore, large-scale hardware implementation and performance validation in real-world deployments are considered future extensions.

1.5 Research Questions

- This research seeks to address the following questions:
- What are the primary security concerns affecting wireless IoT networks?
- In what ways can Software-Defined Networking (SDN) strengthen IoT defense and enhance threat detection?
- Which Machine Learning (ML) techniques provide the most reliable detection and mitigation of DDoS and MiTM attacks?
- How can the proposed system be assessed with respect to accuracy, scalability, and operational efficiency?

2 Design of SDN Network

The architecture of the proposed system relies on a Software-Defined Networking (SDN) structure that strengthens the security of wireless IoT communications. As illustrated in Fig. 1, the topology was constructed within Mininet-WiFi and orchestrated by the Ryu controller through Python scripts. The OpenFlow switch acts as a bridge between the control and data planes, granting centralized visibility over all traffic. This modular design provides scalability and rapid reconfiguration while enabling the controller to react quickly to abnormal network behavior [22].

2.1 Proposed System

The proposed framework integrates SDN technology with Machine Learning (ML) algorithms to protect wireless IoT networks against MiTM and DDoS attacks. The centralized SDN controller observes traffic in real time and applies adaptive flow rules that restrict suspicious communication paths. Statistical traffic features are processed through a Random Forest (RF) model to recognize malicious patterns, while a dynamic MAC-to-port mapping table maintains verified device associations and minimizes controller workload. This configuration achieves fast mitigation, dependable scalability, and effective isolation of hostile traffic within dense IoT environments.

2.2 Methodology and Implementation: DDOS Attack

1. A dataset was collected from an emulated wireless topology created in Mininet-WiFi under Ubuntu 18.04 with the Ryu controller. Two operating conditions were emulated: ordinary traffic and DDoS traffic created through ICMP flooding via the *hping3* tool. The recorded flow statistics were exported as CSV files and merged into a unified dataset containing 22 initial features, such as source and destination IP addresses, flow duration, packet counts, and byte totals.
2. To improve the efficiency of model training, Random Forest feature-importance scores were used to rank all attributes, and the 17 most discriminative features were retained. Reducing dimensionality decreased processing overhead and improved scalability during high-traffic simulations.
3. Six supervised learning methods—Logistic Regression (LR), Naïve Bayes (NB), Support Vector Machine (SVM), K-Nearest Neighbors (KNN), Decision Tree (DT), and Random Forest (RF)—were compared through tenfold evaluations to ensure stable performance. Metrics such as accuracy, true/false positives, false negatives, and computation time were used to assess their effectiveness.
4. The SDN controller continuously gathered flow statistics and submitted them to the trained RF model for classification. When the model flagged traffic as malicious, the controller dynamically installed a high-priority OpenFlow rule on the affected switch to discard packets originating from the attacker's IP or ingress port. Offending hosts were quarantined for 120 s, and timers were refreshed automatically if abnormal activity persisted. Legitimate flows continued uninterrupted, ensuring efficient and lightweight mitigation

2.3 Implementation: First Scenario #: DDoS Attack

In the experimental testbed, two attacker nodes (H2 and H5) launched ICMP flood traffic targeting the IoT device at 10.0.0.8. The simulation was conducted in Mininet-WiFi under the Ryu controller on Ubuntu 18.04.

Attack traffic was generated using the following command: hping3–1 —flood -p 80 10.0.0.8 which continuously sent ICMP packets to exhaust the victim's resources.

The controller collected ongoing flow statistics and evaluated them through the RF classifier. Once malicious behavior was detected from H2 or H5, OpenFlow rules were installed to block packets from the identified sources. Each rule had a hard timeout of 120 s and was automatically extended if the attack persisted. The controller logged both attacker and victim identifiers for analysis.

Port-level blocking was also executed when the ingress interface itself was identified as the attack origin—immediately isolating the compromised port while maintaining normal operations for other devices. As a result, attack traffic was contained, throughput returned to normal levels, and controller resource consumption remained minimal.

2.4 Methodology and Implementation: Man-in-the-Middle Attack

MiTM exploits vulnerabilities in the Address Resolution Protocol (ARP) to redirect network packets by falsifying IP-to-MAC mappings. Using *Ettercap*, the attacker generated counterfeit ARP packets to poison legitimate ARP tables and intercept communication between IoT hosts.

The proposed detection algorithm monitors all ARP exchanges and verifies each mapping against a maintained MAC-to-port table. When inconsistencies are observed, the controller identifies the attacker's MAC address and blocks it directly at the switch layer.

Implementation: Second Scenario #: ARP Poisoning.
In this setup, host H5 acted as the attacker attempting to deceive the victim (10.0.0.7) and redirect packets intended for the server (10.0.0.1). Fake ARP responses allowed the attacker to masquerade as the legitimate server, compromising confidentiality and integrity. Once the controller detected the forged mapping, it immediately blacklisted the attacker's MAC address, preventing further interception. The victim's ARP table remained correct, ensuring secure communication with the server and maintaining data integrity throughout the session.

2.5 Results of Scenario #: ARP Poisoning

The outcomes of the ARP poisoning experiment show that the proposed mitigation method introduces minimal computational burden on the SDN controller. Measurements of CPU utilization and control-plane latency revealed no significant increase during activation of the mitigation flow entries.

Throughput testing, performed using *iPerf* and *NMON*, compared network performance before, during, and after the attack. Figure 1 depicts the results, where TCP throughput dropped sharply during the ARP attack but quickly recovered to baseline once the blocking rule was enforced. This recovery demonstrates that the system can restore normal communication almost instantly after malicious activity is detected.

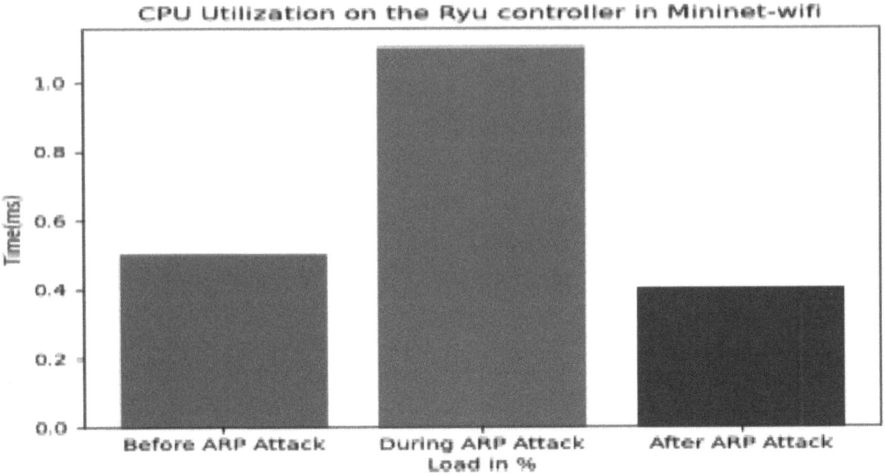

Fig. 1. CPU Utilization on Ryu Controller

To assess network performance under varying operational states, throughput measurements were obtained for the topology illustrated in Fig. 1 using the *iPerf* [30] and *NMON* [32] utilities. The experiment quantified TCP transmission rates across three distinct phases—prior to the attack, during active interference, and following mitigation. In this context, throughput represents the total volume of data successfully transferred across the network within a specific period. Figure 1 visualizes these results, showing the fluctuation in TCP rates throughout the experiment. By contrasting the observed throughput across the three phases, the system's resilience and data-handling capability under normal and adversarial conditions can be effectively evaluated.

2.6 Results of Machine Learning Algorithms

Table 1 summarizes the classification outcomes for all tested machine-learning algorithms. Ten repeated trials ensured consistent results across runs. Decision Tree (DT) and Random Forest (RF) achieved the best overall accuracy, while RF offered the best balance between execution time and detection fidelity (Table 2).

The Random Forest model achieved perfect separation under the tested conditions, with no false positives or negatives, and very low computation time—making it the optimal choice for SDN-based DDoS detection (Fig. 2).

2.7 DDoS Experiment Results

The DDoS scenario was monitored through *sFlow-RT* while the Mininet-WiFi nodes generated attack traffic. The RF classifier successfully recognized ICMP flood behavior

Table 1. Results of Machine Learning Algorithms

Results of Supervised Models for DDoS Detection

Classifier	Accuracy (%)	TP	FP	FN	TN	Time (s)
LR	99.56	1263	0	11	1226	0.1620
KNN	99.04	1263	0	24	1213	0.9657
SVM	85.96	1263	0	351	886	5.1944
NB	60.32	271	0	0	1237	0.0495
DT	100.00	1263	0	0	1237	0.5090
RF	100.00	1263	0	0	1237	0.4319

Table 2. Confusion Matrix for Random Forest

Random Forest Confusion Matrix

	Predicted: Normal	Predicted: DDoS
Actual: Normal	1237	0
Actual: DDoS	0	1263

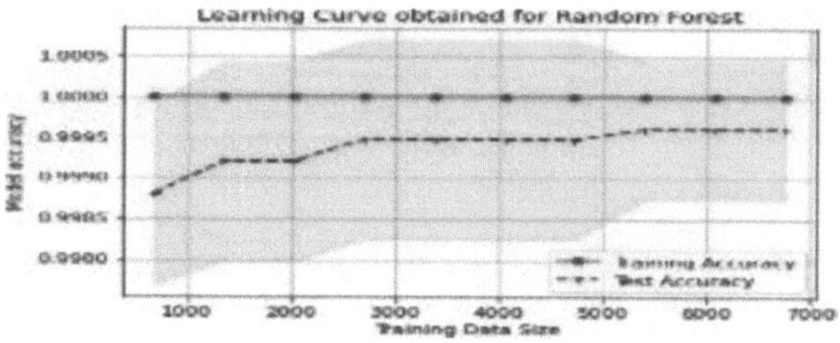

Fig. 2. Result of RF Machine Learning Model

and triggered appropriate mitigation rules at the switch level. CPU and memory utilization on the controller remained within nominal limits, confirming that the process imposed negligible system overhead.

Following mitigation, legitimate flows were immediately restored, and throughput returned to pre-attack values. This validated the efficiency of the control mechanism and confirmed that the system can maintain reliable service continuity even during large-scale distributed attacks.

2.8 Comparative Analysis

For performance benchmarking, the proposed system was compared with prior studies, its outcomes were compared with the study conducted by Naman et al. (2023), which employed similar ML techniques for DDoS detection in SDN environments. The redesigned framework presented here achieved equivalent or improved accuracy using fewer monitored attributes and lighter controller interactions. This optimization reduced control-plane traffic while preserving rapid detection—an essential advantage for scalable IoT implementations.

3 Conclusion and Future Work

3.1 Conclusion

This research presented a hybrid SDN–ML framework for securing IoT networks against Man-in-the-Middle and Distributed Denial-of-Service attacks. The system leverages ML-based classification to recognize DDoS patterns and an adaptive ARP-verification mechanism to prevent spoofing. Both modules were integrated within an SDN architecture that enables centralized control, automatic mitigation, and minimal performance degradation.

Experiments conducted in a Mininet-WiFi environment under Ubuntu 18.04 confirmed that the framework provides accurate, real-time detection while maintaining low overhead on the Ryu controller. Findings confirm that ML-based defensive logic can be efficiently integrated into lightweight IoT infrastructures to improve reliability and resilience against evolving network threats.

3.2 Future Work

Further development may explore the incorporation of deep-learning models and real-time feedback mechanisms to enhance adaptability against emerging attack vectors. Expanding the framework to operate in physical IoT testbeds and large-scale deployments could also provide deeper insight into system scalability and robustness. Future efforts will aim to refine detection latency and optimize feature selection for even faster response in heterogeneous IoT environments. grained QoS requirements.

References

1. Gan, G., Lu, Z., Jiang, J.: Internet of things security analysis. In: Proceedings of the IEEE Conference on Internet of Things, pp. 1–4 (2011)
2. Open Networking Foundation, Software-Defined Networking: The New Norm for Networks (2013)
3. Suh, M., Park, S., Lee, B., Yang, S.: Building firewall over the software-defined network controller. In: Proceedings of the IEEE International Conference on Advanced Communication Technology (ICACT), pp. 744–748 (2014)
4. Chippalkatti, O., Nimbhorkar, S.U.: An approach for detection of attacks in software defined networks. In: International Conference on Innovations in Information, Embedded and Communication Systems (ICIIECS) (2017)

5. Sood, K., Yu, S., Xiang, Y.: Software-defined wireless networking: opportunities and challenges for internet-of-things – a review. IEEE Internet Things J. **2**(6), 453–463 (2015)
6. J. Kodi: Open Source Home Theatre Software (2016)
7. AbdelSalam, A.M., El-Sisi, A.B., Reddy, V.: Mitigating ARP spoofing attacks in software-defined networks. In: Proceedings 25th International Conference on Computer Theory and Applications (ICCTA), pp. 126–131, Alexandria, Egypt (2015)
8. Hayajneh, A.A., Bhuiyan, M.Z.A., McAndrew, I.: Improving internet of things (IoT) security with software-defined networking (SDN). IEEE Trans. Ind. Inform. **103**, 2–6 (2020)
9. Bhunia, S.S., Gurusamy, M.: Dynamic attack detection and mitigation in IoT using SDN. In: Proceedings 27th Int. Telecommunication Networks and Applications Conference (ITNAC), pp. 1–6, Melbourne, Australia (2017)
10. Karmakar, K.K., Varadharajan, V., Nepal, S., Tupakula, U.: SDN enabled secure IoT architecture. IEEE Internet Things J. **5**, 2–5 (2020)
11. Saritakumar, N., Anusuya, V.K., Balasaraswathi, B.: Detection and mitigation of MITM attack in software defined networks. In: Proceedings of the International Conference on Combinatorial Optimization (ICCAP) (2021)
12. Mohsin, M.A., Hamad, A.H.: Performance evaluation of SDN DDoS attack detection and mitigation based on random forest and K-nearest neighbors. Revue d'Intelligence Artificielle. **36**(2), 233–240 (2022)
13. Ndayizigiye, J.P.: Utilizing machine learning algorithms for early detection of DDoS attacks in software-defined networking. J. Emerg. Technol. Innov. Res. (JETIR). **10**(5), 1–13 (2023)
14. Sonthalia, N., Reddy, E.V.A., Pagaria, H., Jayasri, G.V.: Using machine learning in software defined networks to recognize and avoid DDoS attacks. Int. J. Res. Appl. Sci. Eng. Technol. (IJRASET). **11**(3), 1045 (2023)
15. Ranger, S.: What Is the IoT? Everything you Need to Know about the Internet of Things Right Now. ZDNet (2020)
16. I. J. Reynolds: Internet of Things (IoT) Architecture, Zibtek Technical Blog (2020)
17. Bizanis, N., Kuipers, F.: SDN and virtualization solutions for the internet of things: a survey. IEEE Internet Things J. **4**, 5591–5606 (2016)
18. Kreutz, D., Ramos, F.M.V., Verissimo, P.E., Rothenberg, C.E., Azodolmolky, S., Uhlig, S.: Software-defined networking: a comprehensive survey. Proc. IEEE. **103**(1), 14–76 (2015)
19. Open Networking Foundation, Open Networking Foundation (ONF) (2016)
20. Imperva, ARP Spoofing: Definition and Prevention, Imperva Security Portal (2023)
21. Cloudflare, What is a DDoS Attack?, Cloudflare Learning Center (2023)
22. Ryu SDN Framework, Ryu SDN Controller (2017)
23. SDNHub, Using the Ryu SDN Controller (2015)
24. Mohammed, M., Khan, M.B., Bashier, E.: Machine Learning: Algorithms and Applications. CRC Press (2016)
25. Sathya, R., Abraham, A.: Comparison of supervised and unsupervised learning algorithms for pattern classification. Int. J. Adv. Res. Artif. Intell. **2**, 34–38 (2013)
26. Géron, A.: Hands-on Machine Learning with Scikit-Learn and TensorFlow. O'Reilly Media (2017)
27. D. Varghese: Comparative Study on Classic Machine Learning Algorithms, Towards Data Science (2018)
28. Mininet Project: Mininet-WiFi Tool (2017)
29. A. Valleri, M. Ornaghi: Ettercap: Comprehensive Network Sniffer and Interceptor (2015)
30. J. Stone: iPerf: The TCP/UDP Bandwidth Measurement Tool (2017)
31. InMon Corp., sFlow-RT: Real-Time Traffic Analysis Software (2017)
32. IBM, Power Systems Performance Monitor (nmon) (2016)
33. Open Networking Foundation, Software-Defined Networking: The New Norm for Networks (2013)

A Review System Formed by Ensuring the Trustworthiness of DID and Acquiring Relationships Based on Verifiable Credentials

Yuki Deai[1(✉)] and Katsumi Wasaki[2]

[1] Graduate School of Science and Technology, Shinshu University, Nagano, Japan
25w6059h@shinshu-u.ac.jp

[2] Department of Electrical and Computer Engineering, Faculty of Engineering, Shinshu University, Nagano, Japan
wasaki@cs.shinshu-u.ac.jp

Abstract. Current review systems face declining credibility due to fraudulent activities such as fake and defamatory reviews. This research proposes a new review system that leverages Verifiable Credentials (VC) and Decentralized Identifiers (DID) to enable third-party verification of relationships between reviewers and reviewed parties By leveraging Universal Profiles (UP) on the LUKSO blockchain, users can securely and selectively manage and share their VC and DID information. During the review process, sharing the VC that demonstrates the relationship enables verification of the review's trustworthiness. Furthermore, introducing a 'relationship list' based on Zero-Knowledge Proofs (ZKP) enables third parties to verify trustworthy relationships while preserving personal confidentiality. This approach aims to ensure review authenticity in decision-making scenarios where trust is critical, such as personnel selection. Furthermore, the system is applicable to other fields requiring evaluation of review reliability based on usage history. Future challenges include establishing disclosure control for information contained within VC and developing models for generating and revoking relationship lists.

Keywords: Review System · Decentralized Identifiers · Verifiable Credentials · Universal Profiles · Zero-Knowledge Proofs

1 Introduction

Currently, evaluation systems like reviews face significant challenges due to a lack of trustworthiness. Problems such as fake reviews generated through mass account creation and defamatory reviews spreading false information are particularly severe, with documented damage across multiple review sites [1]. However, while existing countermeasures exist to verify whether such reviews were created by legitimate entities or contain accurate information, they are insufficient, making confirmation difficult. Against this backdrop, we propose a framework

H. R. Arabnia et al. (Eds.): CSCI 2025, CCIS 2970, pp. 432–439, 2027.
https://doi.org/10.1007/978-3-032-28466-2_33

to enhance review reliability by leveraging Verifiable Credentials (VC), which enable the digital storage and presentation of verifiable qualification information. This proposal clarifies the relationship between reviewers and those being reviewed (hereafter: reviewers and reviewed parties) using VC, establishing a mechanism to verify this relationship when viewing reviews. This approach eliminates reviews from unrelated parties, enabling the identification of fraudulently created accounts and reviews.

2 Background Technology

2.1 Verifiable Credentials(VC)

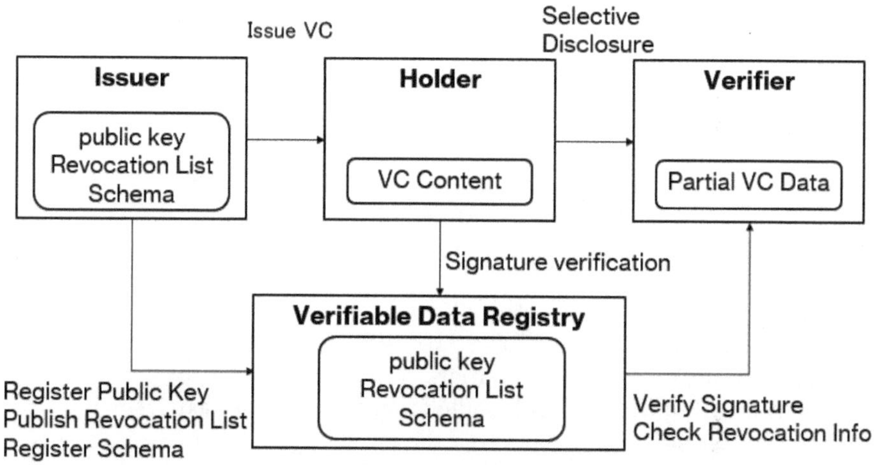

Fig. 1. Example of a VC Architecture: Interaction Between Issuer, Holder, and Verifier based on a Data Registry for Issuance, Possession, and Verification of VC [2]

Verifiable Credentials (VC) provide a secure and reliable framework for handling digital credentials such as diplomas and medical records [2].

The VC model consists of three primary entities, shown in Fig. 1. The Issuer creates and issues VC by attaching digital credential data and guaranteeing its authenticity. The Holder receives the VC, stores it, and manages its use, including the ability to disclose specific parts of the credential selectively. The Verifier checks the received VC to confirm its integrity and validity. The Schema defines the structure and attributes of a VC. This shared template allows Issuers and Verifiers to interpret credentials consistently. Verifiers use the Schema to apply validation rules that determine whether the VC includes the required attributes and preserves structural integrity. The Verifiable Data Registry manages metadata needed for VC verification. This registry contains the Issuer's public keys, credential Schemas, and revocation lists. The registry does not store the VC itself. During the verification process, Verifiers access the registry to verify digital signatures and check for revocation, as indicated in Fig. 1.

2.2 Decentralized Identifier(DID)

A Decentralized Identifier (DID) is a globally unique, cryptographically verifiable identifier that allows individuals to manage their own digital identity without relying on centralized authorities, providing a foundation for privacy and trust in decentralized environments [2]. Within the Verifiable Credentials (VC) framework, DID function as unique references to the issuer, holder, and verifier. This role is essential for validating the authenticity and origin of credentials, thereby establishing trust in the VC ecosystem.

Our proposed system builds upon this integration by utilizing DID and VC to represent an entity's roles and attributes. This approach facilitates decentralized, reliable verification without the need for intermediaries.

2.3 Universal Profiles(UP)

A mechanism for securely and selectively managing DID and VC to represent the relationship between reviewers and reviewed parties is required in this research context. To fulfill this requirement, Universal Profiles (UP)—a smart contract system on LUKSO, an Ethereum-compatible blockchain—are adopted [3]. UP, which manages assets such as VC and NFTs linked to DID, ensures secure storage and dynamic asset control through standard protocols such as ERC725 for metadata (including VC) and ERC20 for token management. Its ability to centrally manage identities and credentials, and to selectively invoke, share, and verify assets, makes UP a suitable foundation for the proposed system.

3 Proposed Review System

3.1 Third-Party Verifiability Through Relationship Introduction

Relationships are established using DID and their associated VC. As described in the previous section, leveraging DID and VC enables credentials to be reliably proven to third parties. For example, regarding relationships within a university, the reviewed party shares a portion of their VC related to the reviewer (e.g., a student ID VC if the reviewed party is a student and the reviewer is a faculty member) with the reviewer. The reviewer then compares the review text with the received VC and their own VC and records this comparison. Consequently, when a third party verifies the review, they can assess the reliability of the review text by validating the authenticity of the VC attached to the review. This mechanism makes it difficult for fraudulent accounts created for malicious or fake reviews to possess highly credible VC, thereby hindering attempts to bypass verification. Furthermore, comparing matching information from VC allows for some assessment of the depth of the relationship, further reinforcing the accuracy and credibility of the review.

3.2 Use Case Examples: Recruitment and Restaurant Reviews

As examples of a relationship-based review system, Fig. 2 shows the flow in a restaurant use case, while Fig. 3 illustrates the process in a recruitment scenario. Based on these diagrams, we explain the system's components and operational steps for real-world deployment in the following sections. Based on these diagrams, we explain the system's components and operational steps for real-world deployment in the following sections.

Restaurant Use Case

First, we explain the procedure of the review process in a restaurant use case.

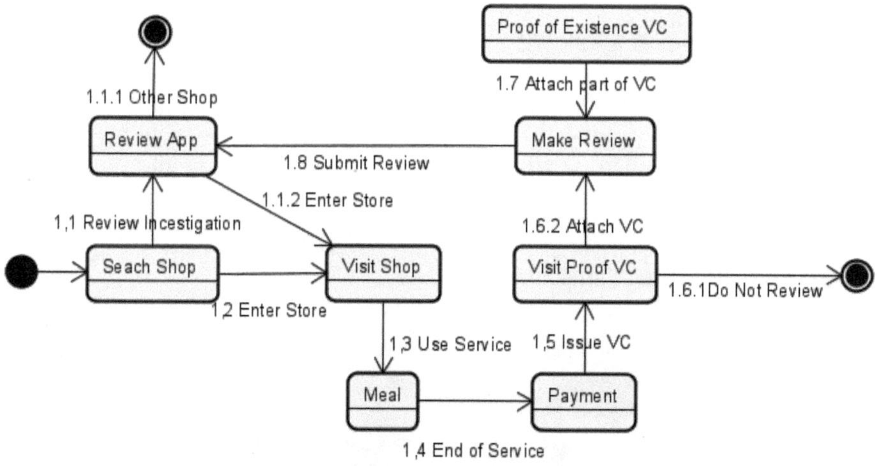

Fig. 2. Review Flow for Restaurant Use Cases (from Store Usage to Review Generation with Verifiable Credentials)

Step 1: Search for Restaurant Information, Decide to Use the Service, and Have a Meal - Searching for and using the store are the same as in a typical case.

Step 2: Issue of a Visit-Proof VC at Payment - A Verifiable Credential (VC) proving the user's visit is issued at checkout, for example, by scanning a QR code displayed in the store or entering a unique code provided at payment. By using this framework, it becomes possible to prove facts such as the user's actual visit to the store, repeat usage, or other relevant context—thereby enhancing the credibility of the review.

Step 3: Issue a Verifiable Credential (VC) that Proves the User's Identity or Presence - To prevent fraudulent activities caused by the widespread distribution of numerous VC, a portion of VC associated with the reviewing DID—specifically those issued by public institutions, such as student or employee VC—is attached to the review. This mechanism makes it possible to prevent fraudulent review submissions using fake DID. Furthermore, by utilizing such VC to verify attributes like age, it becomes feasible to conduct demographic analysis and apply accurate filtering to the reviews.

Step 4: Registration of a Review with an Attached VC - Record reviews to which a DID verification VC and a visit-proof VC are attached. To ensure the trustworthiness of these credentials, store their metadata on a blockchain or similar infrastructure.

This system has several advantages over conventional review platforms, including making it difficult for unrelated individuals to submit reviews and preventing manipulation through multiple identities by the same person. However, since reviews are intended to be viewed by an unspecified number of users, careful consideration must be given to the handling of personal information contained in the VC.

Personnel Selection Use Case

Second, we explain the procedure of the review process in a Personnel Selection.

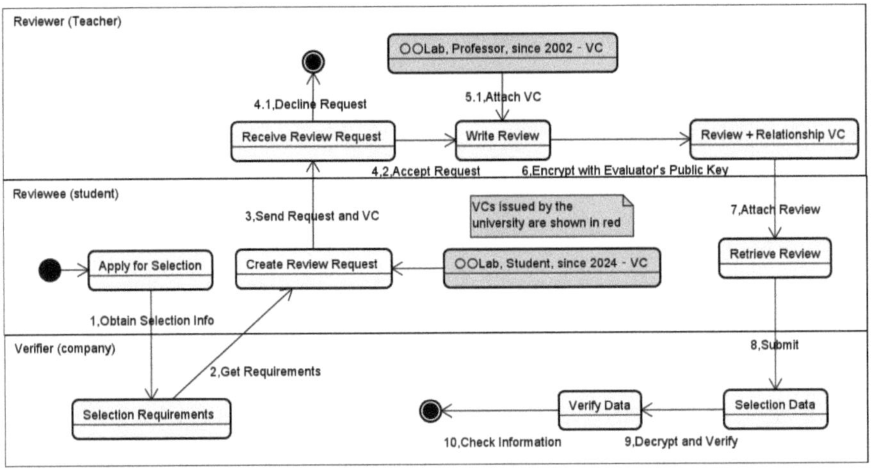

Fig. 3. Review Flow for Personnel Selection and Candidate Requirement Assessment (From Generation to Verification of Review Items)

Step 1: Confirming Required Reviews from Selection Information (Reviewee) - Investigate the relationships required for reviews by the selecting entity (e.g., professors or internship supervisors).

Step 2: Presenting Review Requirements (Recruiter) - Present the required review content and the VC requirements for the reviewer.

Step 3: Requesting a Review (Reviewee) - Request a review from someone who meets the requirements. When doing so, hide information from your VC that is unnecessary for establishing the relationship, sharing only the minimum necessary information, such as affiliation and signature details.

Step 4: Review Creation (Reviewer) - Describe the reviewed party's character and conduct. Attach relevant VC information (e.g., University Name, Lab Name, Enrollment Date, Graduation Date) and store both the review and VC details on the blockchain. Encrypt your VC and the review using the selection committee's public key.

Step 5: Review Submission (Reviewed Individual) - Compile the required information and submit it to the selection body. Encrypt your own VC using the selection body's public key.

Step 6: Review Verification (Selection Body) - Decrypt the review and VC, then verify them using the blockchain.

This example illustrates several key features of the system For example, in this scenario, the audience allowed to view the review is limited and trustworthy. Unlike restaurant reviews that are publicly visible and must minimize the disclosure of verifiable relationships, this recruitment case does not face such restrictions. As a result, more contextual information—such as the affiliations of both the reviewer and the subject, and the duration of their relationship—can be shared. This enables the expression of deeper and more meaningful relationships, thereby improving the credibility of the review and expanding its applicability across a wider range of evaluation scenarios.

4 ZKP-Derived Relationship List Associated with a Single DID

One framework to ensure review reliability is to verify whether a reviewer's DID has multiple trustworthy VC. However, publishing VC carries risks of personal identification and misuse. To address these issues, we are exploring "Relationship Lists (Fig. 4)". A Relationship List utilizes ZKP (Zero-Knowledge Proofs: a framework to prove the truth of certain information without revealing its content [4]) to exchange minimal information from VC with individuals or organizations sharing close relationships. This enables proving information such as attending the same university, being in the same research lab, or sharing the same affiliation for a certain period to third parties. Combining this proof mechanism with partial information concealment enables a system that displays the relationships associated with a DID. This approach demonstrates trustworthy relationships using minimal information, reducing the risk of personal identification and making misuse difficult. By partially disclosing relationships linked to a DID, it could serve as a means to establish trust in the DID. For instance, if a certain number of relationships exist between VC issued by trusted institutions, users could trust the DID.

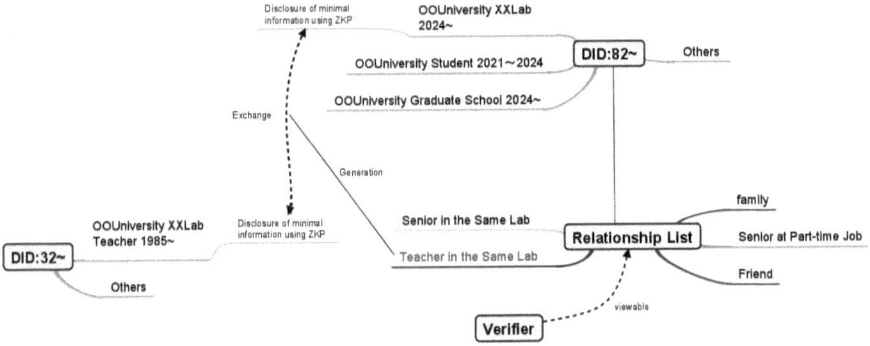

Fig. 4. Example. Relationship List Using DID and ZKP

5 Conclusion

This research presents a framework combining DID and VC to enable third-party verification of reviews, proposing a technical approach to overcome the decline in trust caused by fraudulent reviews. This addresses existing challenges and enables highly trustworthy reviews in contexts such as personnel selection. Future challenges include handling personal information contained within VC and refining the relationship list. For the latter, future work will detail the generation, issuance, and revocation of Relationship Lists, develop a secure model, and integrate it into the review system.

References

1. AP News: "The internet is rife with fake reviews. Will AI make it worse?" (2023) https://apnews.com/article/fake-online-reviews-generative-ai-40f5000346b1894a778434ba295a0496 (2025)
2. W3C: Verifiable credentials data model v2.0 (2025). https://www.w3.org/TR/vc-data-model-2.0/

3. Universal Profiles: LUKSO developer docs (2025). https://docs.lukso.tech/learn/overview
4. Zero Knowledge Proof (ZKP) in Blockchain: A Beginner's Guide (2025). https://www.gate.com/ja/blog/10554/zero-knowledge-proof-ZKP-in-blockchain-a-beginners-guide

Computational Biology

Application of Artificial Intelligence (AI) to Predict Genetic Biomarkers of Brain Tumors from Transcriptomic Datasets

Tijesunimi Adeyemi[1], Skylar Sawyer[1], Adewale Adeniyi[1], Arpana Basnet[1], Ato Bhatta[1], Richard Williams[2], and Roshan Paudel[1](✉)

[1] Morgan State University, Baltimore, MD 21251, USA
roshan.paudel@morgan.edu
[2] Baltimore City Public School, Baltimore, MD 21202, USA

Abstract. Brain tumors pose significant diagnostic and therapeutic challenges, particularly in distinguishing between malignant and benign cases at an early stage. In this study, we integrated gene expression data from multiple publicly available brain tumor datasets, comprising 1,100 samples and 16,384 genes, to enhance classification accuracy and clinical relevance. Counts per million (CPM) normalization and log fold-change analysis were applied to reduce dimensionality, resulting in 2,154 significant differentially expressed genes (DEGs). Among these, H4C14 was identified as a potential novel biomarker strongly associated with malignant tumors. Two machine learning models, XGBoost, was implemented for classification, demonstrating high predictive performance. The RNA-seq–based approach offers advantages over conventional MRI by detecting molecular changes that precede structural alterations, thereby enabling earlier diagnosis. Our findings highlight the promise of combining multi-cohort genomic data with advanced machine learning techniques to accelerate biomarker discovery and improve brain tumor classification. Further validation in independent cohorts and experimental studies is warranted to confirm the diagnostic utility of H4C14 and other identified DEGs.

Keywords: biomarkers · rna-seq · differentially expressed genes (DEGs) · machine learning · gene expression · brain tumor

1 Introduction

According to the Global Cancer Observatory (GLOBOCAN) 2020 estimates, brain and central nervous system cancer constitutes a significant portion of the global disease burden, ranking 19th among the most prevalent malignancies (1.9% of all cancers) and 12th among the primary causes of cancer mortality (2.5% of all cancers) [2]. In Iraq in 2020, brain and central nervous system cancer was the fourth highest cause of death among all cancers for both men and women across all age groups [2]. In 2017, the Global Burden of Disease (GBD)

H. R. Arabnia et al. (Eds.): CSCI 2025, CCIS 2970, pp. 443–458, 2027.
https://doi.org/10.1007/978-3-032-28466-2_34

Study identified brain and central nervous system cancer as the 8th significant cause of absolute Years of Life Lost (YLLs) among all cancers for both genders [3]. From 2008 to 2017, the incidence of brain and central nervous system cancers in the United States declined annually by 0.8% [4]. Nonetheless, while the incidence trend among adults (aged 20 and above) was declining, the trends in children and adolescents were significantly rising [4].

The brain is a highly complex, multifunctional, and delicate organ in the human body, coordinating with billions of neurons and cells to respond to many stimuli, including environmental influences and stress. The brain features a complex anatomy consisting of the cerebrum, cerebellum, and brain-stem, all protected by the skull. The research by Ruchi et al. [6] illustrates that understanding gene expression in various brain regions is essential for clarifying the complex molecular processes of the brain. Brain tumors are a varied category of disorders characterized by varying aspects, including origin location, malignancy, and age of formation, along with complex genomic or transcriptome profiles [7].

The usual treatment protocol for brain tumors encompasses chemotherapy and radiotherapy, alongside a surgical removal [8]. However, these methodologies are not always applicable, as tumors occasionally arise in sensitive or inoperable areas of the brain [9]. A novel approach under investigation for treating brain tumor patients involves the integration of cytotoxic and immune-stimulatory gene therapy utilizing an adenoviral vector [10]. A recent clinical experiment shown that combined cytotoxic and immune-stimulatory gene therapy exhibited acceptable toxicity and a notable enhancement in T cell levels, resulting in a median overall survival of 21 months [11]. However, another clinical trial failed to demonstrate enough tumor suppression or enhancement in overall survival [12]. Therefore, to implement the suitable immunotherapy and enhance its efficacy, it is essential to investigate and compare the transcriptome characteristics among diseases.

Previous research has demonstrated the effectiveness of machine-learning–based gene-expression analysis for biomarker identification across various cancer types. In our prior work [25], we applied this framework to colon cancer, identifying key over- and under-expressed genes using differential analysis and predictive modeling. Likewise, Abdul et al. [26] developed a machine-learning pipeline to identify tumor-associated genes in breast cancer RNA-Seq data. Collectively, these studies confirm the cross-cancer applicability of machine-learning–driven genomic analysis and support the methodological framework employed in the present brain tumor investigation.

The study combines gene expression data from several brain tumor forms and classifies them into two primary categories (malignant and benign), instead of examining each tumor subtype separately. This methodology aims to improve the efficiency and clinical significance of tumor identification. By simplifying the categorization process to a binary problem, the method facilitates the design of a machine learning model that can generate quicker and more actionable predictions. Benign tumors are non-malignant (low-grade, WHO Grade I and II), slowly proliferating, and arise within the brain. Conversely, malignant tumors (high-grade, Grade III and IV) are aggressive, rapidly proliferating, and frequently reveal ill-defined margins. Malignant tumors can originate in the brain (primary/metastasize) to the brain from other regions of the body (secondary/metastatic) [5].

2 Methods and Material

2.1 Datasets

The dataset utilized in this research was obtained from the Gene Expression Omnibus (GEO) repository. The datasets acquired for this study include GSE50161, GSE43378, GSE36245, GSE108476, and GSE68848. After proper cleaning and transposition, GSE108476 was found to have 420 samples, GSE50161 had 160, GSE43378 included 50, GSE68848 comprised 332, and GSE26245 had 46 samples. The dataset includes gene expression profiles of brain tumor samples, encompassing 16,384 genes across 1,100 samples, with 654 malignant and 446 benign cases. This results in a moderate class imbalance, where approximately 59.5% of the samples are malignant and 40.5% are benign (see Fig. 1).

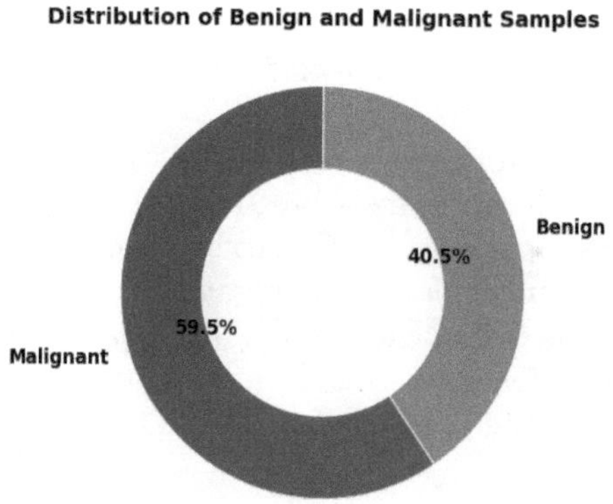

Fig. 1. Percentage difference of malignant and benign tumor samples.

Given the large dimensionality of gene expression data, a systematic preprocessing and feature selection pipeline was established to mitigate noise and remove unnecessary genes that could impede tumor classification efficacy. The data were initially standardized using Counts Per Million (CPM) normalization to adjust for variations in sequencing depth among samples. Subsequent to normalization, low-variance genes were excluded to preserve only informative signals. Subsequently, differential gene expression analysis was conducted to differentiate between tumor and normal tissues. This approach found 5,327 differentially expressed genes (DEGs), which were further classified into over-expressed, under-expressed, and non-significant categories according to their log2 fold change values. The application of log fold change not only emphasized biologically significant genes but also functioned as a dimensionality reduction method by concentrating on genes exhibiting meaningful expression patterns. Genomic datasets frequently encounter the high-dimensional data issue, characterized by a substantial disparity between the number of characteristics (genes) and the number of observations (samples).

This presents major challenges for machine learning, such as higher computational complexity, difficulties in model interpretation, and an elevated risk of overfitting. According to [13], statistical methodologies need to tackle these problems to derive significant insights. Consequently, the use of dimensionality reduction and feature selection methods is essential for enhancing the efficiency and precision of machine learning algorithms in the analysis of extensive genomic data.

2.2 Methods

Count Per Million (CPM). A normalization method in which raw read counts assigned to a transcript are divided by the total number of sequencing reads in the sample and multiplied by one million. It standardizes RNA-seq data for sequencing depth, excluding gene length considerations. The CPM normalization was conducted using the following equation.

$$\mathrm{CPM}_{ij} = \frac{x_{ij}}{\sum_{k=1}^{n} x_{kj}} \times 10^6$$

where x_{ij} is the raw count of the ith gene in the jth sample

After normalizing the raw gene expression data using CPM, a differential expression analysis was performed to identify genes that significantly distinguish malignant from benign brain tumor samples. The average expression of each gene was computed within each group. The log_2 fold change (log_2FC) was subsequently calculated for each gene to measure the extent and direction of differential expression between the two groups. The log_2 fold change (FC) was computed as:

$$\log_2 \mathrm{FC}_i = \log_2 \left(\frac{\mu_{i,\text{Malignant}} + \varepsilon}{\mu_{i,\text{Benign}} + \varepsilon} \right)$$

where $\mu_{i,\text{Malignant}}$ and $\mu_{i,\text{Benign}}$ are the mean CPM-normalized expressions of gene in malignant and benign samples, respectively, and $\varepsilon = 1 \times 10^{-9}$ is a small constant to avoid division by zero.

Welch's t-test was employed on a gene-by-gene basis to evaluate statistical significance by contrasting expression values between malignant and benign groups, accounting for different variances. The obtained p-values were adjusted for multiple comparisons using the Benjamini-Hochberg method to control the False Discovery Rate (FDR). A moderate threshold was utilized to identify significant differentially expressed genes (DEGs), with an absolute log_2 fold change surpassing 0.5 and a false discovery rate (FDR) under 0.5. Genes with positive log_2FC are over-expressed in malignant situations, while genes with negative

log_2FC are under expressed in malignant conditions. The complete workflow for gene identification and tumor classification is shown in Fig. 2.

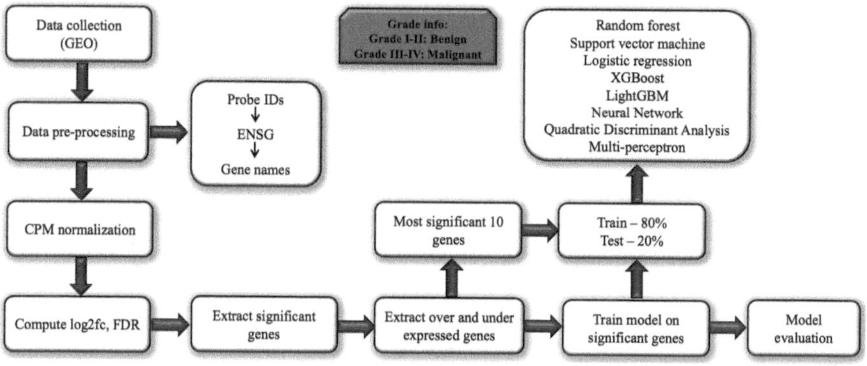

Fig. 2. Workflow of the proposed method for gene identification and tumor classification.

Machine Learning Preparation. Following the identification of the relevant DEGs, the dataset was organized for supervised classification. The feature matrix X comprised the expression values of the chosen genes, whereas the target vector y included the binary class labels, with benign and malignant recorded as 0 and 1, respectively. Before model training, the features were standardized using Z-score normalization through the traditional scaler approach, which is an essential step prior to model implementation. This procedure standardized all gene characteristics to a single scale, with a mean of 0 and a standard deviation of 1, which is crucial for enhancing the performance of numerous machine learning methods. The standardized dataset was then divided into training and testing sets in an 80:20 ratio. This division guaranteed that the model could be trained on the majority of the data while retaining an unobserved segment for assessing its generalization efficacy. No resampling procedures were employed, as the class imbalance was minor, and model evaluation was conducted directly on the original data distribution to accurately represent real-world diagnostic conditions.

We originally examined various supervised machine learning models to assess their capability in classifying brain cancers using gene expression data. These encompassed Random Forest, Support Vector Machine, Logistic Regression, XGBoost, LightGBM, Neural Networks, Quadratic Discriminant Analysis, and Multi-layer Perceptron (MLP). For this investigation, we concentrated on two ensemble models, XGBoost and Random Forest, owing to their robust performance, interpretability, and efficacy in managing high-dimensional biological datasets. The following paragraph describes the operational concepts of both approaches.

Random Forest. A supervised learning technique that is applicable to regression and classification problems [14]. This is commonly known as an ensemble learning classification technique, which generates a series of CART models by bootstrap sampling of the initially provided data [15]. Each individual tree is constructed utilizing a random subset of the model characteristics and use bagging to generate an uncorrelated forest of trees following the same purpose. The tree optimizes the categorization criteria for each iterative node. In numerous instances, it exceeds many of its parametric counterparts and is also less computationally demanding. The following outlines the technical methodology of the Random Forest algorithm as derived from [16].

Algorithm 1. XGBoost Algorithm

1. Given training data with N samples and p predictor variables.
2. Initialize model with a constant value:

$$F_0(x) = \arg\min_{\gamma} \sum_{i=1}^{N} L(y_i, \gamma)$$

3. For $m = 1$ to M do:
   (a) Compute the pseudo-residuals:

$$r_{mi} = -\left[\frac{\partial L(y_i, F(x_i))}{\partial F(x_i)}\right]_{F(x)=F_{m-1}(x)}$$

   (b) Fit a regression tree $h_m(x)$ to $\{(x_i, r_{mi})\}_{i=1}^{N}$
   (c) Compute optimal leaf weights w_j and regularized objective:

$$\Omega(h_m) = \gamma T + \frac{1}{2}\lambda \sum_{j=1}^{T} w_j^2$$

   (d) Update the model:

$$F_m(x) = F_{m-1}(x) + \nu \cdot h_m(x)$$

4. Output: Final prediction model $F_M(x)$

Gradient Boosting. An ensemble learning method that develops a robust predictive model by sequentially integrating several weak learners, usually shallow decision trees [17]. In each iteration, a new model is developed to estimate the residual errors (pseudo-residuals) of the existing ensemble by minimizing a differential loss function. This additive method progressively enhances the overall predictive function. The gradient boosting method formally constructs a function $F_m(x) = F_{m-1}(x) + \rho_m h_m(x)$, where ρ_m represents the learning rate, and

$h_m(x)$ denotes the newly incorporated weak learner, often a decision tree, fitted to the pseudo-residuals at iteration m. The residuals are calculated as the negative gradients of the loss function about the current model's predictions.

This study employed XGBoost (Extreme Gradient Boosting), an efficient and scalable variant of gradient boosting designed for performance and efficiency. XGBoost improves the conventional gradient boosting framework by integrating second-order derivatives into the loss function and adding a specific regularization term to manage tree complexity. The objective function in XGBoost comprises both the loss and a regularization term $\Omega(h) = \gamma T + \frac{1}{2}\lambda\|w\|^2$, where T denotes the number of leaves in the decision tree, and w represents the vector of output values linked to those nodes. The parameter γ imposes a penalty on the inclusion of new leaves, thus promoting the development of simpler and more generalizable models.

XGBoost incorporates randomization methods to enhance generalization and training efficiency, such as sub-sampling of rows and columns at both the tree and node levels. It utilizes a sparsity-aware method that effectively manages missing data, together with a histogram-based split-finding strategy to reduce computational cost by evaluating only a portion of potential splits.

Algorithm 2. Random Forest Algorithm

1. Given training data with N samples and p predictor variables.
2. Construct a Random Forest model consisting of B decision trees.
3. For $x = 1$ to B do:
   (a) Draw a bootstrap sample S^* of size N (with replacement) from the training data.
   (b) Grow a decision tree T_x on S^* using the following steps:
      i. At each node, select m variables randomly from the p predictor variables ($m \ll p$).
      ii. Find the best split among the m selected variables using a chosen splitting criterion (e.g., Gini impurity or information gain).
      iii. Split the node into two child nodes.
   (c) Continue splitting nodes recursively until a stopping criterion is met (e.g., minimum node size or maximum depth).
4. Combine the B trees to form an ensemble.
5. For a new sample, pass it through all B trees:
   - For classification: Use majority voting to determine the predicted class.
   - For regression: Use the average prediction from all trees.
6. Output: Final prediction for the new sample.

3 Results

The brain tumor dataset consists of 1,100 observations and 5,327 genes, making it a high-dimensional dataset. To accurately predict the classification of genes as cancerous or benign, a statistical technique was utilized. The log_2 fold change and corrected p-values (FDR) were calculated to identify differentially expressed genes (DEGs). This technique reduces the 5,327 genes to the most significant predictive features for predicting the outcome variable. We implemented eight distinct machine learning models on the differentially expressed genes (DEGs). The dataset was divided into training (80%) and testing (20%), and machine learning models were employed to determine which model was best.

We performed differential gene expression analysis utilizing a CPM-normalized dataset to find genes substantially correlated with brain tumor malignancy. A Welch's t-test was utilized to evaluate gene expression levels between benign and malignant samples, then implementing multiple hypothesis testing correction via the Benjamini-Hochberg false discovery rate. Genes were defined as substantially differentially expressed if they satisfied two criteria: an absolute log_2FC over 0.5 and an adjusted $FDR < 0.5$, as illustrated in Fig. 3.

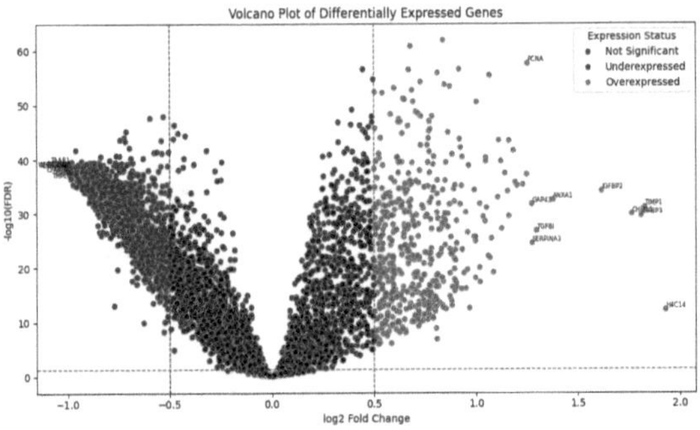

Fig. 3. Distribution of DEGs. The x-axis represents the log_2fold change, while the y-axis shows the -log_{10}(FDR)

This analysis identified a total of 2,154 differentially expressed genes (DEGs). These genes are classified into three categories according to their expression characteristics. These groups are over-expressed, under-expressed, and non-significant. Genes that were over-expressed were characterized by a positive log_2FC in malignant samples relative to benign ones, whereas under-expressed genes exhibited a negative log_2FC. Genes that failed to satisfy the statistical thresholds were categorized as non-significant.

Numerous genes with established associations to tumor biology were identified as highly over-expressed and under-expressed. Genes including H4C14, TIMP1, IGFBF3, CHI3L1, IGFBP2, ANXA1, TGFBI, SERPINA3, GAP43, and PCNA w ere significantly higher in malignant samples, indicating their possible involvement in cell proliferation, invasion, and tumor growth. Conversely, genes including KRT13, TRPC7, CPA6, RGS9BP, SRY, TAAR1, CFAP65, WFDC108, TLR8-AS1, and GPR50 exhibited significant under-expression, suggesting potential tumor suppressor functions or down-regulation in malignant conditions, as illustrated in Fig. 4

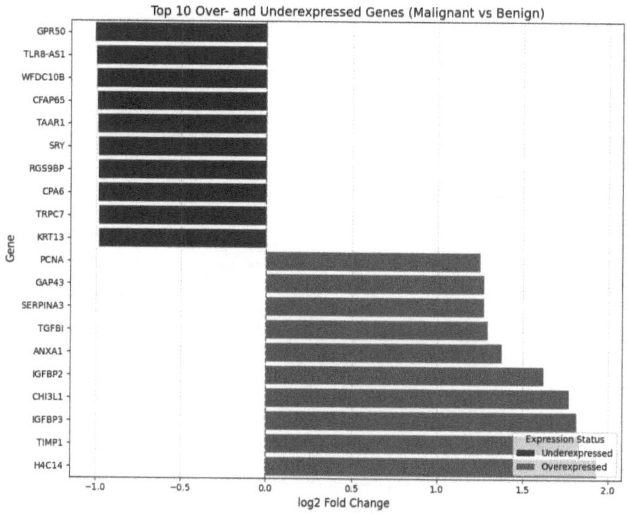

Fig. 4. Top differential genes of over-expressed and under-expressed genes

To gain more insight into the most relevant genes, we concentrated on the top differentially expressed genes according to their log_2FC and statistical significance. Figure 4 illustrates the magnitude and direction of expression changes. This gene will be discussed further as we investigate each gene. Genes exhibiting strong positive log_2FC demonstrate elevated expression levels in malignant tumors compared to benign cases. Conversely, the gene demonstrating a negative log_2FC indicates suppression in malignant tissue.

Genes such as IGFBP2, TIMP1, IGFB3, CHI3L1 and H4C14 demonstrated consistently high expression in a substantial number of samples, indicating their overexpression in malignant tumors, as illustrated in Fig. 5a. Conversely, GPR50, TLR8-AS1, WFDC10B, CFAP65, and TAAR1 exhibit a reduced expression in the malignant group, consistent with their classification as underexpressed.

Figure 5b illustrates the quantitative summary of gene expression differences and supports Fig. 5a by confirming consistent trends at the overall level. The observed differences in mean expression highlight the discriminatory capacity of the selected genes, strengthening their importance for tumor classification and the development of potential biomarkers.

To assess the statistical significance of the top 10 differentially expressed genes identified as potential biomarkers, a Mann-Whitney U (MWU) test was performed to compare their expression levels between malignant and benign samples. Figure 6 illustrates the mean expression levels of each gene categorized by sample type, along with their respective p-values.

All genes exhibited statistically significant differences in expression between malignant and benign tissues. The genes IGFB2, TIMP1, CHI3L1 and IGFB3 exhibited significantly elevated expression in malignant samples, with p-values of $p < 1\exp{-40}$, providing strong statistical evidence against the null hypoth-

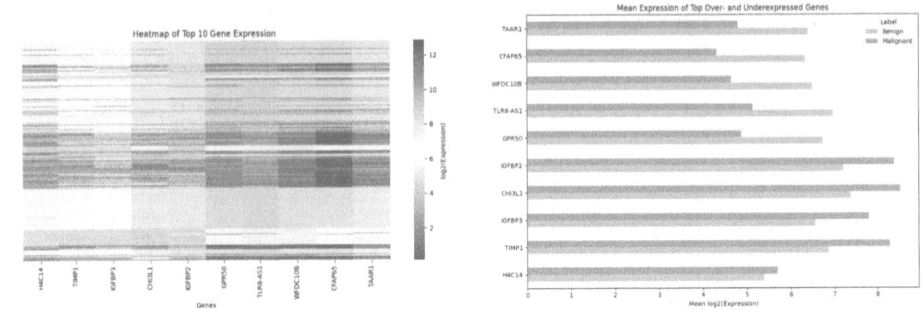

(a) Heatmap of top 10 gene expression (b) Mean expression by tumor type

Fig. 5. Comparison of heatmap and mean gene expression across tumor types

esis. These results confirm the previous over-expression findings identified in the volcano plot and bar charts. In contrast, genes including TAAR1, CFAP65, WFDC10B, TLR8-AS1 and GPR50 exhibited significantly decreased expression in malignant samples, indicating their potential role as down-regulated tumor suppressor markers. H4C14, while still significant ($p = 0.0389$), exhibited a comparatively modest p-value, indicating that its expression pattern may be less distinct between tumor classes or subject to greater variance between samples.

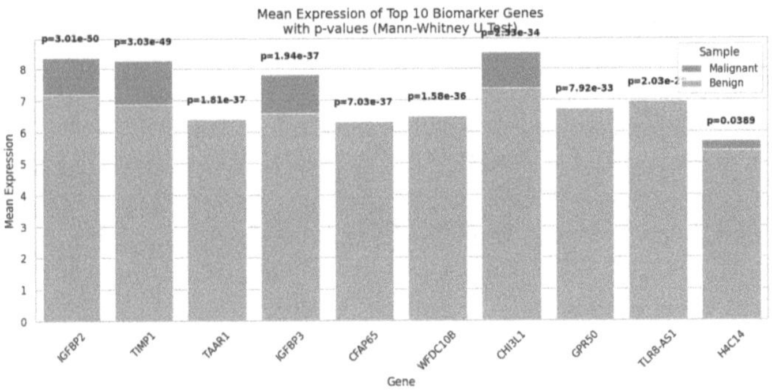

Fig. 6. Mean expression weighted by p-value

Figure 7 indicates that most genes are positioned near the diagonal line, suggesting strong agreement between the two tests. Certain genes, including IGFBP2, TIMP1, and IGFBP3, demonstrate significantly stronger results in the t-test, indicating a potential sensitivity of this method to the assumptions of normality and equal variance. In contrast, genes such as H4C14 exhibited increased significance when analyzed with the MWU test, a non-parametric method that does not assume normality.

Among the differentially expressed genes identified in this study, H4C14 stands out as a significant biomarker. Although several prominent genes, including IGFBP2, TIMP1, and CHI3L1, have previously been associated with malignancies, H4C14 has not been extensively documented in relation to brain tumor classification.

The analysis indicated that H4C14 is markedly over-expressed in malignant samples relative to benign samples. The findings were consistently supported by the log_2FC analysis and the Mann-Whitney U test, both of which produced a statistically significant p-value. The expression pattern of the gene, illustrated through this test, demonstrates increased levels in malignant tumors, indicating its possible role in oncogenic pathways.

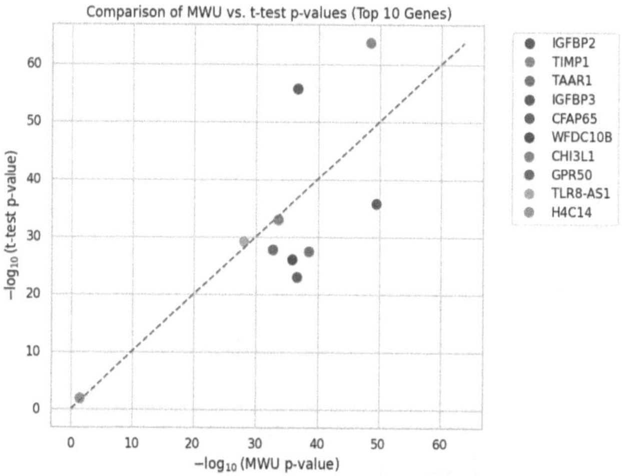

Fig. 7. Statistical validation of top biomarker genes

We assessed the effectiveness of eight classifiers in predicting tumor types based on gene expression data. The performance metrics analyzed are accuracy, precision, recall, and F1-score.

Figure 8 illustrates that XGBoost is the top-performing model, demonstrating a better overall balance across all metrics. The model achieved 80% accuracy, 81% precision, 85% recall, and 81% F1-score, demonstrating robust sensitivity and precision in the classification of tumor samples. The random forest exhibited a recall of 77% and a precision of 82%, yielding an identical F1-score of 81%. Both ANN and QDA achieved a recall rate of 100%, indicating that they accurately classified all malignant cases. However, their precision of 59% and accuracy of 59% indicate a potential over-classification of positive cases, resulting in elevated false positive rates.

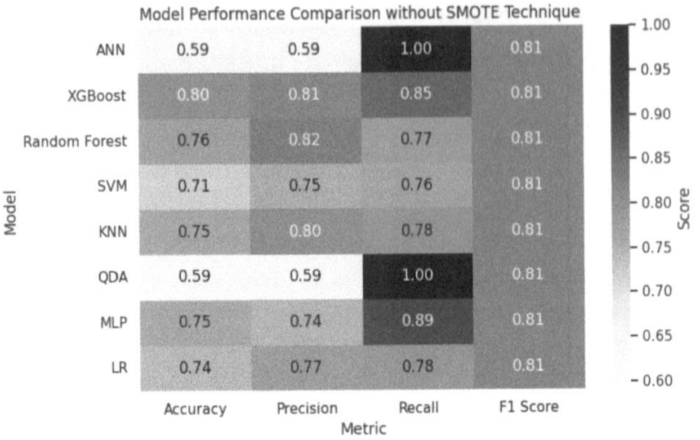

Fig. 8. Assessment of predictive models on transcriptomic profiles

4 Conclusion

In this study, we combined gene expression data from multiple brain tumor datasets and classified them into two clinically relevant categories: malignant and benign. By applying CPM normalization, differential expression analysis, and feature selection, we identified 2,154 significant DEGs, among which H4C14 emerged as a potential novel biomarker strongly associated with malignant tumors. The use of RNA-seq data provided a molecular-level view of tumor biology that overcomes certain limitations of conventional MRI imaging, enabling earlier and more precise detection of malignancy. For predictive modeling, we evaluated several machine learning algorithms, with XGBoost emerging as the top-performing model, achieving robust classification accuracy. These models demonstrated the potential of integrating genomic profiling with machine learning for rapid and accurate tumor classification, which aligns with real-world clinical priorities of determining tumor aggressiveness for timely intervention. While our findings suggest promising diagnostic utility, particularly for H4C14, further validation through independent cohorts and experimental assays is essential before clinical translation. Overall, this work underscores the value of combining multi-source genomic data with advanced computational approaches to accelerate biomarker discovery and improve brain tumor diagnostics.

5 Discussion

In this study, we conducted an integrated analysis of gene expression data to identify biomarkers capable of discriminating between benign and malignant brain tumors. Using multiple statistical and machine learning technique.

While 2,154 differentially expressed genes (DEGs) were discovered from 1,100 samples, the high dimensionality of gene expression data in relation to the sample size could increase the risk of overfitting. To reduce the risk, we implemented a dual method. A differential expression analysis utilizing log_2 FC and false discovery rate (FDR) thresholds was used to exclude genes with insignificant expression differences. This approach minimized noise and ensured that only biologically significant genes influenced classification.

Among the multiple models evaluated, including Support Vector Machines (SVM), K-Nearest Neighbors (KNN), Logistic Regression (LR), Multi-Layer Perceptron (MLP), and Quadratic Discriminant Analysis (QDA), XGBoost consistently yielded strong performance across accuracy, precision, and F1-score metrics. Despite its slightly lower recall compared to MLP and QDA, its balanced overall metrics made it the most suitable model for our feature ranking. Interestingly, both ANN and QDA exhibited perfect recall (1.00), but their poor accuracy and precision values suggest that they over-classified malignant cases. To ensure the robustness of our results, we applied statistical techniques such as Mann–Whitney U tests and Welch's t-tests to mitigate overfitting and account for sample size differences. The comparison between MWU and t-test p-values across the top 10 genes confirmed consistent statistical significance.

IGFBP2 is a member of the secreted IGFBP family that functions by interacting with circulating IGFs to modulate IGF-mediated signaling. The recognition of the diverse functions and cellular locations of members of the IGFBP family has challenged this traditional view of IGFBP2 activities. IGFBP2 has been previously established as a driver of glioma progression to a higher grade. Lynette et al. [18], determined whether IGFB-over-expressing tumors are dependent on continued oncogene expression and whether IGFBP2 is a viable therapeutic target in glioma, and also it was established that IGFBP2 is considered as one of the strongest biomarkers of aggressive behavior in glioblastoma (GBM). It was shown in there result that IGFBP2 also impaired tumor cell spread. IGFBP2 is said to be one of the most potential glioma oncogenes and functions as a hub of oncogenic signaling pathways by regulating pro-tumorigenic signals of tumor initiation and progression [19]. Moreover, the clinical utility of both IGFBP2 and IGFBP3 as biomarkers has been supported by Abdolhoseinpour et al. [22], who demonstrated that serum levels and tissue expression of these two genes were significantly associated with GBM progression and patient survival. There findings showed that the preoperative plasma levels of IGFBP-2 and IGFBP-3 were markedly higher in glioblastoma patients ($mean \pm$ SD: 521.5 $\pm$ 164.2 ng/ml; 402.4 $\pm$ 126 ng/ml) compared with healthy controls (301.28 $\pm$ 73.12 ng/ml; 244 $\pm$ 89.5 ng/ml; $p < 0.001$). Immunohistochemical results indicated that the median H score for glioblastoma tissues was higher than that of normal tissues. There data suggest that plasma levels and tissue levels of IGFBP-2 and IGFBP-3 may be as potential biomarkers for predicting the progression and survival in patients with GBM.

Groft et al. [20] reported that TIMP1 is involved in the progression of gliomas. Significantly increased TIMP1 levels were detected in patients with astrocytomas and glioblastomas. In addition, TIMP1 expression was positively correlated with the gene expression of two predominant components of the extracellular matrix, tenascin C and fibronectin, both of which were also highly expressed in wild-type IDH gliomas. String analysis, co-immunoprecipitation, and immunofluorescence revealed a significant correlation between TIMP1 and CHI3L1. Furthermore, we found, both in vivo and in vitro, that TIMP1 promoted gliomagenesis via CHI3L1 over-expression as well as NF-κB activation. TIMP1 expression correlated with tumor immune infiltration and immune checkpoint-related gene expression. In addition, TIMP1 resulted in immunosuppressive macrophage polarization. In summary, TIMP1 or CHI3L1 might be perceived as a diagnostic marker and an immunotherapy target for gliomas [21].

H4C14 encodes a histone H4 variant, and histones are known to play a crucial role in chromatin remodeling and gene regulation. Aberrations in histone expression and modification are increasingly recognized as contributors to tumor progression through epigenetic mechanisms. The over-expression of H4C14 in malignant samples may reflect such epigenetic dysregulation, which could influence the transcriptional landscape of cancerous cells and contribute to uncontrolled proliferation and survival. Given its significant differential expression and the lack of prior documentation in brain tumor studies, H4C14 represents a promising novel biomarker that warrants further investigation.

Trace amine-associated receptor 1 (TAAR1) has emerged as a promising therapeutic target for neuropsychiatric disorders due to its ability to modulate monoaminergic and glutamatergic neurotransmission. In particular, agonist compounds have generated interest as potential treatments for schizophrenia and other psychoses due to TAAR1-mediated regulation of dopaminergic tone [23]. TAAR1 is a neuro-regulator with emerging evidence suggesting a role in immunomodulation [24].

The predictive model created in this work, including machine learning methods like XGBoost, exhibited significant promise for differentiating tumor from non-tumor samples through gene expression profiles. In addition to statistical validation, these found biomarkers could function as molecular signatures that improve precision diagnostics when integrated with imaging data, such as MRI. Integrating radiological characteristics with genomic predictions via a radiogenomic framework would enable doctors to noninvasively deduce molecular abnormalities from imaging phenotypes, hence enhancing early detection and risk stratification.

Acknowledgments. We would like to acknowledge the Center for Equitable Artificial Intelligence and Machine Learning Systems (CEAMLS) for its support throughout this research.

Disclosure of Interests. The authors declare that they have no competing interests.

References

1. Bagley, J.A., Reumann, D., Bian, S., Lévi-Strauss, J., Knoblich, J.A.: Fused cerebral organoids model interactions between brain regions. Nat. Methods **14**(7), 743–751 (2017). https://doi.org/10.1038/nmeth.4304

2. Ferlay, J., Colombet, M., Bray, F.: Cancer incidence in five continents, CI5plus: IARC CancerBase No. 9. International Agency for Research on Cancer, Lyon, France (2018). http://ci5.iarc.fr. Accessed 02 Aug 2025
3. Global Burden of Disease Cancer Collaboration: Global, regional, and national cancer incidence, mortality, years of life lost, years lived with disability, and disability-adjusted life-years for 29 cancer groups, 1990 to 2017: a systematic analysis for the global burden of disease study. JAMA Oncol. **5**(12), 1749–1768 (2019). https://doi.org/10.1001/jamaoncol.2019.2996
4. Miller, K.D., et al.: Brain and other central nervous system tumor statistics. CA Cancer J. Clin. **71**(5), 381–406 (2021). https://doi.org/10.3322/caac.21693
5. Cancer Research UK: Grades of brain and spinal cord tumours (2023). https://www.cancerresearchuk.org/about-cancer/brain-tumours/grades. Accessed 02 Aug 2025
6. Yadav, R., Sharma, A., Prakash, J.: Prediction and identification of signature genes expressed in different brain regions through RNA-Seq data analysis. Biomed. Pharmacol. J. **15**(4) (2022)
7. Sanai, N., Alvarez-Buylla, A., Berger, M.S.: Neural stem cells and the origin of gliomas. N. Engl. J. Med. **353**(8), 811–822 (2005)
8. Louis, D.N., et al.: The 2016 world health organization classification of tumors of the central nervous system: a summary. Acta Neuropathol. **131**(6), 803–820 (2016)
9. Ricard, D., Idbaih, A., Ducray, F., Lahutte, M., Hoang-Xuan, K., Delattre, J.-Y.: Primary brain tumours in adults. The Lancet **379**(9830), 1984–1996 (2012)
10. Faisal, S.M., Castro, M.G., Lowenstein, P.R.: Combined cytotoxic and immune-stimulatory gene therapy using Ad-TK and Ad-Flt3L: translational developments from rodents to glioma patients. Mol. Ther. **31**(10), 2839–2860 (2023)
11. Umemura, Y., et al.: Combined cytotoxic and immune-stimulatory gene therapy for primary adult high-grade glioma: a phase 1, first-in-human trial. Lancet Oncol. **24**(9), 1042–1052 (2023)
12. Rainov, N.G.: A phase III clinical evaluation of herpes simplex virus type 1 thymidine kinase and ganciclovir gene therapy as an adjuvant to surgical resection and radiation in adults with previously untreated glioblastoma multiforme. Hum. Gene Ther. **11**(17), 2389–2401 (2000)
13. Johnstone, I.M., Titterington, D.M.: Statistical challenges of high-dimensional data. Phil. Trans. R. Soc. A **367**(1906), 4237–4253 (2009)
14. Breiman, L.: Random forests. Mach. Learn. **45**(1), 5–32 (2001)
15. Maroco, J., Silva, D., Rodrigues, A., Guerreiro, M., Santana, I., De Mendonça, A.: Data mining methods in the prediction of Dementia. BMC. Res. Notes **4**(1), 299 (2011)
16. Yekkala, I., Dixit, S.: Prediction of heart disease using random forest and rough set based feature selection. Int. J. Big Data Analytics Healthc. **3**(1), 1–12 (2018)
17. Chen, H.-H.: Understanding gradient boosting classifier: training, prediction, and the role of γ_j. arXiv preprint arXiv:2410.05623 (2024)
18. Phillips, L.M., et al.: Glioma progression is mediated by an addiction to aberrant IGFBP2 expression and can be blocked using anti-IGFBP2 strategies. J. Pathol. **239**(3), 355–364 (2016)
19. Kalya, M., Kel, A., Wlochowitz, D., Wingender, E., Beißbarth, T.: IGFBP2 is a potential master regulator driving the dysregulated gene network responsible for short survival in glioblastoma multiforme. Front. Genet. **12**, 670240 (2021). https://doi.org/10.3389/fgene.2021.670240

20. Groft, L.L., Shridhar, T.T., Matrisian, R.M., Murray, D.W.: Differential expression and localization of TIMP-1 and TIMP-4 in human gliomas. Br. J. Cancer **85**(1), 55–63 (2001). https://doi.org/10.1054/bjoc.2001.1854
21. Xu, J., et al.: TIMP1/CHI3L1 facilitates glioma progression and immunosuppression via NF-κB activation. Biochimica et Biophysica Acta Mol. Basis Dis. **1870**(3), 167041 (2024). https://doi.org/10.1016/j.bbadis.2024.167041
22. Abdolhoseinpour, H., Mehrabi, F., Shahraki, K., et al.: Investigation of serum levels and tissue expression of IGFBP-2 and IGFBP-3 as potential biomarkers for predicting progression and survival in glioblastoma multiforme patients. J. Neurol. Sci. **366**, 202–206 (2016). https://doi.org/10.1016/j.jns.2016.05.018
23. Dedic, N., Dworak, H., Zeni, C., Rutigliano, G., Howes, O.D.: Therapeutic potential of TAAR1 agonists in schizophrenia: evidence from preclinical models and clinical studies. Int. J. Mol. Sci. **22**(24), 13185 (2021). https://doi.org/10.3390/ijms222413185
24. Barnes, D.A., Galloway, D.A., Hoener, M.C., Berry, M.D., Moore, C.S.: TAAR1 expression in human macrophages and brain tissue: a potential novel facet of MS neuroinflammation. Int. J. Mol. Sci. **22**(21), 11576 (2021). https://doi.org/10.3390/ijms222111576
25. Adeyemi, T., et al.: Prediction of genetic biomarkers from RNA-Seq dataset of colon cancer. In: 2023 International Conference on Computational Science and Computational Intelligence (CSCI), pp. 1378–1385 (2023). https://doi.org/10.1109/CSCI62032.2023.00226
26. Abdul, A., Paudel, R., Rahman, M.M.: Using machine learning algorithms to find novel biomarkers for breast cancer using RNA-Seq dataset. Preprints **2023**, 1–15 (2023). https://doi.org/10.20944/preprints202309.0006.v1

Graph Structure Matters: Graph Design and Feature Selection for Accurate MIC Prediction in *Salmonella* Species

Ganga Gireesan[1(✉)], Moses B. Ayoola[1], Athish Ram Das[1], Bindu Nanduri[1], and Mahalingam Ramkumar[2]

[1] Department of Comparative Biomedical Sciences, College of Veterinary Medicine, Mississippi state University, Starkville, MS, USA
{gg733,mba185,ar2903}@mssate.edu, bnanduri@cvm.msstate.edu

[2] Department of Computer Science and Engineering, Mississippi state University, Starkville, MS, USA
ramkumar@cse.msstate.edu

Abstract. Antimicrobial resistance in *Salmonella* poses a major health challenge, making rapid determination of bacterial susceptibility essential. The Minimum Inhibitory Concentration (MIC) provides precise quantitative information, but standard laboratory methods for determining MIC are slow. We present a graph-based framework to predict MIC directly from bacterial genomes. Each genome is modeled as a graph with k-mers as nodes and edges defined by sequence similarity, capturing complex genomic relationships. We evaluated three graph neural network architectures - Graph Convolutional Networks (GCN), Graph Attention Networks (GAT), and Graph SAmple and aggreGatE (GraphSAGE) - and show that GAT achieves the highest predictive accuracy across 15 antibiotics. Our results highlight the importance of graph design in leveraging GNNs (Graph Neural Networks) for biological prediction and demonstrate the potential for rapid, interpretable MIC prediction to support clinical treatment and public health surveillance.

Keywords: Graph Structures · Graph Neural Networks · MIC prediction · k-mers · Antimicrobial Resistance

1 Introduction

Bacteria that are resistant to antibiotic treatments pose one of the greatest challenges in modern medicine. When common infections become difficult to treat, we risk returning to a pre-antibiotic era with much higher mortality rates (Talukder et al. [23]). Foodborne pathogens such as *Salmonella* are particularly concerning, as they spread through our global food supply and are increasingly developing resistance to antibiotics. *Salmonella* causes approximately 93.8 million foodborne illnesses and 155,000 deaths each year worldwide (Majowicz et al. [17]). The burden of these resistant infections falls heavily on developing regions where

H. R. Arabnia et al. (Eds.): CSCI 2025, CCIS 2970, pp. 459–473, 2027.
https://doi.org/10.1007/978-3-032-28466-2_35

health care resources are limited. Recent research found that over half (51.35%) of raw meat samples contained *Salmonella* strains resistant to multiple antibiotics, including 100% resistance to erythromycin and more than 80% resistance to several other antibiotics (Fatima et al. [8]).

Physicians need to know sensitivity of bacterial pathogens to different antibiotics to select the best treatment. Antimicrobial resistance (AMR) is now one of the biggest global health threats, leading to hundreds of thousands of deaths each year and making many treatments less effective (Murray et al. [19]). Due to this, accurate antimicrobial susceptibility tests have become essential for both patient care and public health monitoring (O'Neill [22]). The MIC is 'the lowest concentration of an antimicrobial that results in no visible bacterial growth after overnight incubation' (Kowalska et al. [14]). Unlike simple labels such as 'resistant' or 'susceptible', MIC values provide detailed quantitative information that helps physicians select the right antibiotic dose (Andrews [1]). Standard lab-based methods to measure MIC are reliable but time consuming, which can delay treatment in urgent cases. Faster approaches, such as predicting MIC values directly from bacterial genomes, could greatly improve patient care and strengthen resistance surveillance (Ardila et al. [3]).

Over the years, computational methods for predicting bacterial AMR have evolved significantly. Traditional approaches focused on finding specific resistance genes or mutations (Zankari et al. [27]), but these methods often missed the complex interactions between multiple genomic elements that together result in resistance. A comprehensive review documented various machine learning approaches that have been applied to this problem, including Logistic Regression (LR), Support Vector Machine (SVM), Random Forest (RF), K-nearest Neighbors (KNN), Decision Tree (DT), and Gradient Boosting Machine (GBM). More recently, researchers have been using Convolutional Neural Networks (CNNs) and other deep learning methods to improve the prediction accuracy (Tang et al. [24]).

In this study, we explore the use of GNNs for MIC prediction in *Salmonella* species. GNNs, which have shown success in domains such as molecular property prediction, protein structure analysis, and social network mining (Zhou et al. [29]; Wieder et al. [26]), are well-suited for modeling complex interactions between genomic elements like k-mers that contribute to antimicrobial resistance. We design graphs where nodes represent k-mer sequences ($k = 21$) relevant to AMR and edges are based on sequence similarity, creating sample-specific subgraphs that capture unique genomic relationships (Kromer et al. [15]). We evaluate several GNN architectures, including GCN (Kipf & Welling [13]), GAT (Velivckovic et al. [25]), and GraphSAGE (Hamilton et al. [12]), to predict MIC values across 15 antibiotics. The main contributions of this work include the development of a graph-based genome representation for MIC prediction, evaluation of multiple GNN architectures, and comparison with existing models for *Salmonella* MIC prediction.

2 Related Work

2.1 Machine Learning for AMR Prediction

Numerous studies have applied machine learning techniques to predict antimicrobial resistance in bacteria. Early approaches detected resistance genes or mutations using reference-based methods (Gupta et al. [10]). While interpretable, these methods depend on prior knowledge of resistance mechanisms. To overcome this, recent studies have applied machine learning and deep learning-based models that learn patterns directly from genomic sequences, enabling prediction of both known and novel resistance mechanisms. For instance, Moradigaravand et al. [18] employed Gradient boosted decision trees achieved up to 91% accuracy in predicting resistance to multiple antibiotics in 1936 *E-coli* strains using whole-genome sequences (WGS). This study highlights that gene content and population structure alone can provide informative features for resistance prediction, demonstrating the potential of machine learning to go beyond traditional reference-based detection.

Traditional machine learning approaches for AMR prediction commonly utilize algorithms such as LR, SVM, RF, k-nearest neighbors(k-NN), DT, and GBM to analyze clinical and microbiological data for predicting AMR profiles (Tang et al. [24]). This systematic review and meta-analysis found that machine learning prediction models generally achieved higher specificity (87%) compared to traditional risk score methods (37%) when predicting AMR.

In a landmark study specifically focused on *Salmonella*, Nguyen et al. [21] developed extreme gradient boosting (XGBoost) machine learning models for predicting MICs for 15 antibiotics using whole-genome sequences from 5,278 nontyphoidal *Salmonella* isolates at an average accuracy of 95% within ±1 2-fold dilution step, demonstrating the potential of machine learning approaches for MIC prediction. The study also showed that highly accurate models could be generated with fewer than 500 diverse genomes and that their approach was stable over time despite fluctuations in antimicrobial resistance gene content.

2.2 Deep Learning Methods

Deep learning approaches have recently gained traction in AMR prediction due to their ability to automatically learn complex features from raw data. CNNs have emerged as powerful tools for predicting AMR from genomic data. Green et al. [9] presented how effective these deep learning approaches can be, with their multi-drug CNN achieving prediction accuracy with AUCs ranging between 82.6–99.5% across 13 antibiotics in *Mycobacterium tuberculosis*, while their single-drug CNNs achieved AUCs between 80.1–97.1%. The authors also used saliency methods to evaluate the contribution of the features of the input sequence in their prediction models, identifying 18 sites in the genome not previously associated with resistance. Similarly, Kuang et al. [16] developed accurate and rapid prediction models for tuberculosis drug resistance using traditional machine learning algorithms and CNNs. Their approach used a training dataset of 10,575 *M. tuberculosis* isolates collected from 16 countries in six continents, demonstrating the scalability of deep learning methods for AMR prediction. Their multi-input CNN architecture incorporated both SNP features and AMR-associated gene information to achieve robust prediction performance. Arango-Argoty et al. [2] introduced DeepARG, a deep learning approach to predict antibiotic resistance genes from metagenomic data. Their method demonstrated high precision > 0.97 and recall > 0.9 across 30 antibiotic resistance categories, highlighting the power of deep learning for the detection of AMR. Their evaluation showed an advantage over typical best-hit approaches, consistently yielding lower false negative rates and higher overall recall > 0.9.

Barros et al. [6] trained neural networks on 7,268 *Salmonella* genomes from the PATRIC database using reference-free k-mer analysis to predict MIC and susceptibility profiles. Their models achieved within ±1–2-fold dilution accuracy of 85–95% for MICs and similar accuracy for susceptibility classification. Kromer-Edwards and Oliveira [15] developed a Recurrent Neural Network (RNN) that integrates bacterial k-mer fingerprints and antibiotic SMILES strings to predict MIC. Their model achieved a test F1 score of 0.82 on ten known antibiotics and showed reasonable generalization to four novel antimicrobial agents. This approach demonstrates the potential of sequence and compound-based embeddings for MIC prediction. In our previous work, we utilized neural networks to predict MIC and identify important genomic features for nontyphoidal *Salmonella*, further demonstrating the efficacy of deep learning (Ayoola et al. [5]).

Recently, GNNs have emerged as powerful tools for analyzing biological data with inherent relational structures. Their ability to learn both from node features and graph topology makes them particularly suitable for biological applications. Previous work from our group, Zhang et al. [28] proposed an innovative approach using GNNs for MIC prediction, providing novel insights into the correlation between gene fragment similarities and MIC values. This approach is particularly relevant for organisms such as non-typhoidal *Salmonella* that show increasing resistance to antimicrobial agents. Although this initial study demonstrated utility of GNNs for MIC prediction, the study has some limitations as follows:

- The same graph structure for every bacterial sample, ignoring the natural diversity in bacterial genomes.
- Usage of shorter DNA sequences or k-mers (k=10) that provide less biological context.
- Model worked well only for a single antibiotic, Cotrimaxole (COT)

The current study is designed to address the limitations identified in the previous work by proposing a new graph structure to represent bacterial WGS where:

- Each bacterial genome has its own graph structure.
- Use longer k-mers ($k = 21$).

- The size and complexity of each graph depend on which specific k-mers are present in that particular WGS.
- Node representation as one-hot encoded vectors and edges are defined based on sequence similarity.

This new approach allows our model to capture the subtle and complex relationships between genomic elements that together determine resistance. The evidence for our method's effectiveness is that it works well across all 15 antibiotics we tested, not just one as described in our previous graph-based approach.

3 Materials and Methods

3.1 Dataset and Preprocessing

Genomic Data and MIC Values. This study utilizes genomic data from the National Antimicrobial Resistance Monitoring System (NARMS), comprising 15,622 assembled *Salmonella* genomes. Each genome is annotated with MIC values for 15 different antibiotics, representing various antibiotic classes. The study used MIC data for 15 antibiotics commonly tested in *Salmonella* species. Table 1 lists the abbreviations, full names, and categories of known resistance mechanisms. The antibiotics represent different classes, including cell wall synthesis inhibitors, protein synthesis inhibitors and DNA synthesis inhibitors, which influence their mode of action and susceptibility profiles.

We created separate datasets for each antibiotic, including only samples with available MIC values for that specific antibiotic. This resulted in 15 distinct datasets, one for each antibiotic.

Our work builds upon the dataset and methodology established by Ayoola et al. [5], who developed a "Genome Feature Extractor Pipeline" for MIC prediction in *Salmonella*. Their work identified crucial 10-mers using Random Forest and expanded them to 20-mers for better genomic region identification. Following their approach of k-mer expansion for capturing broader genomic contexts, we utilize 21-mers derived from their identified significant sequences. This process resulted in a set of 896 significant 21-mers used as features for our models. For each genome, we recorded the presence or absence of each 21-mer, creating a binary feature vector.

To stabilize training and reduce the impact of label skewness, we applied a logarithmic transformation to the MIC values prior to model training. All reported results were computed using these log-transformed labels, and predictions were transformed back when needed for evaluation.

Graph Construction. We developed a graph-based representation for each *Salmonella* genome, capturing its unique k-mer composition. Each graph node corresponds to a specific 21-mer sequence, represented as an 896-dimensional one-hot vector. Edges are established between nodes when the Longest Common Subsequence (LCS) similarity exceeds a threshold of 0.6, balancing connectivity and graph complexity.

Table 1. NARMS tested antibiotics against *Salmonella* and the annotated resistance mechanism

Abbr.	Antibiotic	Mechanism
AMP	Ampicillin	Cell-wall synthesis inhibitor (β-lactam)
AUG	Amoxicillin-Clavulanic Acid	Cell-wall synthesis inhibitor (β-lactam)
AXO	Ceftriaxone	Cell-wall synthesis inhibitor (β-lactam)
FOX	Cefoxitin	Cell-wall synthesis inhibitor (β-lactam)
TIO	Ceftiofur	Cell-wall synthesis inhibitor (β-lactam)
AZI	Azithromycin	Protein synthesis inhibitor (Macrolide)
CHL	Chloramphenicol	Protein synthesis inhibitor (Phenicol)
GEN	Gentamicin	Protein synthesis inhibitor (Aminoglycoside)
KAN	Kanamycin	Protein synthesis inhibitor (Aminoglycoside)
STR	Streptomycin	Protein synthesis inhibitor (Aminoglycoside)
TET	Tetracycline	Protein synthesis inhibitor (Tetracycline)
CIP	Ciprofloxacin	DNA synthesis inhibitor (Quinolone)
NAL	Nalidixic Acid	DNA synthesis inhibitor (Quinolone)
COT	Trimethoprim-Sulfamethoxazole	DNA synthesis inhibitor (Sulfonamide)
FIS	Sulfisoxazole	DNA synthesis inhibitor (Sulfonamide)

Formally, each sample graph is defined as

$$G = (V, E, X),$$

where V is the set of k-mer nodes, E contains edges based on sequence similarity, and X represents node features. Alternative feature embeddings, including transformer embeddings and binary codes, were evaluated, but one-hot encoding provided the best performance and was used in the final model.

LCS measures the longest ordered subsequence shared between two sequences, providing a numerical measure of similarity used to define edges.

The final dataset consists of 15,622 graphs (one per sample), each with nodes representing the 21-mers present in specific WGS data, edges based on sequence similarity, and one-hot encoded node features, with 15 MIC values as target labels. This graph construction demonstrates improved performance across the NARMS antibiotic panel.

The graph construction process is described in Algorithm 1:

Algorithm 1. Graph Construction

Require: Set of 21-mer sequences S for each sample, similarity threshold $\tau = 0.6$
Ensure: Graph $G_i = (V, E, X)$ with node and edge features

for each sample i in the dataset **do**
    Initialize an empty graph G_i
    for each present 21-mer m in sample i **do**
        Create node m with features $f = f_{\text{one-hot encoding}}$
        Add node m to G_i
    end for
    for each pair of nodes (u, v) in G_i **do**
        Compute similarity: $\text{sim} = \text{LCS\_similarity}(u, v)$
        if $\text{sim} > \tau$ **then**
            Add edge (u, v) to G_i
        end if
    end for
    Validate graph G_i for self-loops and isolated nodes
    Assign label y_i to graph G_i
end for

3.2 Graph Neural Network Framework

Overview of GNNs. We implemented and evaluated several GNN architectures for MIC prediction. A generic GNN framework can be described using three main components:

- **Propagation**: how information flows through the network,
- **Sampling**: how neighborhood nodes are selected,
- **Pooling**: how node-level information is aggregated.

Among the different architectures we experimented with, GCN, GAT and GraphSAGE produced the most promising results. A comparison of these models in terms of their core components is presented in Table 2.

3.3 Model Architecture

We implemented three GNN models for regression: GAT, GCN, and GraphSAGE. Each model uses three stacked graph convolutional layers of a single type (GAT: three GATConv layers; GCN: three GCNConv layers; GraphSAGE: three

SAGEConv layers). ReLU activations follow each convolution. Batch normalization is applied after the first two layers in the GAT and GraphSAGE models to stabilize training. Global mean pooling aggregates node embeddings into a graph-level representation, which is then mapped to MIC predictions by a final fully connected layer; dropout (0.5) is applied before this layer to reduce overfitting. This layered design, shown in Fig. 1, allows the models to progressively capture local and global similarities within the constructed graphs, enabling more accurate MIC prediction from k-mer representations.

Table 2. Comparison of different GNNs based on propagation, sampling, and pooling.

Model	Propagation	Sampling	Pooling
GCN	Matrix multiplication with normalized adjacency	All neighbors	Matrix-based averaging
GAT	Attention-weighted message passing	All neighbors	Matrix-based averaging
GraphSAGE	Neighborhood aggregation (mean, max, LSTM)	Fixed-size sampled neighbors	Mean/max pooling

3.4 Training Procedure and Optimization

Train-Test Split. The dataset was split into training (80%), validation (10%), and test (10%) sets using random sampling with a fixed seed for reproducibility. Each graph represents a *Salmonella* sample constructed from k-mer features and associated MIC values, with the validation set used for hyperparameter tuning. Because not all antibiotics had MIC labels for every sample, the effective dataset size varied across antibiotics, may influence model performance.

Loss Function. This system uses the mean squared error between predictions and targets

$$\text{MSE} = \frac{1}{N} \sum_{i=1}^{N} (y_i - \hat{y}_i)^2 \tag{1}$$

Optimization Strategy. The training process employs an optimization strategy that combines mean squared error loss with L2 regularization to prevent overfitting while maintaining model performance. Optimization is performed using the Adam optimizer with an initial learning rate of 0.001 and a weight decay coefficient of 1e-5. To handle plateaus in training and ensure better convergence, a learning rate scheduler reduces the learning rate by a factor of 0.1 if the validation loss does not improve for three consecutive epochs. This adaptive scheduling enables the model to escape local minima and achieve more stable convergence.

4 Model Evaluation Metrics

MIC Prediction Accuracy. Model performance in MIC prediction is evaluated according to CLSI guidelines [7], considering a prediction successful if it falls within a two-fold dilution of the laboratory-measured MIC. This accounts for the inherent variability in experimental MIC measurements. Previous studies have adopted the same evaluation strategy (Ayoola et al. [5]; Nguyen et al. [21]; Kromer-Edwards and Oliveira [15]). Based on this criterion, the prediction accuracy within the two-fold dilution is defined as the fraction of predictions that fall within two-fold dilution step of the laboratory-measured MIC values:

$$\mathrm{Acc}_{\text{2-fold}} = \frac{1}{N}\sum_{i=1}^{N} \mathbb{1}\left[\frac{y_i}{2} \le \hat{y}_i \le 2y_i\right],$$

where N is the total number of predictions.

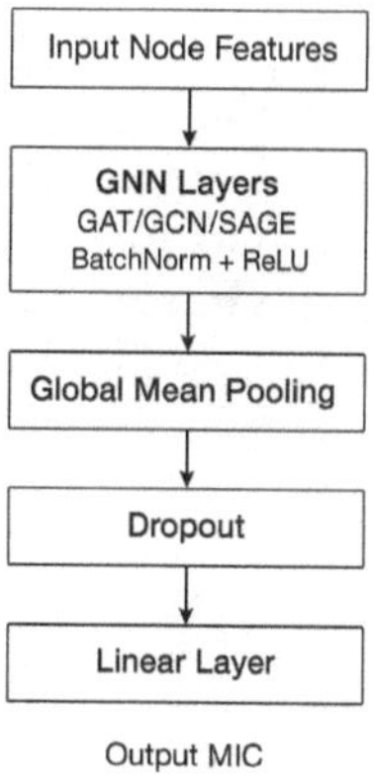

Fig. 1. GNN architecture for MIC prediction: node features pass through GNN layers with BatchNorm and ReLU, followed by global mean pooling, dropout, and a linear layer to output MIC values.

R^2-score To provide a more complete understanding of our model's performance, we complement accuracy-based metrics with the coefficient of determination (R^2-score), which quantifies the proportion of variance in MIC values explained by the model. As a scale-independent metric, R^2 allows us to fairly compare model performance across different antibiotics, even when their MIC values vary widely.

The R^2 score is calculated as:

$$R^2 = 1 - \frac{\sum_i (y_i - \hat{y}_i)^2}{\sum_i (y_i - \bar{y})^2}$$

where y_i are the observed MIC values, $\hat{y}_i$ are the predicted values, and $\bar{y}$ is the mean of the observed values.

The combination of accuracy and R^2-score offer a comprehensive evaluation framework that captures both clinical relevance and statistical robustness. Accuracy reflects the practical utility of predictions, while R^2 highlights the model's predictive power. Together, these metrics provide complementary insights, ensuring that our evaluation aligns with both scientific rigor and practical considerations.

5 Results and Analysis

Table 3. Comparison of best-performing GNN model vs. baseline MLP across 15 antibiotics. Δ values indicate performance improvement of GNN over MLP.

Antibiotic	BestGNN(Accuracy/R^2)	MLP (Accuracy/R^2)	ΔAccuracy	ΔR^2
Ampicillin	GAT 96.03/0.8560	96.19/0.7870	−0.16	+0.0690
Amoxicillin-Clavulanic Acid	GAT 97.38/0.9254	96.59/0.8153	+0.79	+0.1101
Ceftriaxone	GAT 95.97/0.6402	95.43/0.6272	+0.54	+0.0130
Azithromycin	GCN 94.50/0.5198	94.06/0.3390	+0.44	+0.1808
Chloramphenicol	GAT 96.48/0.8062	96.52/0.7892	−0.04	+0.0170
Ciprofloxacin	GAT 98.14/0.4352	98.13/0.4079	+0.01	+0.0273
Trimethoprim-Sulfamethoxazole	SAGE 94.95/0.3581	94.81/−0.9509	+0.14	+1.3090
Sulfisoxazole	GAT 91.80/0.9312	91.80/0.8649	0.00	+0.0663
Cefoxitin	GAT 93.60/0.8846	93.50/0.8577	+0.10	+0.0269
Gentamicin	GAT 90.53/0.7870	89.37/0.7376	+1.16	+0.0494
Kanamycin	GAT 95.16/0.7586	95.11/0.7076	+0.05	+0.0510
Nalidixic Acid	GAT 97.00/0.8884	96.97/0.8352	+0.03	+0.0532
Streptomycin	GAT 85.86/0.8308	86.39/0.7542	-0.53	+0.0766
Tetracycline	GAT 98.53/0.9547	97.42/0.8867	+1.11	+0.0680
Ceftiofur	SAGE 96.71/0.9148	96.68/0.8780	+0.03	+0.0368

We evaluated several GNN variants, including GAT, GCN, and GraphSAGE, against the Multi-Layer Perceptron (MLP) as baseline across 15 antibiotics. Table 3 presents the best-performing GNN for each antibiotic compared to the MLP baseline, while Fig. 2 provides a visual comparison of accuracy and R^2 improvements. GNNs demonstrated substantially improved predictive power compared to the MLP baseline, achieving higher R^2 values in all 15 antibiotics.

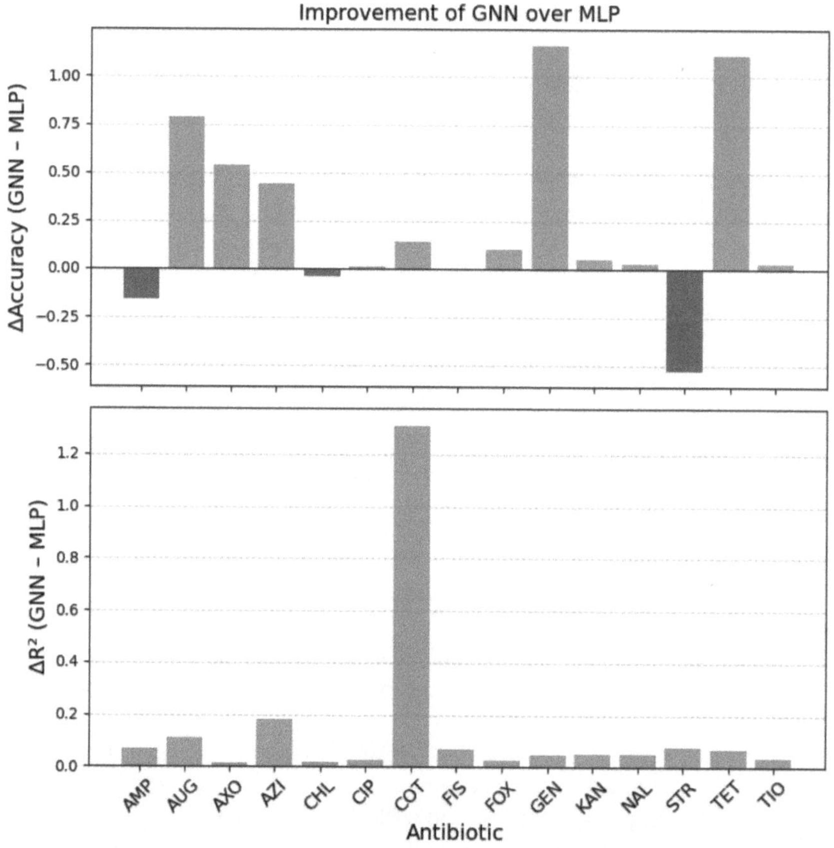

Fig. 2. Comparison of best-performing GNN variants with MLP baseline across 15 antibiotics.

We employed a graph-based approach because genomic sequences exhibit interdependencies and structural relationships that cannot be captured by treating k-mers as independent features. Unlike MLPs, which treat k-mers as isolated features, our GNN represents each k-mer as a node and connects nodes using sequence similarity edges, enabling the propagation of contextual information across related genomic regions. This design is particularly well-suited for biological data, which is inherently structured and networked rather than flat. Genes, proteins, and resistance determinants often interact in pathways, co-resistance networks, and regulatory hierarchies. By aggregating information from neighboring nodes, GNNs capture these contextual dependencies, for example, the influence of a resistance gene may depend on neighboring variants or plasmid-borne elements(Nair et al. [20], Arzanlou et al. [4]). This design allows the model to capture cooperative effects among genetic elements that contribute to antimicrobial resistance. Studies on multidrug-resistant Escherichia coli have demonstrated that resistance is often driven by clusters of coexisting genes, including virulence factors, iron acquisition systems, and metabolic pathways, which collectively influence resistance phenotypes (Haley et al. [11])

When evaluated on identical train–validation–test splits, the GNN consistently achieved higher R^2 values than the MLP across all antibiotics ($\Delta R^2 > 0$ for every case), demonstrating its ability to capture relational information between k-mers. The largest gain was observed for Cotrimaxole ($\Delta R^2 = 1.3090$), reflecting the GNN's capacity to model complex interactions that the MLP cannot capture. Substantial improvements were also seen for Azithromycin (+0.1808), Amoxicillin-Clavulanic Acid (+0.1101), and Streptomycin (+0.0766), while moderate gains were noted for Ampicillin, Sulfisoxazole, Tetracycline, Nalidixic Acid and Kanamycin ($\Delta R^2 \approx 0.05$–0.07). Smaller but consistent improvements were detected for the remaining antibiotics ($\Delta R^2 = 0.01$–0.05). Importantly, even modest increases indicate that the GNN leveraged cooperative k-mer relationships to explain variance in MIC beyond the MLP baseline.

In terms of two-fold accuracy, GNN models performed on par with MLPs, with only minor differences across antibiotics. For Amoxicillin-Clavulanic Acid (+0.79), Gentamicin (+1.16), and Tetracycline (+1.11), the GNN achieved slightly higher accuracy. Several antibiotics showed near-identical results (e.g., Sulfisoxazole, Ciprofloxacin, Nalidixic Acid, Ceftiofur), while a few (Ampicillin, Chloramphenicol, Streptomycin) showed marginally lower values. Overall, the GNN maintained accuracy comparable to MLP baseline while delivering significantly higher R^2, ensuring predictive reliability alongside improved variance explanation.

From a clinical perspective, improved R^2 translates to more reliable MIC predictions, supporting targeted antimicrobial therapy and reducing treatment failure risk. The large gain for Cotrimaxole suggests that the GNN effectively captures complex resistance pathways involving multiple genes, offering early-warning potential for emerging mechanisms. This network-based framework is particularly valuable for outbreak surveillance, where identifying clusters of resistance determinants before phenotypic confirmation could inform proactive interventions. Methodologically, our results show that simple one-hot encoding of k-mer nodes combined with similarity-based edges provides a biologically meaningful representation, outperforming independent k-mer treatment. Beyond antimicrobial resistance, this framework can be extended to phenotypes driven by multi-gene interactions, including virulence traits, metabolic adaptations, and horizontal gene transfer. Moreover, the interpretability afforded by attention weights may help identify novel resistance-associated motifs, offering actionable insights for genomic epidemiology

GAT achieved optimal performance for 12 out of 15 antibiotics, establishing it as the most effective architecture for this task. GraphSAGE excelled for Cotrimaxole and Ceftiofur, while GCN proved optimal for Azithromycin. This performance distribution demonstrates that attention mechanisms successfully identify the most relevant k-mers for resistance prediction, consistently outperforming traditional aggregation methods.

A critical component of our framework is the design of the graph, including the selection of node and edge features. In our setup, nodes correspond to k-mers represented as one-hot encoded vectors, while edges reflect similarity relationships. One-hot encoding preserves the exact identity and position of each k-mer, which is crucial for the GNN to propagate information accurately across the graph. Alternative embeddings, such as transformer-based or binary coding added complexity without improving performance, suggesting that relational information between k-mers is more important than individual feature richness. In contrast to an MLP, which treats each sample as a fixed-length vector and ignores relationships between k-mers, the GNN leverages the graph structure, making precise identity encoding via one-hot vectors particularly effective. These results highlight that careful graph construction and feature selection are essential for capturing underlying biological relationships, leading to more accurate and interpretable MIC predictions with enhanced reliability for clinical decision-making.

6 Conclusion

In this study, we used a graph-based framework for predicting the MIC in *Salmonella* species. Our approach constructs graphs from k-mer sequences, where nodes are represented using **one-hot encoded** features and edges are defined by sequence similarity through the Longest Common Subsequence. Experimental results show that this design effectively captures sequence relationships, enabling graph neural networks to achieve accurate MIC predictions, with GAT achieving the best performance among the evaluated variants. These findings highlight the critical role of graph construction, particularly the choice of node and edge features, in the use of GNNs for biological prediction tasks. While our approach demonstrates strong predictive performance, it currently relies on a fixed k-mer length and sequence similarity thresholds, which may limit generalizability to other bacterial species. Future work will focus on analyzing feature importance, especially identifying key k-mers via GAT's attention mechanism, to provide deeper biological insights into antimicrobial resistance and enhance model interpretability.

Acknowledgments. This work was supported by USDA-ARS NACA #58-6064-3-017 and USDA-ARS CA 59-6064-4-004

AI tools, such as ChatGPT and Claude, were used to assist in writing. All content was carefully reviewed by the authors to ensure accuracy and alignment with the study's objectives. The manuscript reflects the authors' original analysis, insights, and contributions.

References

1. Andrews, J.M.: Determination of minimum inhibitory concentrations. J. Antimicrob. Chemother. **48**(suppl_1), 5–16 (2001)
2. Arango-Argoty, G., Garner, E., Pruden, A., Heath, L.S., Vikesland, P., Zhang, L.: DeepARG: a deep learning approach for predicting antibiotic resistance genes from metagenomic data. Microbiome **6**, 1–15 (2018)
3. Ardila, C.M., Yadalam, P.K., González-Arroyave, D.: Integrating whole genome sequencing and machine learning for predicting antimicrobial resistance in critical pathogens: a systematic review of antimicrobial susceptibility tests. PeerJ **12**, e18213 (2024)
4. Arzanlou, M., Chai, W.C., Venter, H.: Intrinsic, adaptive and acquired antimicrobial resistance in gram-negative bacteria. Essays Biochem. **61**(1), 49–59 (2017)
5. Ayoola, M.B., Das, A.R., Krishnan, B.S., Smith, D.R., Nanduri, B., Ramkumar, M.: Predicting salmonella mic and deciphering genomic determinants of antibiotic resistance and susceptibility. Microorganisms **12**(1), 134 (2024)
6. Barros, C.C.: Neural network-based predictions of antimicrobial resistance in salmonella spp. using k-mers counting from whole-genome sequences. bioRxiv, pp. 2021–08 (2021)
7. Clinical, Institute, L.S.: Performance standards for antimicrobial susceptibility testing. CLSI Suppl. M100 **31** (2021)
8. Fatima, A., Saleem, M., Nawaz, S., Khalid, L., Riaz, S., Sajid, I.: Prevalence and antibiotics resistance status of salmonella in raw meat consumed in various areas of Lahore, Pakistan. Sci. Rep. **13**(1), 22205 (2023)
9. Green, A.G., et al.: A convolutional neural network highlights mutations relevant to antimicrobial resistance in mycobacterium tuberculosis. Nat. Commun. **13**(1), 3817 (2022)
10. Gupta, S.K., et al.: ARG-ANNOT, a new bioinformatic tool to discover antibiotic resistance genes in bacterial genomes. Antimicrob. Agents Chemother. **58**(1), 212–220 (2014)
11. Haley, B.J., Kim, S.W., Salaheen, S., Hovingh, E., Van Kessel, J.A.S.: Genome-wide analysis of escherichia coli isolated from dairy animals identifies virulence factors and genes enriched in multidrug-resistant strains. Antibiotics **12**(10), 1559 (2023)
12. Hamilton, W., Ying, Z., Leskovec, J.: Inductive representation learning on large graphs. In: Advances in Neural Information Processing Systems, vol. 30 (2017)
13. Kipf, T.: Semi-supervised classification with graph convolutional networks. arXiv preprint arXiv:1609.02907 (2016)
14. Kowalska-Krochmal, B., Dudek-Wicher, R.: The minimum inhibitory concentration of antibiotics: metholds, interpretation, clinical relevance. Pathogens **10**(2), 165 (2021)
15. Kromer-Edwards, C., Oliveira, S.: Compound RNN to predict mics using k-mer fingerprints and antibiotic smiles. In: Proceedings of the 14th ACM International Conference on Bioinformatics, Computational Biology, and Health Informatics, pp. 1–5 (2023)
16. Kuang, X., Wang, F., Hernandez, K.M., Zhang, Z., Grossman, R.L.: Accurate and rapid prediction of tuberculosis drug resistance from genome sequence data using traditional machine learning algorithms and CNN. Sci. Rep. **12**(1), 2427 (2022)
17. Majowicz, S.E., et al.: The global burden of nontyphoidal salmonella gastroenteritis. Clin. Infect. Dis. **50**(6), 882–889 (2010)

18. Moradigaravand, D., Palm, M., Farewell, A., Mustonen, V., Warringer, J., Parts, L.: Prediction of antibiotic resistance in Escherichia coli from large-scale pangenome data. PLoS Comput. Biol. **14**(12), e1006258 (2018)
19. Murray, C.J., et al.: Global burden of bacterial antimicrobial resistance in 2019: a systematic analysis. Lancet **399**(10325), 629–655 (2022)
20. Nair, R.R., Andersson, D.I., Warsi, O.M.: Antibiotic resistance begets more resistance: chromosomal resistance mutations mitigate fitness costs conferred by multi-resistant clinical plasmids. Microbiol. Spectr. **12**(5), e04206-23 (2024)
21. Nguyen, M., et al.: Using machine learning to predict antimicrobial mics and associated genomic features for nontyphoidal salmonella. J. Clin. Microbiol. **57**(2), 10–1128 (2019)
22. O'neill, J.: Antimicrobial resistance: tackling a crisis for the health and wealth of nations. Rev. Antimicrob. Resist. (2014)
23. Talukder, H., Roky, S.A., Debnath, K., Sharma, B., Ahmed, J., Roy, S.: Prevalence and antimicrobial resistance profile of salmonella isolated from human, animal and environment samples in south asia: a 10-year meta-analysis. J. Epidemiol. Glob. Health **13**(4), 637–652 (2023)
24. Tang, R., Luo, R., Tang, S., Song, H., Chen, X.: Machine learning in predicting antimicrobial resistance: a systematic review and meta-analysis. Int. J. Antimicrob. Agents **60**(5–6), 106684 (2022)
25. Veličković, P., Cucurull, G., Casanova, A., Romero, A., Lio, P., Bengio, Y.: Graph attention networks. arXiv preprint arXiv:1710.10903 (2017)
26. Wieder, O., et al.: A compact review of molecular property prediction with graph neural networks. Drug Discov. Today Technol. **37**, 1–12 (2020)
27. Zankari, E., et al.: Identification of acquired antimicrobial resistance genes. J. Antimicrob. Chemother. **67**(11), 2640–2644 (2012)
28. Zhang, Z., et al.: Leveraging graph neural networks for mic prediction in antimicrobial resistance studies. In: 2024 46th Annual International Conference of the IEEE Engineering in Medicine and Biology Society (EMBC), pp. 1–4. IEEE (2024)
29. Zhou, J., et al.: Graph neural networks: a review of methods and applications. AI Open **1**, 57–81 (2020)

Comparing SMILES and SELFIES Embeddings and ECFP4 Fingerprints for Molecular Similarity Search

Amey Karmarkar and Ramon Lawrence(✉)

University of British Columbia Okanagan 3333 University Way, Kelowna, BC V1V 1V7, Canada
ameykarmarkar@rocketmail.com, ramon.lawrence@ubc.ca

Abstract. We compare SMILES and SELFIES embeddings generated by two transformer models (ChemBERTa, SELFormer) against 1024-bit ECFP4 fingerprints for molecular similarity search. Using thirty chemically diverse query molecules, we retrieve the top-k nearest neighbours $k \in \{5, 8, 10, 15, 20, 25, 35\}$ via cosine similarity for all embedding representations and Tanimoto's coefficient for fingerprints. Hits are evaluated on chemical similarity (Hamming distance) and structural similarity (Graph Edit Distance, GED). Statistical significance is assessed with one-way ANOVA and Tukey's HSD. Our results confirm that ECFP4 with Tanimoto yields the lowest mean Hamming distances (mean = 12.5 at k = 35) as well as the lowest mean GEDs (mean = 18.9). SELFIES embeddings consistently outperform SMILES embeddings yet remain inferior to fingerprints.

Keywords: Molecular Similarity Search · SELFIES and SMILES Embeddings · Graph Edit Distance (GED) · ECFP4 Fingerprints

1 Introduction

Molecular similarity search lies at the heart of modern cheminformatics and underpins critical applications such as drug design, computer-aided drug discovery, virtual screening, and the de novo design of molecular libraries. By quantifying how "close" two molecules are, whether in terms of backbone scaffold, functional groups, or physicochemical properties, researchers can mine large chemical databases for lead compounds, predict off-target effects, or explore novel bio isosteric replacements.

1.1 String Representations: SMILES and SELFIES

The Simplified Molecular Input Line Entry System (SMILES) [1] is a line-notation for chemicals in which atoms and bonds are encoded as ASCII characters in a depth-first traversal of the molecular graph. While SMILES is compact

H. R. Arabnia et al. (Eds.): CSCI 2025, CCIS 2970, pp. 474–486, 2027.
https://doi.org/10.1007/978-3-032-28466-2_36

and human-readable, it suffers from issues of invalid syntax and ambiguity under small perturbations, which can disrupt machine-learning pipelines.

In contrast, SELF-referencing Embedded Strings (SELFIES) [2] are an alternative where every SELFIES string decodes to a chemically valid molecule. SELFIES replace atomic symbols and branching rules with self-contained tokens and grammar automata, eliminating invalid representations and improving model stability when learning chemical language models.

1.2 Transformer Embeddings: ChemBERTa vs. SELFormer

Recent advances have applied Transformer architectures to molecular strings:

- **ChemBERTa** [3] is a RoBERTa-style Transformer pretrained on SMILES corpora to produce continuous, high-dimensional embeddings that capture chemical semantics.
- **SELFormer** [4] is a Transformer pretrained on SELFIES instead of SMILES. By leveraging SELFIES' guaranteed validity, SELFormer embeddings often exhibit greater sensitivity to structural changes and improved performance on property-prediction benchmarks.

Although both models generate vector representations that can drive similarity search via cosine similarity, their pretrained token vocabularies and input grammars differ fundamentally, leading to potentially divergent notions of molecular similarity.

1.3 Molecular Fingerprints: ECFP4 vs. MACCS

Before the rise of learned embeddings, similarity search often relied on fixed molecular fingerprints [5], bit strings designed to flag the presence or absence of common substructures. One of the oldest and most widely used is the MACCS166 key [6], which tests for 166 predefined structural features (e.g. aromatic rings, specific atom pairs). However, MACCS can miss finer details, since its limited key set forces very different fragments into the same bit. Morgan's circular fingerprints (ECFP4) [7] improve on this by hashing every atom's local environment (radius = 2) into a much larger bit space, capturing more unique substructures. Numerous benchmarks have shown ECFP4's higher resolution and diversity lead to better scaffold-level and bio-activity similarity retrieval than MACCS, which is why we use ECFP4 in this work. Tanimoto coefficient is the conventional metric for comparing binary fingerprints in contrast to cosine similarity, typically used with embeddings for distance calculations.

1.4 Types of Molecular Similarity

We distinguish three distinct flavours of molecular similarity search:

- **Structural Similarity:** Two molecules are structurally similar if they have similar shapes, i.e. their molecular graphs are similar to one another. We use GED [8] as a baseline metric to measure the structural similarity between two molecules. Structural similarity has important applications in areas like predicting enzyme-substrate binding and receptor-ligand binding.
- **Chemical Similarity:** Two molecules are chemically similar if they have the same chemical groups: atoms, functional groups, and bonds. Two molecules may have the exact same shape (molecular graphs) but have different bond types and constituent atoms. We use the Hamming distance [9] as a baseline metric to measure the chemical similarity between two molecules. Chemical similarity has applications in designing de-novo molecular libraries, in computer-aided drug discovery and materials discovery. If two molecules have a particular chemical group in common, they may share the same function brought about by that chemical group.
- **Functional Similarity:** Two molecules are functionally similar if they share the same biological activity or pharmacophore features. Functional similarity is not evaluated in this study. While structural and chemical similarity often correlate with shared function [10], we make no claims about functional similarity here and reserve its evaluation for future work.

1.5 Similarity Metrics: GED and Hamming

To compare methods, we employ two orthogonal distance metrics:

- **Graph Edit Distance (GED)** quantifies the minimum number of edit operations (node/edge insertions, deletions, substitutions) required to transform one molecular graph into another. GED excels at capturing overall scaffold and connectivity similarity, but it cannot detect the difference between two iso-structural graphs that differ only in atom identity: for example, a benzene ring where one carbon is swapped for chlorine still has the same graph shape, so GED will rate them as identical even though their chemical properties differ. In this case, the Hamming distance can distinguish between these two structurally similar but chemically different molecules.
- **Hamming Distance** computes bit-wise differences between two ECFP4 fingerprint vectors, capturing chemical similarity at the substructure level. However, it cannot distinguish stereoisomers or changes in 3D shape, two molecules with identical substructure counts but different spatial arrangements will have identical fingerprints. In this case, GED can distinguish between two chemically similar but structurally different molecules like isomers.

As these two metrics focus on different aspects of similarity, we treat structural similarity (GED) and chemical similarity (Hamming) as separate axes of analysis.

1.6 Statistical Analysis

To assess significance, we conduct one-way ANOVA tests across all methods for both GED and Hamming distributions, followed by post-hoc Tukey HSD comparisons to identify pairwise differences.

1.7 Research Questions

This study addresses three core questions:

1. Fingerprints vs. Embeddings: Do traditional ECFP4 fingerprints outperform learned embeddings on structural (GED) and chemical (Hamming) metrics?
2. SELFIES vs. SMILES: Do SELFIES based embeddings outperform their SMILES counterparts in retrieving structurally or chemically similar molecules?
3. SELFormer vs. ChemBERTa: Does pretraining on SELFIES confer an advantage over SMILES trained ChemBERTa for similarity search?

1.8 Related Work

String and fingerprint based molecular representations have become foundational in cheminformatics. Early SMILES-based methods dominated virtual screening workflows, leveraging rule-based fingerprints such as ECFP4 and Tanimoto similarity.

2 Methods

2.1 Dataset Preparation

A set of 500,000 compounds (PubChem CIDs 1,000,000,001 to 1,000,500,000) were downloaded from PubChem in SDF format. SMILES strings were extracted and validated using RDKit v2023.09.1. SELFIES representations were generated from validated SMILES with the selfies Python package (v2.1.1). All records were stored in a single CSV file containing CID, SMILES, and SELFIES columns.

2.2 Molecular Representations

Two classes of molecular descriptors were computed for each compound:

- **ECFP4 fingerprints:** 1024-bit Morgan fingerprints (radius = 2) were generated via RDKit's GetMorganFingerprintAsBitVect.
- **Transformer embeddings:** using two pretrained models from Hugging Face:
  - *chemBERTa-zinc-base-v1*, trained on SMILES.
  - *HUBioDataLab/SELFormer*, trained on SELFIES.

For each model, we tokenized the corresponding string representation and extracted a 768-dimensional vector.

These particular models were chosen because they are fully open-source and readily available with pretrained weights on Hugging Face, support SMILES or SELFIES inputs, and integrate seamlessly with FAISS for large-scale vector search. While newer architectures (e.g. MolFormer, ChemGPT, graph-based GNN embeddings) have been developed, their pretrained weights are not yet publicly released or are scattered across private repositories, preventing their use in this study.

2.3 Similarity Retrieval

Five FAISS (v1.7.4) IndexFlatIP indexes were built and populated locally: four for each embedding type and one for ECFP4 fingerprints. Thirty chemically diverse query molecules were selected to span a wide range of chemical space. Specifically:

- Aromatic scaffolds (8): simple (e.g., 4-chloropyridine) and polyaromatic systems (e.g., tetraphenylbenzene), including fused heterocycles (quinoline-like cores).
- Heterocycles (6): nitrogen- and sulfur-containing rings, including bicyclic azaheterocycles and thiadiazoles.
- Aliphatic chains (6): linear (C12–C23) and unsaturated alkenes, plus long fluorinated chains.
- Ionic salts (4): nitro- and iodonium-bearing aromatics, dicationic nitro-diols.
- Organometallics & main-group (3): antimony and mercury centers bound to alkyl/aryl groups, phosphine derivatives.
- Multifunctional polycyclics (3): ester-lactone macrocycles, amide-rich polycycles, azo-linked systems.

We retrieved the top k nearest neighbours for $k \in \{5, 8, 10, 15, 20, 25, 35\}$ by cosine similarity in the four embedding indexes, and by Tanimoto's coefficient in the fingerprints index. We L2-normalize all 768-dimensional transformer embeddings prior to FAISS indexing. While binarizing learned embeddings could in principle permit a truly "binary-to-binary" comparison, we opted to preserve their continuous representation to retain semantic granularity. Exploring embedding binarization or other normalization schemes is a promising direction for future studies.

2.4 Similarity Metrics

For every query–hit pair, two distance metrics were computed:

- **Hamming distance:** Defined as the bitwise difference between ECFP4 fingerprints, quantifying chemical similarity.

– **Graph edit distance (GED):** Computed on the molecular graphs with unit edit costs (node/edge insertions, deletions, substitutions) and a 300 s timeout using NetworkX's graph_edit_distance, quantifying structural similarity. We note that GED uses only the 2D graph and ignores stereochemistry or 3D conformation. For example, two enantiomers have GED=0. As GED computation is expensive [20], a 300 s cutoff was imposed per pair, which may truncate the calculation for very large or complex molecules.

2.5 Data Cleaning and Aggregation

Hits returning invalid SMILES or SELFIES (unparsable by RDKit) were excluded from analysis (identified by Hamming = -1 or GED = NaN). For each method and each k, distances were pooled across all thirty queries, yielding approximately 2,500 measurements per method.

2.6 Statistical Analysis

We performed one-way ANOVA on pooled Hamming and GED values to test for a method effect, followed by Tukey's HSD for pairwise comparisons ($\alpha = 0.05$). All analyses were conducted in Python using Pandas and NumPy.

3 Results

We have five distinct groups:

1. chemBERTa SMILES embeddings
2. chemBERTa SELFIES embeddings
3. SELFormer SMILES embeddings
4. SELFormer SELFIES embeddings
5. ECFP4 Fingerprints

For each of the first four groups, we retrieved the top k nearest neighbours per query by cosine similarity. For the last group (ECFP4 Fingerprints), we instead ranked hits by the Tanimoto's coefficient. We then computed, for every query–hit pair across all thirty queries, both the bit-wise Hamming distance of their ECFP4 fingerprints and the unit-cost graph edit distance (GED) of their molecular graphs.

3.1 Hamming Distance

Figure 1 displays the number of top k hits on the x-axis and the mean Hamming Distance on the y-axis. Each line plotted corresponds to one of the five methods being evaluated.

ECFP4 (Tanimoto) achieves the lowest distances at all k (mean = 12.9 at $k = 35$), indicating superior chemical similarity. SELFIES based embeddings outperform SMILES based ones (e.g. SELFormer SELFIES mean = 33.3 vs. SELFormer SMILES = 46.8), but remain well above fingerprint performance.

ANOVA indicates a significant method effect on Hamming (F = 228.9, $p < 10^{-15}$). Post-hoc tests confirm that ECFP4 (Tanimoto) is significantly lower than all other methods ($p < 0.001$), and that SELFIES is less than SMILES within each model ($p < 0.01$).

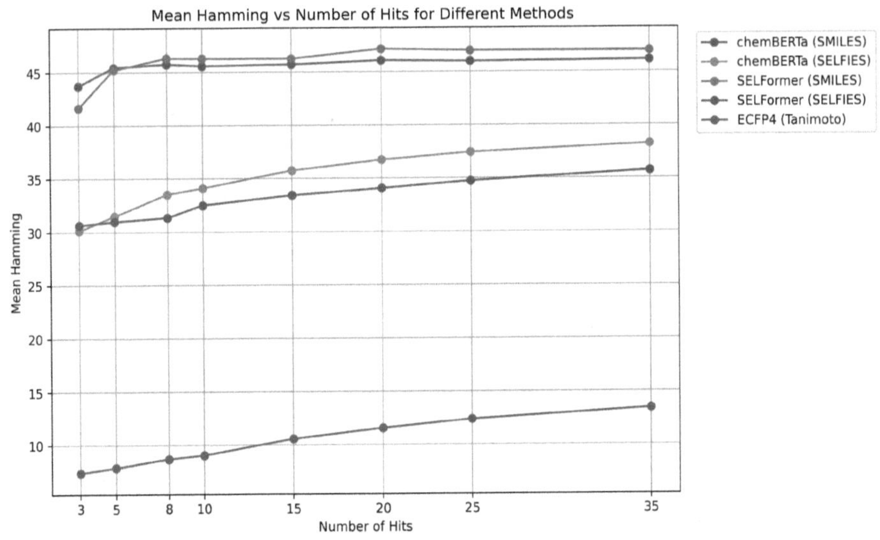

Fig. 1. Mean Hamming Distance for Varying Number of Top-k Hits (lower is better)

Table 1. Hamming Distance Summary (mean, median and standard deviation across all hits)

Method	Mean	Median	Std	N
chemBERTa (SMILES)	46.17	46.00	14.67	412
chemBERTa (SELFIES)	36.18	32.00	19.91	419
SELFormer (SMILES)	46.77	45.00	15.13	382
SELFormer (SELFIES)	33.31	33.00	18.71	418
ECFP4 (Tanimoto's)	12.92	13.00	8.58	420

3.2 Graph Edit Distance

Figure 2 displays the number of top k hits on the x-axis and the mean GED on the y-axis. Each line plotted corresponds to one of the five methods being evaluated.

ECFP4 (Tanimoto) yields the closest structural matches (mean = 29.6 at $k = 35$). All embeddings have lower performance, with SELFormer SELFIES having the largest GED (54.8).

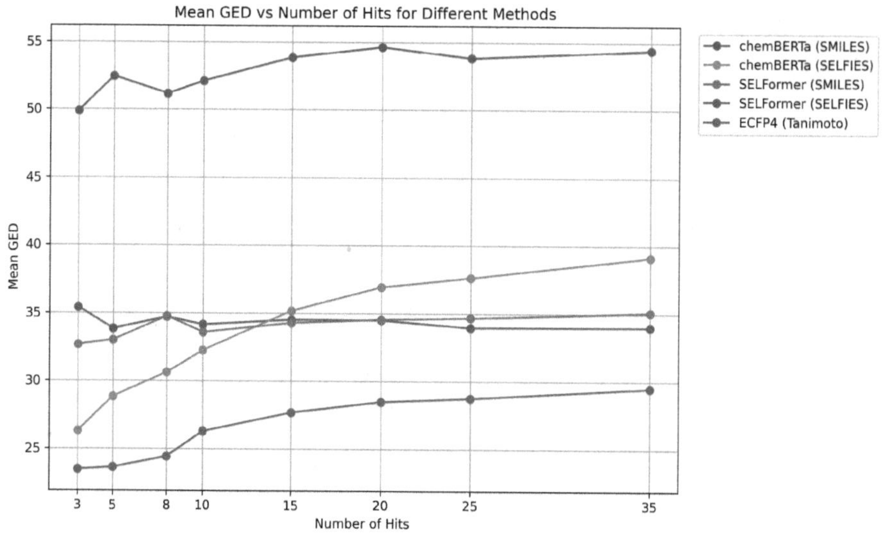

Fig. 2. Graph Edit Distance for Varying Number of Top-k Hits (lower is better)

Table 2. GED Summary (mean, median and standard deviation across all hits)

Method	Mean	Median	Std	N
chemBERTa (SMILES)	34.76	32.00	17.10	412
chemBERTa (SELFIES)	40.16	33.00	33.98	419
SELFormer (SMILES)	35.21	31.00	21.92	382
SELFormer (SELFIES)	54.76	49.00	38.62	418
ECFP4 (Tanimoto's)	29.60	26.00	18.91	420

Method effect on GED is highly significant (F = 85.3, $p < 10^{-82}$). Tukey HSD shows ECFP4 (Tanimoto) is superior to all other methods ($p < 0.001$).

4 Discussion

Our comparison of five retrieval methods: two transformer models (chemBERTa, SELFormer) with SMILES or SELFIES inputs and the ECFP4 based approach, reveals clear distinctions in chemical and structural similarity performance.

4.1 Chemical Similarity (Hamming Distance)

ECFP4 fingerprints ranked by Tanimoto's coefficient consistently yield the lowest Hamming distances (mean 12.9 at $k = 35$, Table 1; Fig. 1), indicating that bit-overlap remains the most reliable proxy for substructural similarity. SELFIES-based embeddings (chemBERTa, SELFormer) outperform their SMILES counterparts by 10–13 units (mean 33.3–36.2 vs. 46.2–46.8), confirming that SELFIES' robust, bijective encoding better preserves chemical information. However, even the best embedding method (SELFormer/SELFIES) remains substantially inferior to ECFP4/Tanimoto. ANOVA and Tukey's tests support these conclusions at $p < 0.01$.

4.2 Structural Similarity (Graph Edit Distance)

When structural similarity is measured via unit-cost GED, ECFP4 with Tanimoto retrieves the closest graph matches (mean 29.6 at $k = 35$, Table 2; Fig. 2). All embedding methods yield higher GEDs (35–55), with SELFormer/SELFIES having the weakest performance. These results, confirmed by significant ANOVA contrasts ($p < 10^{-15}$), indicate that ECFP4 fingerprint vectors, ranked by Tanimoto's coefficient, effectively approximates global scaffold similarity. Transformer embeddings do not yet capture connectivity information as precisely.

4.3 GED Limitations

It is to be noted that GED has known limitations: it operates on 2D graphs and ignores stereochemistry or 3D geometry. Thus, two stereoisomers are treated as identical graphs (GED=0), potentially underestimating actual differences. Additionally, GED computation is expensive, and our 300 s timeout per comparison may terminate before finding a true minimum in highly branched or large graphs. These factors could bias the distances for complex molecules and should be considered when interpreting structural similarity results.

4.4 Model and Representation Insights

Within the transformer methods, SELFIES encoding consistently improves chemical similarity but does not enhance structural retrieval. Differences between chemBERTa and SELFormer (within the same string type) are not significant, suggesting that pretrained model choice is secondary to string representation. Notably, SELFormer/SELFIES underperforms on GED, perhaps reflecting training objectives that emphasize local chemical features over global graph topology.

4.5 Implications for Molecular Search

Our findings reaffirm the enduring utility of ECFP4 fingerprints in similarity search. For applications prioritizing chemical substructure matching, such as scaffold hopping or lead analog identification, ranking by Tanimoto coefficient on ECFP4 is recommended. Additionally, ranking by Tanimoto coefficient on ECFP4 is also recommended for structural motif retrieval, important in shape-based screening. Transformer embeddings may serve as complementary filters to capture orthogonal semantic features, but should not replace fingerprints in primary screening.

In conclusion, ECFP4/Tanimoto remains the most effective method for both chemical and structural similarity search. SELFIES offers modest gains over SMILES for embedding based chemical similarity, but transformer methods do not yet surpass classic fingerprints in either metric.

4.6 Limitations

This study is limited to 500,000 PubChem compounds and thirty queries, constrained by local compute and memory. The unit-cost GED computation (300 s timeout) remains a bottleneck, restricting throughput and potentially biasing against larger or highly branched structures. Embedding and fingerprint indexes each require substantial RAM, limiting dataset size in a single host. Future efforts should prioritize approximate or learned graph-distance measures, cloud-based or GPU-accelerated indexing for scale, and expanded query libraries to validate generality.

4.7 Future Work

Building on these findings, we plan to develop a hybrid search engine that partitions large libraries into parallel FAISS batches, retrieves candidates via both ECFP4/Tanimoto and SELFormer/SELFIES–cosine, and then merges and re-ranks hits by Hamming distance to balance chemical precision and semantic breadth. We also plan to scale up the query set to encompass diverse chemotypes (e.g. natural products, peptidomimetics) and explore approximate graph-distance algorithms to reduce GED computation time.

5 Conclusion

In this study, we benchmarked transformer-based embeddings (chemBERTa, SELFormer) against 1024-bit ECFP4 fingerprints for molecular similarity search, using both Hamming distance (chemical) and graph edit distance (structural) metrics. ECFP4 fingerprints, ranked by Tanimoto's coefficient, consistently outperformed all embedding methods (mean Hamming = 12.9 vs. 33–47; mean GED = 19.2 vs. 35–55) for both chemical as well as structural similarity search. SELFIES encoding modestly improved embedding performance over SMILES, but neither chemBERTa nor SELFormer matched fingerprint accuracy. These results reaffirm ECFP4's utility as the primary descriptor in similarity retrieval. For comprehensive search tools, we recommend a hybrid pipeline that combines ECFP4/Tanimoto screening with SELFormer/SELFIES embeddings to capture complementary semantic features.

Data Availability Statement. Data sets and code for experiment reproducibility are available at: https://github.com/ameykarmarkar02/MolSimSearch

A Appendix A

Top 35 hits retrieved for a sample Query Molecule: ``c1ccc2cc3c(ccc4cccnc43)cc2c1'' using the chemBERTa SMILES method (Table 3)

Table 3. Mol1: Top-35 nearest neighbours ranked by cosine distance. GED: Graph Edit Distance; Hamming: Hamming Distance.

#	Molecule ID	SMILES	Distance	GED	Hamming
0	9186	O=C(CF)Sc1ccccc1	0.913306	28	35
1	9185	O=[N+]([O-])c1ccc(Nc2cccc(C(F)(F)F)c2)cc1	0.911452	19	47
2	9176	CC(C)S[C@H]1O[C@H](CO)[C@H](O)[C@H](O)[C@H]1O	0.886004	25	48
3	123037	O=c1ccc(Nc2c3ccccc3nc3ccccc23)ccc1O	0.884637	30	40
4	9164	Cc1c(C)c2c(c(C)c1O)CCC(C)(CCCCC(C)CCCC(C)CCCC...)	0.882073	43	62
5	9168	CN(C)c1ccc(-c2ccc(N(C)C)cc2)cc1	0.880770	15	32
6	9188	O=C(NN=C1CCCCC1)C(=O)NN=C1CCCCC1	0.880425	19	42
7	9187	O=C(Nc1ccccc1)OCC(F)(F)F	0.880004	24	42
8	9179	Nc1nc(Nc2ccccc2)nc(C(F)(F)F)n1	0.879970	25	40
9	9178	Fc1nc(F)nc(C(F)(F)F)n1	0.873959	40	33
10	9163	Nc1c(F)cc(F)cc1[N+](=O)[O-]	0.873817	25	42
11	9169	CNNCc1ccc(C(=O)NC(C)C)cc1.Cl	0.872413	16	49
12	123038	Cc1ccc(C(=O)NCCN(C)C)c2nc3ccccc3cc12	0.871472	24	42
13	136230	Cc1ccn2c(c1)nc1c(N)cc(C)nc12	0.869793	18	42
14	139693	CSCC[C@H](Nc1c2ccccc2nc2cccc([N+](=O)[O-])c12)...	0.858947	31	58
15	9177	F[Si](F)(F)c1ccccc1	0.856451	25	26
16	43771	Cc1cc(N(CCO)CCO)ccc1/N=N/c1c(Cl)cc([N+](=O)[O-]...)	0.853779	39	64
17	9183	N#[N+]c1ccccc1	0.853577	24	28
18	123040	CNC(=O)c1cccc2c(Nc3ccc(NS(C)(=O)=O)cc3N(C)C)c3...	0.852634	57	64
19	9104	FC(F)(F)C1(F)C(F)(F)C(F)(F)C(F)(F)C(F)(F)C1(F)F	0.852002	36	36
20	142086	Cc1cccc2c3c(ccc12)C1C(C)C(C)CCC1(C)CC3	0.851336	17	53
21	136231	Cc1ccc2nc3ccccn3c2n1	0.847425	17	36
22	9105	FC(F)(F)C(F)(F)C(F)(F)C(F)(F)F	0.847051	38	31
23	95719	C=CC(=O)OCC(C)OC(=O)Nc1ccc(C)c(NC(=O)OC(C)COC(...	0.844028	42	59
24	9123	CC=CC(O)(CC[N+](C)(C)C)c1ccc(OC)cc1.[I-]	0.842143	17	48
25	9200	O=C(O)CCCCCF	0.841506	27	41
26	264429	CC=C(C)C(C)(C#N)Nc1ccccc1	0.840558	18	43
27	9120	C[N+](C)(C)CCC(O)(c1ccccc1)c1cccs1	0.839670	16	47
28	9167	CNC(C)C(O)c1ccc(O)cc1	0.839560	22	41
29	9125	FC(F)(F)C(F)(F)OC(F)(F)C(F)(F)F	0.838116	43	34
30	139713	COc1cc(NS(C)(=O)=O)ccc1Nc1c2ccccc2nc2c(C(=O)N(...	0.837669	54	61
31	9165	CCN(CC)C(=O)CF	0.834127	25	39
32	123036	CC(O)CN(CCO)N=O	0.830843	26	45
33	264422	CC(C#N)(Cc1ccccc1)Nc1ccccc1	0.829175	13	42
34	9153	CCCC(=O)Nc1ncnc2c1ncn2[C@@H]1O[C@@H]2COP(=O)(O)...	0.828089	51	74

References

1. Weininger, D.: SMILES, a chemical language and information system. 1. Introduction to methodology and encoding rules. J. Chem. Inf. Comput. Sci. **28**(1), 31–36 (1988). https://doi.org/10.1021/ci00057a005
2. Krenn, M., Häse, F., Nigam, A., Friederich, P., Aspuru-Guzik, A.: Self-referencing embedded strings (SELFIES): a 100% robust molecular string representation. Mach. Learn. Sci. Technol. **1**(4), 045024 (2020). https://doi.org/10.1088/2632-2153/aba947
3. Chithrananda, S., Grand, G., Ramsundar, B.: ChemBERTa: Large-Scale Self-Supervised Pretraining for Molecular Property Prediction (2020). arXiv:2010.09885. https://arxiv.org/abs/2010.09885
4. Yüksel, A., Ulusoy, E., Ünlü, A., Doğan, T.: SELFormer: Molecular Representation Learning via SELFIES Language Models (2023). arXiv:2304.04662. https://arxiv.org/abs/2304.04662
5. Yang, J., Cai, Y., Zhao, K., Xie, H., Chen, X.: Concepts and applications of chemical fingerprint for hit and lead screening. Drug Discov. Today **27**(11), 103356 (2022). https://doi.org/10.1016/j.drudis.2022.103356
6. C. Mellor, R. et al.: Molecular fingerprint-derived similarity measures for toxicological read-across: Recommendations for optimal use. Regul. Toxicol. Pharmacol. **101**, 121–134 (2019). https://doi.org/10.1016/j.yrtph.2018.11.002
7. Rogers, D., Hahn, M.: Extended-connectivity fingerprints. J. Chem. Inf. Model. **50**(5), 742–754 (2010). pMID: 20426451. https://doi.org/10.1021/ci100050t
8. Lv, J., Zhang, L., Huang, Y., Huang, J., Chen, S.: Graph Edit Distance Learning via Different Attention (2023). arXiv:2308.13871. https://arxiv.org/abs/2308.13871
9. Kulshrestha, A.: On the Hamming Distance between base-n representations of whole numbers (2012). arXiv:1203.4547. https://arxiv.org/abs/1203.4547
10. Bamberger, N.D., et al.: Beyond simple structure-function relationships: the interplay of geometry, electronic structure, and molecule/electrode coupling in single-molecule junctions. J. Phys. Chem. C **126**(15), 6653–6661 (2022). https://doi.org/10.1021/acs.jpcc.2c00761
11. Honda, S., Shi, S., Ueda, H.R.: SMILES Transformer: Pre-trained Molecular Fingerprint for Low Data Drug Discovery (2019). arXiv:1911.04738. https://arxiv.org/abs/1911.04738
12. Hu, J.: Ensemble Model With Bert, Roberta and Xlnet For Molecular property prediction (2024). arXiv:2406.06553. https://arxiv.org/abs/2406.06553
13. Bjerrum, E.J.: SMILES Enumeration as Data Augmentation for Neural Network Modeling of Molecules (2017). arXiv:1703.07076. https://arxiv.org/abs/1703.07076
14. Krenn, M., et al.: SELFIES and the future of molecular string representations. Patterns **3**(10), 100588 (2022). https://doi.org/10.1016/j.patter.2022.100588
15. Reiser, P., et al.: Graph neural networks for materials science and chemistry (2022). arXiv:2208.09481. https://arxiv.org/abs/2208.09481
16. Riba, P., Fischer, A., Lladós, J., Fornés, A.: Learning Graph Edit Distance by Graph Neural Networks (2020). arXiv:2008.07641. https://arxiv.org/abs/2008.07641
17. Wu, C., Wu, F., Huang, Y.: DA-Transformer: Distance-aware Transformer (2021). arXiv:2010.06925. https://arxiv.org/abs/2010.06925

18. White, M.D., Atkinson, M.D., Plowman, A.J., Shanthraj, P.: 3D variational autoencoder for fingerprinting microstructure volume elements (2025). arXiv:2503.17427. https://arxiv.org/abs/2503.17427
19. Kosonocky, C.W., Feller, A.L., Wilke, C.O., Ellington, A.D.: Using alternative SMILES representations to identify novel functional analogues in chemical similarity vector searches. Patterns **4**(12), 100865 (2023). https://doi.org/10.1016/j.patter.2023.100865
20. Blumenthal, D.B., Gamper, J.: On the exact computation of the graph edit distance. Pattern Recogn. Lett. **134**, 46–57 (2020). applications of Graph-based Techniques to Pattern Recognition. https://doi.org/10.1016/j.patrec.2018.05.002

Synthetic Riboswitch Sequences for CVB3 Detection across Salt Concentrations

Jonathan Auyong, Marelyn Negrete, William Andreopoulos, and Leila Khatib(✉)

San Jose State University, San Jose 95192, USA
leila.khatib@sjsu.edu

Abstract. The detection of viruses in complex environments often relies on methods that are both time-intensive and dependent on specialized equipment. While tools such as PCR, qPCR, and RT-LAMP have advanced sensitivity and improved time to detection, these methods are challenged by equipment requirements, training personnel, and financial constraints faced by public health organizations. Synthetic riboswitches, which are engineered RNA sequences capable of altering conformation upon ligand binding, offer a promising route to real-time, field-deployable sensing. By coupling target recognition with riboswitches that control gene expression, the synthetic riboswitch can serve as a biosensor that does not require specialized equipment. In this study, we explore an approach that predicts riboswitch performance across a range of salt concentrations and Coxsackievirus B3 RNA polymerase sequences. The platform aims to optimize synthetic riboswitches for virus detection across a range of salt concentrations. The approach integrates free energy modeling to predict the stability of the riboswitch's secondary structure, coupled with iterative changes to the ligand-binding region. The output produces candidate designs with improved stability and specificity. Our results demonstrate the impact of varying salt concentrations and iterative changes to the ligand-binding region on riboswitch function, as well as a computational approach that reduces the need for the in vivo screening of riboswitch biosensors. This framework lays the foundation for the rapid development of molecular diagnostics, reducing the need for trial-and-error in vivo approaches.

Keywords: Computational · Virus detection · Riboswitch

1 Introduction

1.1 A Subsection Sample

Portable, low-maintenance, and adaptable diagnostic platforms are necessary for robust virus monitoring in regions with limited equipment and financial infrastructure, as well as increased risk from virus exposure [1–3]. Synthetic riboswitches have emerged as a compelling option for the rapid detection of viral targets, with several studies demonstrating their potential for real-time sensing. Riboswitches are structured non-coding RNA molecules that regulate gene expression by altering their secondary or tertiary structure upon binding specific ligands such as metabolites [4]. In organisms, conformational changes can influence transcription termination, translation initiation, or

H. R. Arabnia et al. (Eds.): CSCI 2025, CCIS 2970, pp. 487–494, 2027.
https://doi.org/10.1007/978-3-032-28466-2_37

mRNA stability. Functionally, a ligand bound to the ligand-binding region destabilizes the riboswitch's secondary structure, shifting it between distinct conformations. In the "on" state, the structure permits expression of the downstream gene, while in the "off" state, gene expression is repressed. Synthetic riboswitches can be tailored to recognize chosen targets and express a linked reporter gene to indicate detection [5, 6]. Current studies have utilized synthetic riboswitches to detect Zika virus RNA, achieving low detection limits in human serum [7, 8].

While past studies have focused on blood and human serum, riboswitch-based diagnostics have the potential to be used for the real-time detection of viruses in environmental samples. Previous work has often required screening thousands of variants to identify optimal performance [9]. Past studies have also shown that synthetic riboswitches can be adapted to flow cell design, enabling real-time detection of viruses without significant infrastructure and equipment [6]. Some computational models have accelerated this process by predicting structural states before laboratory validation; however, these models do not consider various environmental parameters, which would enable broader application [10, 11].

Current computational platforms such as NuPack, ViennaRNA, and Ribologic predict the free energy of the RNA secondary structure under the assumption of 1 M NaCl [9, 11–13]. Understanding the free energy and secondary structure that governs riboswitch switching is essential for both natural biology and synthetic RNA circuit design. Predicting the two-state model computationally is challenging due to conformational heterogeneity, especially in the region of ligand or target binding [14]. Additionally, riboswitch functionality and conformation lead to different variations in efficiency of riboswitch control; therefore, artificial riboswitch development requires platforms that can predict iterative changes to the ligand-binding region to determine an optimized synthetic riboswitch. Environmental water samples can vary widely in salinity, creating challenges for virus detection by synthetic riboswitches, as they destabilize the secondary structure [14, 15]. The ionic strength of a matrix can alter the performance of a synthetic riboswitch, and past studies have modeled the impacts of salt concentration on nucleic acid secondary and tertiary structures [16, 17]. To narrow the range of synthetic riboswitches that detect viruses in water, this study presents a platform for identifying optimal synthetic riboswitch sequences with iterative changes to the ligand-binding region across varied salt concentrations. The calculated minimum free energy of the synthetic sequence provides a parameter for determining RNA secondary structure stability, which could more accurately predict the optimal riboswitch sequence for virus detection in water.

2 Materials and Methods

Three riboswitches designated A, B, and C were used as part of a model to predict minimum free energy with the iterative changes to the ligand-binding region [6]. The riboswitch and linker sequences were published previously [6, 18]. Riboswitch complexes were designated as the ligand-binding region, riboswitch, linker sequences, and the first 10 bp of the mCherry coding sequence [19]. CVB3 was selected as a representative non-polio enteric virus found in water and wastewater, serving as the model viral RNA sequence. The Coxsackievirus B3 (CVB3) Nancy strain genome from the American Type Cultures Collection (ATCC) was used to produce 30 bp sequences representative of the entire CVB3 RNA polymerase region. The 30 bp fragments were used to make iterative changes to the ligand-binding region.

Riboswitch A:

GGGUCUUAUCUUAUCUAUCUCGUUUAUCCCUGCAUACAGAAACA-GAGGAGAUAUGCAAUGAUAAACGAGAACCUGGCGGCAGCGCAAAAGA-UGCGUAAA

Riboswitch B:

GGGUCAAUAAGGCGGAGUUCGUCGAGGUGCCUGAGCAGAGGAGA-CAGGCAAUGCGACGAACUAACCUGGCGGCAGCGCAAAAGA-UGCGUAAA

Riboswitch C:

GGGUAUGAAUUGAUGUCGAUUGUUAUGUCUUGAGACAGAGGAGA-UCAAGAAUGAACAAUCGAAACCUGGCGGCAGCGCAAAA-GAUGCGUAAA.

The computational framework developed in this study integrates the ViennaRNA open-source simulator with custom control code to evaluate riboswitch performance under ligand binding across various salt concentrations. The ViennaRNA package was employed to calculate the minimum free energy for each riboswitch design: the complex's minimum free energy of the ligand–riboswitch sequences, the minimum free energy associated with the riboswitch alone, and the target energy reflecting binding efficiency. To ensure robustness, these calculations were repeated systematically over 151 iterations for each of the three riboswitches, where each iteration sampled different ligand-binding sequences and riboswitch sequences. At the same time, the linker region and a fixed segment of ten base pairs corresponding to the mCherry reporter gene remained unchanged. This strategy enabled an exploration of riboswitch minimum free energy predictions for synthetic riboswitches.

Following the energy calculations, the raw simulation outputs were exported and analyzed using MATLAB R2025a [20]. The MATLAB scripts were designed to parse the ViennaRNA data files, extract the minimum free energy values for each synthetic riboswitch variant, and calculate switching probabilities by quantifying the fraction of conformations adopting the alternative structural state. In particular, the probabilities between the minimum free energy of the complex and the minimum free energy of the ligand were calculated and plotted using a cumulative density function. The cumulative density functions (CDFs) for the complex energy of the riboswitch and the ligand sequence energy of CVB3 were plotted over a 10-fold range of salt concentrations (0.1–1 M NaCl) using the Vienna RNA simulator. The cumulative distribution function (CDF) for the ligand-bound riboswitch was determined by

$$\Delta G = E_{\text{complex}} - E_{\text{switch}} \tag{1}$$

In contrast, the target energy E_{Target} CDF was calculated to determine the probability of target binding and conformational changes occurring simultaneously. For each plot, the average cumulative frequency count of the CDF plot was then plotted for the three riboswitches across the range of salt concentrations. The method provides a foundation for rational riboswitch design, enabling researchers to compare alternative constructs based on quantitative energy calculations, thereby narrowing the range of possible synthetic sequences.

3 Results

Two sets of results were obtained to quantify riboswitch performance across the range of salt concentrations. The first set is shown in Fig. 1, which visualizes the cumulative density functions (CDFs) of two free energy parameters: the free-energy difference ΔG (red) and the target binding energy E_{Target} (black). Figure 1 represents one distribution plot at 0.5 M NaCl. The horizontal axis corresponds to energy values, while the vertical axis denotes the fraction of cases (0–1) where the value is less than or equal to the corresponding energy.

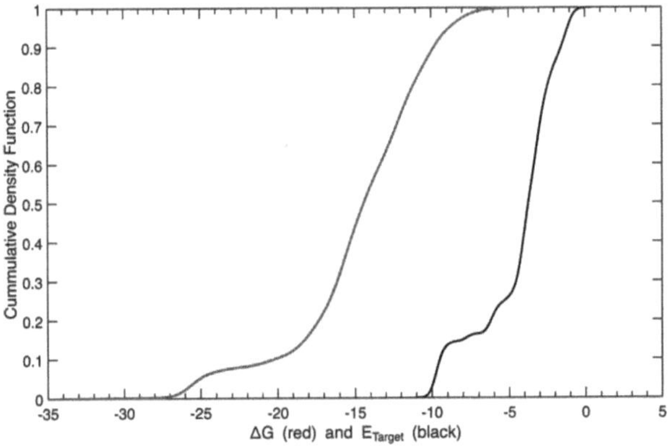

Fig. 1. Coupled CDF analysis of ΔG and E_{Target} reveals riboswitch switching probabilities at 0.5 M NaCl. The red curve represents the cumulative distribution function (CDF) of $\Delta G = E_{\text{complex}} - E_{\text{switch}}$, while the black curve shows the CDF of the target energy E_{Target}. The intersection or relative alignment of these two CDFs provides a direct, probabilistic measure of switching events under a specific salt concentration.

In the CDF plots, a steep slope implies that most values fall within a narrow energy band, whereas a shallow slope suggests a wider spread. The overlap and relative positioning of the ΔG and E_{Target} curves provide a probabilistic measure of switching likelihood, with the closer the alignment, the higher the probability of riboswitch activation. This coupled CDF approach, therefore, visualized the variations in switching threshold and salt concentration.

Coupled CDF analysis of ΔG and E_{Target} provides a statistical view of riboswitch energetics under 0.5 M salinity. The red curve represents the cumulative distribution function (CDF) of

$$\Delta G = E_{\text{complex}} - E_{\text{switch}} \tag{2}$$

that shows for each MFE the fraction of riboswitch sequence variants with ΔG values less than or equal to that threshold. Since ΔG represents the MFE of switching, the CDF indicates that the probability of switching is highly likely when the ligand MFE is equivalent to the MFE complex energy. The black curve corresponds to the CDF of the target binding energy, E_{Target}, which characterizes the interaction strength between the riboswitch and its ligand. Overlaying these distributions highlights regions where conformational stability and binding affinity align, which identifies the highest probability of functional switching. This coupled CDF analysis thus enables the quantification of the proportion of conditions favorable for riboswitch activation, providing a predictive framework for comparing designs and assessing the robustness of the synthetic riboswitch across various environmental salinities.

The second set of results is displayed in Fig. 2, which shows the probability of riboswitch function as a function of salt concentration for three riboswitch designs. The black curve corresponds to the A riboswitch, the blue curve to the B riboswitch, and the red curve to the C riboswitch. The plot reveals clear differences in salinity tolerance among the three designs, with the C riboswitch maintaining consistently high switching probability across the salt concentrations. In comparison, the other two designs show steep declines with increasing salt concentrations. The divergence from the C riboswitch suggests that salt possibly destabilizes the ligand-bound state by disrupting the secondary structure of the riboswitch. This disruption suggests a suboptimal riboswitch design, indicating that particular riboswitch sequences are more suitable for environmental applications.

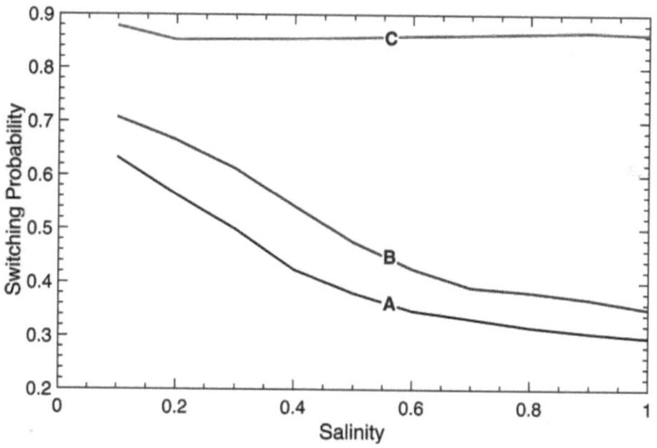

Fig. 2. Switching probability of three riboswitch designs (A, B, and C) as a function of salinity. Design A (black) and design B (blue) exhibit strong sensitivity to salinity, with switching probability decreasing sharply as salinity increases. In contrast, design C (red) demonstrates robust performance, maintaining a consistently high switching probability across the full salinity range calculated.

4 Results

4.1 Riboswitch Dynamics: ON and OFF States

As illustrated in Fig. 3, the riboswitch operates as a ligand dependent genetic switch that regulates transcriptional outcomes through RNA secondary structures. In the *OFF state*, the RNA folds into a hairpin loop or other secondary structures, thereby preventing translation. Upon ligand binding, the conformation shifts to favor the *ON state*, allowing for translation of the reporter gene. RNA conformation and transcription exemplify how riboswitches couple ligand sensing and gene expression.

4.2 Discussion of Simulation Results

The results offer several insights into the performance of riboswitches under various environmental conditions. First, the coupled CDF framework in Fig. 1 demonstrates that riboswitch minimum free energy can be represented in a probabilistic manner, allowing direct visualization of switching likelihood. This probabilistic perspective highlights both the effects across varying salt concentrations and the probability of switching (as indicated by the slope in the overlap region). Such information is crucial for predicting the reliability of synthetic riboswitches under varied environmental conditions, such as high salt concentrations.

Second, the comparative analysis in Fig. 2 underscores the advantage of orthogonal design strategies from previous studies [6]. The C riboswitch exhibits stable performance ($\approx$0.85–0.9 probability) across a range of salt concentrations, making it suitable for virus detection in environments with elevated salt

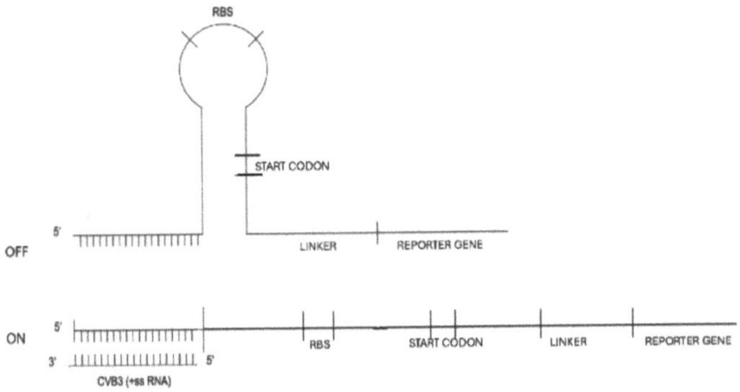

Fig. 3. Schematic illustration of riboswitch operational dynamics. In the OFF state (top), the ribosome binding site (RBS) is sequestered within a stem–loop structure, preventing ribosome access and translation initiation at the start codon. In the presence of the cognate trigger RNA (CVB3 + ssRNA), base-pairing unfolds the hairpin, exposing the RBS and allowing translation of the downstream reporter gene, thereby switching the riboswitch ON (bottom).

levels, such as recycled water, marine biosensing, and industrial processes. In contrast, the A and B riboswitches exhibit marked salt sensitivity, losing nearly half their functional probability as salt concentrations increased. These designs may therefore be restricted to low-salinity applications where specific functional attributes are prioritized over robustness.

Third, the relative ordering of riboswitch designs suggests the importance of structural independence and orthogonality in maintaining secondary structure stability under salt stress. The riboswitch C design likely avoids competing structural constraints that destabilize switching under ionic stress, and suggests that the A and B riboswitch designs appear more vulnerable to such destabilization. The robust performance of the C riboswitch suggests that its sequence and subsequent folding are resilient as compared with A and B. Additional studies will need to examine the riboswitch sequence differences to determine which features of the sequence increase salt resilience in the secondary structure of riboswitches. This relative ordering underscores the importance of structural independence and orthogonality in shaping environmental resilience.

Using this predictive platform, this study sought to understand the relationship between riboswitch design and salt concentrations. The current implementation, however, was limited to modeling NaCl as a representative monovalent salt and conformational changes in the secondary structure used thermodynamic equilibrium measurements but did not consider kinetic rate parameters. Prior studies have demonstrated that divalent ions, such as Mg^{2+}, exert a significantly stronger stabilizing effect on RNA structure due to their higher charge density and ability to influence a broader region of RNA. This broader electrostatic influence reduces phosphate–phosphate repulsion, promoting more stable folding. Incorporating such effects will be an important next step, and future iterations of the platform should extend to models that account for multiple ion types and variable concentrations to more accurately capture riboswitch behavior under diverse environmental conditions. Thus, future studies will need to incorporate models that address various ion types in different concentrations and kinetic parameters.

5 Conclusion

These results underscore the importance of integrating environmental variability into riboswitch engineering pipelines. Designs that appear effective under laboratory conditions may fail in the field if they are not robust to factors such as ionic strength. Therefore, salinity, which has a strong effect on RNA folding, may cause riboswitches to not produce valid results. Overall, the analysis demonstrates that, while some riboswitch variants are highly sensitive to changes in salt concentration, others can maintain function across diverse salt conditions. This finding is critical for applications in environmental or field-based diagnostics, where ionic strength can vary widely. By identifying points of divergence between stable and unstable designs, this method reduces reliance on trial-and-error screening, narrows the range of possible riboswitch designs, and accelerates the development of riboswitch-based methods for detecting environmental viruses. The predictive platform and coupled-CDF approach provide a powerful quantitative framework for anticipating such failures by comparing the free energy distributions of switching and binding interactions across various salt concentrations. This platform, coupled with CDF analysis of synthetic riboswitches, is one approach toward bridging the gap between laboratory optimization and environmental applications.

Disclosure of Interests. The authors have no competing interests to declare that are relevant to the content of this article.

References

1. Becker, W.C., LeChevallier, M., Rosario-Ortiz, F., Stanford, B., Walker, T.: Assessing Water Quality Monitoring Needs, Tools, Gaps, and Opportunities for Potable Water Reuse. Webcast, The Water Research Foundation (2023) Available: https://www.waterrf.org/resource/
2. Jiang, S.C., et al.: Integrating virus monitoring strategies for safe non-potable water reuse. Water. **14**(8), 1187 (2022). https://doi.org/10.3390/w14081187
3. Bogler, A., et al.: Rethinking wastewater risks and monitoring in light of the COVID-19 pandemic. Nat. Sustain. **3**(12), 981–990 (2020). https://doi.org/10.1038/s41893-020-00605-2
4. Wieland, M., Hartig, J.S.: Artificial riboswitches: synthetic mRNA-based regulators of gene expression. Chembiochem. **9**(12), 1873–1878 (2008)
5. Pardee, K., et al.: Rapid, low-cost detection of zika virus using programmable biomolecular components. Cell. **165**(5), 1255–1266 (2016). https://doi.org/10.1016/j.cell.2016.04.059
6. Pardee, K., et al.: Paper-based synthetic gene networks. Cell. **159**(4), 940–954 (2014)
7. Amaral, C., et al.: A molecular test based on RT-LAMP for rapid, sensitive and inexpensive colorimetric detection of SARS-CoV-2 in clinical samples. Sci. Rep. **11**(1), 16430 (2021)
8. Park, S., Lee, J.W.: Detection of coronaviruses using RNA toehold switch sensors. Int. J. Mol. Sci. **22**(4) (2021)
9. Wu, M.J., Andreasson, J.O.L., Kladwang, W., Greenleaf, W., Das, R.: Automated design of diverse stand-alone riboswitches. ACS Synth. Biol. **8**(8), 1838–1846 (2019)
10. Clote, P.: Computational prediction of riboswitches. Methods Enzymol. **553**, 287–312 (2015)
11. Hofacker, I.L., Fontana, W., Stadler, P.F., Bonhoeffer, L.S., Tacker, M., Schuster, P.: Fast folding and comparison of RNA secondary structures. Monatsh. fur¨ Chemie. / Chem. Monthly. **125**(2), 167–188 (1994)
12. Lorenz, R., et al.: Viennarna package 2.0. Algorithms Mol. Biol. **6**(1), 26 (2011)
13. Zadeh, J.N., et al.: Nupack: analysis and design of nucleic acid systems. J. Comput. Chem. **32**(1), 170–173 (2011). https://doi.org/10.1002/jcc.21596
14. Furtig, B., Nozinovic, S., Reining, A., Schwalbe, H.: Multiple conformational states of riboswitches fine-tune gene regulation. Curr. Opin. Struct. Biol. **30**, 112–124 (2015)
15. Badelt, S., Hammer, S., Flamm, C., Hofacker, I.L.: Thermodynamic and kinetic folding of riboswitches. Methods Enzymol. **553**, 193–213 (2015)
16. Chheda, U., Pradeepan, S., Esposito, E., Strezsak, S., Fernandez-Delgado, O., Kranz, J.: Factors affecting stability of RNA – temperature, length, concentration, ph, and buffering species. J. Pharm. Sci. **113**(2), 377–385 (2024). https://doi.org/10.1016/j.xphs.2023.11.013
17. Tan, Z.-J., Chen, S.-J.: Salt contributions to RNA tertiary structure folding stability. Biophys. J. **101**(1), 176–187 (2011)
18. Takahashi, M.K., et al.: A low-cost paperbased synthetic biology platform for analyzing gut microbiota and host biomarkers. Nat. Commun. **9**(1), 3347 (2018)
19. Shaner, N.C., Campbell, R.E., Steinbach, P.A., Giepmans, B.N.G., Palmer, A.E., Tsien, R.Y.: Improved monomeric red, orange and yellow fluorescent proteins derived from *Discosoma* sp. red fluorescent protein. Nat. Biotechnol. **22**(12), 1567–1572 (2004)
20. The MathWorks, Inc., *MATLAB R2025a*, The MathWorks, Inc., Natick, Massachusetts, United States version R2025a (2025). https://www.mathworks.com/products/matlab.html

Word2Vec-Enhanced Ensemble Learning for Protein Succinylation Site Prediction Using CNN-BiLSTM

Mohammad Hassanzadeh[1,4](✉), Love Fadia[2], Alioune Ngom[3], and Mohammad El Yabroudi[4]

[1] Math and Computer Science, Lawrence Technological University, Southfield, USA
mhassanza@ltu.edu
[2] Electrical and Computer Engineering, University of Windsor, Windsor, Canada
fadial@uwindsor.ca
[3] School of Computer Science, University of Windsor, Windsor, Canada
angom@uwindsor.ca
[4] Electrical and Computer Engineering, Lawrence Technological University, Southfield, USA
melyabrou@ltu.edu

Abstract. Lysine succinylation is a new post-translational modification implicated in diverse biological processes, including erythropoiesis. Based on new findings, we propose a machine learning model that employs Word2Vec-based feature representation with varied vector sizes in conjunction with a CNN-BiLSTM ensemble network for the prediction of succinylation sites. Our approach strictly examines various embedding dimensions and selects the top three ranked models using an ensemble strategy. On 10-fold cross-validation, the model achieves 76.0% accuracy, Matthews correlation coefficient (MCC) of 0.53, sensitivity of 81.0%, and specificity of 71.0%, better than the existing methods. On an independent test set, the model performs equally well as the best existing predictors. These results highlight the effectiveness of multi-view embedding methodologies and deep learning ensembles in driving the computational identification of lysine succinylation sites further, and offering a useful tool for future biological research.

Keywords: Succinylation site prediction · CNN-BiLSTM · Word2vec Embeddings · Ensemble Leaning

1 Introduction

Proteins are basic bio-molecules with key functions in almost all cellular processes. Proteins are assembled in a strictly controlled process referred to as translation, where the information in messenger RNA (mRNA) is translated by ribosomes into polypeptide chains [1]. After being produced, these chains fold into distinctive three-dimensional structures to be fully functional proteins.

H. R. Arabnia et al. (Eds.): CSCI 2025, CCIS 2970, pp. 495–510, 2027.
https://doi.org/10.1007/978-3-032-28466-2_38

But protein synthesis is not the end; proteins often have other chemical modifications that refine their function, stability, and interactions [2]. These chemical alterations, or post-translational modifications (PTMs), add diversity to the proteome beyond what is found in the genome. PTMs include numerous types of chemical modifications such as Phosphorylation, Glycosylation, Acetylation, Ubiquitination, Methylation, and Succinylation, to name a few [3]. Each of these modifications has the potential to profoundly change the activity, localization, or interaction with other molecules of a protein. One particularly intriguing PTM is Succinylation, a non-enzymatic modification resulting from the covalent attachment of fumarate a key intermediate in the citric acid cycle—to protein cysteine residues. This modification leads to the formation of S-(2-succinyl)cysteine (2SC), which can serve as a biomarker for metabolic dysregulation [4]. The classification of Succinylation is essential for several reasons. First, it sheds important light on the molecular basis of several diseases, including diabetes and cancer, where Succinylation of proteins play a significant role [5]. Second, comprehension of the patterns and effects of Succinylation on protein function might give rise to new therapeutic targets and biomarkers for early diagnosis. Despite its importance, Succinylation remains underexplored compared to other PTMs, partly due to the complexity of its biochemical pathways and subtle structural alterations it imparts on proteins [6]. Over the years, machine learning techniques have been increasingly applied to the classification of PTMs. Techniques such as support vector machines, random forests, and neural networks and multiview graph convolutional network have shown promise in identifying and classifying various modifications by learning patterns from protein sequences and structural data [7,8]. However, these methods often face limitations when addressing the multifaceted nature of Succinylation, where multiple layers of biological data—from sequence motifs to structural contexts—interact in complex ways [9]. To tackle the challenge, our method employs a hybrid CNN-BiLSTM architecture augmented with word2vec embeddings. In this approach, protein sequences are first transformed into dense vector representations using word2vec, effectively capturing the underlying semantic relationships. The CNN component then extracts local spatial features from these embeddings, while the BiLSTM module captures long-range dependencies in both forward and backward directions. By integrating these techniques, our method leverages the strengths of both convolutional and recurrent neural networks, resulting in a more robust and precise Succinylation classification of proteins. This CNN-BiLSTM framework not only bridges existing gaps in the analysis of protein post-translational modifications but also opens up new avenues for a more holistic understanding of protein functionalities.

2 Related Work

Protein lysine succinylation is an important post-translational modification (PTM) that is implicated in many metabolic and regulatory processes. Prediction of succinylation sites has been a valuable computational tool, aiding functional annotation of proteins and experimental design. A variety of methods have been proposed over the years, from ensemble learning schemes to advanced deep learning models. This review focuses on methodological aspects of representative studies, highlighting feature extraction techniques, model architecture, and data-handling. Pokharel *et al.* [10] employed a protein language model (Prott5-XL-Uniref50) to represent contextual embeddings of amino acid sequences. By feeding these embeddings into a deep learning model, they achieved improved performance metrics, including accuracy and Matthews correlation coefficient (MCC). These language model embeddings capture evolutionary signals and long-range dependencies most accurately and hence are well-suited for PTM site prediction. Jia *et al.* [11] introduced pSuc-EDBAM, which utilizes an ensemble of dense blocks and an attention layer for predicting lysine succinylation sites. Multi-scale features of protein sequences are extracted by dense blocks, while the attention module selectively highlights essential residues. The combination approach provided higher predictive precision compared to standard classifiers, evidencing the advantage of utilizing convolutional layers integrated with attention mechanisms. Hu *et al.* [12] developed SuccSite, making use of amino acid composition and k-spaced amino acid pairs to represent local and long-range residue contacts. Systematic sampling of different k values by this approach captures weak sequence patterns that are correlated with succinylation. The use of spaced-out pairwise features introduces a further layer of insight beyond position-specific scores. Thapa *et al.* [13] proposed DeepSuccinylSite, a CNN-based framework that predicts local sequence features relevant to succinylation. Convolutional filters scan the input sequences, enabling the model to learn discriminative motifs. With stringent validation, DeepSuccinylSite showed improved performance compared to approaches based on handcrafted features alone. Liu *et al.* [14] introduced Deep_KsuccSite, which employs convolutional and recurrent layers, along with an attention mechanism, to predict succinylation sites. The CNN layers identify short-range patterns, and the recurrent layers identify sequential dependencies. The attention module then focuses on the most informative spots to yield a flexible architecture able to capture local and global context. Xu *et al.* [15] pioneered the application of pseudo amino acid composition (PseAAC) for predicting lysine succinylation sites. Their iSuc-PseAAC tool is an addition to ordinary amino acid composition with position-specific propensities. Inspired by this concept, Jia *et al.* [16] developed iSuc-PseOpt, to combat data imbalance issues by optimizing training sets and accounting for sequence-coupling effects. These methods generally employ support vector machines or other traditional classifiers, emphasizing well-designed features. Jia *et al.* [11] also proposed pSuc-Lys, which integrates PseAAC-based descriptors with ensemble random forest. The ensemble strategy combines predictions from multiple decision trees, removing variance and making the overall robustness improved. By cross-validation, pSuc-Lys was of high accuracy and specificity, justifying the potential of random forest ensembles in succinylation site prediction. Hasan and Kurata [17] developed GPSuc, a global prediction system that incorporates various sequence features—ranging from residue composition to motif-based signatures. GPSuc also evolves features to specific species, which implies that pathways for succinylation could be organism-specific. By employing a random forest classifier, GPSuc balances generalizability and organism-specific accuracy in harmony. Hoang *et al.* [18] advanced the field of succinylation site identification by integrating deep learning with large experimental data. The work emphasizes the importance of high-quality, large-scale data since deep learning algorithms typically require many training examples to master complex sequence patterns. The proposed framework emphasizes the collaboration between modern neural networks and biologically curated databases. Tran *et al.* [19] proposed a framework integrating convolutional neural networks (CNN) and bidirectional long short-term memory (Bi-LSTM) models for predicting protein succinylation sites, using techniques like n-gram modeling. Through the use of CNN to detect local residue patterns and Bi-LSTM to detect longer-range dependencies, the authors reported better predictive accuracy than existing methods.

Table 1 provides an overview of the datasets used in some of the prediction studies of succinylation sites, such as the number of proteins, positive and negative site distribution for training and testing. The datasets are specific to a given prediction tool and differ in sample size, class balance, and evaluation strategy. The SuccinSite 2.0 dataset has 2,192 proteins in the training set, 4,750 positive and 9,500 negative sites, with a positive to negative site ratio of approximately 1:2. The test set is much smaller with 14 proteins, with 254 positive and 2,977 negative sites. GPSuc has the same dataset structure as SuccinSite 2.0 and is thus comparable in terms of training and test data distributions. HybridSucc has a considerably greater training set comprising 7,415 proteins and 21,770 positives along with 165,071 negatives, providing substantially more unbalanced data than in the previous tools. There are 1,415 proteins for the test set, yet there are no particular positive and negative site numbers presented. DeepSuccinylSite and LMSuccSite both use exactly the same breakdown of a set comprising 2,192 proteins for training each with 4,750 positives and 4,750 negatives, making data balanced. Test set is maintained constant with 14 proteins, 254 positive sites, and 2,977 negative sites. pSuc-EDBM differs in dataset structure by incorporating a much larger number of negative sites in the training process. It contains 2,198 proteins, with 4,755 positive and 50,549 negative samples. Test dataset is structured similarly to other tools with 14 proteins, 254 positive sites, and 2,977 negative sites. MDCAN-Lys uses a relatively larger dataset, comprising 2,778 training proteins and 5,005 positive and 64,140 negative sites, rendering the dataset extremely imbalanced. The test dataset consists of 309 proteins with 684 positive and 6,709 negative sites. We utilized CBiLSuccSite, which uses a structured dataset strategy as other tools use. The training set consists of 2,192 proteins, 4,750 positive and 4,750 negative sites, with balanced class distribution. The test set consists of 124 proteins with 254 positive and 2,977 negative sites. Adopted dataset structure here follows previous research to ensure a proper comparison while permitting stringent model assessment. The table provides an overview of dataset diversity distribution used in succinylation site prediction tools. While others favor more extensive datasets with highly imbalanced classes to achieve better generalization, others prefer a balanced dataset to achieve greater sensitivity and specificity. The dataset employed in CBiL-SuccSite takes a systematic approach to allow for the proper assessment of its predictive strength compared to other models that are currently available.

Table 1. Statistics of Datasets Used in Succinylation Site Studies

Available Tools	Dataset	Number of proteins	Positive	Negative
SuccinSite 2.0	Training	2192	4750	9500
SuccinSite 2.0	Testing	14	254	2977
GPSuc	Training	2192	4750	9500
GPSuc	Testing	14	254	2977
HybridSucc	Training	7415	21,770	165,071
HybridSucc	Testing	1415	–	–
DeepSuccinylSite	Training	2192	4750	4750
DeepSuccinylSite	Testing	14	254	2977
LMSuccSite	Training	2192	4750	4750
LMSuccSite	Testing	14	254	2977
pSuc-EDBM	Training	2198	4755	50,549
pSuc-EDBM	Testing	14	254	2977
MDCAN-Lys	Training	2778	5005	64,140
MDCAN-Lys	Testing	309	684	6709
CBiLSuccSite	Training	2192	4750	4750
CBiLSuccSite	Testing	124	254	254
CBiLSuccSite	Testing	124	254	2977
Proposed Method	Training	2192	4750	4750
Proposed Method	Testing	124	254	254

3 Proposed Methodology

In this study, a hybrid deep learning model integrating a Convolutional Neural Network (CNN) with a Bidirectional Long Short-Term Memory (BiLSTM) network is proposed for protein succinylation site prediction. Unlike traditional sequence-based methods that rely on predefined feature extraction, our model employs Word2Vec embeddings to represent protein sequences in a dense vector space, capturing contextual relationships among amino acids. To explore the impact of different embedding sizes, we train Word2Vec models with dimensions ranging from 500 to 1500, incremented by 100, allowing for a comparative analysis of varying feature representations. Once this feature representations are extracted then they are fed to the model for classification. The next few subsections discusses all the necessary parts of the methodology in detail.

3.1 Word2Vec Embedding for Protein Sequence Representation

Word2Vec is a powerful word embedding technique initially developed for natural language processing (NLP) to map words into a continuous vector space where similar words have closer vector representations. This method was introduced by Mikolov *et al.* [20] and has been widely used in various domains, including bioinformatics, where protein sequences can be treated as textual data. The primary advantage of Word2Vec over traditional encoding methods, such as one-hot encoding, is its ability to capture contextual relationships between sequence elements, preserving meaningful biological patterns. Word2Vec operates using two main architectures: Continuous Bag-of-Words (CBOW) and Skip-Gram. CBOW predicts a word based on its surrounding context, making it efficient for large datasets, whereas Skip-Gram predicts surrounding words based on a given central word, making it more effective in capturing rare and long-range dependencies. In this study, we employ the Skip-Gram model due to its ability to capture intricate relationships between amino acids, which is crucial for protein sequence analysis [21] (Fig. 1).

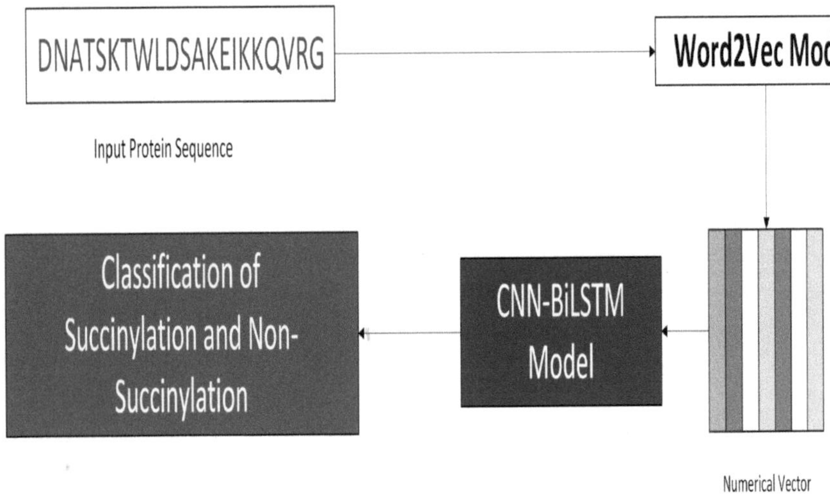

Fig. 1. Proposed Methodology

For this work, each protein sequence is treated as a sentence, and individual amino acids are considered as words. A dictionary of 21 tokens is constructed, comprising the 20 standard amino acids and an additional pseudo amino acid ('X') to handle missing or unknown residues. The sequences are tokenized using a 1-g technique, ensuring that each amino acid is treated as a distinct word in the sequence. This allows the model to learn numerical vector representations for each amino acid, where the positioning and relationship between residues influence the learned embeddings. To extract meaningful representations, a custom

Word2Vec model is trained on the protein dataset using the Skip-Gram approach. The following hyperparameters are set to optimize embedding learning:

- **Window size:** 7, capturing meaningful amino acid interactions within a biologically relevant neighborhood.
- **Minimum word occurrence:** 3, filtering out rare sequence patterns that might introduce noise.
- **Embedding dimensions:** 500 to 1500, incremented by 100, enabling a systematic comparison of feature space richness.

Once trained, the Word2Vec embeddings are used to initialize the embedding layer in the deep learning model. This representation ensures that amino acids with similar biochemical properties have similar vector representations, improving the model's ability to generalize across different protein sequences. Unlike one-hot encoding, which results in sparse high-dimensional representations, Word2Vec compresses sequence information into dense, low-dimensional feature vectors that better reflect underlying sequence properties. Word2Vec is particularly effective in this study because it allows the CNN-BiLSTM model to extract both local sequence motifs and long-range dependencies within protein sequences. The CNN branch identifies conserved short motifs that might correspond to biochemical signals of succinylation, while the BiLSTM branch captures sequential dependencies and broader contextual patterns. By embedding amino acids into a numerical space where semantic similarity is preserved, Word2Vec significantly enhances feature extraction, leading to improved classification accuracy. The effectiveness of Word2Vec in protein sequence representation is evident in its ability to encode biochemical and evolutionary relationships implicitly. Amino acids with similar biochemical functions tend to have closer embeddings, allowing the model to leverage these inherent properties during classification. This method also helps mitigate sparsity issues common in protein sequence analysis, providing a more robust foundation for deep learning-based succinylation site prediction.

3.2 CNN-BiLSTM Model

The architecture of the proposed CNN-BiLSTM model shown in Fig. 2 consists of two parallel branches that process the embedded protein sequence. Unlike traditional approaches that rely on predefined feature extraction, our model employs Word2Vec embeddings with varying dimensions, ranging from 500 to 1500 with increments of 100. These embeddings provide dense, continuous vector representations of amino acids, capturing contextual relationships and improving feature representation. The embedded sequence is then processed by two complementary branches: a Convolutional Neural Network (CNN) branch for extracting local motifs and a Bidirectional Long Short-Term Memory (BiLSTM) branch for capturing long-range dependencies.

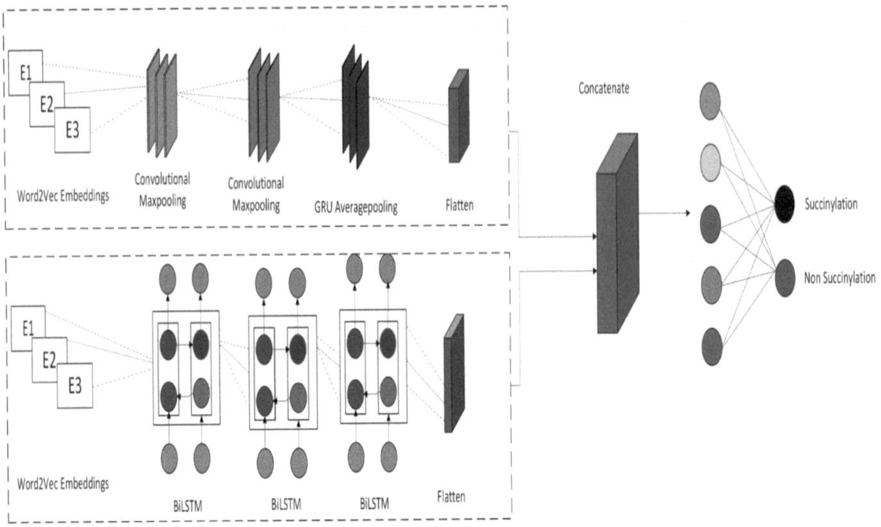

Fig. 2. Proposed Architecture

CNN Branch. In the first branch, the input sequence (padded to a fixed length and embedded into a $L \times d$ matrix, where L is the sequence length and d is the chosen Word2Vec embedding dimension) is passed through a one-dimensional CNN that extracts local sequence motifs. The CNN employs multiple convolutional filters (kernels) of size 3, each activated using the rectified linear unit (ReLU) function. These filters slide over the sequence to detect biologically relevant motifs surrounding the central lysine, potentially indicative of succinylation. Two consecutive convolutional layers are used, each with 64 filters of size 3. Each convolutional layer is followed by a max-pooling layer (pool size = 2), which reduces the sequence length while retaining the most salient features. To prevent overfitting and enhance generalization, a dropout layer (dropout rate = 0.4) is applied after the final pooling operation. The extracted features are then flattened into a fixed-length feature vector [22]. This CNN branch is responsible for capturing short-range dependencies and identifying sequence motifs that are characteristic of succinylation sites.

BiLSTM Branch. The second branch processes the same embedded sequence in parallel using a bidirectional LSTM network, which captures long-range dependencies and sequential patterns that might not be detected by CNN alone. A BiLSTM layer with 128 hidden units is employed, where each LSTM cell processes the sequence in both forward and backward directions. The bidirectional processing ensures that each amino acid's representation is enriched with information from both flanking regions. A dropout layer (dropout rate = 0.4) is applied to the BiLSTM inputs to introduce regularization and prevent overfitting. Unlike conventional LSTM architectures that rely on global pooling to condense features, we flatten the entire BiLSTM output, retaining position-specific information across all sequence positions. This approach proved beneficial in our preliminary experiments, as it preserved finer-grained contextual dependencies [23].

Feature Fusion and Classification. After processing through both branches, the extracted feature vectors from the CNN and BiLSTM are concatenated to form a unified representation of the sequence. This merged feature vector integrates both local motif patterns from CNN and long-range dependencies from BiLSTM, ensuring a comprehensive encoding of succinylation-relevant patterns. The concatenated features are passed through a fully connected dense layer with 128 neurons and ReLU activation, allowing for further refinement of the extracted sequence representations. A dropout layer (rate = 0.3) is applied to introduce additional regularization. Finally, the model outputs a probability score via a single neuron in the final layer with a sigmoid activation function, indicating the likelihood of the input lysine being a succinylation site. A threshold of 0.5 is used for binary classification: if the output probability is ≥ 0.5, the site is classified as *succinylated*, otherwise, it is classified as *non-succinylated.*

This hybrid CNN-BiLSTM framework enables the model to leverage CNN's capability of identifying conserved local motifs and BiLSTM's strength in capturing sequence dependencies, leading to a more effective prediction system for protein succinylation sites.

3.3 Ensemble Learning

To further enhance prediction accuracy and improve generalization, we adopted an ensemble learning approach. Instead of relying on a single best-performing model, we leveraged the top three models that achieved the highest validation performance during cross-validation to form an ensemble. These selected models corresponded to the three Word2Vec embedding dimensions that yielded the best cross-validation metrics (Accuracy and MCC). By combining multiple models, the ensemble approach ensures that variations learned by individual models are collectively utilized, thereby increasing predictive robustness.

The ensemble strategy works as follows: Each of the three selected models is independently applied to a test sample, generating an output probability indicating the likelihood of the central lysine being a succinylation site. To produce a final prediction, we compute the average probability across these three models. This averaged probability is then thresholded at 0.5 to yield the final binary classification—if the probability is ≥ 0.5, the site is classified as *succinylation*; otherwise, it is classified as *non-succinylation*. Since we selected three models, this approach effectively functions as a majority-vote ensemble, where a site is predicted as succinylated if at least two of the models agree.

The motivation behind ensembling is that each model, trained with different embedding sizes and initialized weights, may capture distinct sequence features. By aggregating their predictions, errors made by any one model are compensated by the others, leading to reduced variance and improved generalization.

Empirical results demonstrated that the ensemble consistently outperformed any single model on validation folds, confirming its effectiveness. The results for the ensemble model, along with those for individual models, are presented in the Results section.

3.4 Training Strategy

We trained and evaluated our models using a k-fold cross-validation approach to ensure robust performance estimation. In our experiments, we used $k = 10$ (10-fold cross-validation), partitioning the dataset into ten equal subsets. In each fold, nine subsets (90% of the data) were used for training and the remaining one subset (10% of the data) was used for validation. This process was repeated such that each subset served as the validation set exactly once, and performance metrics were averaged across all ten folds. All models were trained using the same training protocol: we used the Adam optimizer with a learning rate of 1×10^{-4}, and the binary cross-entropy loss function for optimization. Each fold model was trained for up to 80 epochs, with early stopping based on validation loss to prevent overfitting. We used a batch size of 32 sequence segments. The model with the best validation performance in each fold was saved for later ensemble use.

4 Results and Discussion

To evaluate model performance, we considered several standard metrics commonly used in binary classification, particularly in the context of post-translational modification (PTM) site prediction: Accuracy (ACC), Matthews Correlation Coefficient (MCC), Sensitivity (SEN), Specificity (SPE), and Area Under the ROC Curve (AUC) [24]. These metrics provide a comprehensive assessment of the model's ability to distinguish between succinylation and non-succinylation sites.

Accuracy is the proportion of correctly classified lysine sites, including both positive and negative cases, and is defined as:

$$\text{Accuracy} = \frac{TP + TN}{TP + TN + FP + FN}, \tag{1}$$

where TP (True Positives) represents correctly identified succinylation sites, TN (True Negatives) represents correctly identified non-succinylation sites, FP (False Positives) denotes non-succinylation sites incorrectly classified as succinylated, and FN (False Negatives) denotes succinylation sites incorrectly classified as non-succinylated.

Sensitivity, also known as recall or the true positive rate (TPR), measures the model's ability to correctly identify succinylation sites and is calculated as:

$$\text{Sensitivity} = \frac{TP}{TP + FN}. \tag{2}$$

Specificity, also known as the true negative rate (TNR), quantifies the proportion of correctly identified non-succinylation sites and is given by:

$$\text{Specificity} = \frac{TN}{TN + FP}. \tag{3}$$

Matthews Correlation Coefficient (MCC) is a balanced metric that accounts for both false positives and false negatives and is particularly useful for evaluating imbalanced datasets [25]. It is defined as:

$$\text{MCC} = \frac{TP \times TN - FP \times FN}{\sqrt{(TP + FN)(TP + FP)(TN + FP)(TN + FN)}}. \tag{4}$$

The MCC ranges from -1 to $+1$, where $+1$ indicates perfect prediction, 0 indicates no better than random guessing, and -1 indicates total disagreement between predictions and true labels.

Finally, Area Under the ROC Curve (AUC) provides a threshold-independent measure of model performance, illustrating the trade-off between sensitivity and specificity. It is computed as:

$$\text{AUC} = \int_0^1 \text{TPR}(x)\, dx\ , \tag{5}$$

where $\text{TPR}(x)$ represents the true positive rate at a given false positive rate x. AUC values range from 0 to 1, where a value of 0.5 indicates random guessing, while a value closer to 1 reflects a stronger discriminative ability.

Each of these metrics was computed for all cross-validation folds, and the average results are reported in the Results section. By utilizing cross-validation and these comprehensive evaluation metrics, we ensure that the performance assessment is robust and not dependent on a specific train-test split.

4.1 Performance of Individual Models (Varying Embedding Dimensions)

The classification performance was analyzed across multiple Word2Vec embedding dimensions, ranging from 500 to 1500. Table 2 presents the accuracy, MCC, sensitivity, specificity, and AUC values obtained for each dimension.

The classification performance was evaluated across different Word2Vec embedding dimensions to determine their impact on model effectiveness. Table 2 presents the results obtained for embedding dimensions ranging from 500 to 1500. The accuracy values remained within a narrow range, fluctuating between 75.86% and 76.24%, suggesting that while embeddings contribute to feature extraction, an excessively large embedding size does not necessarily enhance model performance. The highest accuracy (76.24%) was observed at an embedding size of 1500, closely followed by 900 (76.20%), indicating that these dimensions provide a more informative representation of sequence data. The Matthews Correlation Coefficient (MCC) values remained relatively stable, with the highest value of 0.5267 corresponding to the 1500-dimensional embeddings. This

Table 2. Performance of different Word2Vec dimensions

Word2Vec Dim	Accuracy	MCC	Sensitivity	Specificity	AUC
500	0.7617	0.5248	0.7968	0.7263	0.8277
600	0.7599	0.5213	0.7960	0.7231	0.8268
700	0.7597	0.5220	0.8040	0.7156	0.8282
800	0.7609	0.5241	0.8070	0.7142	0.8283
900	0.7620	0.5253	0.8021	0.7201	0.8288
1000	0.7616	0.5251	0.7923	0.7302	0.8287
1100	0.7581	0.5176	0.7964	0.7183	0.8287
1200	0.7618	0.5246	0.8008	0.7216	0.8279
1300	0.7598	0.5209	0.7951	0.7240	0.8281
1400	0.7586	0.5187	0.7832	0.7343	0.8256
1500	0.7624	0.5267	0.8004	0.7246	0.8278

stability suggests that increasing dimensionality beyond a certain point yields diminishing returns.

Sensitivity and specificity results highlight an interesting trade-off. The highest sensitivity (80.7%) was recorded for the 800-dimensional embeddings, suggesting that this model configuration maximized the ability to detect true succinylation sites. However, specificity peaked at 1400 dimensions (73.43%), demonstrating that models trained with higher embeddings were more precise in distinguishing non-succinylation sites. The area under the ROC curve (AUC) remained consistently high across all configurations, ranging between 0.8256 and 0.8288, confirming that the model effectively discriminates between positive and negative instances. Given these findings, the selected embedding dimensions significantly impact classification efficiency, and future work will further optimize their selection. It was also observed that these models scored the highest cross-validation scores across different metrics at average against the best state of the art predictors in the exisiting literature.

4.2 Comparison with Existing Predictors

The performance of the proposed model was benchmarked against existing succinylation site prediction tools using an independant test set, as summarized in Table 3.

To validate the robustness of the proposed approach, its performance was compared with existing succinylation site prediction tools, as summarized in Table 3. The proposed model achieved an accuracy of 76.0%, which is highly competitive with the CBiLSuccSite model (76.4%) and outperformed all other benchmarked tools. It also achieved an MCC of 0.53, which is notably higher than GPSuc (0.35), DeepSuccinylSite (0.48), and pSuc-EDBAM (0.40), emphasizing its effectiveness in handling class imbalances. While CBiLSuccSite recorded the

Table 3. Comparison of the proposed model with existing predictors

Method	Accuracy	MCC	Sensitivity	Specificity
GPSuc	67.0%	0.35	66.0%	68.0%
DeepSuccinylSite	70.0%	0.48	79.0%	69.0%
LMSuccSite	74.0%	0.51	76.0%	73.0%
pSuc-EDBAM	69.9%	0.40	74.8%	65.0%
MDCAN-Lys	70.7%	0.42	76.8%	64.6%
CBiLSuccSite	76.4%	0.53	80.3%	72.4%
Proposed Model	76.0%	0.53	81.0%	71.0%

highest specificity (72.4%), the proposed model demonstrated a higher sensitivity (81.0%), which is crucial in biological applications where missing true succinylation sites could lead to incomplete functional analyses. This suggests that the proposed approach excels at capturing true positive cases but requires further refinement to enhance specificity. Despite the competitive performance of the model, future work will focus on refining feature representations and adjusting model hyperparameters to improve classification efficiency. Although the results are closely aligned with CBiLSuccSite, the proposed model exhibited higher cross-validation stability, suggesting its potential for further refinement. Overall, while the current findings establish a strong baseline, continued optimizations will be undertaken to enhance generalization performance.

4.3 Discussion and Future Work

The results presented above provide several key insights into the effectiveness of the CNN-BiLSTM hybrid model for protein sequence classification. First, the integration of convolutional and recurrent neural network architectures proves to be highly effective. The CNN branch captures conserved short motifs, which may correspond to biochemical sequence patterns essential for distinguishing succinylation sites. Simultaneously, the BiLSTM branch enables the model to capture long-range dependencies in sequences, ensuring that position-specific patterns extending beyond the receptive field of CNNs are incorporated. This synergy allows the hybrid model to outperform standalone CNN or LSTM-based approaches.

The role of Word2Vec embeddings in enhancing model performance was also evident. The contextual encoding provided by these embeddings enables the model to discern amino acid similarities in a biologically meaningful manner. Interestingly, the results indicate that an embedding size of around 1500 dimensions provided the highest classification accuracy, although gains beyond 900 dimensions were marginal. This suggests that while increased dimensionality helps in learning richer representations, excessive embedding size can lead to overparameterization without substantial improvements in predictive power. The ability to achieve high accuracy without drastically increasing dimensionality is an encouraging result, as it keeps the model size and computational complexity manageable.

Despite the strong performance observed, this research remains an ongoing effort with several areas for further exploration. One immediate direction is the incorporation of attention mechanisms in the BiLSTM branch. Attention layers can help highlight the most informative residues within a sequence, improving interpretability and allowing the model to focus on the most relevant regions for prediction.

Additionally, future work will explore the use of more advanced protein embeddings derived from state-of-the-art transformer-based protein language models such as ESM and ProtT5. These embeddings incorporate global contextual knowledge from vast protein databases, potentially improving succinylation site prediction at the cost of increased computational demand. A comparative analysis between our current Word2Vec-based embeddings and transformer-derived embeddings will be conducted to assess trade-offs between performance and efficiency.

5 Conclusion

In conclusion, the proposed CNN-BiLSTM ensemble model demonstrates competitive performance in protein sequence classification, effectively combining local feature extraction with global sequence pattern learning. The results validate the significance of hybrid architectures and word embedding representations in biological sequence modeling. Future improvements will focus on integrating attention mechanisms, testing transformer-based embeddings, and refining the handling of class imbalance. These advancements will not only further improve predictive accuracy but also provide deeper biological insights into sequence determinants of succinylation and related modifications.

References

1. Jové, M., et al.: Succination of protein thiols in human brain aging. Front. Aging Neurosci. **12**, 52 (2020)
2. Hillmann, K.B., Goethel, M.E., Erickson, N.A., Niehaus, T.D.: Identification of a S-(2-succino)cysteine breakdown pathway that uses a novel S-(2-succino) lyase. J. Biol. Chem. **298**(12), 102639 (2022)
3. Schmidt, C., Sciacovelli, M., Frezza, C.: Fumarate hydratase in cancer: a multifaceted tumour suppressor. Semin. Cell Dev. Biol. **98**, 15–25 (2020)
4. Zheng, L., et al.: Circulating succinate-modifying metabolites accurately classify and reflect the status of fumarate hydratase-deficient renal cell carcinoma. J. Clin. Invest. **133**(10), e165028 (2023)
5. Meng, L., et al.: Mini-review: recent advances in post-translational modification site prediction based on deep learning. Comput. Struct. Biotechnol. J. **20**, 3522–3532 (2022)

6. Tasmia, S.A., Ahmed, F.F., Mosharaf, P., Hasan, M., Mollah, N.H.: An improved computational prediction model for lysine succinylation sites mapping on Homo sapiens by fusing three sequence encoding schemes with the random forest classifier. Curr. Genom. **22**(2), 122–136 (2021)
7. Song, L., Xu, Y., Wang, M., Leng, Q.: PreCar_Deep: a deep learning framework for prediction of protein carbonylation sites based on Borderline-SMOTE strategy. Chemom. Intell. Lab. Syst. **218**, 104428 (2021)
8. Shah, V., Hassanzadeh, M., Ahmadi, M.: Integrating graph signal processing with graph convolutional networks for N - and O-glycosylation site prediction. In: IEEE 68th International Midwest Symposium on Circuits and Systems (MWSCAS), Lansing/E. Lansing, MI, USA, pp. 793–797 (2025). https://doi.org/10.1109/MWSCAS53549.2025.11244593
9. Phloyphisut, P., Pornputtapong, N., Sriswasdi, S., Chuangsuwanich, E.: MHCSeqNet: a deep neural network model for universal MHC binding prediction. BMC Bioinform. **20**(1), 270 (2019)
10. Pokharel, S., et al.: Improving protein succinylation sites prediction using embeddings from protein language model. Sci. Rep. **12**, 16933 (2022)
11. Jia, J., Wu, G., Li, M., Qiu, W.: pSuc-EDBAM: predicting lysine succinylation sites in proteins based on ensemble dense blocks and an attention module. BMC Bioinform. **23**(1), 450 (2022)
12. Hu, J.N., et al.: SuccSite: incorporating amino acid composition and informative k-spaced amino acid pairs to identify protein succinylation sites. Genom. Proteom. Bioinform. **18**(2), 168–178 (2020)
13. Thapa, N., et al.: DeepSuccinylSite: a deep learning based approach for protein succinylation site prediction. BMC Bioinform. **22**(Suppl 3), 153 (2021)
14. Liu, X., et al.: Deep_KsuccSite: a novel deep learning method for the identification of lysine succinylation sites. Front. Genet. **13**, 1007618 (2022)
15. Xu, Y., et al.: iSuc-PseAAC: predicting lysine succinylation in proteins by incorporating peptide position-specific propensity. Sci. Rep. **5**, 10184 (2015)
16. Jia, J., et al.: iSuc-PseOpt: Identifying lysine succinylation sites in proteins by incorporating sequence-coupling effects into pseudo components and optimizing imbalanced training dataset. Anal. Biochem. **497**, 48–56 (2016)
17. Hasan, M.M., Kurata, H.: GPSuc: global prediction of generic and species-specific succinylation sites by aggregating multiple sequence features. PLoS ONE **13**(10), e0200283 (2018)
18. Huang, K.-Y., Hu, J.B.-K., Lee, T.-Y.: Characterization and identification of lysine succinylation sites based on deep learning method. Sci. Rep. **9**(1), 16175 (2019)
19. Tan, T.-X., Khanh, N.Q., Nguyen, V.-N.: Integrating CNN and Bi-LSTM for protein succinylation sites prediction based on natural language processing technique. Comput. Biol. Med. **152**, 106954 (2025)
20. Mikolov, T., Chen, K., Corrado, G., Dean, J.: Efficient estimation of word representations in vector space. In: Proceedings of the Workshop at ICLR (2013). https://arxiv.org/abs/1301.3781
21. Mikolov, T., Sutskever, I., Chen, K., Corrado, G.S., Dean, J.: Distributed representations of words and phrases and their compositionality. Adv. Neural Inf. Process. Syst. (NIPS) **26**, 3111–3119 (2013)
22. Kim, Y.: Convolutional neural networks for sentence classification. In: Proceedings of the 2014 Conference on Empirical Methods Natural Language Processing (EMNLP), pp. 1746–1751 (2014)

23. Graves, A., Schmidhuber, J.: Bidirectional LSTM networks for improved phoneme classification and recognition. In: Proceedings of the IEEE International Conference on Acoustics, Speech, Signal Processing (ICASSP), pp. 1–4 (2005)
24. Matthews, B.W.: Comparison of the predicted and observed secondary structure of T4 phage lysozyme. Biochim. Biophys. Acta Protein Struct. **405**(2), 442–451 (1975)
25. Sokolova, M., Lapalme, G.: A systematic analysis of performance measures for classification tasks. Inf. Process. Manage. **45**(4), 427–437 (2009)

Beyond Words: How Well Can ChatGPT Classify Pneumonia from Chest X-Rays?

Pragna Prahallad(✉) and Pranathi Prahallad

Emerald High School, Dublin, CA, USA
pragnaprahallad@gmail.com

Abstract. This study evaluates the capability of OpenAI's `gpt-4o` model to classify chest X-rays as `NORMAL` or `PNEUMONIA` without fine-tuning. Using a balanced test set of 400 images, we assessed how four prompt designs—from minimal instructions to reasoning-based formats—affect diagnostic accuracy. Concise, feature-focused prompts achieved the highest accuracy of 74%, while detailed reasoning reduced performance. These results indicate that ChatGPT shows promise for zero-shot medical image interpretation but remains limited in reliability. Enhancing visual understanding and domain-specific adaptation will be essential before such multimodal models can be applied safely in clinical image analysis.

Keywords: Zero-Shot Learning · Medical Image Classification · Pneumonia Detection · Prompt Engineering · Multimodal Large Language Models

Research Track: Computational Biology (CSCI-RTCB)

1 Pneumonia - A Health Challenge

Pneumonia is an infection that causes inflammation in the air sacs of the lungs, called alveoli, and can occur in one or both lungs. The air sacs in the lungs can fill with fluid or pus, resulting in coughs accompanied by phlegm or pus. The filling of the alveoli (air sacs) can make it harder for oxygen to reach the bloodstream [1].

Pneumonia can be life-threatening to infants, people over the age of 65, and those with weakened immune systems. Viruses, bacteria, fungi, and bacteria-like organisms are all leading causes of pneumonia. Additionally, there are three other ways pneumonia can be acquired: Hospital-acquired pneumonia, Healthcare-acquired pneumonia, and aspiration pneumonia. Key symptoms of pneumonia include chest pain when breathing or coughing, fatigue, sweating, fever, body chills (signs of infection), nausea, vomiting, diarrhea, and shortness of breath, confusion or changes in mental awareness, and lower than normal body temperature in those who are over the age of 65.

H. R. Arabnia et al. (Eds.): CSCI 2025, CCIS 2970, pp. 511–518, 2027.
https://doi.org/10.1007/978-3-032-28466-2_39

Pneumonia remains one of the biggest health challenges in the world. In the United States alone, 1.4 million emergency room visits were a result of pneumonia. Additionally, over 40,000 deaths were due to pneumonia [2]. According to the World Health Organization (WHO), in 2017, pneumonia killed over 808,000 children under the age of 5, accounting for 15% of deaths in children under 5 [3]. The high mortality rates of pneumonia tie back to delayed diagnoses, lack of equipment in low-resource areas, and the difficulty of differentiating pneumonia from other respiratory conditions. These statistics emphasize the need for a faster, more efficient, and cost-effective method to diagnose pneumonia, an area that requires more research and testing, including in AI-assisted imaging.

1.1 Detection of Pneumonia in Clinics

Pneumonia is most commonly detected by a doctor asking for a medical history and using a stethoscope to listen for bubbling or crackling sounds when placed on your lungs. If a doctor suspects pneumonia, there are four main tests to pinpoint the cause of the infection and prevent its spread. First, a blood test can be used to confirm the presence of an infection and attempt to identify the type of organism (virus, fungus, bacteria, or bacteria-like organisms) causing the infection. [4] A drawback of blood tests is that they are not specialized in precise identification. Second, a doctor might conduct a chest X-ray, which can help determine the extent and location of the pneumonia. Third, a pulse oximetry can be conducted to measure the level of oxygen in the blood. Lastly, a sputum test can aid in zeroing in on the cause of infection.

2 The Aid of AI in Detecting Pneumonia

Fast, accurate, and efficient diagnoses of pneumonia are essential for patient health, especially for patients over the age of 65. A misdiagnosis of pneumonia can have devastating effects on their health, leading to long-term complications. A study titled "Inappropriate Diagnosis of Pneumonia Among Hospitalized Adults", twelve percent of all pneumonia diagnoses are misdiagnosed [5]. AI has the potential to aid in faster, more consistent interpretation of chest scans and X-rays.

2.1 Zero-Shot Learning for Image Classification

Zero-shot learning (ZSL) is a paradigm in machine learning where a model is capable of recognizing and classifying new categories that it has never encountered during training. Unlike conventional supervised models that rely on large labeled datasets, ZSL leverages the model's ability to associate visual inputs with semantic or textual representations learned during pretraining [6]. In the context of medical imaging, zero-shot learning allows an AI model—such as a large vision-language model—to analyze radiographs or scans and infer diagnostic labels based on semantic understanding, rather than direct exposure to labeled examples of each condition.

2.2 The Role of Prompts in Zero-Shot Inference

In zero-shot image classification using large multimodal models, prompts act as the interface between human intent and model behavior. They define the task, provide linguistic context, and specify the expected output format. Effective prompt design aligns the model's general visual–linguistic knowledge with the target task, emphasizing relevant features while reducing ambiguity.

Prompts vary in their level of specificity. Some may request a concise classification, while others guide the model to explain its reasoning or use domain-specific terminology. By carefully structuring these instructions, prompts enable zero-shot models to transfer learned knowledge to new domains without explicit supervision.

Furthermore, prompts can influence the model's reasoning process. Including instructions to focus on key diagnostic features or to provide step-by-step explanations can improve interpretability and reduce errors arising from incomplete or inconsistent reasoning. Overall, prompt design plays a critical role in directing.

2.3 Image Classification in Health Care

Recent research has increasingly explored the integration of zero-shot learning into medical diagnosis and analysis. A 2025 study in the *Journal of Medical Internet Research* demonstrated how prompt engineering, chain-of-thought reasoning, and retrieval-augmented generation can enhance diagnostic accuracy for pulmonary diseases [7]. Although this work focused on textual radiology reports rather than images [8], it emphasized how prompt design and reasoning strategies directly influence model performance and interpretability.

Building on this, *HealthPrompt: A Zero-Shot Learning Paradigm for Clinical Natural Language Processing* introduced a framework for classifying medical texts through prompt-based learning, further underscoring the importance of linguistic structure and task formulation in achieving accurate predictions.

In the visual domain, studies have extended zero-shot and few-shot prompting methods to medical imaging tasks. For instance, *Improving Pneumonia Localization via Cross-Attention on Medical Images and Reports* leveraged both image and textual features to enhance pneumonia detection [9]. Similarly, the *PM2: Prompting Multi-Modal Model for Few-Shot Medical Image Classification* framework combined textual and visual cues for improved classification accuracy [10]. More recently, *A ChatGPT-Aided Explainable Framework for Zero-Shot Medical Image Classification* utilized large language models such as ChatGPT to directly interpret and classify medical images, highlighting the growing potential of zero-shot approaches in healthcare [11].

2.4 Previous Work in Pneumonia Image Classification

Machine learning has played a pivotal role in automating medical image diagnosis, including the detection of pneumonia from chest X-rays. Early studies, such as that by Mujahid et al. [12], employed Inception-based architectures to distinguish between normal and pneumonia-infected lungs. Their work demonstrated that transfer learning could achieve high accuracy even with limited training data. Similarly, El Asnaoui and Chawki [13] developed an ensemble deep learning framework to improve reliability and minimize classification errors. Kundu et al. [14] further showed that combining multiple deep learning models enhanced both recall and robustness compared to individual networks.

More recently, Singh et al. [15] introduced Vision Transformers (ViTs) for pneumonia detection, leveraging attention mechanisms to focus on critical image regions. Mabrouk et al. [16] extended this approach by integrating ViTs with other deep learning models, achieving further improvements in classification accuracy.

Despite these advances, most existing approaches rely on fully supervised learning with large labeled datasets. In contrast, emerging large multimodal models, such as ChatGPT with vision capabilities, can interpret and classify medical images through natural-language prompts without task-specific retraining. This shift introduces new possibilities for zero-shot image classification in healthcare.

In this study, we investigate the use of OpenAI models for classifying chest X-rays as normal or pneumonia-affected using zero-shot learning guided by prompt engineering, and we evaluate their diagnostic accuracy.

3 Experimental Methodology

3.1 Database Used

This study used the publicly available *Chest radiographic Images (Pneumonia)* data set from Kaggle [17], which is widely used to benchmark pneumonia detection models. The data set consists of a total of 5,863 anterior chest X-ray images, categorized into two classes: *Normal* and *Pneumonia*. The images were taken from pediatric patients aged one to five years at the Guangzhou Women and Children's Medical Center, China.

The data set is organized into three subsets:

- **Training set:** 5,216 images
- **Test set:** 624 images
- **Validation set:** 16 images

Among the 624 test images, 234 show normal chest radiographs, while 390 show radiographic evidence of pneumonia. Since the goal of this work was to evaluate the performance of zero-shot image classification, only the **test set** was utilized in our experiments. All images were used in their original form without fine-tuning or supervised training, aligning with the zero-shot learning framework.

3.2 Sample Chest X-Ray Images

To illustrate the dataset, Fig. 1 shows two representative chest X-ray images from the test set. The first image corresponds to a pneumonia-affected lung, while the second image represents a normal chest X-ray. These examples highlight the visual contrast between infected and healthy lungs that the model aims to classify.

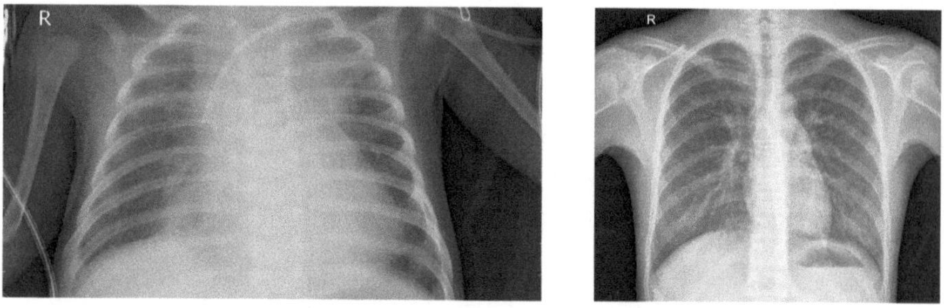

Fig. 1. Example chest X-ray images from the test set in the *Chest X-Ray Images (Pneumonia)* dataset [17]. (a) Pneumonia X-ray (b) Normal X-ray.

4 Model and Prompt Design

4.1 Model and Image Preprocessing

We used the `gpt-4o` model (also called "GPT-4 Omni"), a version of ChatGPT that can understand both pictures and text together. This means that it can look at a chest X-ray while reading written instructions (called prompts) to decide if the image shows a healthy chest (`NORMAL`) or one with pneumonia (`PNEUMONIA`).

Before each X-ray was sent to the model, the images were prepared so that they were clear but not too large. Each image was opened in color mode, resized so that its longest side was no more than 2048 pixels, and saved in high quality JPEG format. This ensured that all images had the same quality and size, helping the model analyze them fairly and accurately.

4.2 Prompt Design

To communicate effectively with the model, each interaction used two types of messages: a **system prompt** and a **user prompt**.

The **system prompt** tells the model how to behave or what role it should take—for example, to act like a medical image labeler and to give answers only in JSON format. It sets the overall tone and rules for the task.

The **user prompt**, on the other hand, gives the actual instruction for that specific image—for example, what kind of answer to return, what format to use, and the image itself. Together, the system and user prompts help the model stay consistent, structured, and focused while analyzing the X-rays.

We tested four prompt types to see how the way we ask the model affects its answers. Each prompt gave the model both the X-ray image and a short instruction. The model always returned results in strict JSON format. The four prompt types are shown below.

4.2.1 Prompt 1: Minimal Output (No Features)

System: `You are labeling chest X-rays as either NORMAL or PNEUMONIA. Output strict JSON only.`
User: `Return ONLY JSON like this: "label":"NORMAL|PNEUMONIA","confidence":0..1. (Image attached)`

This was the simplest version. The model only had to pick a label and confidence value, without explaining its choice.

4.2.2 Prompt 2: Including Features in Output

System: `You are labeling chest X-rays as either NORMAL or PNEUMONIA. Output strict JSON only.`
User: `Return ONLY JSON like this: "features":"...","label":"NORMAL|PNEUMONIA","confidence":0..1. (Image attached)`

This version asked the model to also list short visual features it noticed in the image, such as "cloudy lung areas" or "clear air spaces."

4.2.3 Prompt 3: Features + Concise Reasoning

System: `You are labeling chest X-rays for dataset use. Be objective and return strict JSON only. Do not include step-by-step reasoning; just give a short justification.`
User: `Return ONLY JSON with keys: "features":"<=160 chars","reason":"<=240 chars (short justification)","label":"NORMAL|PNEUMONIA","confidence":0..1. (Image attached)`

This version added a short reason (one or two sentences) to explain why the model made its choice, while staying brief and direct.

4.2.4 Prompt 4: Features + Step-by-Step Reasoning

System: `You are labeling chest X-rays for dataset use. Be objective and return strict JSON only. Include step-by-step reasoning.`
User: `Return ONLY JSON with keys: "features":"<=160 chars","reason":"<=512 chars (step-by-step)","label":"NORMAL|PNEUMONIA","confidence":0..1. (Image attached)`

This final version asked the model to describe its reasoning process in more detail before deciding.

4.3 Summary of Results

Each prompt was tested on 400 X-rays (200 labeled as `NORMAL` and 200 as `PNEUMONIA`). The results are summarized in Table 1.

Table 1. Accuracy Comparison of Prompt Variants

Description	Acc.(%)
Minimal Output (No features)	64.50
With feature output	74.00
Features + concise reasoning	71.50
Features + step-by-step reasoning	70.75

Adding short visual features (Prompt 2) helped the model better identify `NORMAL` images. However, asking for longer reasoning (Prompt 4) did not improve results further, suggesting that brief, focused reasoning works best for image-based classification tasks.

5 Conclusion

This study demonstrates that while ChatGPT (`gpt-4o`) is capable of classifying chest X-rays into `NORMAL` and `PNEUMONIA` using carefully designed prompts, its diagnostic accuracy remains limited. Even with the best-performing prompt, the overall accuracy reached only 74%, indicating that the model still struggles to consistently recognize subtle medical image features.

The results also show that prompts requiring longer or step-by-step reasoning did not improve performance—in fact, they slightly reduced accuracy. This suggests that reasoning-oriented large language models, while powerful for text-based tasks, are not yet optimized for visual diagnostic reasoning [18]. In image-based classification, especially in medical settings, concise and feature-focused prompts appear more effective than reasoning-heavy instructions [19].

Overall, this experiment highlights the potential of multimodal models like ChatGPT in medical imaging, but also emphasizes the need for future improvements in visual understanding, specialized medical fine-tuning, and integration with domain-specific datasets to achieve clinical reliability.

References

1. Mayo Clinic Staff. Pneumonia: Symptoms & causes (2023). Accessed 04 Oct 2025
2. American Lung Association. Learn about pneumonia (2023). Accessed 04 Oct 2025
3. World Health Organization. Pneumonia (2024). Accessed 04 Oct 2025
4. Mayo Clinic Staff. Pneumonia: Diagnosis & treatment (2023). Accessed 04 Oct 2025

5. Gupta, A.B., Vaughn, M.G., et al.: Inappropriate diagnosis of pneumonia among hospitalized adults. JAMA Intern. Med. (2024). Accessed 04 Oct 2025
6. Liu, J., Hu, T., Zhang, Y., Gai, X., Feng, Y., Liu, Z.: A chatGPT-aided explainable framework for zero-shot medical image diagnosis. arXiv preprint arXiv:2307.01981 (2023). Accessed 04 Oct 2025
7. Zhang, X., Li, Y., Chen, H., Wang, J.: Enhancing pulmonary disease prediction using large language models with feature summarization and hybrid retrieval-augmented generation: Multicenter methodological study based on radiology report. J. Med. Internet Res. **27**(1), e72638 (2025)
8. Zhang, Y., Liu, H., Kim, S.: HealthPrompt: a zero-shot learning paradigm for clinical natural language processing. PubMed (2023)
9. Zhao, L., Huang, X., Sun, Q.: Improving pneumonia localization via cross-attention on medical images and reports. arXiv preprint arXiv:2110.03094 (2021)
10. Li, F., Wang, R., Zhou, T.: PM2: a new prompting multi-modal model paradigm for few-shot medical image classification. arXiv preprint arXiv:2404.08915 (2024)
11. Chen, M., Patel, R., Singh, A.: A chatGPT-aided explainable framework for zero-shot medical image classification. OpenReview (2024)
12. Mujahid, M., et al.: Pneumonia classification from X-ray images with inception. Int. J./PMC (open access) (2022). Accessed 05 Oct 2025
13. El Asnaoui, K., Chawki, Y.: Design ensemble deep learning model for pneumonia classification. Journal/PMC (open access) (2021). Accessed 05 Oct 2025
14. Kundu, R., Das, R., Geem, Z.W., Han, G.-T., Sarkar, R.: Pneumonia detection in chest x-ray images using an ensemble of deep learning models. PLoS ONE **16**(9), e0256630 (2021). Accessed 05 Oct 2025
15. Singh, S., Kumar, M., Kumar, A., Kumar, B.: Efficient pneumonia detection using vision transformers on chest X-rays. Sci. Rep. **14**, 2487 (2024). Accessed 05 Oct 2025
16. Mabrouk, A., Díaz Redondo, R.P., Dahou, A., Abd Elaziz, M., Kayed, M.: Pneumonia detection on chest X-ray images using ensemble of deep convolutional neural networks. Appl. Sci. **12**(13), 6448 (2022)
17. Mahmoud, P.T., Mooney: Chest x-ray images (pneumonia) (2018). Accessed 04 Oct 2025
18. Jiang, Y., Chen, C., Nguyen, D., Mervak, B.M., Tan, C.: GPT-4v cannot generate radiology reports yet (2024)
19. Liu, R., Geng, J., Wu, A.J., Sucholutsky, I., Lombrozo, T., Griffiths, T.L.: Mind your step (by step): chain-of-thought can reduce performance on tasks where thinking makes humans worse (2025)

Quantum Sensing Applications in Drug Discovery and Development

Lennex Annor(✉) and Mohammad Atif

Department of Computer Science, Capitol Technology University, Laurel, USA
lannor@captechu.edu

Abstract. Drug discovery, along with development processes, benefits from the advanced technology of quantum sensing, which operates through quantum mechanics principles. The research investigates quantum sensing methods used to boost the accuracy and productivity of drug discovery operations throughout their target detection molecular bonding evaluation and pharmacokinetic measurements. By using quantum properties like superposition and entanglement, quantum sensors have high sensibility and accuracy, which makes it possible to identify slight molecular bond interactions and structural alterations. This paper explores the possibilities of quantum sensing, underlining their efficiency in revolutionizing the drug discovery process, shortening the timetable of drug development, and enhancing the success rate of the treatments. The further development of QS is in line with the use of artificial intelligence and machine learning for the construction of predictive models in drug development.

Keywords: Quantum Sensing · Drug Discovery · Molecular Interaction · Pharmacokinetics · Quantum Mechanics · Artificial Intelligence

1 Introduction

The high sensitivity with precise molecular detection capability is necessary because drug research and development processes take both a long time and a large financial investment. Clinical methods, including mass spectrometry, together with fluorescence-based techniques, struggle to identify very small biomolecular connection points. Quantum sensing that is deployed with superposition, entanglement, and quantum coherence has better precision and efficiency in analytically analyzing molecules. These technologies allow the observation of molecular behavior in real-time, improvement of drug delivery systems, and increase the efficiency of biomarkers discovery. The paper examines quantum sensing fundamentals and shows its pharmaceutical research applications together with its AI-based integration. The article examines future trends and technical obstacles that quantum sensing technologies will face in pharmaceutical drug development processes. Illustrative figures together with diagrams explain quantum sensing mechanisms. Also NV-center-based spectroscopy and AI integration applications are discussed throughout this presentation.

H. R. Arabnia et al. (Eds.): CSCI 2025, CCIS 2970, pp. 519–526, 2027.
https://doi.org/10.1007/978-3-032-28466-2_40

2 Quantum Sensing: Principles and Technologies

Quantum sensing is an emerging sensing technology that applies quantum mechanical phenomena to enhance sensing capabilities when determining certain physical parameters. According to Degen et al. (2017), quantum systems can perform superposition, which lets them concomitantly exist in various states and entanglement creates state correlations between particles throughout long distances. Because of their quantum properties, these sensors precisely measure micrometer shifts in fields ranging from magnetic to electric fields, temperature, and additional physical changes. Quantum sensors supersede classical sensors through quantum coherence enhancement, because they use quantum coherence to exceed fundamental measurement boundaries as established by thermal noise and the uncertainty principle (Ye and Zoller 2024). Fundamental physics, together with biomedical investigation, benefits from quantum sensing due to its distinct operational capability (Figs. 1, 2, 3, 4, 5, 6).

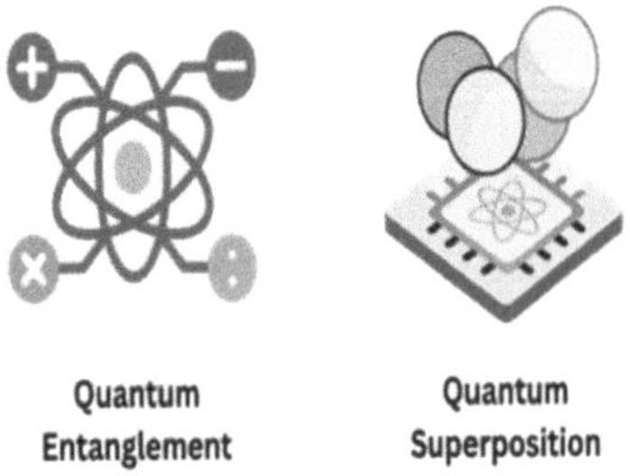

Fig. 1 Quantum entanglement and Quantum Superposition

Recent developments and practical use of multiple quantum sensor systems have occurred in drug discovery research. Tropical Diamond research is actively developing Nitrogen-vacancy centers that provide maximum sensitivity for magnetic field readings at nanometer dimensions. The magnetic character of single molecules becomes detectable due to NV centers, which enables vital analysis of protein structures and molecular bond processes (Rugar et al. 2021). Superconducting quantum interference devices (SQUIDs) function as an essential technology that operates to monitor weak magnetic fields. The bio magnetic fields and molecular dynamics require analysis with SQUIDs to provide information about drug-target interactions (Bao et al. 2023). Scientists continue to study atomic clocks as time-measurement tools because their atom-based vibration capabilities enable precise time measurements for ultrafast molecular research. Nano-sized semiconductor particles known as quantum dots serve extensively for imaging and sensing applications where drug delivery tracking and cell process monitoring represent their main uses.

The superiority of quantum sensing techniques exceeds traditional approaches to a remarkable extent. Quantum sensors entail circle values, making it achievable to discover the molecular interaction and the structural change that occurs on the atomic level (Crawford et al. 2021). Quantum sensors deliver quick results that decrease the duration of vital drug discovery experimental tasks. Quantum sensors can study drug binding

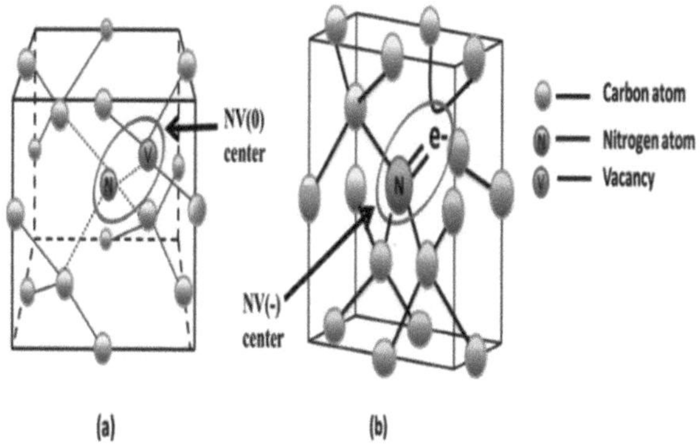

Fig. 2 Nitrogen-vacancy center

evaluation and dynamic structural elements simultaneously while delivering precise drug performance and selectivity information (Kairys et al. 2019). Moreover, the connection of quantum sensing with other innovative technologies, such as AI and ML, can improve predictive thinking and analytics. However, multiple advantages cannot outweigh several limitations like high costs, technical complexity, and specialized expertise (Lee et al. 2021). The continuous progress in quantum technology development aims to solve existing hurdles, thus increasing the usability of quantum sensing methods across drug research activities.

3 Applications in Drug Discovery

Quantum sensing is a clear level of advancement in drug discovery and development as it offers the possibility of improving a variety of elemental steps, from target identification to the method of action of individual ligands, pharmacokinetics, and structure biology. Quantum sensors supply unprecedented structural and dynamic characteristics of drug targets, such as proteins and nucleic acids, for target identification and validation procedures. According to Ali et al. (2023), using NV centers contained within diamonds enables magnetic property detection of proteins to achieve structural and functional information. Scientists use this capability to better determine drug targets and validate them with greater precision, so that researchers can minimize their pursuit of wandering prospects (Zhang et al. 2023). Using quantum sensing in biological systems can assist in the study of the functionality and structure of disease progress and therapeutics. These are efficaciously advanced by raising the possibility of discovering successful targets.

Quantum sensing creates a strong analytical tool that enables scientists to study drug-target binding dynamics together with their specific interaction strength as well as alterations in molecular configuration. For instance, superconducting quantum interference devices (SQUIDs) can record biomagnetic activities originating from molecular interactions and offer information regarding the binding and occurrence of the event in real time (Vettoliere and Granata 2023). The optimization of drug candidates heavily depends on this essential information, because researchers need to know precisely how tightly and selectively drugs attach to their targets. Quantum sensing also allows for the

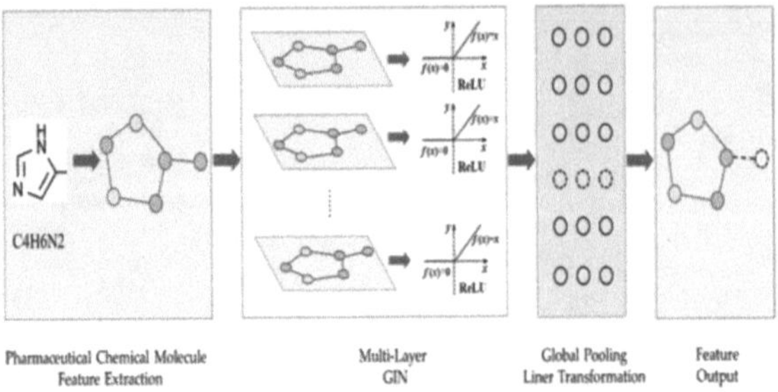

Fig. 3 Drug-target interaction quantum sensing process

investigation of depths, such as protein folding and changes in conformation, besides other procedures which are very hard by normal techniques (Oh et al. 2024). Quantum sensing will help design better drugs since it helps in understanding interactions at the molecular level.

The research field of pharmacokinetics, together with drug delivery, receives meaningful advancements through quantum sensing technology. According to Ahmed et al. (2024), the basic understanding of drug absorption as well as distribution and metabolism and excretion patterns in the body remains a key requirement for drug safety inspections and overall delivery quality. The drug-monitoring capability of quantum dots serves as fluorescent probes that track drug movement inside cells and tissues (Ding et al. 2022). These nanoscale sensors provide precise, real-time insights into drug deposition and metabolism, enabling experts to design optimized drug delivery systems and improve treatment efficacy. Quantum sensing technology enables continuous monitoring of drug release dynamics as well as drug formulation stability and ensuring that drugs arrive at their designated targets as both timed and concentrated correctly.

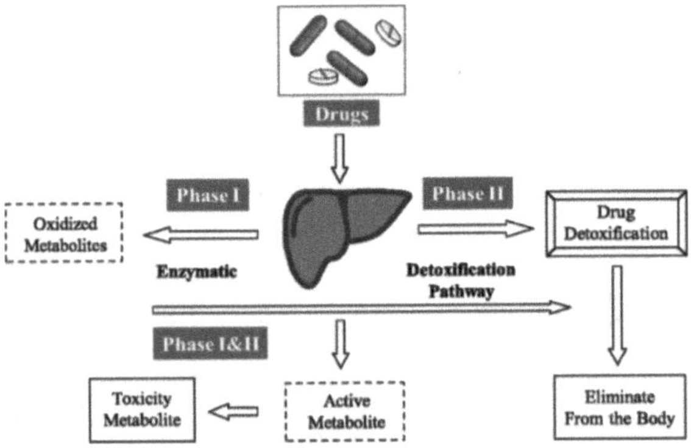

Fig. 4 General pathway of drug metabolism

Finally, quantum sensing is revolutionizing structural biology and drug discovery by providing molecular architecture to biological macromolecules. Forger (2024) observes that atomic clocks measure time with unmatched precision, so scientists use them to study atomic vibrations in proteins through which they gain structural and dynamic information. The knowledge obtained establishes a crucial foundation for designers of rational drugs because it enables them to create drugs that focus on biomolecules' active locations. Degen et al. (2017) observe that quantum sensing helps researchers study membrane proteins and nucleic acids as well as complex biological systems through analysis that would be difficult to achieve with conventional scientific methods. Quantum sensing, therefore, provides an enhancement in knowledge about structures of molecules in the body to enhance the discoveries of newer and precise therapies.

4 Future Directions

The incorporation of the quantum sensor with the help of artificial intelligence (AI) and machine learning (ML) is the way forward for the drug discovery process. Quantum sensors obtain massive high-quality data, and through AI and ML specialists, it is possible to read a pattern and determine an outcome that would have beenimpossible to accomplish using classical methods (Vernuccio et al. 2022). The binding affinity prediction of drug candidates becomes faster through quantum sensing data analysis by AI models, resulting in decreased experimental trial expenses. Policy design optimization through ML algorithms allows researchers to assess quantum sensor-derived structural and dynamic data for developing targeted and effective therapies (Naskar et al. 2025). Quantum sensing, together with AI/ML methods, presents a transformative pathway to enhance drug discovery while it becomes a predictive data-dependent method that simultaneously cuts down development times and expenses.

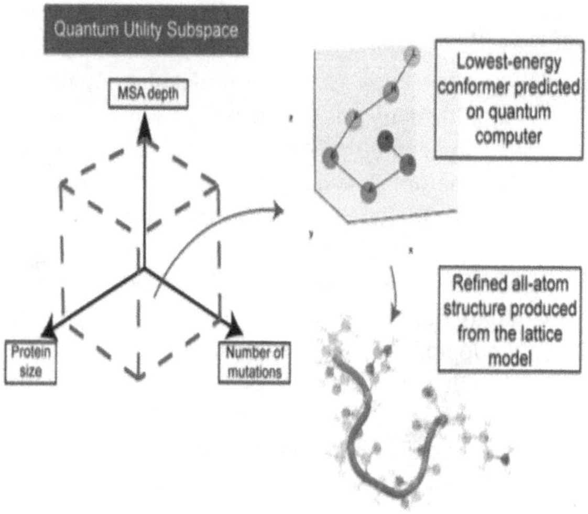

Fig. 5 Protein structure quantum sensing

A further promising area of research is compact quantum sensors for continuous monitoring of the state of health and various industries. The functionality of modern

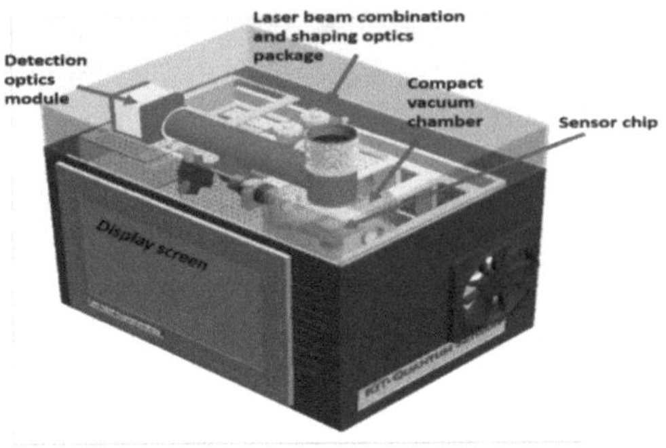

Fig. 6 A Portable quantum sensor

quantum sensing demands both dedicated facilities and specialized instruments, which reduces their practical use. Shukla (2024) asserts that modern miniaturization and quantum hardware development enable the creation of portable sensors that will operate in research facilities, medical facilities, and primary care locations. These portable devices can be used to monitor interactions, pharmacokinetics, and patient response, which helps in enhancing the efficacy and speed of drug development and the effectiveness of medicines. For example, a portable NV center-based sensor can be used to monitor drug targeting in a real-time manner and thus offer instant feedback for drug formulation processes. The pharmaceutical industry's adoption of accessible and easy-to-use portable quantum sensors will expand due to their projected market growth, which will enable widespread use of this innovative technology.

Hence, the integration of quantum sensing (QS) in drug discovery and development holds immense potential for future advancements. By leveraging quantum properties like superposition and entanglement, quantum sensors can achieve unprecedented sensitivity and accuracy, enabling the detection of subtle molecular interactions and structural changes. Future work should focus on enhancing quantum sensor design, exploring new materials, and optimizing configurations to further improve detection capabilities. Additionally, integrating quantum sensing with AI and machine learning can lead to the development of predictive models that accelerate drug discovery and improve the success rate of treatments.

To fully realize the benefits of quantum sensing in drug development, future research should also address scalability and standardization, ensuring that quantum sensing technologies can be widely adopted across the pharmaceutical industry. Interdisciplinary collaboration between quantum physicists, biochemists, and pharmacologists will be crucial in uncovering new applications and overcoming regulatory challenges. Furthermore, exploring the clinical applications of quantum sensing, such as real-time monitoring of drug efficacy, can lead to personalized treatment plans and improved patient outcomes. By pursuing these avenues, researchers can revolutionize the drug discovery process, making it more efficient and effective.

5 Conclusion

Quantum sensing may be the key enabler to change the practice of drug discovery and development by providing better accuracy, sensitivity, and time efficiency when it comes to studying molecular dynamics, structural biology, and pharmacokinetics. Superposition and entanglement of quantum are fundamental principles that help researchers study and model systems of biology and related phenomena beyond the atomic and subatomic level as compared with the available standard techniques. The drug development process benefits from quantum sensing through target identifications and molecular relationship examinations that combine with drug delivery enhancement methods to drive drug advancement and result in better therapeutic performance. The prospect of quantum sensing includes the combination of artificial intelligence, machine learning, portable sensor development, and utilization for personalized medicine to maximize its value. Quantum technology developments continue to tackle cost and technical hurdles so the procedure can become practical and accessible for applications. The future of drug discovery depends on quantum sensing because this technology will establish itself as a foundational modern pharmaceutical discovery approach for global patient care innovation.

References

Ahmed, N., Prajapati, V., Mishra, P., Mishra, N.T.: Dosage form design: from concept to compliance-navigating regulatory standards and patient needs. Int. J. Drug Regul. Aff. **12**(1), 24–35 (2024)

Ali, M.K., et al.: Exploring the multifunctional roles of quantum dots for unlocking the future of biology and medicine. Environ. Res. **232**, 116290 (2023)

Bao, B., Hua, Y., Wang, R., Li, D.: Quantum-based magnetic field sensors for biosensing. Adv. Quantum Technol. **6**(5), 2200146 (2023)

Crawford, S.E., et al.: Quantum sensing for energy applications: review and perspective. Adv. Quantum Technol. **4**(8), 2100049 (2021)

Degen, C.L., Reinhard, F., Cappellaro, P.: Quantum sensing. Rev. Mod. Phys. **89**(3), 035002 (2017)

Ding, R., et al.: Recent advances in quantum dots-based biosensors for antibiotics detection. J. Pharm. Anal. **12**(3), 355–364 (2022)

Forger, D.B.: Biological clocks, rhythms, and oscillations: the theory of biological timekeeping (2024).

Kairys, V., Baranauskiene, L., Kazlauskiene, M., Matulis, D., Kazlauskas, E.: Binding affinity in drug design: experimental and computational techniques. Expert Opin. Drug Discov. **14**(8), 755–768 (2019)

Lee, C., Lawrie, B., Pooser, R., Lee, K.G., Rockstuhl, C., Tame, M.: Quantum plasmonic sensors. Chem. Rev. **121**(8), 4743–4804 (2021)

Naskar, S., et al. (2025). The biomedical applications of artificial intelligence: an overview of decades of research. J. Drug Target., (just-accepted), 1–85.

Oh, E., et al.: Perspective on quantum sensors from basic research to commercial applications. AIAA J. **62**(11), 4029–4053 (2024)

Rugar, A.E., et al.: Quantum photonic interface for tin-vacancy centers in diamond. Phys. Rev. X. **11**(3), 031021 (2021)

Shukla, S.K.: Nanotechnology in Miniaturization: an Emerging Trend to Fabricate Future Devices (2024).

Vernuccio, F., et al.: Artificial intelligence in classical and quantum photonics. Laser Photonics Rev. **16**(5), 2100399 (2022)

Vettoliere, A., Granata, C.: A highly sensitive tunable magnetometer is based on a superconducting quantum interference device. Sensors. **23**(7), 3558 (2023)

Ye, J., Zoller, P.: Essay: quantum sensing with atomic, molecular, and optical platforms for fundamental physics. Phys. Rev. Lett. **132**(19), 190001 (2024)

Zhang, Y., Deng, Z., Xu, X., Feng, Y., Junliang, S.: Application of artificial intelligence in drug-drug interactions prediction: a review. J. Chem. Inf. Model. **64**(7), 2158–2173 (2023)

Proposition on Quantum Computing Applications in Drug Discovery and Development

Lennex Annor(✉) and Atif Mohammad

Department of Computer Science, Capitol Technology University, Laurel, USA
lannor@captechu.edu

Abstract. Quantum computing is emerging as a transformative technology poised to redefine the future dynamics of drug discovery and development. This paper proposes how quantum computing potentials could influence pharmaceutical research, from molecular modeling to drug optimization. It employs contemporary literature to elaborate on two quantum algorithms: quantum annealing, which is used for optimization problems, and variational quantum eigensolvers, which are used for solving eigenvalue problems. It discusses their potential to outperform classical computing techniques. Under this paper's themes of strengths, weaknesses, and direction for improvement, the capabilities and weaknesses of applying quantum computing to drug discovery are assessed, along with the steps required to overcome these weaknesses. Additionally, this paper describes various application areas, quantum learning models, and how quantum computing can transform pharmaceutical value chains. The paper comprehensively reviews existing research, highlights practical quantum computing applications in pharmaceutical development, such as drug-target interaction studies, and offers suggestions for advancement.

Keywords: Quantum Computing · Drug Discovery · Quantum Chemistry · Quantum Machine Learning · Molecular Optimization · Drug Development

1 Introduction

Quantum computing (QC) is emerging as a revolutionary technology with the potential to significantly transform the landscape of drug discovery and development. By leveraging the principles of quantum mechanics, QC can address complex computational challenges that are currently beyond the reach of classical computers. This paper delves into how quantum computing can impact various stages of pharmaceutical research, ranging from the initial modeling of molecules to the optimization of drug candidates. Two key quantum algorithms are highlighted: quantum annealing and variational quantum eigensolvers. Quantum annealing is particularly effective for optimization problems, which are crucial in drug design where the goal is to find the most stable and effective molecular configurations. Variational quantum eigensolvers, on the other hand, are instrumental in solving eigenvalue problems, which are essential for understanding the energy states of molecules and their interactions.

H. R. Arabnia et al. (Eds.): CSCI 2025, CCIS 2970, pp. 527–533, 2027.
https://doi.org/10.1007/978-3-032-28466-2_41

The paper critically evaluates the strengths and weaknesses of applying quantum computing to drug discovery, providing a balanced perspective on its current capabilities and limitations. It identifies areas where quantum computing shows promise, such as in drug-target interaction studies, where understanding the binding affinities between drugs and their targets is vital. However, the paper also acknowledges the challenges and weaknesses, including the need for more robust quantum hardware and error correction methods. To overcome these obstacles, the paper suggests several directions for improvement, such as advancing quantum algorithms and integrating quantum computing with classical methods to create hybrid approaches. Additionally, the paper explores various application areas and quantum learning models, illustrating how quantum computing can transform current value chains in the pharmaceutical industry. By reviewing existing research and offering practical insights, the paper aims to guide future advancements in this burgeoning field, ultimately paving the way for more efficient and innovative drug discovery processes.

Drug discovery is a lengthy and costly process, and the computational challenges of predicting molecular interactions often limit its efficiency. While classical computational systems are reliable in some contexts, they struggle to solve problems related to molecular structures and interactions at the quantum level. Quantum computing, however, offers a promising alternative. Using qubits, which can exist in superposition, quantum computers significantly accelerate computations compared to classical systems [1].

This paper argues that quantum computing has the potential to revolutionize drug discovery by enhancing molecular simulations, improving drug design, and enabling faster screening of therapeutic compounds. The aim is to demonstrate that employing quantum computing in drug discovery is not only feasible but also offers valuable insights, identifies limitations, and uncovers new opportunities. This work synthesizes and extends previous research from both academic and industrial perspectives.

2 The Role of Quantum Computing in Drug Discovery

Quantum computing leverages the principles of quantum mechanics to solve problems in complex systems that are beyond the reach of classical computing. This is particularly advantageous in drug discovery, where the interaction of molecules can be highly intricate and difficult for classical algorithms to manage [2]. The paper below highlights three key areas where quantum computing can directly influence the drug discovery pipeline: molecular simulations, optimization of molecular structures, and quantum machine learning (QML).

A. Quantum Chemistry and Molecular Simulations

The core of drug discovery lies in the ability to model molecules and their interactions at various levels with sufficient precision. Quantum computers, in particular, can simulate quantum mechanical systems far more effectively than classical computers. Two of the most promising quantum algorithms currently under consideration for this purpose are variational quantum eigensolvers (VQE) and quantum phase estimation (QPE) [3]. These algorithms enable more accurate solutions to the Schrödinger equation, which governs molecular behavior at the quantum level [4]. As a result, the molecular properties predicted by these algorithms are more precise than those obtained through physical simulations in quantum chemistry.

However, one of the challenges in applying quantum computing to drug discovery is the limited number of qubits that can be processed by current quantum computers. This limitation makes it difficult to simulate large molecules effectively. The primary obstacles to progress in this area are quantum error correction and qubit coherence [5]. To address these issues, new hybrid quantum-classical approaches are being developed, where certain aspects of the simulation are handled by classical devices while the most computationally intensive tasks are performed by quantum processors [6] (Figs. 1, 2).

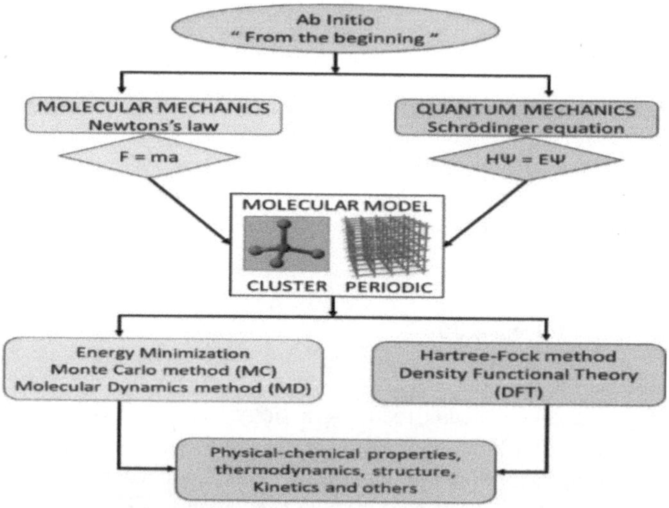

Fig. 1. Diagram of molecular mechanics and quantum mechanics methods(https://www.researchgate.net/publication/338167568_Trends_in_Modeling_Design_and_Optimization_of_Multiphase_Systems_in_Minerals_Processing?_tp=eyJjb250ZXh0Ijp7ImZpcnN0UGFnZSI6Il9kaXJlY3QiLCJwYWdlIjoiX2RpcmVjdCJ9fQ)

B. Quantum Machine Learning for Drug Design

Over the years, machine learning has been widely used in drug discovery to predict the interaction between drugs and targets, analyze molecular data, and optimize potential drug compounds. It furthermore extends this into quantum machine learning (QML), where data throughput and pattern recognition are improved through quantum-enhanced algorithms. There are quantum generative adversarial networks (QGANs) and quantum support vector machines (QSVM) that are being researched to help design and create new molecules that possess desirable characteristics quickly [7]. There is also increased implementation of QML for drug repurposing, that is, identifying drugs in the existing M&M database, which may be helpful for new targets, thus lowering the time and cost of new treatments.

Wang et al. [8] have stressed that QML models make a difference when it comes to analyzing big data, which researchers can use to come up with new patterns in molecular interfaces. As [9] pointed out; however, most of QML's possibilities are still in the concept stage because quantum computing is not yet mature enough to realize its full potential.

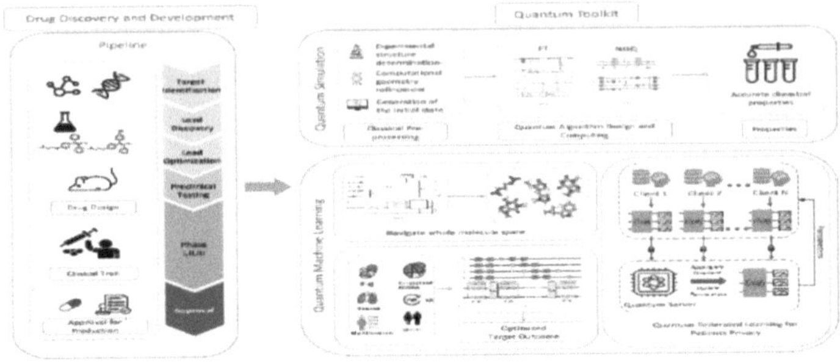

Fig. 2. The process includes target identification, lead discovery, lead optimization, preclinical testing, clinical trials, and approval. Quantum Simulation and QML enhance this process by providing accurate chemical properties, navigating molecular space, and optimizing clinical trials through advanced quantum algorithms. Quantum federated learning ensures secure data handling across multiple clients (https://arxiv.org/html/2408.13479v2).

C. Optimization of Molecular Structures

Optimization is the final stage of drug discovery, focusing on enhancing the binding of the drug to its target molecule, reducing toxicity, and improving pharmacokinetic properties. Classical algorithms often struggle to optimize these parameters simultaneously due to the complexity of the problem [10]. Quantum algorithms such as Quantum Approximate Optimization Algorithms (QAOA) and quantum annealing are considered well-suited to solving these complex optimization problems.

For example, quantum annealers, like those created by D-Wave Systems, are already being tested to solve combinatorial optimization problems (COPs) more efficiently than classical algorithms. However, scalability remains a limitation, as current systems can only accommodate small molecular datasets [6]. This aligns with the work of [11], who suggested that as quantum processors grow in power, they may provide polynomial speedups, offering new solutions to previously challenging optimization problems in drug design.

3 Opportunities for Quantum Computing in Drug Discovery

Given the current state of quantum computing applications in drug discovery, several areas show promise for positive impact. These opportunities align with the pharmaceutical industry's need for faster, more effective drug discovery methods.

A. Accelerated Molecular Screening

One of the most time-consuming tasks in drug discovery is screening large compound databases for promising leads. Quantum computing can expedite this process through quantum parallelism, enabling the simultaneous testing of multiple candidate molecules. This could reduce screening time from months or years to days [12]. Additionally, quantum-based molecular docking methods could predict the binding behavior of a drug to its receptor more accurately than traditional approaches, allowing researchers to identify the best drug candidates early in the development process, minimizing the risk of failed trials [9].

B. Quantum-Enhanced Drug Design

Developing a new drug requires fine-tuning multiple parameters simultaneously, including potency, selectivity, and stability. Quantum computers, with their ability to efficiently explore large solution spaces, offer significant advantages in this area. By applying quantum algorithms, researchers can design drugs with a higher likelihood of success in clinical trials, reducing costs associated with late-stage failures [1]. Furthermore, quantum computing can support the development of precision medicine by enabling drugs to be tailored to individual patients' genetic profiles. Even though quantum computers deal with molecular interactions, simulations could help design effective treatments for patients based on their unique molecular makeup [13].

4 Challenges and Limitations

Several obstacles need to be overcome for quantum computing to reach its full potential in the pharmaceutical industry. The most significant challenge remains hardware related. Current quantum computers, known as noisy intermediate-scale quantum (NISQ) devices, are error-prone and can only handle a limited number of qubits. These limitations reduce the scale of simulations and hamper the efficiency of quantum algorithms [13]. Developing quantum processors that are resilient to computational errors is a critical step forward.

Moreover, quantum algorithms require specialized knowledge of both quantum mechanics and computer science. As the field matures, there will be a growing demand for more user-friendly programming interfaces, allowing scientists—particularly in drug discovery—to apply these algorithms without needing deep expertise in quantum mechanics [14]. Making these tools accessible to non-specialists will be essential for maximizing the impact of quantum computing in drug discovery.

While current quantum systems can simulate small molecules, scaling these simulations to accommodate larger and more complex molecules remains difficult. Quantum algorithms must be robust enough to handle this added complexity in large-scale pharmaceutical research. Until quantum computers are fully error-resistant, hybrid quantum-classical approaches will remain necessary for many applications [5].

Drug-Target Interaction Studies Another challenge is determining how well a drug molecule binds to its target (e.g., a protein receptor) requires analyzing countless configurations and energy landscapes. Quantum computing can help calculate binding energies and simulate drug-target interactions with greater accuracy, allowing pharmaceutical companies to identify the most promising candidates faster.

5 Future Directions

Despite these challenges, the future of quantum computing in drug discovery appears promising. To accelerate the adoption of quantum computing, researchers and pharmaceutical companies should focus on several key areas. As quantum hardware improves, the ability to simulate larger and more complex molecules with greater accuracy will be possible. Quantum processors capable of correcting errors while maintaining functionality will be essential for this advancement [1].

Given current hardware constraints, exploring hybrid quantum-classical architectures is a practical approach. Combining the strengths of classical and quantum computers can address drug discovery challenges that neither system can solve alone [2]. Further research into quantum machine learning (QML) is also crucial to advancing its application in drug discovery. Integrating QML with classical models may provide significant advantages in drug-target interaction prediction and the optimization of drug candidates [7]. These advancements will be key to developing personalized medicine and improving patient outcomes.

6 Conclusion

Emerging technologies like quantum computing offer exciting opportunities for drug discovery and development by accelerating molecular simulations and improving drug design and screening processes. As researchers unlock the potential of quantum algorithms and address the limitations of current quantum hardware, the pharmaceutical industry is on the brink of a revolution that will significantly accelerate and reduce the cost of drug development. This innovation will require close collaboration between academia, industry, and quantum computing developers.

[1] S. M. K. A. P. F. a. S. A. K. Dhanalekshmi Unnikrishnan Meenakshi, 2024.

[2] K. Hartman-Caballero, 2021.

[3] F. D. P. B. J. E. L. B. a. L. F. M. Zinner, 2021.

[4] C. V. M. F. S. N. P. R.-V. a. A. P. T. Cova, 2021.

[5] P. P. a. B. C. B. Bonde, 2023.

[6] R. L. a. M. M. H. Salloum, 2024.

[7] S. C. a. A. Rajawali, 2023.

[8] J.-H. C. Y.-Y. Y. C. L. a. Y. J. T. P.-H. Wang, 2023.

[9] P. Sharma, 2024.

[10] A. P. e. al., 2023.

[11] A. G. a. R. G. M. F. Bova, 2021.

[12] P. Hassanzadeh, 2020.

[13] M. Raparthi, 2022.

[14] P. S. M. R. M. A. M. a. S. V. 1R. Mishra, 2024.

References

1. Bonde, B., Patil, P., Choubey, B.: The future of drug development with quantum computing. High Perform. Comput. Drug Discov. Biomed., 153–179 (2023). https://doi.org/10.1007/978-1-0716-3449-3_7
2. Bova, F., Goldfarb, A., Melko, R.G.: Commercial applications of quantum computing. EPJ Quantum Technol. **8**(1), 2 (2021). https://doi.org/10.1140/epjqt/s40507-021-00091-1
3. Chen, S., Rajawali, A.: Leveraging the power of quantum computing and machine learning to disrupt drug development. Q. J. Comput. Technol. Healthc., 8(4), 1–11 (2023). https://vectoral.org/index.php/QJCTH/article/view/53. Accessed 17 Oct. 2024
4. Cova, T., Vitorino, C., Ferreira, M., Nunes, S., Rondon-Villarreal, P., Pais, A.: Artificial intelligence and quantum computing as the next pharma disruptors. Artif. Intell. Drug Des., 321–347 (2022). https://doi.org/10.1007/978-1-0716-1787-8_14
5. Hassanzadeh, P.: Towards the quantum-enabled technologies for development of drugs or delivery systems. J. Control. Release. **324**, 260–279 (2020). https://doi.org/10.1016/j.jconrel.2020.04.050
6. Meenakshi, D.U., Kesavan, S.M., Francis, A.P., Khan, S.A.: Quantum computational concepts and approaches in drug discovery, development and delivery. Drug Deliv. Syst. Using Quantum Comput., 1–25 (2024). https://doi.org/10.1002/9781394159338.ch1
7. Mishra, R., Mishra, P.S., Mazumder, R., Mazumder, A., Varshney, S.: Quantum computing and its promise in drug discovery. Drug Deliv. Syst. Using Quantum Comput., 57–92 (2024). https://doi.org/10.1002/9781394159338.ch3
8. Pyrkov, A., et al.: Quantum computing for near-term applications in generative chemistry and drug discovery. Drug Discov. Today. **28**(8), 103675 (2023). https://doi.org/10.1016/j.drudis.2023.103675
9. Quantum Computing and Drug Discovery (2021). https://cklixx.people.wm.edu/teaching/math400/Kylee-2.pdf. Accessed 17 Oct. 2024
10. Raparthi, M., et al.: Harnessing quantum computing for drug discovery and molecular modelling in precision medicine: exploring its applications and implications for precision medicine advancement. Adv. Deep Learn. Tech. **2**(1), 27–36 (2022) https://thesciencebrigade.com/adlt/article/view/136
11. Salloum, H., Lukin, R., Mazzara, M.: Quantum computing in drug discovery: a review of quantum annealing and gate-based approaches. In: International Conference on Computational Optimization. https://openreview.net/forum?id=PHRRdOdUya. Accessed 17 Oct. 2024
12. Sharma, P.: Quantum computing in drug design: enhancing precision and efficiency in pharmaceutical development. Sage Sci. Rev. Appl. Mach. Learn. **7**(1), 1–9 (2024) https://journals.sagescience.org/index.php/ssraml/article/view/120
13. Wang, P.H., Chen, J.H., Yang, Y.Y., Lee, C., Tseng, Y.J.: Recent advances in quantum computing for drug discovery and development. IEEE Nanotechnol. Mag. **17**(2), 26–30 (2023). https://doi.org/10.1109/MNANO.2023.3249499
14. Zinner, M., Dahlhausen, F., Boehme, P., Ehlers, J., Bieske, L., Fehring, L.: Quantum computing's potential for drug discovery: early stage industry dynamics. Drug Discov. Today. **26**(7), 1680–1688 (2021). https://doi.org/10.1016/j.drudis.2021.06.003

The World's First LSD Discrimination Theory Using RIP (Combinatorial Optimization LDF)

Shuichi Shinmura(✉)

Seikei University, Tokyo, Japan
sshinmura2@gmail.com

Abstract. Since 1971, we have been developing combinatorially optimized **LDF** (**RIP**: Revised IP-optimal LDF) to minimize data-specific misclassification. Through six empirical studies, we have completed the world's first discrimination theory for linearly separable data (**LSD**). By identifying and excluding misclassified cases, all discriminant data becomes LSD. By applying the same algorithm as Yamanaka's four genes, we select the smallest LSD (Basic Gene Set). **BGS** solves many problems in science, engineering, industry, and business, et al.

Keywords: 169 microarrays · vital BGS of Multivariate Oncogene · Exam's pass/fail determination · Authenticating Swiss banknotes · Legal classification of industrial products · Species and BGSs of plants and animals · Case selection method of misclassified medical data (**Method4**)

1 Introduction

Fisher developed a linear discriminant function (**LDF**) based on the hypothesis that two groups are normally distributed with different means and evaluated it using three species of iris with four variables. The matrix scatterplot in Fig. 2 shows that the variables (x1, x2), x3, and x4 separate the two groups into triplicates with an LSD. The minimum number of misclassification (**MNM**) for the triplicates is 0. We call the LSD with the minimum variables as Basic Gene Set. BGS plays the same role as Yamanaka's four factors. (x1, x2) is a bivariate BGS, while x3 and x4 are two univariate BGSs. In decision trees (**DT**), x3 and x4 have NM = 0. The NMs of LDFs such as SVM [41] are larger than 1. These are worse than the NM of DT. Furthermore, various LDFs have different NMs for the same data, making them unreliable. This deficiency is the first serious problem in discriminant theory [21]. MNM is the only statistic in LSD. In multiple regression analysis, the dependent variable (y) is predicted using the explanatory variables. It is correct to assume a normal distribution, as discovered by Gauss. However, LDF is a classification method for determining whether an external criterion is correct. Because NM can only take n integer values from 0 to (n-1), "discriminant analysis is a combinatorial problem." For example, the LDF discrimination of **CPD** is a descriptive method for assessing whether physicians can correctly discriminate between external criteria (180 natural births and 60 cesarean sections) using 19 variables [6, 10, 13–16, 25, 38]. Results can be evaluated using **NM** or error rate (**ER**) [5] using k-fold CV [18].

H. R. Arabnia et al. (Eds.): CSCI 2025, CCIS 2970, pp. 534–555, 2027.
https://doi.org/10.1007/978-3-032-28466-2_42

SAS RSQUARE [3] selected 16 variables without multicollinearity from 2^{19} models. The combinatorial optimization LDF (RIP) found an optimal convex polyhedron (**OCP**) with MNM 2 from domains of 2^{240} convex polyhedra (**CPs**) [38]. Removing two pregnant women resulted in an LSD. Next, a 14-variable BGS without multicollinearity was found from 2^{19} discriminant models. This was selected using the same elimination method as Yamanaka's four-factor approach for 238 women. This BGS was significantly superior to a 16-variable model with two misclassifications. Clinicians could investigate the reasons for the misclassified women, and RIP was better at finding the optimal model than RSQUARE. Furthermore, RIP using the Branch & Bound method of integer programming analyzed 169 microarrays (maximum n = 357 * p = 22,278) and found 169 LSD. Because they are LSD, they become better evaluation problems for quantum computers than the traveling salesman problem (**TSP**). Until quantum computers are completed, the B&B algorithm by LINGO [9] is the best for combinatorial optimization problems. Even after quantum computers are developed, LSD can be solved quickly, enabling high-quality genetic analysis at low cost.

We have conducted many empirical studies on cancer gene diagnoses using 169 DNA/microarrays [26–40], pass/fail determination for the National Center Test for University Admissions, my exams [20, 25], classification of industrial products as determined by law, and authenticity detection of Swiss banknotes [2, 25]. Using the world's first LSD discriminant theory [12–40], we can identify the minimum causal variables and clarify the cause of defects, such as defective and normal processes of factories. LSD discriminant theory is to become one of the most essential statistical methods, with a wide range of applications. Assuming basic knowledge of high school mathematics, students can easily understand the LSD theory and its applications.

2 The world's First LSD Discriminant Theory Using RIP

2.1 ChatGPT's Wrong Information on LSD Research

LSD is a fundamental concept in science and technology, yet no one has studied it except for us. ChatGPT suggests to us the survey of the LSD study in SVM. Because H-SVM can separate only LSD and stop for other data, most people ignore it. Maximized soft-margin SVM accepts the sum of the distances from the discriminant hyperplane of cases misclassified by the two SVs multiplied by a penalty c. It can separate between the two classes if by trial and error and selecting an appropriate penalty-c value from the interval [0, 1]. Ultimately, the success of SVM lies in the ability of kernel-SVM to handle both LDFs and nonlinear discriminant functions. Still, no one confirms that it can always find LSD. This paper explains the simple, perfect RIP that can discover the MNM = k unique for data. In six empirical studies, we show that LSD theory is a revolutionary method with a wide range of applications.

When a problem occurs, you classify the subjects into two classes and separate them. Suppose you want to find the cause of a faulty process. You can analyze variables that indicate deviations from a normal process and identify the cause by finding the BGS. When repeating experiments, you can discriminate between successful and unsuccessful results after each experiment and determine the cause of success using BGS.

RIP splits the discriminant coefficient space into 2^n CPs using the n linear discriminant hyperplanes Hi shown in Fig. 1 and finds the OCP with MNM = k. k = 0 results in LSD. If k is not 0, there are unique misclassified cases. After omitting these k $\mathbf{x}_i$, all discriminated data becomes LSD. This can be achieved by using binary integers $e_i = 1/0$ for misclassified/correctly classified $\mathbf{x}_i$. Using the same backward stepwise method used to find the Yamanaka four genes, the BGS of the minimum-dimensional LSD from a 2^p-pair discriminant model is found. Until quantum computers are active, only the B&B algorithm can solve combinatorial optimization problems. Because RIP can quickly see that 169 arrays [1, 4] are LSDs, the cost-performance is better than that of high-cost quantum computers. Furthermore, empirical studies on five kinds of LSDs were conducted [38], demonstrating a practical and straightforward LSD theory that had not previously been studied. RIP and 169 arrays offer many more correct evaluation problems based on LSD for quantum computers.

2.2 RIP that Can Discriminate both LSD and Misclassified Data

RIP defined in (1) minimizes NM (MIN Σe_i) under (n*p) simultaneous inequality constraints and finds the OCP corresponding to MNM. Figure 1 shows that RIP identifies OCP (SM1: Small Matryoshka1) as the domain of RIP among many CPs in LDF's coefficient space. RIPs can quickly solve large-scale combinatorial problems for LSD.

$$\text{MNM} = \text{MIN}(\Sigma ei);\ yi * (\ \mathbf{xi}\ \mathbf{b} + 1) \geq 1 - 1000 * ei;\ \ldots\ldots.\ \text{for } i = 1, .., n. \quad (1)$$

The constant term c in LDF (= **xb** + c) can be fixed at 1 because dividing LDF by c does not change the discriminant result. If LDF > 0 for class1 and LDF < 0 for class2, it becomes LSD (MNM = 0). To avoid using two inequalities, two classes are expressed as an external criterion, $y_i = 1/-1$. If LDF = $y_i * (\mathbf{xb} + 1) > 0$, it becomes LSD. However, because we cannot strictly decide whether LDF = 0 through numerical calculation, we define LSD when LDF ≥ 1. There is no $\mathbf{x}_i$ in the open interval (−1, 1). If MNM ≥ 1, H-SVM cannot compute and immediately terminates. Many SVM users avoid H-SVM and do not study LSD for this reason. S-SVM is a complex two-objective optimization LDF that multiplies the sum of the distances of misclassified $\mathbf{x}_i$ from the discriminant hyperplane of SV with a penalty c. Choosing an appropriate c through trial and error can discriminate LSD by chance. Vapnik [41] and his followers focus on kernel SVMs. They use it to accommodate LDF and nonlinear discriminant functions, but they must always confirm to find LSD using RIP results.

RIP is superior to both H-SVM and S-SVM. The constant term (1–1000* e_i) can analyze both classified cases for $e_i = 0$ and misclassified cases for $e_i = 1$. The $e_i = 1/0$ is binary integer variables. This allows for the identification of misclassified $\mathbf{x}_i$ unique to the data. IP finds the MNM by minimizing the sum of misclassified $\mathbf{x}_i$ in the objective function. In Japanese high school mathematics, Linear Programming is taught through the max/min problems of the domain. To solve many problems in the world, actual data, scientists need knowledge of both statistical and IP software. The e_i is the most critical variable in RIP. With these advantages, RIP can, for the first time, investigate the causes of misclassifications, achieving the true purpose of LSD theory.

2.3 Fact1: the Relation of MNM and OCP on P-Dimensional Coefficient Space

Consider three bivariate cases, $\mathbf{x}_i$. Substituting the three $\mathbf{x}_i$ in LDF, $\mathbf{x}_i$ b + 1 = 0 forms a straight line in the two-dimensional discriminant coefficient space. We call them three linear discriminant hyperplanes Hi. Hi divides the discriminant coefficient space into 7 (=2^3–1) CPs. '-1' is the null model with no variables. H1, which slopes downward to the right, divides it into the lower-left + half-plane, where the discriminant score (**DS**) of $\mathbf{x}_1$ is positive, and the upper-right - half-plane, where the DS of $\mathbf{x}_1$ is negative. The infinite points in the lower + half-plane correspond to infinite LDFs that correctly discriminate $\mathbf{x}_1$ in the data space. The upper half-plane corresponds to infinite LDFs that misclassify $\mathbf{x}_1$. The numbers 0 through 2 written in Fig. 1 represent NM. Because NM can only take three integers, 0, 1, and 2, discriminant theory should be considered in terms of combinatorial theory rather than normal distribution. NM = 3 is the same as NM = 0. Only RIP finds an OCP where MNM = k. This data becomes an LSD because a triangular OCP is k = 0. If k is not 0, the data is misclassified. Although Fig. 1 is simple, there are 2^{240} CPs in the CPD data (240*19). Furthermore, we must consider other combinations of 1 to 19 variables. Fig. 1 explains the relation between NM and LDF, which becomes a very important **Fact1** for LSD study.

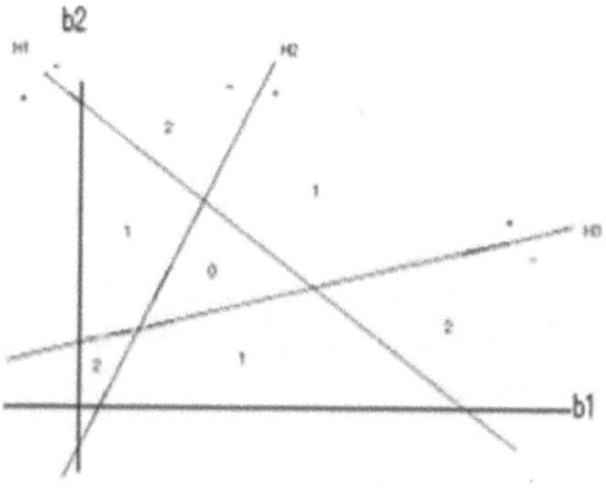

Fig. 1 7-CPs' NM of 3 cases and 2 variables (Fact1)

RIP is a combinatorial optimization LDF that finds the OCP with MNM among 2^n CPs using (1–1000 * e_i). MNM = 0 is the only statistic that defines LSD. Other LDFs, such as H-SVM, S-SVM, RDA, and Lasso, also find one CP among seven CPs. We have published a summary of LSD research from 1971 to 2024 on Irises (150 * 4), Swiss banknotes (200 * 6) [25], pass/fail judgments on numerous exams [17–21], and 29 regular cars and 19 compact cars (48 * 6) [25]. From 2015, we separated six first-generation US arrays (maximum 173 subjects * 12,625 genes) measured up to 2004 [4, 22, 26–29] and found that they were all LSDs. Program3 splits them into many SMs of n or fewer genes. Each SMs included the smallest LSDs (BGSs), which explained the Matryoshka dolls (**Fact3**) [26–39]. Furthermore, in 2024, we confirmed that 163 second-generation new arrays (maximum 357 subjects * 54,676 genes) measured after 2004 were all LSDs and completed the LSD Theory [31–40]. RIP could quickly analyze the optimal combination problems of 2^n and 2^p pairs using the B&B algorithm because of LSD. These 169 results provide good evaluation data for quantum computers because RIP is well-defined and arrays are open to everyone.

2.4 Comparison of RIP and Two SVMs in LINGO Programs

We developed four programs [38] with different purposes using LINGO, an MP software developed at the University of Chicago [9]. They enable selective execution of four LDFs (RIP, H-SVM, S-SVM, and LP-OLDF) as defined in the SUMMODEL section. This section explains the difference between RIP and SVM. We canceled my Amazon Kindle contract to distribute a free Japanese translation of LINGO [9] download from LINDO Japan (https://lindo.jp/). You can download the original book from LINDO (https://lindo.com/). The SUB MODEL RIP: to END SUBMODEL defines RIP by setting BigM = 1000. Within the CALC section, "@SOLVE (RIP);" executes RIP, while "@SOLVE (SVM);" executes H-SVM or S-SVM. The next "MIN = "statement calculates the MNM using the objective function described in (1), and the next statement is the n simultaneous inequalities constraint with p variables. "@FREE(VARK(J1)));" specifies coefficient **b**, which is a free variable with no restriction on the discriminant coefficient $(-\infty, \infty)$. "@FOR(N(i):@BIN(E(i));" specifies a binary integer variable e_i.

```
SUBMODEL RIP:
MIN = @SUM(N(i):E(i));
@FOR(N(i):@SUM(P1(J1):IS(i,J1)*VARK(J1)*CHOICE(J1)) > 1- BigM *E(i));
@FOR(P1(J1):@FREE(VARK(J1))); @FOR(N(i):@BIN(E(i)));
ENDSUBMODEL
```

The following SVM becomes an H-SVM if BigM is set to 0 in the DATA section. Instead of the method described in Sect. 2.2, the discriminant coefficient **b** is calculated using Vapnik's Quadratic Programing model. However, it cannot handle misclassifications. S-SVM can handle both types of data if the BigM (penalty c) value is chosen appropriately, but it has the serious drawback of not being able to identify data-specific misclassifications because it summarizes misclassified cases as the sum of their distances from the discriminant hyperplane using @SUM(N(i):E(i)). However, the complex explanations of H-SVM and S-SVM using QP are difficult for high school students. SVM researchers are satisfied with those explanations and haven't conducted empirical research as seriously as we have.

```
SUBMODEL SVM:
MIN = ER; ER = @SUM(P1(J):VARK(j)^2)/2 + BigM*@SUM(N(i):E(i));
@FOR(N(i):@SUM(P1(j):IS(i,j)*VARK(j)*CHOICE(j)) > = 1-E(i));
@FOR(P1(j):@FREE(VARK(j))); ENDSUBMODEL
```

2.5 Four LINGO Programs

We made four LINGO programs to execute RIP, H-SVM, S-SVM, and LP-OLDF.

Program 1 of RIP separates between two classes and selects the OCP with MNM = k from the 2^n CPs. If k = 0, it means LSD. If k is over 1, RIP finds k misclassifications specific to the data. One of the actual goals of discriminant theory is that the analyst, who has set **external criteria** $\mathbf{y_i}$, examines the reasons for the k misclassifications and improves the quality of the data. After that, omitting k cases results in LSD for all discriminant data. This "case selection Method4" infinitely expands the scope of application of LSD theory. When RIP separated the **Shipp** array (77*7129) [40] on Oct. 28, 2015, only 32 RIP coefficients were nonzero, and 7,097 became zero in 1 s because of LSD. We need no other Feature Selection (**FS**) and test statistics. We call it the 32-gene SM1 (one of OCPs), which can separate cancer and normal classes or two different cancers. We uploaded 15 technical reports to **ResearchGate** [22] by Dec. 25, 2015. We developed the Matryoshka feature selection method (**Method2**) to find all exclusive SMs automatically, which became **Program3**.

Program2 evaluates the selected models using the average error rate (**M2**) on the validation sample [18]. While we used a 100-fold CV for ordinary data ($n \geq p$), we switched to a ten-fold CV midway through the microarray analysis. In many cases, BGS1 becomes the **best model** with the smallest M2. The best model is the first successful model satisfying **Occam's razor**. Many studies use the wrong k-fold CV, violating statistical basic principles. It cannot find the best model. Analyst copies the original data k-times. It is the **pseudo-population** (**validation sample**). Sort it and split the k training samples. We discriminate k training samples and obtain k RIPs. Next, we discriminate the validation sample by k RIPs and obtain the k ERs. M2 is the average of these k ERs. This is the correct k-fold CV. We select the best BGS1 with the lowest M2 and confirm that BGS1 with M2 is the best model among all models.

For ordinary LSD ($n \geq p$), **Program3** finds several SMs with **p or fewer variables**. For high-dimensional LSD ($n < < p$), it finds many SMs with **n or fewer variables**. Statisticians and engineers wrongly claim that to find multivariate oncogenes separating cancer from normal is difficult because they are buried in a massive amount of noise. In reality, because we found many multivariate oncogenes candidates, we needed to develop five validations to select true cancer genes from among many SMs.

Program4 finds BGSs, which are the smallest LSDs among all LSDs. Until now, no one has found BGSs. New FS is performed based on the criterion of whether or not an LSD exists, the same as in iPS research, which determines whether pluripotent stem cells can be produced or not. In iPS research, Prof. Yamanaka identified 24 genes activated in rat ES cells from about 30,000 human genes using the gene DB. Also, Program1 found the first SM1 (32 genes) from the Shipp array. It achieves both successes for high-dimensional and ordinary data. Next, 24 sets of 23 genes were co-cultured; if pluripotent cells were formed, the omitted genes were deemed useless. Finally, the Yamanaka 4 genes were found to be the minimum set required to create pluripotent cells. Program4 uses this algorithm to perform the backward stepwise method. The statistical stepwise using R^2 cannot find Yamanaka's four genes and BGSs. The fact that pluripotent cells can be formed and two classes become LSDs are the same obvious facts without test statistics. Both studies are based on the same gene expression, also.

In 2024, we completed the LSD discriminant theory [39] and subsequently discovered two critical points:

1) We discovered that wide-column LSDs ($n < p$), including high-dimensional LSDs ($n < <p$), have SMs with n or fewer variables [28]. Furthermore, each SMs includes one of the BGSs. We refer to this structure as the Matryoshka doll structure (Fact3) and illustrate it using the Iris data. While ordinary LSDs ($n > p$) have Fact3 consisting of

SMs with p or fewer variables. Both types of SMs include several BGSs. However, a Matryoshka doll includes only one BGS.

2) Around 1978, we were asked to statistically examine the correctness of the Suzumura method for determining the delivery method for pregnant women with CPD data (240*19) [6, 10, 13, 16]. It consists of 19 variables (age, number of births, number of the sacrum, and 16 X-ray measurements) for 180 natural births and 60 cesarean sections. Over six months, we used **SAS RSQUARE** to perform 2^{19} multiple regression models, selecting 342 (=19*18) models with the top 19 highest R^2 values for one to 18 variables. We then selected 38 models using the backward and forward stepwise methods with large R^2 and selected 16 variables without multicollinearities using PCA and test statistics such as AIC. On the other hand, MNM monotonically decreased from MNM = 20 for one variable to MNM = 2 for 19 variables. While writing [39], we excluded these two cases and used 238 subjects to find 14-variable BGS without multicollinearities using Program4. We realized that this "case selection Method4" will expand the scope of application of discriminant analysis after 2025.

3 Fisher's LDF and Iris Data (150 Cases * 4 Variables)

3.1 Three LSDs Confirmed by Three BGSs

Fisher formulated the LDF based on Fisher's hypothesis and verified it using the Iris data. The matrix scatterplot in Fig. 2 shows that 50 cases each of Sctosa G1 (red), Versicolor G2 (blue), and Virginica G3 (green) form three LSDs (G1&G2, G1&G3, and G1&G23). The scatterplot easily finds that G2 and G3 are not LSD. Two univariate X3 and X4 can find the three LSD. One bivariate (x1, x2) finds G1 and G23 as LSD.

Multiple regression analysis ($y_i = b*x_i + c$) is a method that treats the measured value y_i as the dependent variable and predicts it using the linear equation. The expected value y_i follows the normal distribution (Gaussian distribution), which led to the development of FS and test statistics. LDF, which uses the same linear equation, is a descriptive method for predicting an external standard y_i, a higher-level concept. FS and test statistics can be obtained by performing multiple regression analysis using the plug-in rule, treating $y_i = 1/-1$ as the dependent variable. However, the Iris data is the only data to satisfy Fisher's hypothesis, making the hypothesis a fairy tale with no applicable data. Furthermore, it is aberrant to believe that more than 10,000 genes satisfy the hypothesis. Unfortunately, many precision medicine products are based on this hypothesis, hindering the development of accurate cancer diagnosis, which humanity desires.

3.2 Iris's LSD Explained by 14 Matryoshka Dolls (Fact3)

We can explain a vital data structure of LSDs (Fact3) using the 14 Matryoshka dolls of Iris. An LSD with four variables contains 14 LSD with three variables, which then contains 14 LSD with two variables. Because two LSDs include X1 and X2, they become two Matryoshka dolls 13 & 14. The other 12 LSD become each six Matryoshka dolls 1–6, including X3, and six Matryoshka dolls 7–12, including X4.

Matryoshka1&2: (X1, X2, X3, X4) ⊃ (X2, X3, X4) ⊃ (X3, X4)/ (X2, X3) ⊃ (X3),

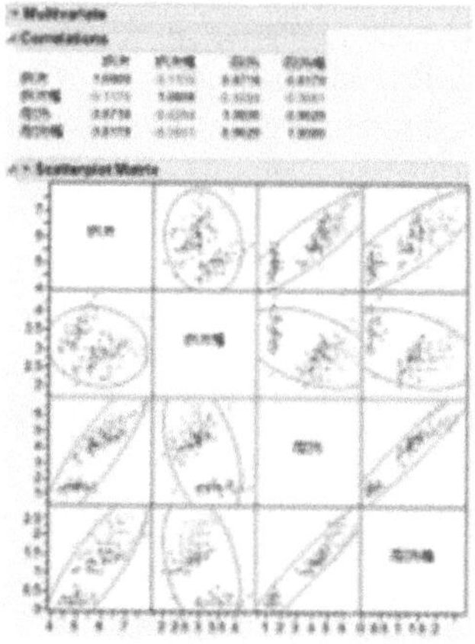

Fig. 2. Matrix scatter plot of Iris

Matryoshka3&4: (X1, X2, X3, X4) ⊃ (X1, X3, X4) ⊃ (X3, X4)/ (X1, X3) ⊃ (X3),
Matryoshka5&6: (X1, X2, X3, X4) ⊃ (X1, X2, X3) ⊃ (X2, X3) /(X1, X3) ⊃ (X3),
Matryoshka7&8 (X1, X2, X3, X4) ⊃ (X2, X3, X4) ⊃ (X3, X4)/ (X2, X4) ⊃ (X4),
Matryoshka9&10: (X1, X2, X3, X4) ⊃ (X1, X3, X4) ⊃ (X3, X4)/ (X1, X4) ⊃ (X4),
Matryoshka11 & 12: (X1, X2, X3, X4) ⊃ (X1, X2, X4) ⊃ (X2, X4)/ (X1, X4) ⊃ (X4),
Matryoshka13 & 14: (X1, X2, X3, X4) ⊃ (X1, X2, X4)/ (X1, X2, X3) ⊃ (X1, X2).

Ordinary LSD ($n \geq p$) has a nested structure, as in Iris, with p variables including each BGS. However, researchers in statistics, machine learning (ML), and AI have pointed out three difficulties with high-dimensional LSD ($n << p$): 1) no high-dimensional analysis method exists, 2) it is difficult to find information buried in high-dimensional noise, and 3) no high-dimensional FS exists. The term "high-dimensional array" is incorrect. Dr. Sall, in his introduction to high-dimensional LDF and PCA developed using singular value decomposition (SVD) at the 2015 JMP Users Conference, referred to this as wide-column data ($n < p$). Engineers have forgotten that, in high-school mathematics, the solution to a system of n simultaneous equations containing p variables ($n < p$) can be obtained by selecting n variables from the p and computing the inverse matrix [29, 34]. This is a problem in science and engineering education. The 169 arrays are LSDs, and SMs would become huge Matryoshka dolls containing smaller BGSs. This finding indicates that the arrays contain a wealth of information on BGSs that distinguish cancer from normal tissue, rather than a small amount of information buried in noise. Therefore, we developed four validation methods to select vital BGSs that are truly useful for cancer gene diagnosis.

3.3 13 M2s = 0 of the 15 Models

Program 2 outputs 15 M1 (the average ER of 15 training samples with ten-fold CV) and 15 M2 (the average ER of 15 validation samples with 15 RIPs calculated using 15 training samples) for all 15 iris models. Table 1 shows the 15 M2. Because M1 monotonically decreases, like MNM, the four-variable M1 is the smallest and is useless for FS. Because

X1 and X2 are not LSDs, M2 is 11% and 17%. The remaining 13 models (including one of the three BGSs) are LSDs, and surprisingly, M2 is 0%. In other words, all 13 models are optimal to the best models. All 13 models, including the original data, the 10 training samples, and one validation sample, are LSDs. A surprising result from the models in Table 1 is that the widths and lengths of the petals and sepals have three BGSs. If we were to examine gene expression in the iris, we would likely find three BGSs for the genes. This means that if the BGS can identify species based on the morphology of animals and plants, it is likely that the BGS of genes also corresponds to these species. At a public seminar at Kyoto University, researchers from the Inuyama Monkey Center reported on the relationships among species, body length, region, and genes in monkeys in China.

Table 1. The surprising result of 15 models

Model	X1	X2	X3	X4	c	M2
1	1	0	0	0	1	11
2	0	1	0	0	1	17
3(BGS1)	0	0	1	0	1	0
4(BGS2)	0	0	0	1	1	0
5(BGS3)	1	1	0	0	1	0
15	1	1	1	1	1	0

4 CPD (240*19) Results by SAS, JMP and RIP

4.1 2^19 Discriminant Models by SAS and JMP

CPD has 17 variables. Prof. Suzumura of Nippon Medical School proposed a diagnostic method using two new variables, X12 (= X13 - X14) and X9 (= X7 - X8), which cause multicollinearities. Sall [9], the vice president of SAS and the founder of JMP, explains multicollinearity. Because the author studied the ER of LDF [5] and a heuristic-OLDF with Prof. Miyake of Nippon Medical School [6, 11], he asked us to write two different papers: 1) verifying the validity of the Suzumura method, 2) covering 320,000 SSM injection patients for the Ministry of Health, Labor, and Welfare's Maruyama vaccine application [11, 12].

Table 2. FS summary for models over 520,000

P	Rank	F/B	AIC	MNM	OCP
1	1	FB12	110	20	1
2	1	FB9	94	13	3
3	1	FB18	93	12	2

(continued)

Table 2. (*continued*)

P	Rank	F/B	AIC	MNM	OCP
14	1	B11	102	3	10
	2	F4	102	3	16
15	1	B19	104	2	1
	2	F11	104	3	1
16	1	F19B16	106	2	2
17	1	FB6	108	2	8
18	1	FB3	111	2	15
19	1	FB8	113	2	16

Since around 1976, we have provided SAS calculation services in Japan. We learned SAS as our private tutor, mastered statistical methods, and published many Japanese papers and books. Using Goodnight's sweep operator [3], RSQUARE in SAS and JMP can output the R^2 values for the top 19 models (1 to 18 variables) using a 19-variable variance-covariance matrix. We selected 37 models by the backward and forward stepwise methods from the 19 variables. Table 2 summarizes the outputs of 343 (= 18*19 + 1) models among 2^{19} models. With p = 1, both methods chose the same x12, so we'll refer to this model as FB12 with AIC (110) and MNM (20). LINGO found one OCP. We selected three models: 1) FB18 with the minimum AIC (93), 2) B19 with minimum MNM (2), 3) F19B16 without the multicollinearities. We chose F19B16 with p (16), MNM (2), and AIC (106) because of no multicollinearity. MNM monotonically decreases from 20 to 2 for 1 to 19 variables. The OCP column indicates the number of OCP. 19 and 16 variables have 16 and 2 OCPs. This table shows how to select the better model statistically from 2^{19} models.

Figure 3 shows that NM is adversely affected by multicollinearity, making it difficult to determine the best model among 2^{19} models ranging from 1 to 19-variables. The left figure shows JMP. The red circle represents the forward stepwise method, and the blue + represents the backward stepwise method. The right one shows SAS. While NM generally tends to decrease as the number of variables increases, MNM monotonically decreases from 20 to 2 for 1 to 19 variables. In writing [38], we realized the importance of omitting two misclassified $\mathbf{x}_i$ (Method4), Table 2 and Fig. 3 become simpler. However, RSQUARE is not supported now.

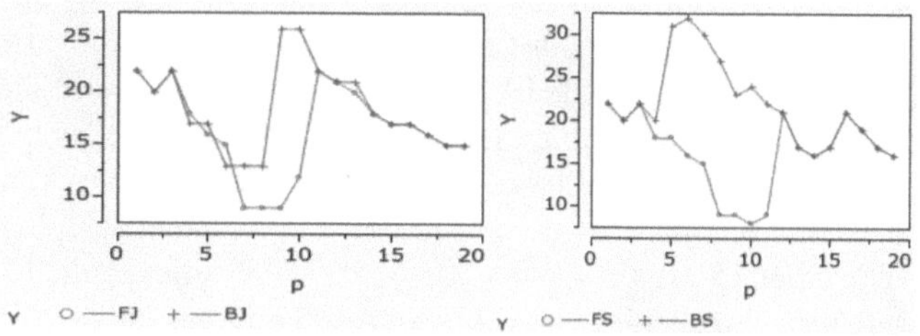

Fig. 3. Forward (red) and backward (blue) NMs in JMP (left) and SAS (right)

4.2 A 14-Variable BGS without Multicollinearity by RIP

After Method4 omits two $\mathbf{x}_i$ unique to the CPD, 238 cases are LSD. MNM = 2 is determined from 2^{240} CPs. MNM = 0 is determined from 2^{238} CPs. Only by examining why two $\mathbf{x}_i$ fall into opposite classes can analysts truly fulfill the mission of discriminant theory. Next, using Program4, we quickly selected a 14-variable BGS1 with MNM = 0 and M2 = 2.44 from the 2^{19} discriminant models. BGS1 is the best model with the minimum M2 among 2^{19} models. M2 = 2.44 shows the difficulty of CPD diagnosis.

In 2024, we discovered the following advantages of BGS: 1) It is easier to find a BGS of 14 variables without multicollinearity using Program4 than to select a 16-variable model without multicollinearity using R^2 and AIC in RSUQARE and stepwise regression. 2) 32 LSDs of the combination of the remaining 5 variables, including BGS, are accepted. We can reject 524,256 non-LSD models without testing. 3) Furthermore, from X12 = X13 - X14, which causes multicollinearity, we found three BGSs with 14 variables that do not include one of X12, X13, or X14, but do include the other two variables. BGS avoids multicollinearity by omitting one of X12, X13, and X14. 4) ten-fold CV finds BGS1, including X13 and X14, is the smallest and best model, satisfying Occam's razor. 5) We thought that multicollinearity would arise from the two Eqs. X12 = X13-X14 and X9 = X7-X8, but only X12 = X13-X14 caused it. We confirmed these facts through a personal conversation with Vice President Sall, who stated that VIF is not a good statistic for detecting multicollinearity. We learned VIF from his book! 6) Initially, calculating the BGS in the order X1 to X19 was time-consuming due to the loop. Therefore, moving the three variables to the end reduced the time. Method4 can be widely applied to human society, various academic fields, and industry, etc.

5 Three Types of Ordinary LSD and 169 Microarrays

5.1 The World's First LSD Study: Swiss Banknotes Data (200*6)

Flury and Rieduyl [17] wrote a book on discriminant analysis of a six-variable consisting of 100 genuine and 100 counterfeit bills. NMs of S-SVM1 (penalty c = 1), LDF, Quadratic Discriminant Function, and RDA are greater than 1. Statistical software provided this data for evaluation, and many discriminant users prefer it to Iris. However, no one, including the authors, found the LSD. We discriminate 63 models using RIP, H-SVM, S-SVM3, and logistic regression. Only RIP could see that 16 of them are LSD. Therefore, RIP can separate 16 signal models and 47 noise models without FS and test statistics. Also, we found that MNM is monotonically decreasing (Fact2: $\mathrm{MNM}_k \geq \mathrm{MNM}_{(k+1)}$). Fact2 and Fact1 (Relation of NM and LDF in Fig. 1) are significant for the LSD study. Table 3 shows 16 LSD using 100-fold CV. The Var column shows the suffix of 16 models. The last model with SN = 16 is a bivariate BGS (X4 and X6). The other 15 models include BGS and become LSD by Fact2. Three columns, min, max, and M1, are the minimum, maximum, and average of ERs of 100 training samples. M1 is the average of 100 ERs of training samples. MIN, MAX, and M2 show the minimum, maximum, and average of 100 ERs of validation sample. Only M2 of BGS1 is 0%. The other 15 M2s range is [0.05%, 2%]. We compare the M2 result with Iris and conclude that the reliability of M2 is weaker than Iris's M2.

Table 3. Results of M1 and M2 for 16 LSDs with 100-fold CV in Program

SN	p	Var.	min	max	**M1**	MIN	MAX	**M2**
1	6	1–6	0	0.2	0.4	0	1	0.2
3	5	1,3–6	0	0.1	0.01	0	0.5	0.05
5	5	1–4,6	0	0.1	0.4	0	0.5	0.2
7	4	2,4–6	0	0.1	0.01	0	0.5	0.05
8	4	2–4,6	0	0.2	0.06	0	1	0.3
12	3	4–6	0	0.2	0.01	0	1	0.1
15	3	2,4,6	20	0.6	0.4	1	3	2
16	2	4,6	0	0	0	0	0	0

We started the world's first LSD study with this data. A 100-fold CV shows that only (X4, X6) became M2 (0%). In contrast, in a ten-fold CV for Iris, 13 M2 values, including three BGSs, were 0%. Although the 100-fold CV is a stricter evaluation than the ten-fold CV, the Iris results tend to be better. Moreover, since the authenticity test does not rely on ink quality or other characteristics of counterfeit bills, but rather on the length of the banknote, which deteriorates over time, it may yield poor results. For these reasons, we believe this result should not be used for authenticity testing.

5.2 Pass/Fail Judgment of Exams

5.2.1 Trivial LDF and LSD [20, 21]

After the first successful LSD research, we realized the importance of LSD and searched the web, but couldn't find it. After thinking about it for a while, we suddenly realized that the exams are good LSDs and become trivial LDFs. Furthermore, if you're a teacher, exams are readily available. We consider a pass/fail judgment with two scores, T1 and T2. If the passing score is 50 points, a trivial LDF (= T1 + T2–49.5) becomes a trivial LSD. It is vital to use 49.5 points instead of 50 points here. The absence of x_i in the open interval (49, 50) means LSD. Although H-SVM can find LSD quickly, it cannot discriminate misclassified cases Even if the score distribution is normal, two classes dose not satisfy Fisher's hypothesi after the H divides two classes. Exams explain well the LSD.

Because no one pays attention to our exams, we decided to borrow three years of research data from the National Center Test for University Admissions. We analyzed the main and preliminary exams of 13 subjects, targeting current first-year students of national universities in the Tokyo area. The Center decided to open the research data in collaboration with the Japanese Society of Applied Statistics. A symposium was held at Seikei University, where we presented our first results as follows. 1) The LDF misclassified 32% in Mathematics IIb for science and engineering. 2) The QDF also revealed a flaw in the generalized inverse matrix of the variance-covariance matrix, when all passing students answered a particular question correctly, but the failing students varied, all passing students are misclassified as failing. Currently, JMP uses the overall variance-covariance matrix when a variable takes a constant value. 3) Mathematics IIb has several BGSs, and English tends to have a single BGS with many questions. Other subjects fall somewhere between these two exams. Exams provide a variety of new insights for LSD. No one has studied that the pass/fail decision is an LSD by BGSs. Each university should conduct research of entrance exam as a topic of education.

5.2.2 Exams of Introduction to Statistics: 100 Questions with 10 Choice

We analyzed three score levels at the 90%, 50%, and 10% and applied three research papers. However, a private university distributed the exam sheets incorrectly. The Asahi Shinbun reported that the Democratic Socialist Party's Kan administration was "considering reorganizing the Center." As a result, the Center requested us to withdraw our three studies to avoid further confusion. Therefore, we analyzed our multiple-choice exams (100 questions) from a semester-long introductory statistics course for about 130 first-year students. The exam had four categories using 8 4-by-2 integer values: T1: Basic statistics and the normal distribution; T2: Scatter plots and correlations; T3: Contingency tables and chi-square tests; and T4: PCA and simple regression. We used my unique best-selling textbook, which uses 8 integers. The 100 questions covered all the content taught. Basic statistics were calculated mentally, and other problems used the JMP output using 8 integer values. After the exam, the results were returned to everyone, but it would be difficult to understand unless students understood the meaning of the distribution.

Table 4. Results in 2012(ten-fold CV results of 90%, 50%, and 10%)

90%	var	M2	MNM	50%	var	M2	MNM	10%	var	M2	MNM
Full	100	10.5	0	Full	100	13.2	0	Full	100	7.3	0
SM1	77	9.2	0	SM1	93	11.4	0	SM1	93	7.6	0
sm2	23	11	6	sm2	7	13	6	sm2	7	5.2	6
BGS1	8	0	0	BGS1	16	5.2	0	BGS1	16	5.3	0
BGS2	11	2.7	0	BGS2	16	5.9	0	BGS2	16	4.7	0
BGS3	13	4.4	0	BGS3	33	11.7	0	BGS3	33	5.5	0
BGS4	13	4.1	0	BGS4	29	11.4	0	BGS4	29	7.8	0
BGS5	18	4.3	0	bgs5	6	33.4	6	BGS5	6	7.7	0
bgs6	37	12	6								

Table 4 shows the results for the three levels in 2012. Three 100 items, three SM1s, and 14 BGSs are LSD. The other three sm2s, bgs6 of 90%, and bgs5 of 50% are not LSD. They are useless for pass/fail judgments. Because 10% has a few correct answers, there is no bgs. Three M2s for the 100 items are 10.5%, 13.2%, and 7.3%, respectively. The worst result in 50% shows that its judgment is difficult because many students are near the discriminant hyperplane. Three SM1 show the same trend as three100 items.

Three BGS1s have only 8, 16, and 16 items. These M2s achieved good results, 0%, 5.2% and 5.3%, respectively. This fact means the most vital role of BGS1. Primarily, only BGS1 at the 90% level takes M2 = 0%. Three BGS1s have fewer items than the 100-item and SM1s, showing the vital role of BGS. Because five BGSs at 10% have 100 items, this indicates the diversity of correct-answer patterns among students with poor performance.

To summarize the above, we must consider the following improvements.

1) We decide on three passing points to correspond to the four grades of excellent, good, acceptable, and fail, which corresponds to the Japanese general university grading system. However, we need the other pass level. For university entrance exams, we need to consider two classes of successful/unsuccessful applicants. For reserved admissions, it is better to select students with the correct answers among the BGS.
2) In the National Center Test, English has a single BGS with many questions, while Mathematics IIB has several BGSs. Comparative research on BGSs can also provide clues to these subject differences. Since each university can decide pass/fail based on the BGSs, we can now elucidate the quality of the questions. Until now, there has been no study of question quality, which affects many students. Programs 3 and 4 can split the exams into SMs and BGSs. The ten-fold CV determines the M2s. They can evaluate all questions, SMs, and BGSs. By comparing M2s, test creators could assess the quality of their questions. We consider the BGS with the smallest M2 to be more trustworthy for the judgment.
3) Various qualification exams have a greater impact on people's lives than entrance exams. The analysis results for each exam should be published alongside the BGS.
4) The Center and the Japanese Society of Applied Statistics should resume their intended research into pass/fail scores, inviting a wide range of researchers.
5) Pass/fail judgements are a template for various evaluation systems. For example, the Altman score used an LDF to distinguish between high-performing and bankrupt companies in the US, although it is used worldwide. RIP can revise this score to suit each country's companies.

5.3 Japanese Compact and Regular Cars with Two Univariate BGSs

5.3.1 Statistical Analysis

Japanese 15 compact cars and 29 standard cars have six variables: emissions (X1, BGS1), price, seating capacity (X3, BGS2), CO2 emissions, fuel consumption, and sales. Compact cars are separated from standard cars by the legal standard of four seats and an engine displacement of 2000 cc or less. We consider that other industrial products may have obvious univariate BGSs decided by a legal standard.

Table 5 shows the forward stepwise regression analysis. The external standards y_i = 1/−1 is used as the dependent variable (Plug-in rule). The two t-values of BGS1 and BGS2 are large at 11.4 and 8.93, respectively. LDF's NM decreases monotonically from 2, 1, 1, 1, 0, and 0. Because MNM = 0 for X1, the other models, including X1, also have MNM = 0. QDF becomes NMs = 0 for up to the first two variables. However, when X3 is included in the model, all cars are misclassified as compact cars because compact cars have four seats. The last two columns are the RDAs. It must select two adjustment terms from the range [0,1], but there is no way to determine the optimal values for each data point. RDA1 produces poor results when both adjustment values are set to 0.8, while RDA2 coincidentally achieves the same results as RIP when set to 0.1. Searching for these adjustment values from 100 grid points in a 10x10 grid of [0, 1] is a waste of time and unscientific.

Table 5. Results between Japanese compact and standard cars

p	Var.	t	LDF	QDF	RIP	RDA1	RDA2
1	Emission (BGS1)	11.4	2	0	0	2	0
2	Price (X2)	5.42	1	0	0	4	0
3	Seat(BGS2)	8.93	1	29	0	3	0
4	CO_2 (X4)	4.27	1	29	0	4	0
5	Fuel (X5)	−4	0	29	0	5	0
6	Sales (X6)	−0.8	0	29	0	5	0

5.3.2 RIP Surprising Results

Because of two univariate BGSs, we analyzed 63 models using Program2. Table 6 shows simple yet surprising results. The 15 models (SNs > = 49) do not include BGSs and exhibit different RIP coefficients. Their M2s are over 1.295%. The 16 models (SN = 1 ~ 16) include both X1 and X3 result in RIP = 5.92*X1–4.893 or RIP = 2*X3–9. The 16 models (SN = 17 ~ 32) include only X1 and result in RIP = 5.92*X1–4.893. The 16 models (SN = 33 ~ 48) include only X3 and result in RIP = 2*X3–9. Of these 48 models, five models have M2 = 0.023% and the other 43 have M2 = 0. The reason for these results is unclear, and we need further analysis using other similar data.

Table 6. Evaluation of 63 automobile models with ten-fold CV

SN	Var.	M2	X1	X2	X3	X4	X5	X6	C
1	1–6	0.023	5.92	0	0	0	0	0	−4.893
2	1–5	0	5.92	0	0	0	0		−4.893
12	1,3,6	0	5.92		0			0	−4.893
13	1–3	0	0	0	2				−9
16	1,3	0	0		2				−9
17	1,2,4–6	0	5.92	0		0	0	0	−4.893
18	1,2,4,5	0.023	5.92	0		0	0		−4.893
19	1,2,4,6	0.023	5.92	0		0		0	−4.893
20	1,2,5,6	0.023	5.92	0			0	0	−4.893
21	1,4–6	0.023	5.92			0	0	0	−4.893

(continued)

Table 6. (*continued*)

SN	Var.	M2	X1	X2	X3	X4	X5	X6	C
32	1	0	5.92						−4.893
33	3,5,6	0			2		0	0	−9
48	3–5	0			2	0	0		−9
49	2,4–6	1.295		0		−46	−199	0	5342.8
63	6	3.205						−0.7	6773.5

5.4 Analysis of Six First Generation and 73 Second Generation Microarrays

5.4.1 Six Old First- Generation Microarrays

On October 28, 2015, we obtained the old 6Array [4] shown in Table 7. Initially, RIP discriminated Shipp (77*7129) [40] and found SM1 (32 genes) from 7,129 genes in 1 s. B&B rapidly identified LSDs. Program 3 split 222 SMs and sm (44 genes). Program 4 split many BGSs in Alon and Tien. All of these are LSDs. However, five projects other than Chiaretti erroneously claimed only one set of about 50 genes as a new multivariate oncogene and non-hierarchical cluster analysis showed that the two groups were roughly separated into two clusters. Chiaretti et al. received a NIH research grant for their study of multiple small gene pairs, presumably combining previously identified univariate oncogenes. As a result, genetic data analysis has been discontinued in new studies conducted after 2004. Ultimately, we recommend learning the basics of medical genetic diagnosis from Golub et al.'s paper and analyzing the new 73 arrays after the quality control bioengineering techniques.

Table 7. Old 6-microarray of the first generation

2016	SMs	Gene-1	Range
Alon (62*2000)	64	1,999	[21, 42]
Singh (77*7129)	215	12,598	[34, 85]
Tien (173*12625)	101	12,566	[100,139]
Shipp (77*7129)	222	7,085	[18, 62]
Chiaretti(128*12625)	155	12,623	[32, 122]
Golub (72*7129)	170	6,348	[16,56]

Although the six arrays and all SM and BGS were LSD, M2 were not 0 (Validation 1). Ward and PCA also failed to separate the two groups into clear clusters (Validation 3). Even with MNM = 0 and M2 = 0, Ward and PCA did not always separate the two groups. It is essential to validate genetic data using these two Validations.

Therefore, we created signal data (77 * 222) using the Discriminant Scores of 222 RIPs as variables instead of 7,129 genes. The 6 arrays resulted in six small samples, which could be analyzed using various statistical methods [29]. The top panel of Fig. 4 shows the results of the SM signal data from Alon et al. (62 * 2000). Ward's hierarchical clustering separated blue (normal) and red (cancer). PCA plots blue normal cases on the negative first principal component (Prin1), while red cancer cases form fan-shaped clusters in the first and fourth quadrants from the origin. The higher the Prin1 value, the more aggressive the cancer. The figure below shows the results of Alon's BGS signal data analyzed in 2025. Ward's algorithm separated the red normal cases at the top and six colored cancer clusters, ranging from green to yellow. PCA places the red normal cases and the six cancer cases on a line extending from the second quadrant to the origin. BGS is more useful than SM for cancer diagnosis. Prof. Yamanaka focused his research on four factors (BGS) rather than 24 genes (SM). We mistakenly believed that identifying SM1 from 7,129 genes was important, and regretted spending too much time analyzing SM. Doctors can use BGS to study whether the yellow cancers in the lower right corner are the most aggressive.

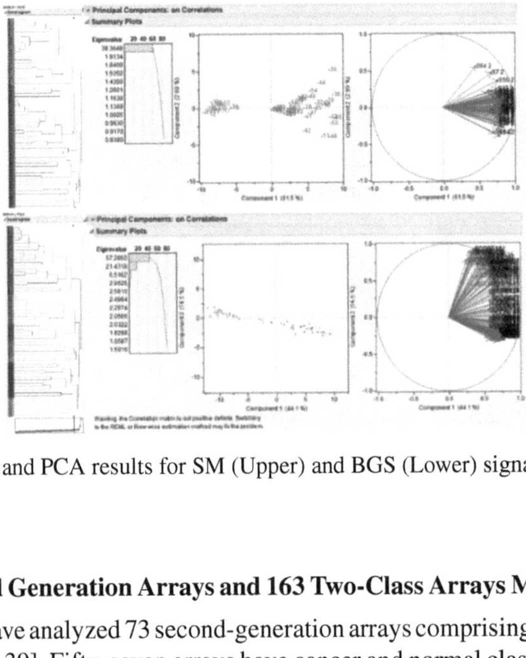

Fig. 4. Ward and PCA results for SM (Upper) and BGS (Lower) signal data in Alon.

5.4.2 73 Second Generation Arrays and 163 Two-Class Arrays Made by 73 Arrays

Since 2019, we have analyzed 73 second-generation arrays comprising over 12,621 genes from CuMiDa [1, 39]. Fifty-seven arrays have cancer and normal classes (Principles 1 & 2). The other 16 arrays have over three groups. We change them to 106 two-group arrays. All 163 arrays were LSD. Analysis of SMs and BGSs revealed that almost the same number of cancer and normal groups (Principles 1 & 2), and fewer than 100 subjects (new Principle 3) performed better. The various studies are summarized below.

The "highest quality expression data" includes many SMs and BGSs. Additionally, scatter plots can reveal heterogeneity in cancer. Figure 5 shows the results of four PCAs. The upper-left scatter plot for Liver3 (357*22,278) satisfies three principles of patient design, but the results are poor. Because it has additional genome cohort samples, it may include different types of research subjects. Because Breast6 (129/10*33,580) has only 10 normal subjects, it violates Principle 2. Two classes are not clearly separated, and there are many misclassifications. Colorectal6 (32 cancer/31 normal*54,676) in the lower left has almost same numbers of cancer and normal (control group) subjects, but the total number of subjects is less than 100. Many empirical studies have shown that arrays with fewer than 100 subjects provide good results because of high-quality of the expression. Colorectal5 (26/26*24,527) in the lower right corner has poor results due to four outliers. However, excluding these four subjects, the remaining 48 subjects have many vital BGSs with fewer than five genes and M2s of 0. However, when the number of subjects of Real1 (GSE66270) has 14 cancer and 14 normal subjects, even if PCA results in a clean separation like Colorectal6, the "stopping rule of M2=50%" will come into effect. In this case, increasing the number of subjects using the same patient design criteria would yield good results similar to those in Colorectal6.

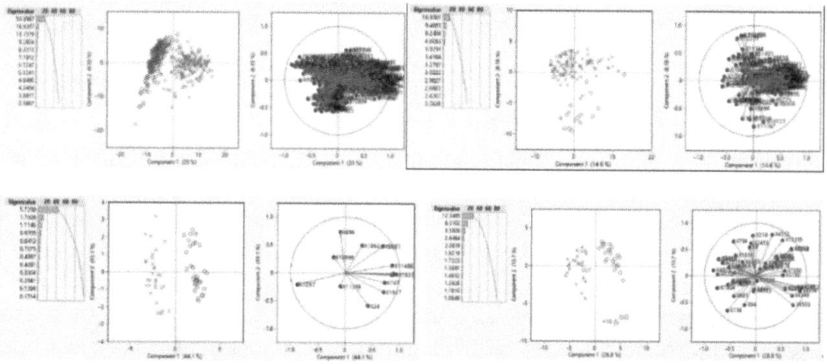

Fig. 5 Four PCA of SM1 (Upper left: Liver3 (GSE14520), Upper right: Breast5 (GSE42568), Lower left: Colorectal6 (GSE8671), Lower right: Colorectal5 (GSE25070).

Lower left: Colorectal6 (GSE8671), Lower right: Colorectal5 (GSE25070)

Table 8 shows the accuracy rates (= 1-ER) of five Weka classifiers (SVM, MLP, DT, NB, and RF) confirmed by three-fold CV [1]. Colorectal5 is the original data, while the modified Col5 data omits four subjects. The predicted values are calculated by multiplying the number of vital BGSs with M2 = 0 among 12,621 genes by 4.3 (= 54,676 / 12,621). Colorectal5 has no vital BGSs, while the modified Col5 predicts 1,743.7 vital BGSs. Cancer research has already identified over 400 univariate oncogenes. We proposed in Validation4 that if these genes are included in the vital BGSs, they represent the characteristics of a multivariate oncogene candidate. Validation4 is more valuable to physicians. However, we can trust 764.2 vital BGSs of Colorectal6 than 1,743.7 vital BGSs because Colorectal6 is the original data.

Table 8. Summary of three studies of 44 microarrays satisfying Principal 1&2

TYPE	N	GENE	SVM	MLP	DT	NB	RF	Forecast
Max	357	54676	1	1	1	1	1	
Min	12	12621	0.52	0.5	0.3	0.5	0.4	
Breast5	129/10	33580	1	1	1	1	1	
Liver 3	181/176	22278	0.97	0.8	0.9	1	1	
Colorectal5	26/26	54676	0.96	0.94	0.81	0.94	0.96	0
Modified Col5	24/24	54676						1743.7
Colorectal6	32/31	54676	1	1	0.9	1	1	764.2

6 Conclusion

Over 52 years, we solved six problems of discriminant theory, devised four methods based on four facts, and conducted empirical studies using five types of LSD. Program1 found LSD (MNM = 0). Program 3 and 4 split LSD into SMs and BGSs. Program2 evaluated them using M2 and specified that most BGS1s became the best model. Because we obtained many BGSs from 169 microarrays, five validations selected vital BGSs. Because Method3 combining both four programs and four validations is the interdisciplinary method many researchers can analyze any gene data. Moreover, we found the good results of around 50 cases with roughly equal numbers of cancer and normal. Because cancer research had already identified more than 400 univariate oncogenes, we found vital BGSs as candidates for multivariate oncogenes, which had M2 = 0 (Validation1) and five or fewer genes. We then proposed that physicians study the characteristics of vital BGSs based on the univariate oncogene characteristics contained within these (Validation4).

Theory1 analyzed four common LSDs. To address Problem1 (NMs are unreliable), we proposed MNMs based on Fact1 (the relationship between LDFs and NMs) and Fact2 ($MNM_k \geq MNM_{(k+1)}$). To find MNMs, we defined the RIP of combinatorial optimization LDFs. In Theory2, since Ward and PCA (Validation3) failed to separate SMs, we proposed signal data. Since 2019, we made 163 two-class data from 73 arrays with over two classes and confirmed that they were LSDs. However, several SMs were not M2 = 0. On the other hand, Fig. 5 and Table 7 showed that the number of cancer and normal classes (Principle1) was roughly equal (Principle2), and that a sample size of fewer than 100 subjects (new Principle3) yielded better results. Liver3 had the largest sample size. This is statistically the best array. However, due to the many genomic cohort, we judged that it included many different research subjects. For the same reason, we believe that many arrays with more than 100 subjects have problems with case collection, resulting in poor results despite high expression quality. For cancer researchers, we developed an easy-to-use two-stage screening method, Method3, which combines four programs and five validations. We hope this will help eradicate cancer, a long-cherished goal for humanity. However, the most vital discovery of LSD theory is the case selection

Method4. Because all data becomes LSD, we can separate between defective products and normal manufacturing processes and identify the cause of the problem using BGS as same as CPD medical diagnoses. Furthermore, by dividing all research subjects into two classes, Method4 can identify the cause of the problem. LSD theory is the simplest and most widely applicable method.

The author proposes five future research themes:

1. The simple formulation of RIP using 169 arrays provides 169 evaluation problems suitable for quantum computers. TSP is useless because the correctness of the results cannot be quickly verified. Everyone can easily confirm 169 LSDs. Furthermore, many SMs and BGSs provide additional plenty problems. Even a low-power PC can analyze all arrays, making it less expensive than a quantum computer.
2. There are many untested issues with the test's pass/fail criteria, and further research is needed. It can also be used for many evaluation problems and risk analysis. For example, the Altman score should be studied using data from various countries.
3. BGS determines the classification of industrial products and provides many topics for statistical research. By comparing defective manufacturing processes with normal ones, the cause of the defects can be identified.
4. Iris can probably be identified its species with three genes BGS. Zoologists believe that creating a list of BGS for the study of every animal species will enable new research frontier.
5. Researcher can analyze new RNA-seq and other genetic data using RIP. LSD theory is an interdisciplinary method, and better results can be obtained with higher-quality data. However, methods that do not reveal the LSD of microarrays will not succeed in new data. Program3 splits 169 wide-column arrays ($n < p$) into many exclusive SMs of n or fewer genes. Each SM became several Matryoshka dolls containing one of the BGSs. Prof. Yamanaka selected 24 genes from about 30,000. Because the 24 genes formed pluripotent cells, he removed one gene from the 24 and discovered the Yamanaka four genes using a new FS. Program4 finds one of BGSs from every SMs. However, if the BGS is univariate, as in Iris, it becomes many Matryoshka dolls, unsuitable for important pluripotent cells. On the other hand, cancer is a runaway of normal cells, and univariate oncogenes are vital candidates for the BGS. We found up to 1743.7 forecast vital BGSs with five or fewer genes for which $M2 = 0$ in Colorectal5 [7], as shown in Table 8. If these BGSs include univariate oncogenes already discovered in cancer research, they will become candidates for multivariate oncogenes (Validation4). We sincerely cooperate with physicians to use our LSD theory to overcome cancer. Please send me an email. We want to work with you to make this happen.

References

1. Bruno, C.F., Eduardo, B.C., Bruno, I.G., Marcio, D.: CuMiDa: an extensively curated microarray database for benchmarking and testing of machine learning approaches in cancer research. J. Comput. Biol. **26**, 1–11 (2019)
2. Flury, B., Riedwyl, H.: Multivariate Statistics: a Practical Approach. Cambridge University Press, New York (1988)

3. Goodnight, J.H.: SAS Technical Report—The Sweep Operator: its Importance in Statistical Computing—R(100). SAS Institute Inc, USA (1978)
4. Jeffery, I.B., Higgins, D.G., Culhane, C.: Comparison and evaluation of methods for generating differentially expressed gene lists from microarray data. BMC Bioinf. **7**(1), 359 (2006)
5. Miyake, A., Shinmura, S.: In: Dombal, F.T., Gremy, F. (eds.) Error Rate of Linear Discriminant Function, pp. 435–445. North-Holland Publishing Company, The Netherland (1976)
6. Miyake, A., Shinmura, S.: An algorithm for the optimal linear discriminant function and its application. Japan Soc. Med. Electron. Biol. Eng. **1815**, 452–454 (1980)
7. Sabates-Bellver, J., Van der Flier, L.G., de Palo, M., Cattaneo, E., et al.: Transcriptome profile of human colorectal adenomas. Mol. Cancer Res. **5**(12), 1263–1275 (2007)
8. Sall, J,P.: SAS regression applications. SAS Institute Inc. USA (1981). (Shinmura translate Japanese)
9. Schrage L.: Optimization Modeling with LINGO. LINDO Systems Inc, Chicago (2006) (Shinmura translates Japanese)
10. Shinmura, S., Miyake, A.: Optimal linear discriminant functions and their application. COMPSAC. **79**, 167–172 (1979)
11. Shinmura, S., Iida, K., Maruyama, C.: Estimation of the effectiveness of cancer treatment by SSM using a null hypothesis model. Inf. Health Soc. Care. **7**(3), 263–275 (1987)
12. Shinmura, S.: A new algorithm of the linear discriminant function using integer programming. New Trends Probab. Stat. **5**, 133–142 (2000)
13. Shinmura S.: Optimal Linear Discriminant Function Using Mathematical Programming. Dissertation, Okayama University, Japan, pp. 1–101 (2000)
14. Shinmura, S.: Overviews of discriminant function by mathematical programming. J. Japan. Soc. Comput. Stat. **20**(12), 59–94 (2007)
15. Shinmura S.: Comparison of revised IP-OLDF and SVM. ISI2009, 1–4 (2007)
16. Shinmura S.: The Optimal Linearly Discriminant Function. Union of Japanese Scientist and Engineer Publishing, Japan (2010). ISBN 978-4-8171-9364-3
17. Shinmura, S.: Problems of discriminant analysis by mark sense test data. Japan. Soc. Appl. Stat. **4012**, 157–172 (2011)
18. Shinmura S.: Evaluation of optimal linear discriminant function by 100-fold cross-validation. ISI2013 CD-ROM, pp. 1–6 (2013)
19. Shinmura S.: End of discriminant functions based on variance-covariance matrices. ICORE2014. 5–16 (2014)
20. Shinmura, S.: A trivial linear discriminant function. Stat. Optim. Inf. Comput. **3**, 322–335 (2015)
21. Shinmura, S.: Four serious problems and new facts of the discriminant analysis. In: Pinson, E., Valente, F., Vitoriano, B. (eds.) Operations Research and Enterprise Systems, pp. 15–30. Springer, Berlin (2015)
22. Shinmura, S.: Matryoshka feature selection method for microarray data. Research Gate. (15), 1–16 (2015)
23. Shinmura, S.: The best model of Swiss bank note data. Stat. Optim. Inf. Comput. **4**, 118–131 (2016)
24. Shinmura, S.: Discriminant analysis of the linearly separable data—Japanese 44 cars. J. Stat. Sci. Appl. **4**(7–8), 165–178 (2016)
25. Shinmura, S.: New Theory of Discriminant Analysis after R. Fisher, First edn. Springer, Singapore (2016)
26. Shinmura, S.: Matryoshka feature selection method for microarray data. Biotechnol, 1–8 (2016) (Best Paper Award)
27. Shinmura S.: First success of cancer gene analysis by microarrays. In: Biocomp'18, pp. 1–7 (2018).

28. Shinmura, S.: Release from the curse of high-dimensional data analysis. In: Big Data, Cloud Computing, and Data Science Engineering (Studies in Computational Intelligence 844), pp. 173–196 (2019)
29. Shinmura, S.: High-Dimensional Microarray Data Analysis–Cancer Gene Diagnosis and Malignancy Indexes by Microarray, First edn. Springer, Singapore (2019)
30. Shinmura, S.: Cancer gene diagnosis of 84 microarrays by rank using 100-fold cross-validation. In: Transactions on Computational Science & Computational Intelligence. Springer Nature (2020)
31. Shinmura, S.: Cancer diagnosis of 78 microarrays registered on GSE from 2007 to 2017. In: Transactions on Computational Science & Computational Intelligence. Springer Nature (2020)
32. Shinmura, S.: First success of cancer gene data analysis of 169 microarrays for medical diagnosis. In: CSCI-ISCB: COMPUTATIONAL BIOLOGY1–7.Transactions on Computational Science & Computational Intelligence. Springer Nature (2020)
33. Shinmura, S.: Twenty-three serious mistakes of cancer gene data analysis because 1995. In: Arabnia, H.R., et al. (eds.) Advances in Computer Vision and Computational Biology, Transaction on Computational Science and Computational Intelligence. Springer Nature Switzerland (2021) in Press
34. Shinmura, S.: First theory of cancer gene data analysis of 169 microarrays and four universal data structures for big data. In: CSCI-ISCB: COMPUTATIONAL BIOLOGY1–14.Transactions on Computational Science & Computational Intelligence. Springer Nature (2021) in Press
35. Shinmura, S.: Practical screening method for cancer gene diagnosis – four universal data structures of discriminant data. In: BIOLOGY1–16.Transactions on Computational Science & Computational Intelligence. Springer Nature (2022) in press
36. Shinmura, S.: Practical screening method for cancer gene Diagnosis-2 –four patients design principles by Method1 and scatter plot. In: BIOLOGY1–17.Transactions on Computational Science & Computational Intelligence. Springer Nature (2022) in press
37. Shinmura S.: Screening Results by Colorectal5 of 169 Microarrays. CSCI22 1–7. IEEE Book and Proceedings (2022)
38. Shinmura, S.: The First Discriminant Theory of Linearly Separable Data – from Exams and Medical Diagnosis with Misclassifications to 169 Microarrays for Cancer Gene Diagnosis, First edn. Springer, Singapore (2024)
39. Shinmura, S.: Practical screening method for cancer gene diagnosis – how to choose cancer and normal patients by four principles. London J. Res. Comput. Sci. Technol. **2**, 1–16 (2024)
40. Shipp, M.A., et al.: Diffuse large B-cell lymphoma outcome prediction by gene-expression profiling and supervised machine learning. Nat. Med. **8**(1.1), 68–74 (2002)
41. Vapnik, V.: The Nature of Statistical Learning Theory. Springer, NY (1995)

Poster Papers

In-Context Learning in Genomic Language Models as an Evaluation Task

Aadit Kapoor(✉) and Wendy Lee

San José State University, San Jose, USA
aadit.kapoor@sjsu.edu

Abstract. We study whether base (non-instruction-tuned) genomic language models (gLMs) exhibit *in-context learning* (ICL) on DNA. Using an adapted NanoGPT trained on multiple *Escherichia coli* references with BPE tokenization, we frame promoter completion as a genomic ICL task: a 1,000 bp upstream context (prompt) conditions autoregressive generation of downstream bases. We introduce an intrinsic evaluation suite that quantifies overall, compositional, structural, and local consistency similarity between generated and ground-truth promoter sequences, alongside loss and GC% diagnostics. Preliminary results suggest the base model learns aggregate nucleotide patterns and motif ordering signals, while position-wise fidelity remains limited. We discuss tokenization–compression tradeoffs, scaling behavior, and cross-species transfer directions for evaluating emergent behavior in genomic models.

Keywords: Genomic language models · In-context learning · Promoter sequences · Tokenization · Intrinsic evaluation

Research Track: CSCI-RTCB

1 Introduction

Large Language Models (LLMs) trained on human language exhibit surprising generalization capabilities, performing tasks not seen during training [1,2]. This emergent property implies that these models capture rich contextual and syntactic knowledge. In contrast, genomic sequences encode the blueprint of life using a four-letter vocabulary {A, T, G, C}, forming a highly non-linear, interdependent system [3,4]. This non-linearity makes them excellent candidates for autoregressive modeling, since genomic contexts—such as gene regulatory regions—resemble long-range semantic dependencies.

However, evaluating LLMs in genomics is non-trivial. Unlike human language, genomic data lacks defined words, grammar, or explicit tasks. Evaluation thus requires intrinsic metrics that probe whether a model genuinely learns the structural and compositional biology underlying the DNA sequences. In this work, we adapt GPT-2/NanoGPT for full-genome training on *E. coli* to test whether the base model (without fine-tuning) demonstrates *In-Context Learning (ICL)*—the ability to generalize from contextual prompts such as promoter regions.

H. R. Arabnia et al. (Eds.): CSCI 2025, CCIS 2970, pp. 559–564, 2027.
https://doi.org/10.1007/978-3-032-28466-2_43

2 Related Work

Language models trained on protein and DNA sequences have achieved notable performance on gene regulation, variant prediction, and expression modeling [5–7]. Enformer [3] and HyenaDNA [4] focus on long-range dependencies, while works such as DNA-GPT and GROVER extend autoregressive modeling to biological sequences. However, most prior work focuses on fine-tuned, task-specific settings rather than base-model evaluation.

Tokenization plays a pivotal role: while NLP models rely on subword units, genomic sequences must balance compression with information retention [8,9]. Prior studies show tokenization impacts model generalization and structural capture capacity. Our work contributes an evaluation framework probing whether a base genomic model can infer biological structure and regulation context through ICL-like behavior.

3 Methods: Genomic ICL Protocol

3.1 Dataset Preparation

We select reference genomes of *E. coli* from NCBI to ensure high diversity while maintaining quality. Each genome is canonicalized by removing ambiguous nucleotides (N, U), ensuring a pure vocabulary of {A, T, G, C}. Only genomes between 4–6 Mbp and GC content between 40–60% were retained to maintain taxonomic similarity (Table 1).

Table 1. Representative *E. coli* genomes used for training.

Strain	Genomic ID	Description
MG1655	NC_000913.3	K-12, Lab strain
O157:H7 EDL933	NC_002655.2	Enterohemorrhagic
UTI89	NC_007946.1	Uropathogenic (UPEC)
CFT073	NC_004431.1	UPEC
REL606	NC_012967.1	Evolved Lab Strain

3.2 Tokenizer Evaluation

To convert raw sequences into model-readable form, we assess four tokenizers: character-level, BPE, WordPiece, and SentencePiece. Each is quantified by a Sequence-to-Compression Ratio (SCR) where $\text{SCR}(\%) = \frac{\text{tokens}}{\text{sequence length}} \times 100$. BPE achieves the optimal tradeoff and is used in all subsequent experiments.

3.3 Model and Training Configuration

We adapt NanoGPT (decoder-only Transformer, $\sim$ 84M parameters) with 12 layers, 12 heads, and embedding dimension 768. A block size of 256 tokens balances efficiency and local context. Training uses AdamW with cosine learning rate decay from 1×10^{-3} to 5×10^{-5}, batch size 32, weight decay 0.1, and mixed precision (bf16). Two RTX 3070 GPUs were used for 100 epochs (Table 2).

Table 2. Comparison of DNA tokenizers.

Tokenizer	Core Idea	Compression Range	Insight
Character	1 token per base	100–120%	Full resolution
BPE	Merges frequent pairs	60–80%	Good balance
WordPiece	Likelihood-optimized merges	50–70%	May duplicate tokens
SentencePiece	Probabilistic unigram modeling	40–60%	High compression, risk of info loss

3.4 Notation for Genomic ICL

We define the notation used in the genomic in-context learning setup as follows:

$$S^{(i)} : \text{Full promoter sequence of sample } i$$
$$L^{(i)} : \text{Total sequence length of } S^{(i)}$$
$$C^{(i)} : \text{Context (upstream) of 1,000 bp used as input prompt}$$
$$G^{(i)} : \text{Generated continuation downstream of the context}$$
$$\hat{S}^{(i)} = C^{(i)} \| G^{(i)} \text{ (Concatenated predicted sequence)}$$

Here, the task is to minimize the divergence between the generated sequence $G^{(i)}$ and the ground-truth continuation using multiple intrinsic similarity metrics. This formulation draws an analogy between sentence completion in NLP and promoter completion in genomics.

3.5 Promoter ICL Setup

We design an *In-Context Learning* paradigm where the start of the promoter acts as a genomic prompt. The model is tasked with predicting the continuation of the sequence beyond this context (Fig. 1).

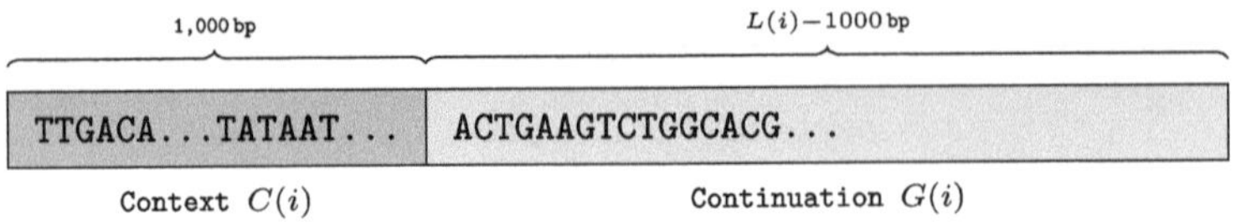

Fig. 1. Compact schematic of promoter ICL setup. Blue: context; green: generated continuation. (Color figure online)

4 Results

5 Metric Formulation

Training Diagnostics. Validation loss and negative log–likelihood (NLL) decreased steadily across epochs, converging to approximately 1.32 for generated sequences. GC content of generated regions aligned closely with true promoter GC% (mean deviation ΔGC ≤ 0.02), indicating compositional awareness.

Negative Log Likelihood of a Sequence. In the context of a genomic language model, predictive uncertainty tells us how confident a model is in its nucleotide predictions (A, T, C, G).

Let $\mathbf{x} = (x_1, x_2, \ldots, x_k)$ be a single DNA sequence of length k bp, where each $x_t \in \{\mathsf{A}, \mathsf{C}, \mathsf{G}, \mathsf{T}\}$ (or the corresponding tokenizer ID). For an autoregressive model with parameters θ, the negative log–likelihood (NLL) is

$$\boxed{\mathrm{NLL}(\mathbf{x};\theta) = -\sum_{t=1}^{k} \log p_\theta\big(x_t \mid x_1, \ldots, x_{t-1}\big)}$$

This can be categorized as follows:

- Lower NLL means that the model assigns high confidence to correct bases.
- Higher NLL suggests that the model struggles to recognize sequence pattern

Similarity Metrics. Table 3 summarizes quantitative results across 500 annotated promoters. Composition similarity exceeded 0.90, indicating accurate nucleotide frequency modeling. Structural similarity using LCS (longest common subsequence)-based approach averaged 0.65, while overall position-wise similarity reached 0.45, showing that while the model captures macro-level nucleotide order, micro-level accuracy remains limited.

NLL Comparison. The negative log-likelihood across different sequence lengths shows decreasing uncertainty with increasing context size (Table 4). This pattern mimics the ICL phenomenon observed in NLP models—longer prompts lead to more context-aware, lower-entropy predictions.

Table 3. Intrinsic metric results for the genomic ICL experiment (with interpretation).

Metric	Mean	Std. Dev.	Min	Max	Interpretation
Overall similarity	0.45	0.08	0.31	0.56	Measures exact position-wise nucleotide matches (higher is better)
Composition similarity	0.91	0.03	0.84	0.95	Quantifies preservation of A/T/G/C frequencies (higher is better)
Structural similarity	0.65	0.07	0.51	0.74	Captures macro-order and motif arrangement (higher is better)
Local consistency	0.58	0.05	0.45	0.69	Evaluates stability of matches across sliding windows (higher is better)

Table 4. NLL across varying context lengths.

Length (bp)	Generated Sequence NLL	True Sequence NLL
100	1.38	1.28
1000	1.32	0.62
10000	1.22	0.48

Analogy with NLP. In natural language, an LLM uses a paragraph of preceding words to predict the next sentence. Similarly, in the genomic setting, the promoter serves as a biological "prompt"—conditioning the model to predict plausible downstream bases. The base model thus encodes contextual regulatory grammar without explicit labels, suggesting an emergent understanding of genomic syntax.

6 Discussion and Conclusion

The results support the idea that even base genomic LMs learn compositional and structural priors analogous to grammatical awareness in NLP models. The observed decline in NLL with longer context and stability in GC% distribution both imply that the model benefits from context-based conditioning, indicative of emergent in-context learning. This work contributes a step toward interpretable evaluation of genomic LMs, framing promoter completion as an ICL probe. This work contributes a step toward interpretable evaluation of genomic LMs, framing promoter completion as an ICL probe.

Disclosure of Interests. The authors have no competing interests to declare that are relevant to the content of this article.

References

1. Radford, A.: Language models are unsupervised multitask learners. OpenAI Blog (2019)
2. Brown, T.B.: Language models are few-shot learners. arXiv:2005.14165 (2020)
3. Avsec, Ž., et al.: Effective gene expression prediction from sequence by integrating long-range interactions. Nat. Methods **18**(10), 1196–1203 (2021)
4. Nguyen, E.: HyenaDNA: Long-range genomic sequence modeling at single-nucleotide resolution. In: NeurIPS, vol. 36 (2024)
5. Sanabria, M., et al.: DNA language model GROVER learns sequence context in the human genome. Nat. Mach. Intell. **6**, 911–923 (2024)
6. Boshar, S., et al.: Are genomic language models all you need? Bioinformatics **40**(9), btae529 (2024)
7. Benegas, G., Batra, S.S., Song, Y.S.: DNA LMs are powerful zero-shot predictors of non-coding variant effects. bioRxiv (2022)
8. Dotan, E., et al.: Effect of tokenization on transformers for biological sequences. Bioinformatics **40**(4), btae196 (2024)
9. Wang, D.: Tokenization matters! arXiv:2405.17067 (2024)
10. Kapoor, A.: Understanding and evaluating genomic language models. MS thesis, San Jose State University (2025)

Author Index

H. R. Arabnia et al. (Eds.): CSCI 2025, CCIS 2970, pp. 565–566, 2027.
https://doi.org/10.1007/978-3-032-28466-2

FSC
www.fsc.org
MIX
Papier | Fördert gute Waldnutzung
FSC® C083411

Kolibri 360 GmbH
Ferdinand-Jühlke-Straße 7
99095 Erfurt, Deutschland
produktsicherheit@kolibri360.de